Lecture Notes in Artificial Intelligence 1711

Subseries of Lecture Notes in Computer Science
Edited by J. G. Carbonell and J. Siekmann

Lecture Notes in Computer Science

Edited by G. Goos, J. Hartmanis and J. van Leeuwen

T0134782

Subseries of Lecture Notes in Computer Science
Edited by J. G. Carbonell and J. Siekmann

Lecture Notes in Computer Science

Springer

Berlin
Heidelberg
New York
Barcelona
Hong Kong
London
Milan
Paris
Singapore
Tokyo

Ning Zhong   Andrzej Skowron
Setsuo Ohsuga (Eds.)

New Directions in Rough Sets, Data Mining, and Granular-Soft Computing

7th International Workshop, RSFDGrC'99
Yamaguchi, Japan, November 9-11, 1999
Proceedings

 Springer

Series Editors

Jaime G. Carbonell, Carnegie Mellon University, Pittsburgh, PA, USA
Jörg Siekmann, University of Saarland, Saarbrücken, Germany

Volume Editors

Ning Zhong
Yamaguchi University, Faculty of Engineering
Department of Computer Science and Systems Engineering
Tokiwa-Dai, 2557, Ube 755, Japan
E-mail: zhong@ai.csse.yamaguchi-u.ac.jp

Andrzej Skowron
Warsaw University, Institute of Mathematics
Banacha 2, 02-097 Warsaw, Poland
E-mail: skowron@mimuw.edu.pl

Setsuo Ohsuga
Waseda University, School of Science and Engineering
Department of Information and Computer Science
3-4-1 Okubo Shinjuku-ku, Tokyo 169, Japan
E-mail: ohsuga@ohsuga.info.waseda.ac.jp

Cataloging-in-Publication data applied for

Die Deutsche Bibliothek - CIP-Einheitsaufnahme

New directions in rough sets, data mining, and granular soft computing : 7th
international workshop ; proceedings / RSFDGrC'99, Yamaguchi, Japan, November 9
- 11, 1999. Ning Zhong ... (ed.). - Berlin ; Heidelberg ; New York ; Barcelona
; Hong Kong ; London ; Milan ; Paris ; Singapore ; Tokyo : Springer, 1999
  (Lecture notes in computer science ; Vol. 1711 : Lecture notes in artificial
intelligence)
  ISBN 3-540-66645-1

CR Subject Classification (1998): I.2, F.4.1, H.3, H.2.4, F.1

ISBN 3-540-66645-1 Springer-Verlag Berlin Heidelberg New York

© Springer-Verlag Berlin Heidelberg 1999
Printed in Germany

Typesetting: Camera-ready by author
SPIN 10705042   06/3142 – 5 4 3 2 1 0   Printed on acid-free paper

Preface

This volume contains the papers selected for presentation at the Seventh International Workshop on Rough Sets, Fuzzy Sets, Data Mining, and Granular-Soft Computing (RSFDGrC'99) held in the Yamaguchi Resort Center, Ube, Yamaguchi, Japan, November 9-11, 1999. The workshop was organized by the International Rough Set Society, the BISC Special Interest Group on Granular Computing (GrC), the Polish-Japanese Institute of Information Technology, and Yamaguchi University.

RSFDGrC'99 provided an international forum for sharing original research results and practical development experiences among experts in these emerging fields. An important feature of the workshop was to stress the role of the integration of intelligent information techniques. That is, to promote a deep fusion of these approaches to AI, soft computing, and database communities in order to solve real-world, large, complex problems concerned with uncertainty and fuzziness. In particular, rough and fuzzy set methods in data mining and granular computing were on display.

The total of 89 papers coming from 21 countries and touching a wide spectrum of topics related to both theory and applications were submitted to RSFDGrC'99. Out of them 45 papers were selected for regular presentations and 15 for short presentations. Seven technical sessions were organized, namely: Rough Set Theory and Its Applications; Fuzzy Set Theory and Its Applications; Non-classical Logic and Approximate Reasoning; Information Granulation and Granular Computing; Data Mining and Knowledge Discovery; Machine Learning; Intelligent Agents and Systems.

The RSFDGrC'99 program was enriched by four invited speakers: Zdzisław Pawlak, Lotfi A. Zadeh, Philip Yu, and Setsuo Arikawa, from soft computing, database, and AI communities. A special session on Rough Computing: Foundations and Applications was organized by James F. Peters.

An event like this can only succeed as a team effort. We would like to acknowledge the contribution of the program committee members and thank the reviewers for their efforts. Many thanks to the honorary chairs Zdzisław Pawlak and Lotfi A. Zadeh as well as the general chairs Setsuo Ohsuga and T.Y. Lin. Their involvement and support have added greatly to the quality of the workshop. Our sincere gratitude goes to all of the authors who submitted papers. We are grateful to our sponsors: Kayamori Foundation of Informational Science Advancement, United States Air Force Asian Office of Aerospace Research and Development, and Yamaguchi Industrial Technology Development Organizer, for their generous support. We wish to express our thanks to Alfred Hofmann of Springer-Verlag for his help and cooperation.

November 1999

Ning Zhong
Andrzej Skowron
Setsuo Ohsuga

RSFDGrC'99 Conference Committee

Honorary Chairs:

Zdzisław Pawlak   Polish Academy of Sciences, Poland
L.A. Zadeh   UC Berkeley, USA

General Chairs:

Setsuo Ohsuga   Waseda University, Japan
T.Y. Lin   San Jose State University, USA

Program Chairs:

Andrzej Skowron   Warsaw University, Poland
Ning Zhong   Yamaguchi University, Japan

Advisory Board:

Setsuo Arikawa   Kyushu U., Japan
Jerzy Grzymala-Busse   U. Kansas, USA
Katsushi Inoue   Yamaguchi U., Japan
T.Y. Lin   San Jose State U., USA
Masao Mukaidono   Meiji U., Japan
Setsuo Ohsuga   Waseda U., Japan
Zdzisław Pawlak   Polish Academy of Sci., Poland
Lech Polkowski   Polish-Japanese Inst. Infor. Tech., Poland
Zbigniew W. Ras   U. North Carolina, USA
Andrzej Skowron   Warsaw U., Poland
Roman Slowinski   Poznan U. Tech., Poland
Hideo Tanaka   Osaka Prefecture U., Japan
Shusaku Tsumoto   Shimane Medical U., Japan
Yiyu Yao   U. Regina, Canada
L.A. Zadeh   UC Berkeley, USA
Wojciech Ziarko   U. Regina, Canada
Ning Zhong   Yamaguchi U., Japan

Program Committee

S.K. Michael Wong   U. Regina, Canada
Xindong Wu   Colorado School of Mines, USA
Tetuya Yoshida   Osaka U., Japan
Philip Yu   IBM T.J. Watson Research Center, USA
Lizhu Zhou   Tsinghua U., China
Jan M. Żytkow   U. North Carolina, USA

Contents

Invited Talks

Rough Computing: Foundations and Applications

Rough Set Theory and Its Applications

Fuzzy Set Theory and Its Applications

Non-classical Logic and Approximate Reasoning

Information Granulation and Granular Computing

Data Mining and Knowledge Discovery

Machine Learning

Intelligent Agents and Systems

Decision Rules, Bayes' Rule and Rough Sets

Zdzisław Pawlak

Institute of Theoretical and Applied Informatics
Polish Academy of Sciences
ul. Bałtycka 5, 44 000 Gliwice, Poland
e-mail:zpw@ii.pw.edu.pl

Abstract. This paper concerns a relationship between Bayes' inference rule and decision rules from the rough set perspective.
In statistical inference based on the Bayes' rule it is assumed that some prior knowledge (prior probability) about some parameters without knowledge about the data is given first. Next the posterior probability is computed by employing the available data. The posterior probability is then used to verify the prior probability.
In the rough set philosophy with every decision rule two conditional probabilities, called *certainty* and *coverage factors*, are associated. These two factors are closely related with the lower and the upper approximation of a set, basic notions of rough set theory. Besides, it is revealed that these two factors satisfy the Bayes' rule. That means that we can use to data analysis the Bayes' rule of inference without referring to Bayesian philosophy of prior and posterior probabilities.

Key words: Bayes' rule, rough sets, decision rules, information system

1 Introduction

This paper is an extended version of the author's ideas presented in [5,6,7,8]. It concerns some relationships between probability, logic and rough sets and it refers to some concepts of Lukasiewicz presented in [3].

We will dwell in this paper upon the Bayesian philosophy of data analysis and that proposed by rough set theory.

Statistical inference grounded on the Bayes' rule supposes that some prior knowledge (prior probability) about some parameters without knowledge about the data is given first. Next the posterior probability is computed when the data are available. The posterior probability is then used to verify the prior probability.

In the rough set philosophy with every decision rule two conditional probabilities, called *certainty* and *coverage factors*, are associated. These two factors are closely related with the lower and the upper approximation of a set, basic concepts of rough set theory. Besides, it turned out that these two factors satisfy the Bayes' rule. That means that we can use to data analysis the Bayes' rule of inference without referring to Bayesian philosophy, i.e., to the prior and posterior probabilities. In other words, every data set with distinguished condition and decision attributes satisfies the Bayes' rule. This property gives a new look on reasoning methods about data.

2 Information System and Decision Table

Starting point of rough set based data analysis is a data set, called an information system.

An information system is a data table, whose columns are labelled by attributes, rows are labelled by objects of interest and entries of the table are attribute values.

Formally by an *information system* we will understand a pair $S = (U, A)$, where U and A, are finite, nonempty sets called the *universe*, and the set of *attributes*, respectively. With every attribute $a \in A$ we associate a set V_a, of its *values*, called the *domain* of a. Any subset B of A determines a binary relation $I(B)$ on U, which will be called an *indiscernibility relation*, and is defined as follows: $(x, y) \in I(B)$ if and only if $a(x) = a(y)$ for every $a \in A$, where $a(x)$ denotes the value of attribute a for element x. Obviously $I(B)$ is an equivalence relation. The family of all equivalence classes of $I(B)$, i.e., partition determined by B, will be denoted by $U/I(B)$, or simple U/B; an equivalence class of $I(B)$, i.e., block of the partition U/B, containing x will be denoted by $B(x)$.

If (x, y) belongs to $I(B)$ we will say that x and y are *B-indiscernible* or indiscernible with respect to B. Equivalence classes of the relation $I(B)$ (or blocks of the partition U/B) are referred to as *B-elementary sets* or *B-granules*.

If we distinguish in an information system two classes of attributes, called *condition* and *decision attributes*, respectively, then the system will be called a *decision table*.

A simple, tutorial example of an information system (a decision table) is shown in Table 1.

Table 1. An example of a decision table

Car	F	P	S	M
1	med.	med.	med.	poor
2	high	med.	large	poor
3	med.	low	large	poor
4	low	med.	med.	good
5	high	low	small	poor
6	med.	low	large	good

The table contains data about six cars, where F, P, S and M denote *fuel consumption, selling price, size* and *marketability*, respectively.

Attributes F, P and S are condition attributes, whereas M is the decision attribute. Each row of the decision table determines a decision obeyed when specified conditions are satisfied.

3   Approximations

Suppose we are given an information system (a datat set) $S = (U, A)$, a subset X of the universe U, and subset of attributes B. Our task is to describe the set X in terms of attribute values from B. To this end we define two operations assigning to every $X \subseteq U$ two sets $B_*(X)$ and $B^*(X)$ called the B-lower and the B-upper approximation of X, respectively, and defined as follows:

$$B_*(X) = \bigcup_{x \in U} \{B(x) : B(x) \subseteq X\},$$

$$B^*(X) = \bigcup_{x \in U} \{B(x) : B(x) \cap X \neq \emptyset\}.$$

Hence, the B-lower approximation of a set is the union of all B-granules that are included in the set, whereas the B-upper approximation of a set is the union of all B-granules that have a nonempty intersection with the set. The set

$$BN_B(X) = B^*(X) - B_*(X)$$

will be referred to as the B-boundary region of X.

If the boundary region of X is the empty set, i.e., $BN_B(X) = \emptyset$, then X is crisp (exact) with respect to B; in the opposite case, i.e., if $BN_B(X) \neq \emptyset$, X is referred to as rough (inexact) with respect to B.

For example, let $C = \{F, P, S\}$ be the set of all condition attributes. Then for the set $X = \{1, 2, 3, 5\}$ of cars with poor marketability we have $C_*(X) = \{1, 2, 5\}$, $C^*(X) = \{1, 2, 3, 5, 6\}$ and $BN_C(X) = \{3, 6\}$.

4   Decision Rules

With every information system $S = (U, A)$ we associate a formal language $L(S)$, written L when S is understood. Expressions of the language L are logical formulas denoted by Φ, Ψ etc. built up from attributes and attribute-value pairs by means of logical connectives $\wedge$ (and), $\vee$ (or), $\sim$ (not) in the standard way. We will denote by $\|\Phi\|_S$ the set of all objects $x \in U$ satisfying Φ in S and refer to as the meaning of Φ in S.

The meaning of Φ in S is defined inductively as follows:

1) $\|(a, v)\|_S = \{v \in U : a(v) = U\}$ for all $a \in A$ and $v \in V_a$,
2) $\|\Phi \vee \Psi\|_S = \|\Phi\|_S \cup \|\Psi\|_S$,
3) $\|\Phi \wedge \Psi\|_S = \|\Phi\|_S \cap \|\Psi\|_S$,
4) $\| \sim \Phi\|_S = U - \|\Phi\|_S$.

A formula Φ is *true* in S if $||\Phi||_S = U$.

A *decision rule* in L is an expression $\Phi \to \Psi$, read *if Φ then Ψ*; Φ and Ψ are referred to as *conditions* and *decisions* of the rule, respectively.

An example of a decision rule is given below

$$(F, med.) \wedge (P, low) \wedge (S, large) \to (M, poor).$$

Obviously a decision rule $\Phi \to \Psi$ is *true* in S if $||\Phi||_S \subseteq ||\Psi||_S$.

With every decision rule $\Phi \to \Psi$ we associate a conditional probability $\pi_S(\Psi|\Phi)$ that Ψ is true in S given Φ is true in S with the probability $\pi_S(\Phi)\frac{card(||\Phi||_S)}{card(U)}$, called the *certainty factor* and defined as follows:

$$\pi_S(\Psi|\Phi) = \frac{card(||\Phi \wedge \Psi||_S)}{card(||\Phi||_S)},$$

where $||\Phi||_S \neq 0$.

This coefficient is widly used in data mining and is called "confidence coefficient".

Obviously, $\pi_S(\Psi|\Phi) = 1$ if and only if $\Phi \to \Psi$ is true in S.

If $\pi_S(\Psi|\Phi) = 1$, then $\Phi \to \Psi$ will be called a *certain decision* rule; if $0 < \pi_S(\Psi|\Phi) < 1$ the decision rule will be referred to as a *possible decision* rule.

Besides, we will also need a *coverage factor*

$$\pi_S(\Phi|\Psi) = \frac{card(||\Phi \wedge \Psi||_S)}{card(||\Psi||_S)},$$

which is the conditional probability that Φ is true in S, given Ψ is true in S with the probability $\pi_S(\Psi)$.

Certainty and coverage factors for decision rules associated with Table 1 are given in Table 2.

Table 2. Certainty and coverage factors

Car	F	P	S	M	Cert.	Cov.
1	med.	med.	med.	poor	1	1/4
2	high	med.	large	poor	1	1/4
3	med.	low	large	poor	1/2	1/4
4	low	med.	med.	good	1	1/2
5	high	low	small	poor	1	1/4
6	med.	low	large	good	1/2	1/2

More about managing uncertainty in decision rules can be found in [2].

5 Decision Rules and Approximations

Let $\{\Phi_i \to \Psi\}_n$ be a set of decision rules such that:

all conditions Φ_i are pairwise mutually exclusive, i.e., $\|\Phi_i \wedge \Phi_j\|_S = \emptyset$, for any

$1 \leq i, j \leq n,\ i \neq j$, and $\qquad\qquad\qquad\qquad\qquad\qquad\qquad\qquad$ (1)

$$\sum_{i=1}^{n} \pi_S(\Phi_i|\Psi) = 1.$$

Let C and D be condition and decision attributes, respectively, and let $\{\Phi_i \to \Psi\}_n$ be a set of decision rules satisfying (1).

Then the following relationships are valid:

a) $C_*(\|\Psi\|_S) = \|\bigvee_{\pi(\Psi|\Phi_i)=1} \Phi_i\|_S,$

b) $C^*(\|\Psi\|_S) = \|\bigvee_{0<\pi(\Psi|\Phi_i)\leq 1} \Phi_i\|_S,$

c) $BN_C(\|\Psi\|_S) = \|\bigvee_{0<\pi(\Psi|\Phi_i)<1} \Phi_i\|_S = \bigcup_{i=1}^{n} \|\Phi_i\|_S.$

The above properties enable us to introduce the following definitions:

i) If $\|\Phi\|_S = C_*(\|\Psi\|_S)$, then formula Φ will be called the *C-lower approximation* of the formula Ψ and will be denoted by $C_*(\Psi)$;

ii) If $\|\Phi\|_S = C^*(\|\Psi\|_S)$, then the formula Φ will be called the *C-upper approximation* of the formula Φ and will be denoted by $C^*(\Psi)$;

iii) If $\|\Phi\|_S = BN_C(\|\Psi\|_S)$, then Φ will be called the *C-boundary* of the formula Ψ and will be denoted by $BN_C(\Psi)$.

Let us consider the following example.
The C-lower approximation of $(M, poor)$ is the formula

$$C_*(M, poor) = ((F, med.) \wedge (P, med.) \wedge (S, med.)) \vee$$
$$((F, high) \wedge (P, med.) \wedge (S, large)) \vee$$
$$((F, high) \wedge (P, low) \wedge (S, small)).$$

The C-upper approximation of $(M, poor)$ is the formula

$$C^*(M, poor) = ((F, med.) \wedge (P, med.) \wedge (S, med.)) \vee$$
$$((F, high) \wedge (P, med.) \wedge (S, large)) \vee$$
$$((F, med.) \wedge (P, low) \wedge (S, large)) \vee$$
$$((F, high) \wedge (P, low) \wedge (S, small)).$$

The C-boundary of $(M, poor)$ is the formula

$$BN_C(M, poor) = ((F, med.) \wedge (P, low) \vee (S, large)).$$

After simplification we get the following approximations

$$C_*(M, poor) = ((F, med.) \land (P, med.)) \lor (F, high),$$
$$C^*(M, poor) = (F, med.) \lor (F, high).$$

The concepts of the lower and upper approximation of a decision allow us to define the following decision rules:

$$C_*(\Psi) \rightarrow \Psi,$$

$$C^*(\Psi) \rightarrow \Psi,$$

$$BN_C(\Psi) \rightarrow \Psi.$$

For example, from the approximations given in the example above we get the following decision rules:

$((F, med.) \land (P, med.)) \lor (F, high) \rightarrow (M, poor),$
$(F, med.) \lor (F, high) \rightarrow (M, poor),$
$((F, med.) \land (P, low) \land (S, large)) \rightarrow (M, poor).$

From these definitions it follows that any decision Ψ can be uniquely discribed by the following two decision rules:

$$C_*(\Psi) \rightarrow \Psi,$$

$$BN_C(\Psi) \rightarrow \Psi.$$

From the above calculations we can get two decision rules

$((F, med.) \land (P, med.)) \lor (F, high) \rightarrow (M, poor),$
$((F, med.) \land (P, low.) \land (S, large)) \rightarrow (M, poor),$

which are associated with the lower approximation and the boudary region of the decision $(M, poor)$, respectively and describe decision $(M, poor)$.

Obviously we can get similar decision rules for the decision $(M, good)$ which are as follows:

$(F, low) \rightarrow (M, good),$
$((F, med.) \land (P, low.) \land (S, large)) \rightarrow (M, good).$

This coincides with the idea given by Ziarko [15] to represent decision tables by means of three decision rules corresponding to positive region the boundary region, and the negative region of a decision.

6 Decision Rules and Bayes' Rules

If $\{\Phi_i \rightarrow \Psi\}_n$ is a set of decision rules satisfying condition (1), then the well known formula for total probability holds:

$$\pi_S(\Psi) = \sum_{i=1}^{n} \pi_S(\Psi|\Phi_i) \cdot \pi_S(\Phi_i). \tag{2}$$

Moreover for any decision rule $\Phi \rightarrow \Psi$ the following Bayes' rule is valid:

$$\pi_S(\Phi_j|\Psi) = \frac{\pi_S(\Psi|\Phi_j) \cdot \pi_S(\Phi_j)}{\sum_{i=1}^{n} \pi_S(\Psi|\Phi_i) \cdot \pi_S(\Phi_i)}. \tag{3}$$

That is, any decision table or any set of implications satisfying condition (1) satisfies the Bayes' rule, without referring to prior and posterior probablities – fundamental in Baysian data analysis philosophy. Bayes' rule in our case says that: if an implication $\Phi \rightarrow \Psi$ is true to the degree $\pi_S(\Psi|\Phi)$ then the implication $\Psi \rightarrow \Phi$ is true to the degree $\pi_S(\Phi|\Psi)$.

This idea can be seen as a generalization of a *modus tollens* inference rule, which says that if the implication $\Phi \rightarrow \Psi$ is true so is the implication $\sim \Psi \rightarrow \sim \Phi$.

For example, for the set of decision rules

$((F,med.) \wedge (P,med.)) \vee (F, high) \rightarrow (M,poor)$,
$((F,med.) \wedge (P,low) \wedge (S,large)) \rightarrow (M,poor)$,
$(F,low) \rightarrow (M,good)$,
$((F,med.) \wedge (P,low) \wedge (S,large)) \rightarrow (M,good)$,

we get the values of ceratinty and coverage factors shown in Table 3.

Table 3. Initial decision rules

Rule	Decision	Certainty	Coverage
certain	poor	1	3/4
boundary	poor	1/2	1/4
certain	good	1	1/2
boundary	good	1/2	1/2

The above set of decison rules can be "reversed" as

$(M,poor) \rightarrow ((F,med.) \wedge (P,med.)) \vee (F, high)$,
$(M,poor) \rightarrow ((F,med.) \wedge (P,low) \wedge (S,large))$,
$(M,good) \rightarrow (F,low)$,

8

$(M,good) \rightarrow ((F,med.) \wedge (P,low) \wedge (S,large))$.

Due to Bayes' rule the certainty and coverage factors for inverted decision rules are mutually exchanged as shown in Table 4 below.

Table 4. Reversed decision rules

Rule	Decision	Certainty	Coverage
certain	poor	3/4	1
boundary	poor	1/4	1/2
certain	good	1/2	1
boundary	good	1/2	1/2

This property can be used to reason about data in the way similar to that allowed by *modus tollens* inference rule in classical logic.

7 Conclusions

It is shown in this paper that any decision table satisfies Bayes' rule. This enables to apply Bayes' rule of inference without referring to prior and posterior probabilities, inherently associated with "classical" Bayesian inference philosophy. From data tables one can extract decision rules – implications labelled by certainty factors expressing their degree of truth. The factors can be computed from data. Moreover, one can compute from data the coverage degrees expressing the truth degrees of "reverse" implications. This can be treated as generalization of *modus tollens* inference rule.

Acknowledgments

Thanks are due to Prof. Andrzej Skowron for his critical remarks.

References

1. Adams, E. W. The logic of conditionals, an application of probability to deductive logic. D. Reidel Publishing Company, Dordrecht, Boston, 1975.
2. Grzymała-Busse, J. Managing Uncertainty in Expert Systems; Kluwer Academic Publishers, Dordrecht, Boston, 1991,
3. Łukasiewicz, J. Die logishen Grundlagen der Wahrscheinilchkeitsrechnung. Krakow (1913). In: L. Borkowski (ed.): Jan Łukasiewicz – Selected Works, North Holland Publishing Company, Amsterdam, London, Polish Scientific Publishers, Warszawa, 1970.

4. Pawlak, Z. Rough Sets – Theoretical Aspects of Reasoning about Data; Kluwer Academic Publishers: Boston, Dordrecht, 1991.
5. Pawlak, Z. Reasoning about data – a rough set perspective. In: L. Polkowski, A. Skowron (eds.), Rough Sets and Current Trends in Computing, Lecture Notes in Artificial Intelligence, 1424 Springer, First International Conference, RSCTC'98, Warsaw, Poland, June, Proceedings, (1998) 25–34
6. Pawlak, Z. Rough Modus Ponens. In: Proceedings of Seventh International Conference, International Processing and Management of Uncertainty in Knowledge-Based (IPMU), Paris, France, July 6–10, 1998 1162–1166.
7. Pawlak, Z. Logic, probability and rough sets (to appear).
8. Pawlak, Z. Inference rules, decision rules and rough sets (to appear).
9. Pawlak, Z., Skowron, A. Rough membership functions. In: R.R. Yaeger, M. Fedrizzi, and J. Kacprzyk (eds.), Advances in the Dempster Shafer Theory of Evidence, John Wiley & Sons, Inc., New York, 1994, 251–271.
10. Polkowski, L., Skowron, A. (eds.). Rough Sets and Current Trends in Computing, Lecture Notes in Artificial Intelligence, 1424 Springer, First International Conference, RSCTC'98, Warsaw, Poland, June, Proceedings, 1998.
11. Polkowski, L., Skowron, A. (eds.). Rough Sets in Knowledge Discovery; Physica-Verlag: Vol. 1, 2, 1998.
12. Skowron, A. Menagement of uncertainty in AI: A rough set approach. In: Proceedings of the conference SOFTEKS, Springer Verlag and British Computer Society. 1994, 69–86.
13. Tsumoto, S., Kobayashi, S., Yokomori, T., Tanaka, H., Nakamura, A. (eds.). Proceedings of the Fourth International Workshop on Rough Sets, Fuzzy Sets, and Machine Discovery (RSFD'96). The University of Tokyo, November 6–8, 1996.
14. Tsumoto, S. Modelling medical diagnostic rules based on rough sets. In: L. Polkowski, A. Skowron (eds.), Rough Sets and Current Trends in Computing, Lecture Notes in Artificial Intelligence, 1424 Springer, First International Conference, RSCTC'98, Warsaw, Poland, June, Proceedings, 1998. 475–482.
15. Ziarko, W. Approximation Region -Based Decision Tables. In: L. Polkowski, A. Skowron (eds.): Rough Sets in Knowledge Discovery, Physica-Verlag, 1998, 1, 2. 178 -185.

A New Direction in System Analysis:
From Computation with Measurements to
Computation with Perceptions

Lotfi A. Zadeh*

Professor in the Graduate School
Director, Berkeley Initiative in Soft Computing (BISC)
Computer Science Division and the Electronics Research Laboratory
Dept. of EECS, University of California, Berkeley, CA 94720-1776
Telephone: 510-642-4959; Fax: 510-642-1712
E-mail: zadeh@cs.berkeley.edu

In one form or another, decision processes play a pivotal role in systems analysis. Decisions are based on information. More often than not, decision-relevant information is a mixture of measurements and perceptions.

It is a long-standing tradition in science to deal with perceptions by converting them into measurements. As is true of every tradition, a time comes when the underlying assumptions cease to be beyond question.

A thesis advanced in our work is that closer analysis leads to the conclusion that in most fields of science – and especially in systems analysis – conversion of perceptions into measurements is, in many cases, infeasible, unrealistic or counterproductive. The alternative is to develop a machinery for computation with perceptions which exploits the vast computational power of modern computers. In essence, this is the aim of the computational theory of perceptions (CTP). Somewhat paradoxically, the source of inspiration for this theory is the remarkable human capability to perform a wide variety of physical and mental tasks without any measurements and any computations. Underlying this capability is the brain's crucial ability to manipulate perceptions – perceptions of time, distance, direction, speed, force, shape, color, similarity, likelihood, intent and truth, among others.

The point of departure in the computational theory of perceptions – the first stage in the reasoning process – is conversion of perceptions into propositions expressed in a natural language, with a proposition viewed as a carrier of information which provides an answer to a question. A key idea in CTP is that the meaning of a proposition may be represented as a generalized constraint on a variable. This idea forms the basis for what is called constraint-centered semantics of natural languages (CSNL).

The second stage in CTP involves translation of propositions in the initial data set into the constraint language GCL, resulting in a collection of antecedent constraints which constitute the initial constraint set ICS. The third stage involves goal-directed propagation of initial constraints augmented with

* Research supported in part by NASA Grant NAC2-1177, ONR Grant N00014-96-1-0556, ARO Grant DAAH 04-961-0341 and the BISC Program of UC Berkeley.

decision-relevant constraints induced by an external knowledgebase. The goal is a terminal constraint set in GCL which upon retranslation – the fourth and last stage – yields the end result of the reasoning process.

Existing theories, especially probability theory, decision analysis and systems analysis, lack the capability to operate on information which is perception-based rather than measurement- based. The primary objective of the computational theory of perceptions is to add such capability to existing theories and thereby enhance their ability to deal with real world problems in an environment of imprecision, uncertainty and partial truth.

On Text Mining Techniques for Personalization

Charu C. Aggarwal and Philip S. Yu

IBM T. J. Watson Research Center, Yorktown Heights, NY 10598

Abstract. The popularity of the Web has made text mining techniques for personalization an increasingly important research topic. We first examine the problem on text mining for building categorization systems. Three different approaches which can be used for building categorization systems are discussed: classification, clustering and partial supervision. We examine the advantages and disadvantages of each approach. Some Web specific enhancements are discussed. Applications of text mining techniques to collaborative filtering have then been examined. Specifically, a content-based collaborative filtering approach is considered.

1 Introduction

The increased amount of online text data on the Web has led to the need for improved text mining techniques for personalization. In this paper we will discuss two important applications:

- **Categorization Systems:** In categorization systems, we wish to provide the ability of classifying documents into categories in an automated way. These categories may either be pre-decided from a training data set or may be generated using a clustering algorithm. Various tradeoffs will be discussed in this paper. Some web specific extensions for categorization are examined.
- **Collaborative Filtering Systems:** Collaborative filtering systems [18, 12, 2] are very applicable to electronic commerce sites in which purchases made by customers can be tracked. The record of purchases or Web pages browsed may be used in order to determine like-minded peer groups and make recommendations for individual customers based on the behavior of their peer groups. In content based collaborative filtering methods [6], a content characterization of the product or Web page is being used on the peer group formation to make these recommendations. Text mining techniques are very effective in using these content characterizations in order to provide recommendations.

2 Categorization Systems

Categorization systems have become increasingly important because of the need to classify large online repositories in a structured way. This can be very useful for personalization applications in which the analysis of textual material browsed on an E-commerce site by an online customer is used in order to make recommendations.

Categorization systems can be built with or without supervision from a pre-existing set of classes in another taxonomy. Several tradeoffs are possible and have been discussed in [1]. These tradeoffs are as follows:

- **Unsupervised systems:** In unsupervised systems, clustering methods are used in order to create sets of classes. These classes are then used in order to perform the recommendations. Examples of such systems include those discussed in [9, 10]. Improved methods for text clustering have also been discussed in [4, 9, 10, 17, 19, 21]. The advantage of this system is that the same measures which are used for clustering may be used for categorization. Thus, this system has 100% accuracy, though the actual quality of categorization is dependent on the nature of the initial clustering. Such systems may not be too useful for personalization because it is not possible to control the range of categories that the system can address. Furthermore, it is difficult to create effective fine grained subject isolation using unsupervised techniques.

- **Supervised Systems:** In supervised systems, a pre-existing sample of documents with the associated classes is available in order to provide the supervision to the categorization system. A training procedure is applied to this sample which models the relationship between the training data and the set of classes. Several text classifiers have recently been proposed [5, 7, 14, 15]. Although these methods seem to work well on structured collections such as the US patent database or the *Reuters* data set, the systems do not work well for heterogeneous collections of documents such as those on the Web. This is primarily because of the varying style, authorship and vocabulary in different documents. For example, it has been shown in [7] that a training procedure on the *Yahoo!* taxonomy achieves only 32% accuracy, whereas the same algorithm achieves much greater success (more that 66%) with the US patent database and Reuters data sets. Clearly, text data on the Web provides special problems in terms of fitting the training data into any particular model.

- **Partially Supervised Clustering:** We have developed a new approach on categorization systems, referred to as the partially supervised approach, where an initial training data set is used in order to partially supervise the creation of a new set of classes. This results in a categorization system in which it is possible to have some control over the range of subjects that one would like the categorization system to address, but with a precise automated definition of how each cluster is defined. The definition of the clusters may then be used for the categorization process. The details of such a categorization system can be found in [1] which uses a projected clustering technique [3] to handle high dimensional clustering, and has shown to be more effective than either purely supervised or unsupervised clustering. This is because the supervision ensures that one is able to create reasonably fine grained subject isolation which is related to the original taxonomy. At the same time, the system is very suited to automated categorization. We have used this system to categorize Web pages for personalized news feed.

It is possible to improve the accuracy of these systems on the Web further by adding certain Web-specific extensions. One interesting method for performing enhanced categorization is by using the information which is latent in hyperlinks [8]. Web pages tends to link with one another based on a proximity in the general subject areas which are discussed in each page. This information can be used in order to improve classifier performance by examining the content of the Web pages which are linked to by the current page. Such a method has been discussed in [8].

3   Content Based Collaborative Filtering Systems

In this section, we will discuss our work on an application of clustering to provide a generalization of the collaborative filtering concept to combine it with content based filtering [16], where recommendations are made on products with similar characteristic to the products likened by a customer. This is referred to as the content based collaborative filtering approach [6]. Content based collaborative filtering systems are useful in providing personalized recommendations at an E-commerce site. In such systems, a past history of customer behavior is available, which may be used for making future recommendations for individual customers. We also assume that a "content characterization" of products is available in order to perform recommendations. These characterizations may be (but are not restricted to) the text description of the products which are available at the Web site. The key here is that the characterizations should be such that they contain attributes (or textual words) which are highly correlated with buying behavior. In this sense, using carefully defined content attributes which are specific to the domain knowledge in question can be very useful for making recommendations. For example, in an engine which recommends CDs, the nature of the characterizations could be the singer name, music category, composer etc., since all of these attributes are likely to be highly correlated with buying behavior. On the other hand, if the only information available is the raw textual description of the products, then it may be desirable to use some kind of feature selection process in order to decide which words are most relevant to the process of making recommendations.

We will now proceed to describe the overall process and method of our approach for performing content-based collaborative filtering. This collaborative filtering process consists of the following sequence of steps, all of which are shown in Figure 1.

(1) **Feature Selection:** It is possible that the initial characterization of the products is quite noisy, and not all of the textual descriptions are directly related to buying behavior. For example, *stop words* (commonly occurring words in the language) in the description are unlikely to have much connection with the buying pattern in the products. In order to perform the feature selection, we perform the following process: we first create a preliminary customer characterization by concatenating the text descriptions for each product bought by the customer. Let the set of words in the lexicon

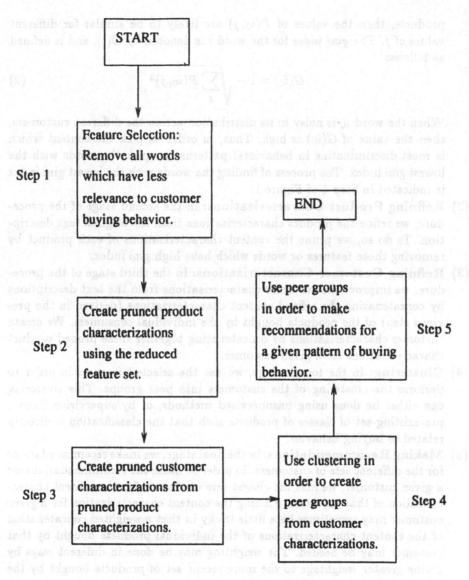

Fig. 1. Content Based Collaborative Filtering

describing the products be indexed by $i \in \{1, \ldots, k\}$, and let the set of customers j for which buying behavior is available be indexed by $j \in \{1, \ldots, n\}$. The frequency of presence of word i in customer characterization j is denoted by $F(i, j)$. The fractional presence of a word i for customer j is denoted by $P(i, j)$ and is defined as follows:

$$P(i, j) = \frac{F(i, j)}{\sum_{j \in \text{All customers}} F(i, j)} \quad (1)$$

Note that when the word $i = i_0$ is noisy in its distribution across the different products, then the values of $P(i_0, j)$ are likely to be similar for different values of j. The *gini index* for the word i is denoted by $G(i)$, and is defined as follows:

$$G(i_0) = 1 - \sqrt{\sum_j P(i_0, j)^2} \qquad (2)$$

When the word i_0 is noisy in its distribution across the different customers, then the value of $G(i_0)$ is high. Thus, in order to pick the content which is most discriminating in behavioral patterns, we pick the words with the lowest gini index. The process of finding the words with the lowest gini index is indicated in Step 1 of Figure 1.

(2) **Refining Product Characterizations:** In the second stage of the procedure, we refine the product characterizations from the original text description. To do so, we prune the content characterizations of each product by removing those features or words which have high gini index.

(3) **Refining Customer Characterizations:** In the third stage of the procedure, we improve the customer characterizations from the text descriptions by concatenating the refined content characterizations (derived in the previous step) of the products bought by the individual consumers. We create customer characterizations by concatenating together these pruned product characterizations for a given customer.

(4) **Clustering:** In the fourth stage, we use the selected features in order to perform the clustering of the customers into peer groups. This clustering can either be done using unsupervised methods, or by supervision from a pre-existing set of classes of products such that the classification is directly related to buying behavior.

(5) **Making Recommendations:** In the final stage, we make recommendations for the different sets of customers. In order to make the recommendations for a given customer, we find the closest sets of clusters for the content characterization of that customer. Finding the content characterization for a given customer may sometimes be a little tricky in that a weighted concatenation of the content characterizations of the individual products bought by that customer may be needed. The weighting may be done in different ways by giving greater weightage to the more recent set of products bought by the customer. The set of entities in this closest set of clusters forms the *peer group*. The buying behavior of this peer group is used in order to make recommendations. Specifically, the most frequently bought products in this peer group may be used as the recommendations. Several variations of the nature of queries are possible, and are discussed subsequently.

We have implemented these approaches in a content-based mining engine for making recommendations, and it seems to provide significantly more effective results than a simple clustering engine which uses only the identity attributes of the products in order to do the clustering.

Several kinds of queries may be resolved using such a system by using minor variations of the method discussed for making recommendations:

(1) For a given set of products browsed/bought, find the best recommendation list.
(2) For a given customer and a set of products browsed/bought by him in the current session, find the best set of products for that customer.
(3) For a given customer, find the best set of products for that customer.
(4) For the queries (1), (2), and (3) above, find the recommendation list out of a pre-specified promotion list.
(5) Find the closest peers for a given customer.
(6) Find the profile of the customers who will like a product the most.

Most of the above queries (with the exception of (6)) can be solved by using a different content characterization for the customer, and using this content characterization in order to find the peer group for the customer. For the case of query (6), we first find the peer group for the content characterization of the current product, and then find the dominant profile characteristics of this group of customers. In order to do so, the quantitative association rule method [20] may be used.

Another related application which we are working on is to provide user profiling on Web browsing patterns by categorizing the Web pages browsed by each person so as to identify the categories of interests to a person. Once the user characterization or profile is built, content-based collaborative filtering can then be applied. Here we have used the categorization system based on the partially supervised clustering approach to categorize the Web pages. Now the product characterization is replaced by the Web page categorization. The user categorization is the concatenation of the categories of Web pages browsed by a user.

4 Conclusions and Summary

In this paper, we discussed some categorization and clustering methods based on text mining, and their applications to content based collaborative filtering systems. With the recent increase in the popularity of the World Wide Web for electronic commerce, such systems are very useful for improving the efficiency of target marketing techniques. Specifically, such methods may be very useful in performing one-to-one sales promotions.

References

1. Aggarwal C. C., Gates S. C., Yu P. S.: On the merits of using supervised clustering for building categorization systems. *Proceedings of the ACM SIGKDD Conference*, (1999).
2. Aggarwal C. C., Wolf J. L., Wu K-L., Yu P. S.: Horting Hatches an Egg: A New Graph-Theorectic Approach to Collaborative Filtering. *Proceedings of the ACM SIGKDD Conference*, (1999).
3. Aggarwal C. C., Procopiuc C., Wolf J. L., Yu P. S., Park J-S.: Fast Algorithms for Projected Clustering. *Proceedings of the ACM SIGMOD Conference*, pages 61–72, (1999).

4. Anick P., Vaithyanathan S.: Exploiting clustering and phrases for context-based information retrieval. *Proceedings of the ACM SIGIR Conference*, pages 314–322, (1997).
5. Apte C., Damerau F., Weiss S. M.: Automated learning of decision rules for text categorization. *IBM Research Report* RC 18879.
6. Balabanovic M., Shoham Y.: Content-Based, Collaborative Recommendation. *Communications of the ACM*, Vol. 40, No. 3, pages 66–72, (1997).
7. Chakrabarti S., Dom B., Agrawal R., Raghavan P.: Using taxonomy, discriminants, and signatures for navigating in text databases. *Proceedings of the VLDB Conference*, (1997). *Extended Version:* Scalable feature selection, classification and signature generation for organizing text databases into hierarchical topic taxonomies. *VLDB Journal*, Vol. 7, pages 163–178, (1998).
8. Chakrabarti S., Dom B., Indyk P.: Enhanced hypertext categorization using hyperlinks. *Proceedings of the ACM SIGMOD Conference*, (1998).
9. Cutting D. R., Karger D. R., Pedersen J. O., Tukey J. W.: Scatter/Gather: A Cluster-based Approach to Browsing Large Document Collections. *Proceedings of the ACM SIGIR Conference*, pages 318–329, (1992).
10. Cutting D. R., Karger D. R., Pedersen J. O.: Constant Interaction-Time Scatter/Gather Browsing of Very Large Document Collections. *Proceedings of the ACM SIGIR Conference*, (1993).
11. Douglas Baker L., McCallum A. K.: Distributional Clustering of words for Text Classification. *Proceedings of the ACM SIGIR Conference*, pages 96–103, (1998).
12. Goldberg D., Nichols D., Oki B. M., Terry D.: Using Collabortive Filtering to Weave an Information Tapestry. *Communications of the ACM*, Vol. 35, No. 12, pages 61–70, (1992).
13. Hearst M. A., Pedersen J. O.: Re-examining the cluster hypothesis: Scatter/Gather on Retrieval Results. *Proceedings of the ACM SIGIR Conference*, pages 76–84, (1996).
14. Koller D., Sahami M.: Hierarchically classifying documents using very few words. *International Conference on Machine Learning*, Vol. 14, Morgan-Kaufmann, (1997).
15. Lam W., Ho C. Y.: Using a Generalized Instance Set for Automatic Text Categorization. *Proceedings of the ACM SIGIR Conference*, pages 81–88, (1998).
16. Lang, K.: Newsweeder: Learning to Filter Netnews. *Proceedings of the 12th Intl. Conference on Machine Learning*, (1995).
17. Schutze H., Silverstein C.: Projections for efficient document clustering. *Proceedings of the ACM SIGIR Conference*, pages 74–81, (1997).
18. Shardanand U., Maes, P.: Social Information Filtering: Algorithms for Automating "Word of Mouth". *Proceedings of the Conference on Human Factors in Computing Systems-CHI'95*, pages 210–217, (1995).
19. Silverstein C., Pedersen J. O.: Almost-constant time clustering of arbitrary corpus sets. *Proceedings of the ACM SIGIR Conference*, pages 60–66, (1997).
20. Srikant R., and Agrawal R.: Mining quantitative association rules in large relational tables. *Proceedings of the ACM SIGMOD Conference*, (1996).
21. Zamir O., Etzioni O.: Web Document Clustering: A Feasibility Demonstration. *Proceedings of the ACM SIGIR Conference*, pages 46–53, (1998).

A Road to Discovery Science

Setsuo Arikawa

Department of Informatics
Kyushu University, Japan

We have started a project: Grant-in-Aid for Scientific Research on Priority Area "Discovery Science" sponsored by Ministry of ESSC, Japan. The "Discovery Science" is a three-year project from 1998 to 2000 that targets to develop new methods for knowledge discovery, install network environments for knowledge discovery, and establish the Discovery Science as a new area of Computer Science. A systematic research has started that ranges over philosophy, logic, reasoning, computational learning, information retrieval and data mining. This lecture will describe the outline of the project, especially how it has been prepared and organized, and will touch on some of the results so far obtained.

Calculi of Granules Based on Rough Set Theory: Approximate Distributed Synthesis and Granular Semantics for Computing with Words

Lech Polkowski[1] and Andrzej Skowron[2]

Polish-Japanese Institute of Information Technology[1]
Koszykowa 86 02008 Warsaw, Poland
& Institute of Mathematics
Warsaw University of Technology[1]
email:polkow@pjwstk.waw.pl
Institute of Mathematics of Warsaw University[2]
Banacha 2 02097 Warsaw Poland
email: skowron@mimuw.edu.pl

Abstract. Rough mereology is a paradigm allowing for a synthesis of main ideas of two potent paradigms for reasoning under uncertainty : fuzzy set theory and rough set theory. In this work, we demonstrate applications of rough mereology to the important theoretical ideas put forth by Lotfi Zadeh [9], [10]: Granularity of Knowledge and Computing with Words.

Keywords reasoning under uncertainty, rough mereology, granule calculi, distributed systems, approximate synthesis, computing with words, synthesis grammars

1 Introduction

We refer the reader to [2], [4], [5], [6], [7] for the basic notions of rough set theory and rough mereology. Here, we begin with the notion of a pre-granule of knowledge. Given either an *information system* or a *decision system* $A = (U, A)$ (resp. $A = (U, A, d)$), we define information sets of objects via $Inf_B(u) = \{(a, a(u)) : a \in B\}$ and we express the indiscernibility of objects as the identity of their information sets : $IND_B(u, w)$ *is TRUE* iff $Inf_B(u) = Inf_B(w)$ for any pair u, w of objects in U. The (boolean) algebra generated over the set of atoms U/IND_B by means of set - theoretical operations of union, intersection and complement is said to be the B-algebra $CG(B)$ *of pre - granules.*

1.1 Granules of Knowledge: Rough set approach

In the language of granules, we may express partial dependencies between sets B, C of attributes by relating classes of IND_B to classes of IND_C. We will call, accordingly, a $(B, C) - granule$ any pair (G, G') where $G \in CG(B)$ and $G' \in CG(C)$. Clearly, given a pre - granule $G \in CG(B)$, there exists a formula (unique in DNF) α_G of the form $\vee_i \wedge_j (a_{i,j} = v_{i,j})$ such that the meaning

$[\alpha_G] = G$; Then, the granule (G, G') may be represented logically as a pair $[\alpha_G], [\alpha_{G'}]$ corresponding to the dependency rule $\alpha_G \implies \alpha_{G'}$ [2]. There are two characteristics of the granule (G, G') important in applications to decision algorithms viz. the characteristic whose values measure what part of $[G']$ is in $[G]$ (the *strength* of the rule $\alpha_G \implies \alpha_{G'}$) and the characteristic whose values measure what part of $[G]$ is in $[G']$ (the *strength of the support* for the rule $\alpha_G \implies \alpha_{G'}$).

A standard choice of an appropriate measure may be based on frequency count; the formal rendering is the *standard rough inclusion function* [3] defined for two sets $X, Y \subseteq U$ by the formula $\mu(X, Y) = \frac{card(X \cap Y)}{card(X)}$ when X is non - empty and $\mu(X, Y) = 1$, otherwise.

To select sufficiently strong rules, we would set a threshold ρ_{cr}. We define then, in analogy with machine learning techniques, two characteristics:

(ρ) $\rho(G, G') = \mu([G], [G'])$; ($\eta$) $\eta(G, G') = \mu([G'], [G])$

and we call an (η, ρ) *granule of knowledge* any granule (G, G') such that (i) $\rho(G, G') \geq \rho_{cr}$; (ii) $\eta(G, G') \geq \rho_{cr}$

This logical model of granulation may not be adequate to practical demands: the relation IND may be too rigid and ways of its relaxation are among most intensively studied topics [7]. Here, we propose to introduce rough mereological approach to the granulation problem in which IND-classes are replaced with mereological classes i.e. similarity classes.

2 Rough mereology

Rough mereology [4], [5], [8] has been proposed and studied as means of clustering in a relational way. Formally, it defines a functor $\mu(r)$ of being a part in degree at least r for each $r \in [0, 1]$. Rough mereology may be introduced conveniently in the logical framework of ontology and mereology proposed by Stanisław Leśniewski [1].

2.1 Mereology

We begin with the notion of a *part* functor. This sets the meaning of "X is a part of Y". We will use the notation of Ontology of Leśniewski $X \varepsilon Y$ (reads "X is Y") which replaces the standard notation of naive set theory as more convenient. The meaning of $X \varepsilon Y$ is specified as:

$$X \varepsilon Y \iff \exists Z. Z \varepsilon X \wedge \forall Z. (Z \varepsilon X \implies Z \varepsilon Y) \wedge \forall U, W. (U \varepsilon X \wedge W \varepsilon X \implies U \varepsilon W)$$

which means that X is an individual, anything which is X is Y and X is non-empty. The symbol V denotes the universe and is defined via

$$X \varepsilon V \iff \exists Y. X \varepsilon Y.$$

We rephrase basic axioms for *pt*.

(ME1) $X \varepsilon pt(Y) \implies \exists Z. Z \varepsilon X \wedge X \varepsilon V \wedge Y \varepsilon V;$

(ME2) $X\varepsilon pt(Y) \wedge Y\varepsilon pt(Z) \Longrightarrow X\varepsilon pt(Z)$;
(ME3) $non(X\varepsilon pt(X))$.

Then $X\varepsilon pt(Y)$ means that the individual denoted X is a proper part (in virtue of (ME3)) of the individual denoted Y. The concept of an improper part is reflected in the notion of an *element el*; this is a name - forming functor defined as follows: $X\varepsilon el(Y) \Longleftrightarrow X\varepsilon pt(Y) \vee X = Y$.

We will require that the following inference rule be valid.

(ME4) $\forall T.(T\varepsilon el(X) \Longrightarrow \exists W.W\varepsilon el(T) \wedge W\varepsilon el(Y)) \Longrightarrow X\varepsilon el(Y)$.

The notion of a collective class i.e. of an object composed of other objects which are its elements may be introduced at this point; this is effected by means of a functor Kl defined as follows.

$$X\varepsilon Kl(Y) \Longleftrightarrow \exists Z.Z\varepsilon Y \wedge \forall Z.(Z\varepsilon Y \Longrightarrow Z\varepsilon el(X)) \wedge$$
$$\forall Z.(Z\varepsilon el(X) \Longrightarrow \exists U,W.U\varepsilon Y \wedge W\varepsilon el(U) \wedge W\varepsilon el(Z)).$$

Thus, the class consists of all objects which have an element in common with an object with the class defining property.

The notion of a class is subjected to the following restrictions

(ME5) $X\varepsilon Kl(Y) \wedge Z\varepsilon Kl(Y) \Longrightarrow Z\varepsilon X$ ($Kl(Y)$ is an individual);
(ME6) $\exists Z.Z\varepsilon Y \Longleftrightarrow \exists Z.Z\varepsilon Kl(Y)$ (the class exists for each non-empty name).

Thus, $Kl(Y)$ is defined for any non-empty name Y and $Kl(Y)$ is an individual object.

2.2 Rough Mereology: first notions

Rough Mereology has been proposed and studied in [4], [5], [8] as a vehicle for reasoning under uncertainty.

The following is a list of basic axiomatic postulates for Rough Mereology. We introduce a graded family μ_r, where $r \in [0,1]$ is a real number from the unit interval, of functors which would satisfy ($\mu_r(X)$ is a new name derived from X via μ_r).

(RM1) $X\varepsilon\mu_1(Y) \Longleftrightarrow X\varepsilon el(Y)$;
(RM2) $X\varepsilon\mu_1(Y) \Longrightarrow \forall Z.(Z\varepsilon\mu_r(X) \Longrightarrow Z\varepsilon\mu_r(Y))$;
(RM3) $X = Y \wedge X\varepsilon\mu_r(Z) \Longrightarrow Y\varepsilon\mu_r(Z)$;
(RM4) $X\varepsilon\mu_r(Y) \wedge s \leq r \Longrightarrow X\varepsilon\mu_s(Y)$;

One may have as an archetypical rough mereological predicate the rough membership function of Pawlak and Skowron [3] defined in an extended form as:
$$X\varepsilon\mu_r(Y) \Longleftrightarrow \frac{card(X \cap Y)}{card(X)} \geq r \text{ in case } X \text{ non-empty, 1 else}$$
where X, Y are (either exact or rough) subsets in the universe U of an information/decision system (U, A).

2.3 Rough mereological component of granulation

The functors μ_r may enter our discussion of a granule and of the relation Gr in each of the following ways:

1. Concerning the definitions $(\eta), (\rho)$ of functions η, ρ , we may replace in them the rough membership function μ with a function μ_r possibly better suited to a given context:

(ρ) $G\varepsilon\mu_{\rho_{cr}}G'$; (η) $G'\varepsilon\mu_{\rho_{cr}}G$.

2. The process of clustering may be described in terms of the class functor of mereology:

(ρ) $G\varepsilon el(Kl_{\rho_{cr}}(G'))$; (η) $G'\varepsilon el(Kl_{\rho_{cr}}(G))$

where $Kl_r(X)$ is the class of objects Z satisfying $Z\varepsilon\mu_r X$. We will adhere to this means of clustering and we denote in the sequel $Kl_r(X)$ with the symbol $gr(X, r)$ read as *the granule of radius r centered at X*.

3 Adaptive Calculus of Granules for Synthesis in Distributed Systems

We construct a mechanism for transferring granules of knowledge among agents by means of transfer functions induced by rough mereological connectives extracted from their respective information systems [5].

We now recall basic ingredients of our scheme of agents [5], [8].

3.1 Distributed Systems of Agents

We assume that a pair (Inv, Ag) is given where Inv is an *inventory of elementary objects* and Ag is a set of inteligent computing units called shortly *agents*. We consider an agent $ag \in Ag$. The agent ag is endowed with tools for reasoning and communicating about objects in its scope; these tools are defined by components of the agent label.

The *label of the agent ag* is the tuple

$$lab(ag) = (\mathbf{A}(ag), \mu(r)(ag), L(ag), Link(ag), O(ag), St(ag)), Unc - rel(ag),$$

$$Unc - rule(ag), Dec - rule(ag))$$

where

1. $\mathbf{A}(ag)$ is an information system of the agent ag.
2. $\mu_r(ag)$ is a functor of part in a degree at ag.
3. $L(ag)$ is a set of unary predicates (properties of objects) in a predicate calculus interpreted in the set $U(ag)$.
4. $St(ag) = \{st(ag)_1, ..., st(ag)_n\} \subset U(ag)$ is the set of *standard objects* at ag.
5. $Link(ag)$ is a collection of strings of the form $ag_1ag_2...ag_kag$ which are elementary teams of agents; we denote by the symbol $Link$ the union of the family $\{Link(ag) : ag \in Ag\}$.

6. $O(ag)$ is the set of operations at ag; any $o \in O(ag)$ is a mapping of the Cartesian product $U(ag_1) \times U(ag_2) \times ... \times U(ag_k)$ into the universe $U(ag)$ where $ag_1 ag_2 ... ag_k ag \in Link(ag)$.

7. $Unc-rel(ag)$ is the set of parameterized uncertainty relations $\rho_i = \rho_i(o_i(ag),$ $st(ag_1)_i, st(ag_2)_i, ..., st(ag_k)_i, st(ag))$ where $ag_1, ag_2, ... ag_k, ag \in Link(ag)$, $o_i(ag) \in O(ag)$ are such that

$$\rho_i((x_1, \varepsilon_1), (x_2, \varepsilon_2), ., (x_k, \varepsilon_k), (x, \varepsilon))$$

holds for $x_1 \in U(ag_1), x_2 \in U(ag_2), .., x_k \in U(ag_k)$ and $\varepsilon, \varepsilon_1, \varepsilon_2, .., \varepsilon_k \in [0, 1]$ iff $x_j \varepsilon \mu(ag_j)_{\varepsilon_j}(st(ag_j))$ for $j = 1, 2, .., k$ and $x \varepsilon \mu(ag)_\varepsilon(st(ag))$ where

$$o_i(st(ag_1), st(ag_2), .., st(ag_k)) = st(ag) \text{ and } o_i(x_1, x_2, .., x_k) = x.$$

Uncertainty relations express the agents knowledge about relationships among uncertainty coefficients of the agent ag and uncertainty coefficients of its children.

8. $Unc - rule(ag)$ is the set of uncertainty rules f_j where $f_j : [0, 1]^k \longrightarrow [0, 1]$ is a function which has the property that
   if $x_1 \in U(ag_1), x_2 \in U(ag_2), .., x_k \in U(ag_k)$ satisfy the conditions

$$x_i \varepsilon \mu(ag_i)_{\varepsilon(ag_i)}(st(ag_i)) \, for \, i = 1, 2, .., k$$

then $\mu_o(o_j(x_1, x_2, ..., x_k), st(ag)) \geq f_j(\varepsilon(ag_1), \varepsilon(ag_2), .., \varepsilon(ag_k))$
where all parameters are as in 7.

9. $Dec - rule(ag)$ is a set of decomposition rules $dec - rule_i$ and

$$(\Phi(ag_1), \Phi(ag_2), .., \Phi(ag_k), \Phi(ag)) \in dec - rule_i$$

where $\Phi(ag_1) \in L(ag_1), \Phi(ag_2) \in L(ag_2), .., \Phi(ag_k) \in L(ag_k), \Phi(ag) \in L(ag)$ and $ag_1 ag_2 .. ag_k ag \in Link(ag))$ iff there exists a collection of standards $st(ag_1), st(ag_2), ..., st(ag_k), st(ag)$ with the properties that $o_j(st(ag_1), st(ag_2)$ $,..., st(ag_k)) = st(ag)$, $st(ag_i)$ satisfies $\Phi(ag_i)$ for $i = 1, 2, .., k$ and $st(ag)$ satisfies $\Phi(ag)$. Decomposition rules are decomposition schemes in the sense that they describe the standard $st(ag)$ and the standards $st(ag_1), ..., st(ag_k)$ from which the standard $st(ag)$ is assembled under o_i in terms of predicates which these standards satisfy.

3.2 Approximate Synthesis of Complex Objects

The process of synthesis of a complex object (e.g. signal, action) by the above defined scheme of agents consists in our approach of the two communication stages viz. the top - down communication/negotiation process and the bottom - up communication/assembling process. We outline the two stages here in the language of approximate formulae.

Approximate logic of synthesis We assume for simplicity that the relation $ag' \leq ag$, which holds for agents $ag', ag \in Ag$ iff there exists a string $ag_1 ag_2$. $..ag_k ag \in Link(ag)$ with $ag' = ag_i$ for some $i \leq k$, orders the set Ag into a tree. We also assume that $O(ag) = \{o(ag)\}$ for $ag \in Ag$ i.e. each agent has a unique assembling operation.

We recall a logic $L(Ag)$ [5], [8] in which we can express global properties of the synthesis process.

Elementary formulae of $L(Ag)$ are of the form $< st(ag), \Phi(ag), \varepsilon(ag) >$ where $st(ag) \in St(ag), \Phi(ag) \in L(ag), \varepsilon(ag) \in [0,1]$ for any $ag \in Ag$. Formulae of $L(ag)$ form the smallest extension of the set of elementary formulae closed under propositional connectives $\vee, \wedge, \neg$ and under the modal operators $\Box, \Diamond$.

The meaning of a formula $\Phi(ag)$ is defined classically as the set $[\Phi(ag)] = \{u \in U(ag) : u$ has the property $\Phi(ag)\}$; we express satisfaction by $u \vdash \Phi(ag)$. For $x \in U(ag)$, we say that x $satifies$ $< st(ag), \Phi(ag), \varepsilon(ag) >$, in symbols:

$$x \vdash < st(ag), \Phi(ag), \varepsilon(ag) >,$$

iff (i) $\boxed{st(ag) \vdash \Phi(ag);}$ and (ii) $\boxed{x \varepsilon \mu(ag)_{\varepsilon(ag)}(st(ag)).}$

We extend satisfaction over formulae by recursion as usual.

By a *selection* over Ag we mean a function sel which assigns to each agent ag an object $sel(ag) \in U(ag)$. For two selections sel, sel' we say that sel *induces* sel', in symbols $sel \rightarrow_{Ag} sel'$ when $sel(ag) = sel'(ag)$ for any $ag \in Leaf(Ag)$ and $sel'(ag) = o(ag)(sel'(ag_1), sel'(ag_2), ..., sel'(ag_k))$ for any $ag_1 ag_2 ... ag_k ag \in Link$.

We extend the satisfiability predicate $\vdash$ to selections: for an elementary formula $< st(ag), \Phi(ag), \varepsilon(ag) >$, we let $sel \vdash < st(ag), \Phi(ag), \varepsilon(ag) >$ iff $sel(ag) \vdash < st(ag), \Phi(ag), \varepsilon(ag) >$.

We now let $sel \vdash \Diamond < st(ag), \Phi(ag), \varepsilon(ag) >$ when there exists a selection sel' satisfying the conditions: $sel \rightarrow_{Ag} sel'$; $sel' \vdash < st(ag), \Phi(ag), \varepsilon(ag) >$.

In terms of $L(Ag)$ it is possible to express the problem of synthesis of an approximate solution to the problem posed to Ag. We denote by $head(Ag)$ the root of the tree $(Ag, \leq)$ and by $Leaf(Ag)$ the set of leaf-agents in Ag. In the process of top - down communication, a requirement Ψ received by the scheme from an external source (which may be called a *customer*) is decomposed into approximate specifications of the form $< st(ag), \Phi(ag), \varepsilon(ag) >$ for any agent ag of the scheme. The decomposition process is initiated at the agent $head(Ag)$ and propagated down the tree. We are able now to formulate the synthesis problem.

Synthesis problem. *Given a formula*

$$\alpha : < st(head(Ag)), \Phi(head(Ag)), \varepsilon(head(Ag)) >$$

find a selection sel over the tree $(Ag, \leq)$ with the property $sel \vdash \alpha$.

A solution to the synthesis problem with a given formula α is found by negotiations among the agents based on uncertainty rules and their succesful result can be expressed by a top-down recursion in the tree $(Ag, \leq)$ as follows: given a local team $ag_1 ag_2 ... ag_k ag$ with the formula $< st(ag), \Phi(ag), \varepsilon(ag) >$

already chosen, it is sufficient that each agent ag_i choose a standard $st(ag_i) \in U(ag_i)$, a formula $\Phi(ag_i) \in L(ag_i)$ and a coefficient $\varepsilon(ag_i) \in [0,1]$ such that

(iii) $\boxed{(\Phi(ag_1), \Phi(ag_2), ... \Phi(ag_k), \Phi(ag)) \in Dec - rule(ag)}$
with standards $st(ag), st(ag_1), ..., st(ag_k)$;

(iv) $\boxed{f(\varepsilon(ag_1), .., \varepsilon(ag_k)) \geq \varepsilon(ag)}$
where f satisfies $unc - rule(ag)$ with $st(ag), st(ag_1), ..., st(ag_k)$ and $\varepsilon(ag_1), ..., \varepsilon(ag_k), \varepsilon(ag)$.

For a formula α, we call an α - *scheme* an assignment of a formula $\alpha(ag)$: $< st(ag), \Phi(ag), \varepsilon(ag) >$ to each $ag \in Ag$ in such manner that (iii), (iv) above are satisfied and $\alpha(head(Ag))$ is $< st(head(Ag)), \Phi(head(Ag)), \varepsilon(head(Ag)) >$. We denote this scheme with the symbol

$$sch(< st(head(Ag)), \Phi(head(Ag)), \varepsilon(head(Ag)) >).$$

We say that a selection sel is *compatible* with a scheme $sch(< st(head(Ag)), \Phi(head(Ag)), \varepsilon(head(Ag)) >)$ in case $sel(ag)\varepsilon\mu(ag)_{\varepsilon(ag)}(st(ag))$ for each leaf agent $ag \in Ag$.

The goal of negotiations can be summarized now as follows.

Proposition 1. *Given a formula* $< st(head(Ag)), \Phi(head(Ag)), \varepsilon(head(Ag)) >$: *if a selection sel is compatible with a scheme* $sch(< st(head(Ag)), \Phi(head(Ag)), \varepsilon(head(Ag)) >)$ *then* $sel \vdash \Diamond < st(head(Ag)), \Phi(head(Ag)), \varepsilon(head(Ag)) >$.

4   Calculi of Granules in (Inv, Ag)

We construct for a given system $(Ag, \leq)$ of agents a granulation relation $Gr(ag)$ for any agent $ag \in Ag$ depending on parameters $\varepsilon(ag), \mu(ag)$. We may have various levels of granulation and a fortiori various levels of knowledge compresion about synthesis; we address here a simple specimen.

4.1   Calculi of pre–granules

For a standard $st(ag)$ and a value $\varepsilon(ag)$, we denote by $gr(st(ag), \varepsilon(ag))$ the pre-granule $Kl_{\varepsilon(ag)}(st(ag))$; then, *a granule selector* sel_g is a map which for each $ag \in Ag$ chooses a granule $sel_g(ag) = gr(st(ag), \varepsilon(ag))$.

We say that $gr(st(ag), \varepsilon(ag))$ satisfies a formula α :$< st(ag), \Phi(ag), \varepsilon(ag) >$ ($gr(st(ag), \varepsilon(ag)) \vdash \alpha$) in case $st(ag) \vdash \Phi(ag)$. Given $ag_1 ag_2 ... ag_k ag \in Link$ and a formula $< st(ag), \Phi(ag), \varepsilon(ag) >$ along with f satisfying $unc - rule(ag)$ with $st(ag), st(ag_1), ..., st(ag_k)$ and $\varepsilon(ag_1), ..., \varepsilon(ag_k), \varepsilon(ag), o(ag)$ maps the product $\times_i gr(st(ag_i), \varepsilon(ag_i))$ into $gr(st(ag), \varepsilon(ag))$. Composing these mappings along the tree $(Ag, \leq)$, we define a mapping $prod_{Ag}$ which maps any set $\{gr(st(ag), \varepsilon(ag)) : ag \in Leaf(Ag)\}$ into a granule $gr(st(head(Ag)), \varepsilon(head(Ag)))$. We say that a selection sel_g is *compatible* with a scheme $sch(< st(head(Ag)), \Phi(head(Ag)), \varepsilon(head(Ag)) >)$ if

$$sel_g(ag_i) = gr(st(ag_i), \varepsilon'(ag_i))\varepsilon el(gr(st(ag_i), \varepsilon(ag_i)))$$

for each leaf agent ag_i. As

$$prod_{Ag}(sel_g) \vdash < st(head(Ag)), \Phi(head(Ag)), \varepsilon(head(Ag)) >$$

we have the pre-granule counterpart of Proposition 1.

Proposition 2. *Given a formula $< st(head(Ag)), \Phi(head(Ag)), \varepsilon(head(Ag)) >$: if a selection sel_g is compatible with a scheme $sch(< st(head(Ag)), \Phi(head(Ag)), \varepsilon(head(Ag)) >)$ then $sel_g \vdash \Diamond < st(head(Ag)), \Phi(head(Ag)), \varepsilon(head(Ag)) >$.*

5   Associated grammar systems: a granular semantics for Computing with Words

We are now in position to present our discussion in the form of a grammar system related to the multi - agent tree $(Ag, \leq)$ [6]. With each agent $ag \in Ag$, we associate a grammar $\Gamma(ag) = (N(ag), T(ag), P(ag))$. To this end, we assume that a finite set $\Xi(ag) \subset [0, 1]$ is selected for each agent ag. We let $N(ag) = \{(s_{\Phi(ag)}, t_{\varepsilon(ag)}) : \Phi(ag) \in L(ag), \varepsilon(ag) \in \Xi(ag)\}$ where $s_{\Phi(ag)}$ is a non–terminal symbol corresponding in a one - to - one way to the formula $\Phi(ag)$ and similarly $t_{\varepsilon(ag)}$ corresponds to $\varepsilon(ag)$. The set of terminal symbols $T(ag)$ is defined for each non–leaf agent ag by letting

$$T(ag) = \bigcup \{\{(s_{\Phi(ag_i)}, t_{\varepsilon(ag_i)}) : \Phi(ag_i) \in L(ag_i), \varepsilon(ag_i) \in \Xi(ag_i)\} : i = 1, 2, ., k\}$$

where $ag_1 ag_2 ... ag_k ag \in Link$.

The set of productions $P(ag)$ contains productions of the form

(v) $\boxed{(s_{\Phi(ag)}, t_{\varepsilon(ag)}) \longrightarrow (s_{\Phi(ag_1)}, t_{\varepsilon(ag_1)})(s_{\Phi(ag_2)}, t_{\varepsilon(ag_2)})...(s_{\Phi(ag_k)}, t_{\varepsilon(ag_k)})}$

where $(o(ag), \Phi(ag_1), \Phi(ag_2), .., \Phi(ag_k), \Phi(ag), st(ag_1), st(ag_2), ..., st(ag_k), st(ag), \varepsilon(ag), \varepsilon(ag_1), \varepsilon(ag_2), ..., \varepsilon(ag_k))$ satisfy **(iii)**, **(iv)**.

We define a grammar system $\Gamma = (T, (\Gamma(ag) : ag \in Ag, ag \text{ non-leaf or } ag = Input), S)$ by choosing the set T of terminals as follows:

(vi) $\boxed{T = \{\{(s_{\Phi(ag)}, t_{\varepsilon(ag)}) : \Phi(ag) \in L(ag), \varepsilon(ag) \in \Xi(ag)\} : ag \in Leaf(Ag)\};}$

and introducing an additional agent $Input$ with non - terminal symbol S, terminal symbols of $Input$ being non-terminal symbols of $head(Ag)$ and productions of $Input$ of the form:

(vii) $\boxed{S \longrightarrow (s_{\Phi(head(Ag))}, t_{\varepsilon(head(Ag))})}$

where $\Phi(head(Ag)) \in L(head(Ag)), \varepsilon(head(Ag)) \in \Xi(head(Ag))$.

The meaning of S is that it codes an approximate specification (requirement) for an object; productions of $Input$ code specifications for approximate solutions in the language of the agent $head(Ag)$. Subsequent rewritings produce terminal strings of the form

(viii) $\boxed{(s_{\Phi\,(ag_1)}, t_{\varepsilon(ag_1)})(s_{\Phi(ag_2)}, t_{\varepsilon(ag_2)})...(s_{\Phi(ag_k)}, t_{\varepsilon(ag_k)})}$

where $ag_1, ag_2, .., ag_k$ are all leaf agents in Ag.

We have

Proposition 3. *Suppose* $(s_{\Phi(ag_1)}, t_{\varepsilon(ag_1)})(s_{\Phi(ag_2)}, t_{\varepsilon(ag_2)})...(s_{\Phi(ag_k)}, t_{\varepsilon(ag_k)})$ *is of the form (viii) and it is obtained from* $S \longrightarrow (s_{\Phi(head(Ag))}, t_{\varepsilon(head(Ag))})$ *by subsequent rewriting by means of productions in* Γ. *Then given any selection sel with* $sel(ag_i)\varepsilon\mu(ag_i)(\varepsilon(ag_i)st(ag_i)$ *for* $i = 1, 2, ..., k$ *we have*

$$sel \models \Diamond < st(head(Ag), \Phi(head(Ag)), \varepsilon(head(Ag) > .$$

Let us observe that each of grammars Γ is a linear context-free grammar. We have thus linear languages $L(\Gamma)$ which provide a semantics for Computing with Words.

Acknowledgement *This work has been prepared under the European ESPRIT Program in CRIT 2 Research Project No 20288, under the Research Grant No 8T11C 024 17 from the State Committee for Scientific Research (KBN) of the Republic of Poland, and under the Research Grant No 1/99 from the Polish-Japanese Institute of Information Technology (Lech Polkowski)*

References

1. S. Leśniewski, 1927, O podstawach matematyki, *Przegląd Filozoficzny*, 30, pp.164-206; 31, pp. 261-291; 32, pp. 60-101; 33, 77-105; 34, 142-170.
2. Z. Pawlak, 1992, *Rough Sets: Theoretical Aspects of Reasoning about Data*, Kluwer, Dordrecht.
3. Z. Pawlak and A. Skowron, 1994, Rough membership functions, in: R.R. Yaeger, M. Fedrizzi, J. Kacprzyk (eds.), *Advances in the Dempster Shafer Theory of Evidence*, John Wiley and S., New York, pp. 251-271.
4. L. Polkowski and A. Skowron, 1994, Rough mereology, in: *Lecture Notes in Artificial Intelligence*, vol. 869, Springer Verlag, Berlin, pp. 85-94.
5. L. Polkowski and A. Skowron, 1996, Rough mereology: a new paradigm for approximate reasoning, *International J. Approximate Reasoning*, 15(4), pp. 333-365.
6. L. Polkowski and A. Skowron, Grammar systems for distributed synthesis of approximate solutions extracted from experience, in: Gh. Paun and A. Salomaa (eds.), *Grammatical Models of Multi-Agent Systems*, Gordon and Breach Sci. Publ., Amsterdam, 1998, pp. 316-333.
7. L. Polkowski and A. Skowron (eds.), 1998, *Rough Sets in Knowledge Discovery*, vols. 18, 19 in the Series: *Studies in Fuzziness and Soft Computing*, J. Kacprzyk (ed.), Physica Verlag (Springer Verlag).
8. A. Skowron and L. Polkowski, Rough mereological foundations for design, analysis, synthesis and control in distributed systems, *Information Sciences. An Intern. J.*, 104(1-2) (1998), pp. 129-156.
9. L.A. Zadeh, 1996, Fuzzy logic = computing with words, *IEEE Trans. on Fuzzy Systems*, 4, pp. 103-111.
10. L. A. Zadeh, 1997, Toward a theory of fuzzy information granulation and its certainty in human reasoning and fuzzy logic, *Fuzzy Sets and Systems*, 90, pp. 111-127.

Discovery of Rules about Complications
– A Rough Set Approach in Medical Knowledge Discovery –

Shusaku Tsumoto

Department of Medicine Informatics, Shimane Medical University, School of Medicine,
89-1 Enya-cho Izumo City, Shimane 693-8501 Japan
E-mail: tsumoto@computer.org

Abstract. One of the most difficult problems in modeling medical reasoning is to model a procedure for diagnosis about complications. In medical contexts, a patient sometimes suffers from several diseases and has complicated symptoms, which makes a differential diagnosis very difficult. For example, in the domain of headache, a patient suffering from migraine, (a vascular disease), may also suffer from muscle contraction headache(a muscular disease). In this case, symptoms specific to vascular diseases will be observed with those specific to muscular ones. Since one of the essential processes in diagnosis of headache is discrimination between vascular and muscular diseases[1], simple rules will not work to rule out one of the two groups. However, medical experts do not have this problem and conclude both diseases. In this paper, three models for reasoning about complications are introduced and modeled by using characterization and rough set model. This clear representation suggests that this model should be used by medical experts implicitly.

1 Introduction

One of the most difficult problems in modeling medical reasoning is to model a procedure for diagnosis about complications. In medical contexts, a patient sometimes suffers from several diseases and has complicated symptoms, which makes a differential diagnosis very difficult. For example, in the domain of headache, a patient suffering from migraine, (a vascular disease), may also suffer from muscle contraction headache(a muscular disease). In this case, symptoms specific to vascular diseases will be observed with those specific to muscular ones. Since one of the essential processes in diagnosis of headache is discrimination between vascular and muscular diseases[2], simple rules will not work to rule out one of the two groups. However, medical experts do not have this problem and conclude both diseases.

In this paper, three models for reasoning about complications are introduced and modeled by using characterization and rough set model. This clear representation suggests that this model should be used by medical experts implicitly.

---

[1] The second step of differential diagnosis will be to discriminate diseases within each group.

[2] The second step of differential diagnosis will be to discriminate diseases within each group[2].

The paper is organized as follows: Section 2 discusses reasoning about complications. Section 3 shows the definitions of statistical measures used for modeling rules based on rough set model. Section 4 presents a rough set model of complications and an algorithm for induction of plausible diagnostic rules. Section 5 gives an algorithm for induction of reasoning about complications. Section 6 discusses related work. Finally, Section 7 concludes this paper.

2 Reasoning about Complications

Medical experts look for the possibilities of complications when they meet the following cases. (1) A patient has several symptoms which cannot be explained by the final diagnostic candidates. In this case, each diagnostic candidate belongs to the different disease category and will not intersect each other (independent type). (2) A patient has several symptoms which will be shared by several diseases, each of which belongs to different disease categories, and which are important to confirm some diseases above. In this case, each diagnostic candidate will have some intersection with respect to characterization of diseases (boundary type). (3) A patient has several symptoms which suggest that his disease will progress into the more specific ones in the near future. In this case, the specific disease will belong to the subcategory of a disease (subcategory type).

3 Probabilistic Rules

3.1 Accuracy and Coverage

In the subsequent sections, we adopt the following notations, which is introduced in [7].

Let U denote a nonempty, finite set called the universe and A denote a nonempty, finite set of attributes, i.e., $a : U \to V_a$ for $a \in A$, where V_a is called the domain of a, respectively. Then, a decision table is defined as an information system, $A = (U, A \cup \{d\})$.

The atomic formulas over $B \subseteq A \cup \{d\}$ and V are expressions of the form $[a = v]$, called descriptors over B, where $a \in B$ and $v \in V_a$. The set $F(B, V)$ of formulas over B is the least set containing all atomic formulas over B and closed with respect to disjunction, conjunction and negation.

For each $f \in F(B, V)$, f_A denote the meaning of f in A, i.e., the set of all objects in U with property f, defined inductively as follows.

1. If f is of the form $[a = v]$ then, $f_A = \{s \in U | a(s) = v\}$
2. $(f \wedge g)_A = f_A \cap g_A;\ (f \vee g)_A = f_A \vee g_A;\ (\neg f)_A = U - f_a$

By the use of this framework, classification accuracy and coverage, or true positive rate is defined as follows.

Definition 1.
Let R and D denote a formula in $F(B, V)$ and a set of objects which belong to a decision d. Classification accuracy and coverage(true positive rate) for $R \to d$ is defined as:

$$\alpha_R(D) = \frac{|R_A \cap D|}{|R_A|} (= P(D|R)), \text{ and}$$

$$\kappa_R(D) = \frac{|R_A \cap D|}{|D|} (= P(R|D)),$$

where $|A|$ denotes the cardinality of a set A, $\alpha_R(D)$ denotes a classification accuracy of R as to classification of D, and $\kappa_R(D)$ denotes a coverage, or a true positive rate of R to D, respectively.

It is notable that these two measures are equal to conditional probabilities: accuracy is a probability of D under the condition of R, coverage is one of R under the condition of D. It is also notable that $\alpha_R(D)$ measures the degree of the sufficiency of a proposition, $R \to D$, and that $\kappa_R(D)$ measures the degree of its necessity.[3]

For example, if $\alpha_R(D)$ is equal to 1.0, then $R \to D$ is true. On the other hand, if $\kappa_R(D)$ is equal to 1.0, then $D \to R$ is true. Thus, if both measures are 1.0, then $R \leftrightarrow D$.

Also, Pawlak recently reports a Bayesian relation between accuracy and coverage[5]:

$$\alpha_R(D)P(D) = P(R|D)P(D) = P(R, D)$$
$$= P(R)P(D|R) = \kappa_R(D)P(R)$$

This relation also suggests that *a priori* and *a posteriori* probabilities should be easily and automatically calculated from database.

3.2 Definition of Rules

By the use of accuracy and coverage, a probabilistic rule is defined as:

$$R \stackrel{\alpha, \kappa}{\to} d \quad s.t. \quad R = \wedge_j \vee_k [a_j = v_k], \ \alpha_R(D) \geq \delta_\alpha,$$
$$\kappa_R(D) \geq \delta_\kappa.$$

This rule is a kind of probabilistic proposition with two statistical measures, which is an extension of Ziarko's variable precision model(VPRS) [12].[4]

It is also notable that both a positive rule and a negative rule are defined as special cases of this rule, as shown in the next subsections.

---

[3] These characteristics are from formal definition of accuracy and coverage. In this paper, these measures are important not only from the viewpoint of propositional logic, but also from that of modelling medical experts' reasoning, as shown later.

[4] This probabilistic rule is also a kind of *Rough Modus Ponens*[4].

4 Rough Set Model of Complications

4.1 Definition of Characterization Set

In order to model these three reasoning types, a statistical measure, coverage $\kappa_R(D)$ plays an important role in modeling, which is a conditional probability of a condition (R) under the decision D ($P(R|D)$).

Let us define a characterization set of D, denoted by $L(D)$ as a set, each element of which is an elementary attribute-value pair R with coverage being larger than a given threshold, δ_κ. That is,

$$L_{\delta_\kappa}(D) = \{[a_i = v_j] | \kappa_{[a_i = v_j]}(D) > \delta_\kappa\}.$$

Then, according to the descriptions in Section 2, three models of reasoning about complications will be defined as below:

1. Independent type: $L_{\delta_\kappa}(D_i) \cap L_{\delta_\kappa}(D_j) = \phi$,
2. Boundary type: $L_{\delta_\kappa}(D_i) \cap L_{\delta_\kappa}(D_j) \neq \phi$, and
3. Subcatgory type: $L_{\delta_\kappa}(D_i) \subseteq L_{\delta_\kappa}(D_j)$.

All three definitions correspond to the negative region, boundary region, and positive region[2], respectively, if a set of the whole elementary attribute-value pairs will be taken as the universe of discourse. Thus, reasoning about complications are closely related with the fundamental concept of rough set theory.

4.2 Characterization as Exclusive Rules

Characteristics of characterization set depends on the value of δ_κ. If the threshold is set to 1.0, then a characterization set is equivalent to a set of attributes in exclusive rules[8]. That is, the meaning of each attribute-value pair in $L_{1.0}(D)$ covers all the examples of D. Thus, in other words, some examples which do not satisfy any pairs in $L_{1.0}(D)$ will not belong to a class D.

Construction of rules based on $L_{1.0}$ are discussed in Subsection 4.4, which can also be found in [9, 10]. The differences between these two papers are the following: in the former paper, independent type and subcategory type for $L_{1.0}$ are focused on to represent diagnostic rules and applied to discovery of decision rules in medical databases. On the other hand, in the latter paper, a boundary type for $L_{1.0}$ is focused on and applied to discovery of plausible rules.

4.3 Rough Inclusion

Concerning the boundary type, it is important to consider the similarities between classes. In order to measure the similarity between classes with respect to characterization, we introduce a rough inclusion measure μ, which is defined as follows.

$$\mu(S,T) = \frac{|S \bigcap T|}{|S|}.$$

It is notable that if $S \subseteq T$, then $\mu(S,T) = 1.0$, which shows that this relation extends subset and superset relations. This measure is introduced by Polkowski and Skowron in their study on rough mereology[6]. Whereas rough mereology firstly applies to distributed information systems, its essential idea is rough inclusion: Rough inclusion focuses on set-inclusion to characterize a hierarchical structure based on a relation between a subset and superset. Thus, application of rough inclusion to capturing the relations between classes is equivalent to constructing rough hierarchical structure between classes, which is also closely related with information granulation proposed by Zadeh[11].

4.4 Rule Induction Algorithm

Algorithms for induction of plausible diagnostic rules (boundary type) are given in Fig 1 to 3, which are fully discussed in [10]. Since subcategory type and independent type can be viewed as special types of boundary type with respect to rough inclusion, rule induction algorithms for subcategory type and independent type are given if the thresholds for μ are set up to 1.0 and 0.0, respectively.

Rule induction(Fig 1.) consists of the following three procedures. First, the characterization of each given class, a list of attribute-value pairs the supporting set of which covers all the samples of the class, is extracted from databases and the classes are classified into several groups with respect to the characterization. Then, two kinds of sub-rules, rules discriminating between each group and rules classifying each class in the group are induced(Fig 2). Finally, those two parts are integrated into one rule for each decision attribute(Fig 3).

5 Induction of Complication Rules

Simple version of complication rules are formerly called disease image, which had a very simple form in earlier versions[8]. Disease image is constructed from $L_{0.0}(D)$, as disjunctive formula of all the members of this characterization set. In this paper, complication rules are defined more effectively to detect complications. This rule is used to detect complications of multiple diseases, acquired by all the possible manifestations of the disease. By the use of this rule, the manifestations which cannot be explained by the conclusions will be checked, which suggest complications of other diseases. These rules consists of two parts: one is a collection of symptoms, and the other one is a rule for each symptoms, which are important for detection of complications.

1. $R \overset{\alpha, \kappa}{\to} \neg d$ s.t. $R = \vee R_{jk} = \vee_j \vee_k [a_j = v_k]$,
   $\alpha_{R_{jk}}(D) = 0$.
2. $R_{jk} \overset{\alpha, \kappa}{\to} d_l$ s.t. $R_{jk} = [a_j = v_k]$,
   $\alpha_{R_{jk}}(D_l) > \eta_\alpha, \; \kappa_{R_{jk}}(D_l) > \eta_\kappa$,

where η denotes a threshold for α and κ.

The first part can be viewed as rules, whose attribute-value pairs belong to $U - L_{0.0}(D_i)$.for each class D_i. On the other hand, the second part can be viewed

procedure *Rule Induction (Total Process)*;
  var
    $i : integer$; $M, L, R : List$;
    $L_D : List$; /* A list of all classes */
  begin
    Calculate $\alpha_R(D_i)$ and $\kappa_R(D_i)$
      for each elementary relation R and each class D_i;
    Make a list $L(D_i) = \{R|\kappa_R(D) = 1.0\}$)
      for each class D_i;
    while $(L_D \neq \phi)$ **do**
      begin
        $D_i := first(L_D)$; $M := L_D - D_i$;
        while $(M \neq \phi)$ **do**
          begin
            $D_j := first(M)$;
            if $(\mu(L(D_j), L(D_i)) \leq \delta_\mu)$
              then $L_2(D_i) := L_2(D_i) + \{D_j\}$;
            $M := M - D_j$;
          end
        Make a new decision attribute D_i' for $L_2(D_i)$;
        $L_D := L_D - D_i$;
      end
    Construct a new table $(T_2(D_i))$ for $L_2(D_i)$.
    Construct a new table $(T(D_i'))$
      for each decision attribute D_i';
      Induce classification rules R_2 for each $L_2(D)$;
      /* Fig.2 */
      Store Rules into a List $R(D)$;
      Induce classification rules R_d
        for each D' in $T(D')$; /* Fig.2 */
      Store Rules into a List $R(D')(= R(L_2(D_i)))$
    Integrate R_2 and R_d into a rule R_D;
      /* Fig.3 */
  end {*Rule Induction* };

Fig. 1. An Algorithm for Rule Induction

as rules, whose attribute-value pairs comes from $L_{\eta_\kappa}(D_j)$ such that $i \neq j$. Thus, complication rules can be constructed from overlapping region of $U - L_{0.0}(D_i)$ and $L_{\eta_\kappa}(D_j)$.

6 Discussion

6.1 Conflict Analysis

It is easy to see the relations of independent type and subcategory type. While independent type suggests different mechanisms of diseases, subcategory type

```
procedure Induction of Classification Rules;
  var
    i : integer;    M, L_i : List;
  begin
    L_1 := L_{er};
    /* L_{er}: List of Elementary Relations */
    i := 1;    M := {};
    for i := 1 to n do
        /* n: Total number of attributes */
      begin
        while ( L_i ≠ {} ) do
          begin
            Select one pair R = ∧[a_i = v_j] from L_i;
            L_i := L_i - {R};
            if   (α_R(D) ≥ δ_α)    and    (κ_R(D) ≥ δ_κ)
              then  do S_{ir} := S_{ir} + {R};
            /* Include R as Inclusive Rule */
              else M := M + {R};
          end
        L_{i+1} := (A list of the whole combination of
        the conjunction formulae in M);
      end
  end {Induction of Classification Rules };
```

Fig. 2. An Algorithm for Classification Rules

does the same etiology. The difficult one is boundary type, where several symptoms are overlapped in each $L_{\delta_\kappa}(D)$. In this case, relations between $L_{\delta_\kappa}(D_i)$. and $L_{\delta_\kappa}(D_j)$ should be examined.

One approach to these complicated relations is conflict analysis[3]. In this analysis, several concepts which shares several attribute-value pairs, are analyzed with respect to qualitative similarity measure that can be viewed as an extension of rough inclusion. It will be our future work to introduce this methodology to analyze relations of boundary type and to develop an induction algorithms for these relations.

6.2 Granular Fuzzy Partition

Coverage is also closely related with granular fuzzy partition, which is introduced by Lin[1] in the context of granular computing.

Since coverage $\kappa_R(D)$ is equivalent to a conditional probability, $P(R|D)$, this measure will satisfy the condition on partition of unity, called BH-partition (If we select a suitable partition of universe, then this partition will satisfy $\sum_\kappa \kappa_R(D) = 1.0.$) Also, from the definition of coverage, it is also equivalent to the counting measure for $|[x]_R \cap D|$, since $|D|$ is constant in a given universe

```
procedure Rule Integration;
  var
    i : integer;    M, L₂ : List; R(Dᵢ) : List;
    /* A list of rules for Dᵢ */
    L_D : List; /* A list of all classes */
  begin
    while(L_D ≠ φ) do
      begin
        Dᵢ := first(L_D); M := L₂(Dᵢ);
        Select one rule R' → D'ᵢ from R(L₂(Dᵢ)).
        while (M ≠ φ) do
          begin
            D_j := first(M);
              Select one rule R → d_j for D_j;
              Integrate two rules: R ∧ R' → d_j.
            M := M − {D_j};
          end
        L_D := L_D − Dᵢ;
      end
  end {Rule Combination}
```

Fig. 3. An Algorithm for Rule Integration

U. Thus, this measure satisfies a "nice context", which holds:

$$|[x]_{R_1} \bigcap D| + |[x]_{R_2} \bigcap D| \leq |D|.$$

Hence, all these features show that a partition generated by coverage is a kind of granular fuzzy partition[1]. This result also shows that the characterization by coverage is closely related with information granulation.

From this point of view, the usage of coverage for characterization and grouping of classes means that we focus on some specific partition generated by attribute-value pairs, the coverage of which are equal to 1.0 and that we consider the second-order relations between these pairs. It is also notable that if the second-order relation makes partition, as shown in the example above, then this structure can also be viewed as granular fuzzy partition.

However, rough inclusion and accuracy do not always hold the nice context. It would be our future work to examine the formal characteristics of coverage (and also accuracy) and rough inclusion from the viewpoint of granular fuzzy sets.

References

1. Lin, T.Y. Fuzzy Partitions: Rough Set Theory, in *Proceedings of Seventh International Conference on Information Processing and Management of Uncertainty in Knowledge-based Systems(IPMU'98)*, Paris, pp. 1167-1174, 1998.
2. Pawlak, Z., *Rough Sets*. Kluwer Academic Publishers, Dordrecht, 1991.
3. Pawlak, Z. Conflict analysis. In: *Proceedings of the Fifth European Congress on Intelligent Techniques and Soft Computing (EUFIT'97)*, pp.1589-1591, Verlag Mainz, Aachen, 1997.
4. Pawlak, Z. Rough Modus Ponens. *Proceedings of IPMU'98* , Paris, 1998.
5. Pawlak, Z. Rough Sets and Decision Analysis, *Fifth IIASA workshop on Decision Analysis and Support*, Laxenburg, 1998.
6. Polkowski, L. and Skowron, A.: Rough mereology: a new paradigm for approximate reasoning. Intern. J. Approx. Reasoning **15**, 333-365, 1996.
7. Skowron, A. and Grzymala-Busse, J. From rough set theory to evidence theory. In: Yager, R., Fedrizzi, M. and Kacprzyk, J.(eds.) *Advances in the Dempster-Shafer Theory of Evidence*, pp.193-236, John Wiley & Sons, New York, 1994.
8. Tsumoto, S. Automated Induction of Medical Expert System Rules from Clinical Databases based on Rough Set Theory *Information Sciences* **112**, 67-84, 1998.
9. Tsumoto, S. Extraction of Experts' Decision Rules from Clinical Databases using Rough Set Model *Journal of Intelligent Data Analysis*, 2(3), 1998.
10. Tsumoto, S., Automated Discovery of Plausible Rules based on Rough Sets and Rough Inclusion, *Proceedings of PAKDD'99*, (in press), LNAI, Springer-Verlag.
11. Zadeh, L.A., Toward a theory of fuzzy information granulation and its certainty in human reasoning and fuzzy logic. *Fuzzy Sets and Systems* **90**, 111-127, 1997.
12. Ziarko, W., Variable Precision Rough Set Model. *Journal of Computer and System Sciences*. 46, 39-59, 1993.

Rough Genetic Algorithms

Pawan Lingras and Cedric Davies

Saint Mary's University
Halifax, Nova Scotia, B3H 3C3, Canada.
e-mail: Pawan.Lingras@StMarys.CA,
WWW: http://cs.stmarys.ca/faculty/home_pages/pawan/

Abstract. This paper proposes rough genetic algorithms based on the
notion of rough values. A rough value is defined using an upper and a
lower bound. Rough values can be used to effectively represent a range or
set of values. A gene in a rough genetic algorithm can be represented using a rough value. The paper describes how this generalization facilitates
development of new genetic operators and evaluation measures. The use
of rough genetic algorithms is demonstrated using a simple document
retrieval application.

1   Introduction

Rough set theory [9] provides an important complement to fuzzy set theory [14]
in the field of soft computing. Rough computing has proved itself useful in the
development of a variety of intelligent information systems [10, 11]. Recently,
Lingras [4–7] proposed the concept of rough patterns, which are based on the
notion of rough values. A rough value consists of an upper and a lower bound.
A rough value can be used to effectively represent a range or set of values for
variables such as daily temperature, rain fall, hourly traffic volume, and daily
financial indicators. Many of the mathematical operations on rough values are
borrowed from the interval algebra [1]. The interval algebra provides an ability
to deal with an interval of numbers. Allen [1] described how the interval algebra
can be used for temporal reasoning. There are several computational issues associated with temporal reasoning based on the interval algebra. van Beek [12] used
a subset of the interval algebra that leads to computationally feasible temporal
reasoning. A rough value is a special case of an interval, where only the upper
and lower bounds of the interval are used in the computations. A rough pattern
consisting of rough values has several semantic and computational advantages in
many analytical applications. Rough patterns are primarily used with numerical
tools such as neural networks and genetic algorithms, while the interval algebra
is used for logical reasoning.

Lingras [7] used an analogy with the heap sorting algorithm and object oriented programming to stress the importance of rough computing. Any computation done using rough values can also be rewritten in the form of conventional
numbers. However, rough values provide a better semantic interpretation of results, in terms of upper and lower bounds. Moreover, some of the numeric com-

putations can not be conceptualized without explicitly discussing the upper and lower bound framework [7].

This paper proposes a generalization of genetic algorithms based on rough values. The proposed rough genetic algorithms (RGAs) can complement the existing tools developed in rough computing.

The paper provides the definitions of basic building blocks of rough genetic algorithms, such as rough genes and rough chromosomes. The conventional genes and chromosomes are shown to be special cases of their rough equivalents. Rough extension of GAs facilitates development of new genetic operators and evaluators in addition to the conventional ones. Two new rough genetic operators, called *union* and *intersection*, are defined in this paper. In addition, the paper also introduces a measure called *precision* to describe the information contained in a rough chromosome. A *distance* measure is defined that can be useful for quantifying the dissimilarity between two rough chromosomes. Both precision and distance measures can play an important role in evaluating a rough genetic population. A simple example is also provided to demonstrate practical applications of the proposed RGAs.

Section 2 provides a brief review of genetic algorithms. Section 3 proposes the notion of rough genetic algorithms and the associated definitions. New rough genetic operators and evaluation measures are also defined in section 3. Section 4 contains a simple document retrieval example to illustrate the use of rough genetic algorithms. Summary and conclusions appear in section 5.

2 Brief Review of Genetic Algorithms

The origin of Genetic Algorithms (GAs) is attributed to Holland's [3] work on cellular automata. There has been significant interest in GAs over the last two decades. The range of applications of GAs includes such diverse areas as job shop scheduling, training neural nets, image feature extraction, and image feature identification [2]. This section contains some of the basic concepts of genetic algorithms as described in [2].

A genetic algorithm is a search process that follows the principles of evolution through natural selection. The domain knowledge is represented using a candidate solution called an *organism*. Typically, an organism is a single *chromosome* represented as a vector of length n:

$$c = (c_i \mid 1 \leq i \leq n), \tag{1}$$

where c_i is called a *gene*.

A group of organisms is called a *population*. Successive populations are called *generations*. A generational GA starts from initial generation $G(0)$, and for each generation $G(t)$ generates a new generation $G(t+1)$ using genetic operators such as *mutation* and *crossover*. The mutation operator creates new chromosomes by changing values of one or more genes at random. The crossover joins segments of two or more chromosomes to generate a new chromosome. An abstract view of a generational GA is given in Fig. 1.

Genetic Algorithm:
    generate initial population, $G(0)$;
    evaluate $G(0)$;
    for(t = 1; solution is not found, t++)
        generate $G(t)$ using $G(t-1)$;
        evaluate $G(t)$;

Fig. 1. Abstract view of a generational genetic algorithm

3   Definition of Rough Genetic Algorithms

In a rough pattern, the value of each variable is specified using lower and upper bounds:

$$x = (\underline{x}, \overline{x}), \tag{2}$$

where $\underline{x}$ is the lower bound and $\overline{x}$ is the upper bound of x. A conventional pattern can be easily represented as a rough pattern by specifying both the lower and upper bounds to be equal to the value of the variable. The rough values can be added as:

$$x + y = (\underline{x}, \overline{x}) + (\underline{y}, \overline{y}) = (\underline{x} + \underline{y}, \overline{x} + \overline{y}), \tag{3}$$

where x and y are rough values given by pairs $(\underline{x}, \overline{x})$ and $(\underline{y}, \overline{y})$, respectively. A rough value x can be multiplied by a number c as:

$$
\begin{aligned}
c \times x = c \times (\underline{x}, \overline{x}) = (c \times \underline{x}, c \times \overline{x}), \text{if } c \geq 0, \\
c \times x = c \times (\underline{x}, \overline{x}) = (c \times \overline{x}, c \times \underline{x}), \text{if } c < 0.
\end{aligned}
\tag{4}
$$

Note that these operations are borrowed from the conventional interval calculus. As mentioned before, a rough value is used to represent an interval or a set of values, where only the lower and upper bounds are considered relevant in the computation.

A *rough chromosome* r is a string of *rough genes* r_i:

$$r = (r_i \mid 1 \leq i \leq n) \tag{5}$$

A *rough gene* r_i can be viewed as a pair of conventional genes, one for the lower bound called *lower gene* $(\underline{r_i})$ and the other for the upper bound called *upper gene* $(\overline{r_i})$:

$$r_i = (\underline{r_i}, \overline{r_i}), \tag{6}$$

Fig. 2 shows an example of a rough chromosome. The value of each rough gene is the range for that variable. The use of a range means that the information conveyed by a rough chromosome is not precise. Hence, an information measure called *precision* given by eq. (7) may be useful while evaluating the fitness of a rough chromosome.

$$precision(r) = - \sum_{1 \leq i \leq n} \left(\frac{\overline{r_i} - \underline{r_i}}{Range_{max}(r_i)} \right). \tag{7}$$

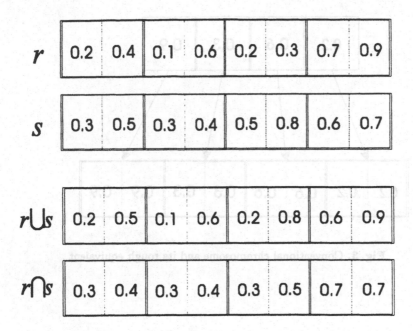

Fig. 2. Rough chromosomes along with associated operators and functions

In eq. (7), $Range_{max}(r_i)$ is the length of maximum allowable range for the value of rough gene r_i.

In Fig. 2,

$$precision(r) = -\frac{(0.4 - 0.2)}{1.0} - \frac{(0.6 - 0.1)}{1.0}$$
$$-\frac{(0.3 - 0.2)}{1.0} - \frac{(0.9 - 0.7)}{1.0}$$
$$= -1.0,$$

assuming that the maximum range of each rough gene is $[0, 1]$.

Any conventional chromosome can be represented as a rough chromosome as shown in Fig. 3. Therefore, rough chromosomes are a generalization of conventional chromosomes. For a conventional chromosome c, $precision(c)$ has the maximum possible value of zero.

New generations of rough chromosomes can be created using the conventional mutation and crossover operators. However, the mutation operator should make sure that $\overline{r_i} \geq \underline{r_i}$. Similarly, during the crossover a rough chromosome should be split only at the boundary of a rough gene, i.e. a rough gene should be treated as atomic.

In addition to the conventional genetic operators, the structure of rough genes enables us to define two new genetic operators called *union* and *intersection*. Let

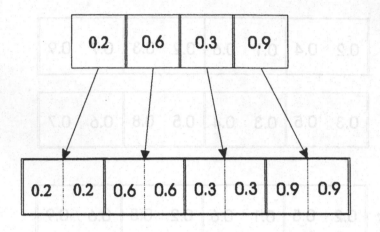

Fig. 3. Conventional chromosome and its rough equivalent

$r = (r_i \mid 1 \leq i \leq n)$ and $s = (s_i \mid 1 \leq i \leq n)$ be two rough chromosomes defined as strings of rough genes r_i and s_i, respectively. The *union* operator, denoted by the familiar symbol $\cup$, is given as follows:

$$r \cup s = (r_i \cup s_i \mid 1 \leq i \leq n), where$$
$$r_i \cup s_i = (min(\underline{r_i}, \underline{s_i}), max(\overline{r_i}, \overline{s_i})) \tag{8}$$

The *intersection* operator, denoted by the the symbol $\cap$, is given as follows:

$$r \cap s = (r_i \cap s_i \mid 1 \leq i \leq n), where$$
$$r_i \cap s_i = \begin{pmatrix} min(min(\overline{r_i}, \overline{s_i}), max(\underline{r_i}, \underline{s_i})), \\ max(min(\overline{r_i}, \overline{s_i}), max(\underline{r_i}, \underline{s_i})) \end{pmatrix}. \tag{9}$$

Fig. 2 illustrates the union and intersection operators.

A measure of similarity or dissimilarity between two chromosomes can be important during the evolution process. The distance between two rough chromosomes is given as follows:

$$distance(r, s) = \sum_{1 \leq i \leq n} \sqrt{\left(\underline{r_i} - \underline{s_i}\right)^2 + \left(\overline{r_i} - \overline{s_i}\right)^2} \tag{10}$$

The distance between rough chromosomes r and s from Fig. 2 can be calculated as:

$$distance(r, s) = \sqrt{(0.2 - 0.3)^2 + (0.4 - 0.5)^2}$$
$$+\sqrt{(0.1 - 0.3)^2 + (0.6 - 0.4)^2}$$
$$+\sqrt{(0.2 - 0.5)^2 + (0.3 - 0.8)^2}$$
$$+\sqrt{(0.7 - 0.6)^2 + (0.9 - 0.7)^2}$$
$$= 1.23.$$

4 An application of rough genetic algorithms

Information retrieval is an important issue in the modern information age. A huge amount of information is now available to the general public through the advent of the Internet and other related technologies. Previously, the information was made available through experts such as librarians, who helped the general public analyze their information needs. Because of enhanced communication facilities, the general public can access various documents directly from their desktop computers without having to consult a human expert. The modern information retrieval systems must assist the general public in locating documents relevant to their needs.

In the traditional approach, user queries are usually represented in a linear form obtained from the user. However, the user may not be able to specify his information needs in the mathematical form, either because he is not comfortable with the mathematical form, or the mathematical form does not provide a good representation of his information needs [8]. Wong and Yao [13] proposed the use of perceptrons to learn the user query based on document preference specified by the user for a sample set. Lingras [8] extended the approach using non-linear neural networks. This section illustrates how rough genetic algorithms can learn the user query from a sample of documents.

Let us consider a small sample of documents a, b, c, and d. Let us assume that each document is represented using four keywords: *Web search, Information Retrieval, Intelligent Agents* and *Genetic Algorithms*. Fig. 4 shows the documents represented as conventional chromosomes. The value $a_1 = 0.6$ corresponds to the relative importance attached to the keyword *Web Search* in document a. Similarly, $a_2 = 0.9$ corresponds to the relative importance attached to the keyword *Information Retrieval* in document a, etc.

As mentioned before, the user may not be able to specify the precise query that could be matched with the document set. However, given a sample set, she may be able to identify relevant and non-relevant documents. Let us assume that the user deemed a and b as relevant. The documents c and d were considered non-relevant to the user. This information can be used to learn a linear query by associating weights for each of the four keywords [13]. However, it may not

a	0.6	0.9	0.4	0.1
b	0.1	0.4	0.5	0.9
c	0.9	0.8	0.1	0.3
d	0.9	0.2	0.9	0.2

Fig. 4. Document set represented as conventional chromosomes

be appropriate to associate precise weights for each keyword. Instead, a range of weights such as 0.3-0.5, may be a more realistic representation of the user's opinion. A rough query can then be represented using rough chromosomes.

The user may supply an initial query and a genetic algorithm may generate additional random queries. The evolution process given by Fig. 1 can be used until the user's preference is adequately represented by a rough chromosome. Fig. 5 shows an objective function which may be used to evaluate the population in such an evolution process.

Objective function:
        repeat for all the relevant documents d
                repeat for all the non-relevant documents d'
                        if $distance(r,d) \leq distance(r,d')$ **then**
                                match++;
        return match;

Fig. 5. An example of objective function for document retrieval

Let us assume that r and s in Fig. 2 are our candidate queries.
In that case,
        $distance(r,a) = 2.53,$
        $distance(r,b) = 1.24,$

$$distance(r, c) = 2.53, \text{ and}$$
$$distance(r, d) = 3.05.$$

Similarly,

$$distance(s, a) = 2.29,$$
$$distance(s, b) = 1.21,$$
$$distance(s, c) = 2.67, \text{ and}$$
$$distance(s, d) = 2.00.$$

Using the objective function given in Fig. 5, rough chromosome r evaluates to 4 and s evaluates to 3. Hence, in the natural selection process, r will be chosen ahead of s. If both of these rough chromosomes were selected for creating the next generation, we may apply genetic operators such as mutation, crossover, union and intersection. The results of union and intersection of r and s are shown in Fig. 2.

The example used here demonstrates a few aspects of a rough genetic algorithm. Typically, we will select twenty candidate queries for every generation. Depending upon a probability distribution, the four different genetic operators will be applied to create the next generation. The evolution process will go on for several generations.

In practice, a document retrieval process will involve hundreds of keywords. Instead of classifying sample documents as relevant or non-relevant, it may be possible to rank the documents. Rough genetic algorithms may provide a suitable mechanism to optimize the search for the user query. Results of applications of RGAs for web searching will appear in a future publication. An implementation of rough extensions to a genetic algorithm library is also currently underway and may be available for distribution in the future.

5 Summary and Conclusions

This paper proposes Rough Genetic Algorithms (RGAs) based on the notion of rough values.

A rough value consists of an upper and a lower bound. Variables such as daily temperature are associated with a set of values instead of a single value. The upper and lower bounds of the set can represent variables using rough values.

Rough equivalents of basic notions such as gene and chromosomes are defined here as part of the proposal. The paper also presents new genetic operators, namely, *union* and *intersection*, made possible with the introduction of rough computing. These rough genetic operators provide additional flexibility for creating new generations during the evolution. Two new evaluation measures, called *precision* and *distance*, are also defined. The precision function quantifies information contained in a rough chromosome, while the distance function is used to calculate the dissimilarity between two rough chromosomes. A simple document retrieval example was used to demonstrate the usefulness of RGAs. Rough genetic algorithms seem to provide useful extensions for practical applications. Future publications will present results of such experimentation.

Acknowledgments

The authors would like to thank the Natural Sciences and Engineering Research Council of Canada for their financial support.

References

1. Allen, J. F.: Maintaining Knowledge about Temporal Intervals. Commnunication of the ACM. **26** (1983) 832-843
2. Buckles, B. P. and Petry, F.E.: Genetic Algorithms. IEEE Computer Press, Los Alamitos, California. (1994)
3. Holland, J.H.: Adaptation in Natural and Artificial Systems. University of Michigan Press, Ann Arbor. (1975)
4. Lingras, P.: Rough Neural Networks. Proceedings of Sixth International Conference on Information Processing and Management of Uncertainty in Knowledge-Based Systems, Granada, Spain (1996) 1445-1450
5. Lingras, P.: Unsupervised Learning Using Rough Kohonen Neural Network Classifiers. Proceedings of Symposium on Modelling, Analysis and Simulation, CESA'96 IMACS Multiconference, Lille, France (1996) 753-757
6. Lingras, P.: Comparison of neofuzzy and rough neural networks, Information Sciences: an International Journal. **110** (1998) 207-215
7. Lingras, P.: Applications of Rough Patterns. In: L. Polkowski and A. Skowron (eds.), Rough Sets in Data Mining and Knowledge Discovery 2, Series Soft Computing, Physica Verlag (Springer). (1998) 369-384
8. Lingras, P.: Neural Networks as Queries for Linear and Non-Linear Retrieval Models, Proceedings of Fifth International Conference of the Decision Sciences Institute, Athens, Greece. (1999) (to appear).
9. Pawlak, Z.: Rough sets. International Journal of Information and Computer Sciences. **11** (1982) 145-172
10. Pawlak Z.: Rough classification. International Journal of Man-Machine Studies. **20** (1984) 469-483
11. Pawlak, Z., Wong, S.K.M. and Ziarko, W.: Rough sets: probabilistic versus deterministic approach. International Journal of Man-Machine Studies. **29** (1988) 81-95
12. van Beek, P.: Reasoning about qualitative temporal information. Artificial Intelligence. **58** (1992) 297-326
13. Wong S.K.M. and Yao Y.Y.: Query Formulation in Linear Retrieval Models. Journal of the American Society for Information Science. (1990) **41(5)** 334-341
14. Zadeh, L.: Fuzzy Sets as a Basis for Theory of Possibility. Fuzzy Sets and Systems. **1** (1978) 3-28

Classifying Faults in High Voltage Power Systems: A Rough-Fuzzy Neural Computational Approach

L.Han, J.F. Peters, S.Ramanna and R.Zhai

Department of Electrical and Computer Engineering
University of Manitoba
Winnipeg, Manitoba, Canada R3T 5V6
e-mail: {liting, jfpeters}@ee.umanitoba.ca

Abstract: This paper introduces an approach to classifying faults in high voltage power system with a combination of rough sets and fuzzy sets in a neural computing framework. Typical error signals important for fault detection in power systems are considered. Features of these error signals derived earlier using Fast Fourier Transform analysis, amplitude estimation and waveform type identification, provide inputs to a neural network used in classifying faults. A form of rough neuron with memory is introduced in this paper. A brief overview of a rough-fuzzy neural computational method is given. The learning performance of a rough-fuzzy and pure fuzzy neural network are compared.

Keywords: Approximation, calibration, classification, faults, fuzzy sets, rough neuron, rough sets, neural network, high voltage power system

1 Introduction

A file of high voltage power system faults recorded by the Transcan Recording System (TRS) a Manitoba Hydro in the past three years provides a collection of unclassified signals. The TRS records power system data whenever a fault occurs. However, the TRS does not classify faults relative to waveform types. To date, a number of power system fault signal readings have been visually associated with seven waveform types. In this paper, a combination of rough set and fuzzy set are used in a neural computing framework to classify faults. Rough neural networks (rNNs) were introduced in 1996 [1], and elaborated in [2]-[4]. This paper reports research-in-progress on classifying power system faults and also introduces the design of neurons in rNNs in the context of rough sets.

This paper is organized as follows. Waveform types of power system faults are discussed in Section 2. The basic concepts of rough sets and design of a rough neural network are presented in Section 3. An overview of a form of rough-fuzzy neural computation is given in Section 4. In this section, the performance comparison between rough-fuzzy neural network and pure-fuzzy neural network is also provided.

2   Power System Faults

Using methods described in [5], a group of 26 pulse signals relative to seven types of waveforms have been selected for this study (see Table 1). Each value in Table 1 specifies the degree-of-membership of a pulse signal in the waveform of a particular fault type. Values greater than 0.5 indicate "definite" membership of a signal in a fault class. Values below 0.5 indicate uncertainty that a signal is of a particular fault type. From line 7 of Table 1, a value of 0.6734 indicates a high degree of certainty that a pole line flasher signal has a type 2 waveform.

Table 1   Sample Power System Faults Relative to Waveform Type

Fault	Degree-of-membership / Waveform Type						
	type1	type2	type3	type4	type5	type6	type7
Value Cab	0.0724	0.0231	0.0381	0.8990	0.0222	0.1109	0.0201
AC filter test	0.0752	0.0270	0.0447	0.1102	0.0259	0.6779	0.0158
Ring counter error	0.1383	0.0446	0.1300	0.0506	0.0410	0.0567	0.0109
500 Kv close	0.0862	0.1234	0.0626	0.2790	0.1224	0.2083	0.8334
pole line flash	0.0369	0.3389	0.0600	0.0251	0.2122	0.0289	0.0214
pole line flash	0.0340	0.6734	0.0573	0.0237	0.1539	0.0271	0.0201
pole line flash	0.0327	0.5836	0.0533	0.0231	0.1537	0.0263	0.0231
pole line flash	0.0337	0.4836	0.0561	0.0211	0.1767	0.0283	0.0221
pole line flash	0.0329	0.5336	0.0582	0.0241	0.1676	0.0275	0.0205
pole line retard	0.0326	0.2056	0.0548	0.0230	0.0854	0.0262	0.0156

3   Classifying Faults

In this paper, the classification of six high voltage power system faults relative to candidate waveforms is carried out with a neural network which combines the use of rough sets and fuzzy sets.

3.1   Basic Concepts of Rough Sets

Rough set theory offers a systematic approach to set approximation [6]-[7].   To begin, let S = (U, A) be an information system where U is a non-empty finite set of objects and A is a non-empty finite set of attributes where $a{:}U \to V_a$ for every a ∈ A.   For each B ⊆ A, there is associated an equivalence relation $Ind_B(A)$ such that

$$Ind_A(B) = \{(x, x') \in U^2 \mid \forall a \in B. \, a(x) = a(x')\}$$

If (x, x') ∈ $Ind_B(A)$, we say that objects x and x' are indiscernible from each other relative to attributes from B. The notation $[x]_B$ denotes equivalence classes of $Ind_B(A)$. For X ⊆ U, the set X can be approximated only from information contained in B by constructing a B-lower and B-upper approximation denoted by $\underline{B}X$ and $\overline{B}X$ respectively, where $\underline{B}X = \{ x \mid [x]_B \subseteq X \}$ and $\overline{B}X = \{ x \mid [x]_B \cap X \neq \varnothing \}$. The objects of $\underline{B}X$ can be classified as members of X with certainty, while the objects of $\overline{B}X$ can only be classified as possible members of X. Let $BN_B(X) = \overline{B}X - \underline{B}X$. A set X is rough if $BN_B(X)$ is not empty. The notation $\alpha_B(X)$ denotes the accuracy of an approximation, where

$$\alpha_B(X) = \frac{|\underline{B}X|}{|\overline{B}X|}$$

where $|X|$ denotes the cardinality of the non-empty set X, and $\alpha_B(X) \in [0, 1]$. The approximation of X with respect to B is precise, if $\alpha_B(X) = 1$. Otherwise, the approximation of X is rough with respect to B, if $\alpha_B(X) < 1$.

3.2 Example

Let PLF denote a pole line fault in a high voltage power system. The set $P = \{ x \mid PLF2(x) = yes \}$ consists of pole line fault readings which are judged to be type 2 waveforms (see Table 2).

Table 2. Sample PLF2 Decision Table

	PLF_α	PLF2
x1	in $[\tau, 1]$	yes
x2	in $[0, \beta)$	no
x3	in $[\tau, 1]$	yes
x4	in $[\beta, \tau)$	yes/no
x5	in $[\tau, 1]$	yes

In effect, PLF2 is a decision attribute whose outcome is synthesized in terms of hidden condition attributes. To see this, let τ, β be thresholds used to assess the candidacy of a fault reading in particular type of waveform and numerical boundary separating the possible approximation region from the rest of the universe, respectively. Recall that a power system fault f is considered to be a waveform of type t if the degree-of-membership of t is greater than or equal to some threshold. Next, we construct a sample decision table for pole line faults of type 2 (see Table 2). From Table 2, we obtain approximation regions $\underline{B}P = \{0.6734, 0.5836, 0.4836, 0.5336\}$ and $\overline{B}P = \{0.3389, 0.6734, 0.5836, 0.4836, 0.5336\}$ relative to condition attributes B (see Fig. 1). The set of pole line fault readings being classified is rough, since the boundary region $BN_B(P) = \{0.3389\}$ in Fig. 1 is non-empty. The accuracy of the approximation is high, since $\alpha_B(P) = 4/5$. We use the idea of approximation regions to design a rough neuron.

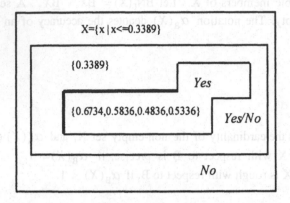

Fig. 1 Approximating the Set of Pole Line Fault Recordings

Notice that six power system faults are represented in Table 1. The degree-of-membership of each fault is computed relative to seven types of waveforms. This leads to 42 different rough sets, seven rough sets for each of the set of fault readings.

3.3 Design of Rough Neurons

A neuron is a processing element in a neural network. Informally, a rough neuron is a processing element designed to construct approximation regions $\underline{B}X$ and $\overline{B}X$ based on the evaluation of its input X on the basis of knowledge in a set of condition attributes B. Let r_m be a rough neuron with memory. Let X, τ be a set of unclassified fault readings and threshold used to assess the candidacy of a fault reading in a particular type of waveform, respectively. Further, let α be a degree-of-membership function such that $\alpha: X \to [0,1]$. Internally, a rough neuron r_m performs the following computation on each $x \in X$.

$$r_m(x) = \begin{cases} \underline{B}X \cup \{x\}, & \text{if } \alpha(x) \geq \tau \\ \overline{B}X \cup \{x\}, & \text{if } \alpha(x) < \tau \\ U - \overline{B}X \cup \{x\}, & \text{if } 0 \leq \alpha(x) < \beta \end{cases}$$

In effect, a rough neuron constructs approximation regions over time. To make this possible, a rough neuron is endowed with memory. During calibratilon, the approximation regions from the previous epoch are recalled and updated during the current epoch. The accuracy of the approximation computed by a rough neuron is measured by $\alpha_B(X)$, which is the output of a rough neuron.

4   Rough-Fuzzy Neural Network Computation

The basic features of a rough-fuzzy neural computing algorithm used to classify the waveforms of power system faults are shown in the flowchart in Fig. 2.

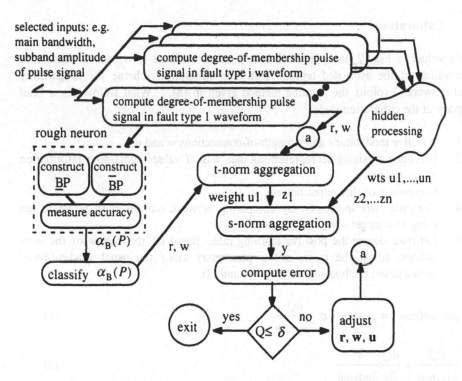

Fig. 2   Rough-Fuzzy Neural Computation

The details of the underlying network have been omitted due to space constraints. The computation in Fig. 2 begins with the initialization of modulator r and strengths-of-connection w, u. During calibration, r, w, u will be adjusted until the error Q is less than some threshold δ.

Let $\alpha: X \rightarrow [0,1]$ be a degree-of-membership function used in Fig. 2. The output of each rough neuron is aggregated with the degree-of-membership values using a t-, s-norm and implication ($\rightarrow$) operations from fuzzy set theory to compute z_j.

$$z_j = \mathop{T}_{i=1}^{n}\left[\left(r_{ij} \rightarrow \alpha(x)_i\right) \text{ s } w_{ij}\right]$$

In the next stage of the neural computation in Fig. 2, the weighted z-values are aggregated with an s-, t-norm operations to compute y.

$$y = \mathop{S}_{j=1}^{n}\left[z_j \text{ t } u_j\right]$$

4.1 Calibration

The flowchart in Fig. 2 has a feedback loop used to calibrate r, w, and u relative to target values of the estimated type-of-fault. The calibration scheme for rough-fuzzy neural networks exploits the learning method given in [8]. What follows is a brief summary of the calibration steps:

1. Initialize modulator r and strength-of-connections w and u.
2. Introduce a training set representing data sets of values containing information of fault signals.
3. Compute y of the output neuron.
4. Compute error indices Q, by comparing network outputs with target values using Q = target − y.
5. Let $\alpha > 0$ denote the positive learning rate. Based on the values of the error indices, adjust the r, w, and u parameters using the usual gradient-based optimization method suggested in (1) and (2).

$$param(new) = param - \alpha \frac{\partial Q}{\partial param} \tag{1}$$

$$\frac{\partial Q}{\partial param} = \frac{\partial Q}{\partial y} \frac{\partial y}{\partial param} \tag{2}$$

4.2 Learning Performance of Two Types of Networks

A rough-fuzzy and pure fuzzy neural network have been calibrated, and compared (see Figures 3, 4, and 5).

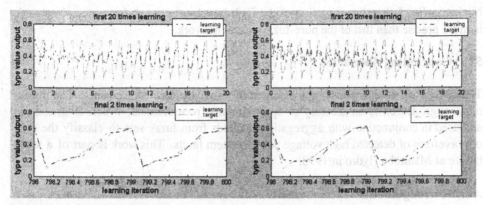

Fig. 3 Rough-fuzzy network performance   **Fig. 4** Performance of fuzzy network

A plot showing a comparison of learning performance of these networks during calibration is given in Figures 3 and 4. It is clear that for the same learning iteration, the performance of rough-fuzzy neural network is better than that of pure-fuzzy neural network. After the calibration of both neural networks, all of the connections relative to the r, w and u parameters have been determined. To test the performance of the sample rough-fuzzy and pure fuzzy neural networks, we utilize an additional 26 data sets of fault signals.

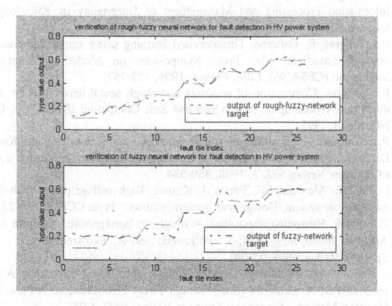

Fig. 5 Verification of rough-fuzzy and pure fuzzy networks

Notice in Fig. 5 that the estimation of the fault type by the rough-fuzzy neural network is more precise than that of the pure-fuzzy neural network.

5   Concluding Remarks

The design of a rough neuron in the context of rough sets has been given.   The output of a rough neuron is an accuracy of approximation measurement, which is granulated and used in conjunction with aggregation methods from fuzzy sets to classify the type of waveform of detected high voltage power system faults. This work is part of a study begun at Manitoba Hydro in 1998.

Acknowledgements

First, we wish to thank Prof. Pawan Lingras, St. Mary's University, Nova Scotia, Canada, for sharing copies of his papers with us.   We also want to thank the reviewers for their comments concerning this paper. We also gratefully acknowledge the support for this research provided by Manitoba Hydro and the National Sciences and Engineering Research Council of Canada (NSERC).

References

1.   P.J. Lingras, Rough neural networks. In: Proc. of the 6th Int. Conf. on Information Processing and Management of Uncertainty in Knowledge-based Systems (IPMU'96), Granada, Spain, 1996, 1445-1450.
2.   P.J. Lingras, P. Osborne, Unsupervised learning using rough Kohonen neural network classifiers. In: Proc. Symposium on Modelling, Analysis and Simulation (CESA'96), Lille, France, 1996, 753-757.
3.   P.J. Lingras, Comparison of neofuzzy and rough neural networks. In: Proc. of the 5th Int. Workshop on Rough Sets and Soft Computing (RSSC'97), Durham, NC, March 1997.
4.   P.J. Lingras, Applications of rough patterns. In: Rough Sets in Knowledge Discovery edited by L. Polkowski and A. Skowron. Physica Verlag, a division of Springer Verlag, vol. 2, 1998, 369-384.
5.   L. Han, R. Menzies, J.F. Peters, L. Crowe, High voltage power fault-detection and analysis system: Design and implementation. Proc. CCECE99, 1253-1258.
6.   Z. Pawlak. Reasoning about data--A rough set persepective. Lecture Notes in Artificial Intelligence 1424, L. Polkowski and A. Skowron (Eds.).   Berlin, Springer-Verlag, 1998, 25-34.
7.   J. Komorowski, Z. Pawlak, L. Polkowski, A. Skowron, Rough sets: A tutorial. In: S.K. Pal, A. Skowron (Eds.), Rough Fuzzy Hybridization: A New Trend in Decision-Making. Singapore: Springer-Verlag, 1999, 3-98.
8.   W. Pedrycz, J.F. Peters, Learning in fuzzy Petri nets, in *Fuzziness in Petri Nets* edited by J. Cardoso and H. Scarpelli. Physica Verlag, a division of Springer Verlag, 1998.

Toward Spatial Reasoning in the Framework of Rough Mereology

Lech Polkowski

Polish-Japanese Institute of Information Technology
Koszykowa 86 02008 Warsaw Poland
& Institute of Mathematics
Warsaw University of Technology
Pl.Politechniki 1 00661 Warsaw Poland
email:polkow@pjwstk.waw.pl

Abstract. Rough mereology is a paradigm allowing to blend main ideas of two potent paradigms for approximate reasoning : fuzzy set theory and rough set theory. Essential ideas of rough mereology and schemes for approximate reasoning in distributed systems based on rough mereological logic were presented in [13], [14], [17]. Spatial reasoning is an extensively studied paradigm stretching from theoretical investigations of proper languages and models for this reasoning to applicational studies concerned with e.g. geographic data bases, satellite image analyses, geodesy applications etc. We propose a rough mereological environment for spatial reasoning under uncertainty. We confront our context with an alternatively studied mereological context defined within Calculus of Individuals [10] by Clarke [5] and developed into schemes for spatial reasoning in [2], [3] where the reader will find examples of linguistic interpretation. We outline how to define in the rough mereological domain the topological and geometrical structures which are fundamental for spatial reasoning; we show that rough mereology allows for introducing notions studied earlier in other mereological theories [2], [3], [5]. This note sums up a first step toward our synthesis of intelligent control algorithms useful in mobile robotics [1], [7], [8].

Keywords rough mereology, mereotopology, spatial reasoning, connection, rough mereological geometry

1 Introduction

Rough mereology has been proposed in [13] and developed into a paradigm for approximate reasoning in [14]. Its applications to problems of approximate synthesis, control, design and analysis of complex objects have been discussed in [17] and in [15] a granular semantics for computing with words was proposed based on rough mereology. We are concerned here with the issues of spatial reasoning under uncertainty. Therefore we study the rough mereological paradigm in a geometric - mereotopological setting (cf. [2], [3]). Spatial reasoning plays an important role in intelligent robot control (cf. [1], [7], [8] and we are aiming at synthesizing a context for control under uncertainty of a mobile robot which may possibly involve natural language interfaces. Rough Mereology is a natural extension of Mereology (cf. [11], [18]) and we give as well a brief sketch of relevant theories of Ontology and Mereology to set a proper language for our discussion.

2 Ontology

Ontological theory of Leśniewski [9], [18] is concerned with the explanation of meaning of phrases like "X is Y" . Naive set theory solves this problem via the notion of an element; in Ontology, the esti symbol $\in$ is replaced by the copula ε (read "is"). Ontology makes use of functors of either of two categories: propositional and nominal; the former yield propositions the latter new names. We begin this very concise outline of Ontology by selecting symbols X, Y, Z .... to denote names (of objects); the primitive symbol of ontology is ε (read "is").

The sole Axiom of Ontology is a formula coding the meaning of ε as follows

2.1 Ontology Axiom

$$X\varepsilon Y \iff \exists Z.Z\varepsilon X \wedge \forall U, W.(U\varepsilon X \wedge W\varepsilon X \implies U\varepsilon W) \wedge \forall T.(T\varepsilon X \implies T\varepsilon Y)$$

This axiom determines the meaning of the formula $X\varepsilon Y$ ("X is Y") as the conjunction of three conditions: $\exists Z.Z\varepsilon X$ ("something is X"); $\forall U, W.(U\varepsilon X \wedge W\varepsilon X \implies U\varepsilon W)$ ("any two objects which are X are identical" i.e. X is an individual name); $\forall T.(T\varepsilon X \implies T\varepsilon Y)$ ("everything which is X is Y").

Therefore the meaning of the formula $X\varepsilon Y$ is as follows: X is a non-empty name of an individual (X is an individual) and any object which is X is also Y.

We introduce a name V defined via : $X\varepsilon V \iff \exists Y.X\varepsilon Y$ being a name for a universal object. The copula ε formalized as above permits to accomodate distributive classes (counterparts of sets in the naive set theory). The next step is to formalize the notion of distributive classes (counterparts of unions of families of sets). This belongs to Mereology.

3 Mereology

Mereology of Leśniewski [11], [19] can be based on any of a few primitive notions related one to another: part, element, class..; here, we begin with the notion of a *part* conceived as a name - forming functor pt on individual names.

3.1 Mereology Axioms

We start with basic axioms for pt.

(ME1) $X\varepsilon pt(Y) \implies \exists Z.Z\varepsilon X \wedge X\varepsilon V \wedge Y\varepsilon V$;

(ME2) $X\varepsilon pt(Y) \wedge Y\varepsilon pt(Z) \implies X\varepsilon pt(Z)$;

(ME3) $non(X\varepsilon pt(X))$.

Then $X\varepsilon pt(Y)$ means that the individual denoted X is a proper part (in virtue of (ME3)) of the individual denoted Y. The concept of an improper part is reflected in the notion of an *element el*; this is a name - forming functor defined as follows:

$X \varepsilon el(Y) \iff X \varepsilon pt(Y) \lor X = Y$.

We will require that the following inference rule be valid.

(ME4) $\forall T.(T \varepsilon el(X) \implies \exists W.W \varepsilon el(T) \land W \varepsilon el(Y)) \implies X \varepsilon el(Y)$.

3.2 Classes

The notion of a collective class may be introduced at this point; this is effected by means of a name - forming functor Kl defined as follows.

$X \varepsilon Kl(Y) \iff$

$$\exists Z.Z \varepsilon Y \land \forall Z.(Z \varepsilon Y \implies Z \varepsilon el(X)) \land$$
$$\forall Z.(Z \varepsilon el(X) \implies \exists U, W.U \varepsilon Y \land W \varepsilon el(U) \land W \varepsilon el(Z)).$$

The notion of a class is subjected to the following restrictions

(ME5) $X \varepsilon Kl(Y) \land Z \varepsilon Kl(Y) \implies Z \varepsilon X$ ($Kl(Y)$ is an individual);

(ME6) $\exists Z.Z \varepsilon Y \iff \exists Z.Z \varepsilon Kl(Y)$ (the class exists for each non-empty name).

Thus, $Kl(Y)$ is defined for any non-empty name Y and $Kl(Y)$ is an individual object. One can also introduce a less restrictive name viz. of a set:

$X \varepsilon set(Y) \iff$

$$\exists Z.Z \varepsilon Y \land \forall Z.(Z \varepsilon el(X) \implies \exists U, W.U \varepsilon Y \land W \varepsilon el(U) \land W \varepsilon el(Z)).$$

Thus, a set is like a class except for the universality property $\forall Z.(Z \varepsilon Y \implies Z \varepsilon el(X))$.

3.3 Mereotopology: first notions

Within mereology one may define (cf. [11]) some functors expressing relative position of objects. The functor *ext* expresses disjointness in terms of parts:

$X \varepsilon ext(Y) \iff non(\exists Z.Z \varepsilon el(X) \land Z \varepsilon el(Y))$.

The notion of a complement is expressed by the functor *comp* :

$X \varepsilon comp(Y, rel Z) \iff Y \varepsilon sub(Z) \land X \varepsilon Kl(el Z | ext Y)$

where $U \varepsilon el Z | ext Y$ iff $U \varepsilon el(Z) \land U \varepsilon ext(Y)$.

4 Rough Mereology

Approximate Reasoning carried out under Uncertainty needs a weaker form of *part* predicate: of being *a part in a degree*. The degree of being a part may then be specified either on the basis of a priori considerations and findings or directly from data [14]. In our construction of rough mereoogical predicate, we are guided by the tendency to preserve Mereology as an exact skeleton of reasoning .

Rough Mereology has been proposed and studied in [13], [14], [17] as a first-order theory. Here, we propose a formalization in the framework of Ontology; hence, rough mereology becomes now a genuine extension of mereology in a unified framework. By virtue of our earlier studies cited above, we may now assume that rough mereology is defined around a certain mereological theory as its extension. We therefore assume that a mereological predicate *el* of an element is given and ε is a symbol for ontological copula as defined above.

4.1 Rough Mereology Axioms

The following is a list of axiomatic postulates for Rough Mereology. We introduce a graded family μ_r, where $r \in [0, 1]$ is a real number from the unit interval, of name–forming functors of an individual name which would satisfy

(RM1) $X\varepsilon\mu_1(Y) \Longleftrightarrow X\varepsilon el(Y)$ (any part in degree 1 is an element);

(RM2) $X\varepsilon\mu_1(Y) \Longrightarrow \forall Z.(Z\varepsilon\mu_r(X) \Longrightarrow Z\varepsilon\mu_r(Y))$ (monotonicity);

(RM3) $X = Y \wedge X\varepsilon\mu_r(Z) \Longrightarrow Y\varepsilon\mu_r(Z)$ (identity of objects);

(RM4) $X\varepsilon\mu_r(Y) \wedge s \leq r \Longrightarrow X\varepsilon\mu_s(Y)$ (meaning of μ_r: a part in degree at least r);

we introduce a following notational convention:

$X\varepsilon\mu_r^+(Y) \Longleftrightarrow X\varepsilon\mu_r(Y) \wedge non(\exists s > r.X\varepsilon\mu_s(Y))$.

In some versions of our approach, we adopt one more axiom

(RM5) $X\varepsilon ext(Y) \Longrightarrow X\varepsilon\mu_0^+(Y)$ (disjointness of objects is fully recognizable)

or its weakened form expressing uncertainty of our reasoning

(RM5)* $X\varepsilon ext(Y) \Longrightarrow \exists r < 1.X\varepsilon\mu_r^+(Y)$ (disjointness is recognizable up to a bounded uncertainty).

4.2 Models

One may have as an archetypical rough mereological predicate the rough membership function of Pawlak and Skowron [12] defined in an extended form as:

$X\varepsilon\mu_r(Y) \Longleftrightarrow \frac{card(X \cap Y)}{card(X)} \geq r$

where X, Y are (either exact or rough) subsets in the universe U of an information/decision system (U, A).

4.3 Mereotopology: Čech Topologies

Topological structures are important for spatial reasoning: setting the interior and the boundary of an object apart, allows for expressing various spatial relations of contact (cf. eg. [2], [3]). We point here that (weak) topologies are immanent to rough mereological structures.

We define an object $Kl_r X$, each $X, r < 1$, as follows:

$Z\varepsilon Kl_r X \Longleftrightarrow Z\varepsilon Kl(\mu_r X)$ where $Z\varepsilon\mu_r X \Longleftrightarrow Z\varepsilon\mu_r(X)$.

Thus $Kl_r X$ is the class of all objects Z such that $Z\varepsilon\mu_r(X)$.

A simplified description of $Kl_r X$ may be provided as follows.

Let $B_r X$ be defined via: $Z\varepsilon B_r X \Longleftrightarrow \exists T.Z\varepsilon el(T) \wedge T\varepsilon\mu_r X$.

Then we have

Proposition 1. $Kl_r X = B_r X$.

Proof. Let $Z\varepsilon el(B_r X)$; there is T such that $Z\varepsilon el(T)$ and $T\varepsilon\mu_r X$. Hence the following is true: $\forall Z.Z\varepsilon el(B_r X) \Longrightarrow \exists U.U\varepsilon el(Z) \wedge U\varepsilon el(Kl_r X)$ and $B_r X\varepsilon el(Kl_r X)$ follows by (ME4). Similarly, for $Z\varepsilon el(Kl_r X)$, we have P, Q with $P\varepsilon el(Z)$, $P\varepsilon el(Q)$, $Q\varepsilon\mu_r(X)$. Hence $P\varepsilon el(B_r X)$ and (ME4) implies that $Kl_r X\varepsilon el(B_r X)$ so finally, $Kl_r X = B_r X$.

There is another property, showing the monotonicity of class operators.

Proposition 2. *For $s \leq r$, $Kl_rX \varepsilon el(Kl_sX)$.*

Indeed, by the previous fact, $Z \varepsilon el(Kl_rX)$ implies that $Z \varepsilon el(T)$ and $T \varepsilon \mu_r X$ for some T hence $T \varepsilon \mu_s X$ and a fortiori $Z \varepsilon el(Kl_sX)$.

Introducing a constant name Λ (the empty name) via the definition:
$$X \varepsilon \Lambda \iff X \varepsilon X \wedge non(X \varepsilon X)$$
and defining the interior $IntX$ of an object X as follows:
$$IntX \varepsilon Kl(int\_X)$$
where $Z \varepsilon int\_X \iff \exists T. \exists r < 1. Z \varepsilon el(Kl_rT) \wedge Kl_rT \varepsilon el(X)$
i.e. $IntX$ is the class of objects of the form Kl_rT which are elements of X, we have

Proposition 3. *(i) $Int\Lambda \varepsilon\ Int\Lambda \iff \Lambda \varepsilon \Lambda$ (the interior of the empty concept is the empty concept);*
(ii) $X \varepsilon el(Y) \implies IntX \varepsilon el(IntY)$ (monotonicity of Int);
(iii) $IntKlV \varepsilon KlV$ (the universe is open).

Properties (i)-(iii) witness that the family of all classes Kl_rT, $r < 1$, is a base for a Čech topology [21]; we call this topology the *rough mereological topology* (*rm-topology*).

5 From Čech mereotopologies to mereotopologies

We go a step further: we make rm-topology into a topology (ie. open sets have open intersections); this comes at a cost: we need a specific model for rough mereology.

5.1 A *t-norm* model

We recall that a *t-norm* is a 2-argument functor $T(x,y) : [0,1]^2 \longrightarrow [0,1]$ satisfying the conditions:
(i) $T(x,y) = T(y,x)$; (ii) $T(x,1) = x$; (iii) $x' \geq x, y' \geq y \longrightarrow T(x',y') \geq T(x,y)$; (iv) $T(x, T(y,z)) = T(T(x,y), z)$
and that the residual implication induced by T, in symbols $\overrightarrow{T}$, is defined via
$\overrightarrow{T}(r,s) \geq t \iff T(t,r) \leq s$.

We apply here the ideas developed in [14] and we define, given a part in degree predicate μ, a new measure of partiality in degree, μ_T, defined as follows
(*) $X \varepsilon \mu_T(r)(Y) \iff \forall Z. (Z \varepsilon \mu(u)(X) \wedge Z \varepsilon \mu(v)(Y) \implies \overrightarrow{T}(u,v) \geq r)$.
It turns out that

Proposition 4. *The functor μ_T satisfies axioms (RM1)-(RM5), (RM5)*.*

Proof. We may check (RM1): $X \varepsilon \mu_T(1)(Y)$ implies that from $Z \varepsilon \mu(u)(X) \wedge Z \varepsilon \mu(v)(Y)$ it follows that $u \leq v$ for each Z hence: $Z \varepsilon el(X) \implies Z \varepsilon el(Y)$ follows for any Z i.e. $X \varepsilon el(Y)$. Similarly, $X \varepsilon el(Y)$ implies via (RM2) for μ that $Z \varepsilon \mu(u)(X) \wedge Z \varepsilon \mu(v)(Y)$ yields $u \leq v$ i.e. $\overrightarrow{T}(u,v) \geq 1$ for any Z thus

$X\varepsilon\mu_T(1)(Y)$. (RM2), (RM3), (RM4) are checked similarly, for (RM5), we begin with the premise $X\varepsilon ext(Y)$ hence $X\varepsilon\mu_0^+(Y)$; assuming $X\varepsilon\mu_T(r)(Y)$ we get by (*) for $Z = X$ that $\overrightarrow{T}(1,0) \geq r$ i.e. $T(r,1) = r \leq 0$. Similar argument handles (RM5)*.

Thus μ_T is a partiality in degree predicate.

Modifying a proof given in ([9], Prop.14), we find that the following deduction rule is valid for μ_T :

(MPR) $\dfrac{X\varepsilon\mu_T(r)(Y), Y\varepsilon\mu_T(s)(Z)}{X\varepsilon\mu_T(T(r,s))(Z)}$.

We denote with the symbol $Kl_{r,T}X$ the class Kl_rX with respect to μ_T.

We may give a new characterization of $Kl_{r,T}X$.

Proposition 5. $Y\varepsilon el(Kl_{r,T}X) \Longleftrightarrow Y\varepsilon\mu_T(r)(X)$.

Indeed, $Y\varepsilon el(Kl_{r,T}X)$ means that $Y\varepsilon el(Z)$ and $Z\varepsilon\mu_T(r)(X)$ for some Z. From $Y\varepsilon\mu_T(1)(Z)$ and $Z\varepsilon\mu_T(r)(X)$ it follows by (MPR) that $Y\varepsilon\mu_T(T(1,r) = r)(X)$.

We may regard therefore $Kl_{r,T}X$ as a "ball of radius r centered at X" with respect to the "metric" μ_T.

Furthermore, we have by the same argument

Proposition 6. $Y\varepsilon el(Kl_{r,T}X)$ and $s_o = \min{}_-\arg(T(r,s) \geq r)$ imply $Kl_{s_0,T}Y\varepsilon el(Kl_{r,T}X)$.

It follows that the family $\{Kl_{r,T}X : r < 1, X\}$ induces a topology on our universe of objects (under the assumption that $T(r,s) < 1$ whenever $rs < 1$). This allows us to define a variety of functors like: Tangential Part, Non-tangential Part etc. instrumental in spatial reasoning (cf. [2], [3]).

6 Connections

We refer to an alternative scheme for mereological reasoning based on Clarke's formalism of connection C [5] in Calculus of Individuals of Leonard &Goodman [10]; see in this respect [3]. This formalism is a basis for some schemes of approximate spatial reasoning (eg. various relations of external contact, touching etc. may be expressed via C) (op.cit.). The basic primitive in this approach is the predicate $C(X,Y)$ (read "X and Y are *connected*") which should satisfy : (i) $C(X,X)$; (ii) $C(X,Y) \Longrightarrow C(Y,X)$; (iii) $\forall Z.(C(X,Z) \Longleftrightarrow C(Y,Z)) \Longrightarrow X = Y$. From C other predicates (as mentioned above) are generated and under additional assumptions (cf. [5]) a topology may be generated from C.

We will define a notion of connection in our model; clearly, as in our model topological structures arise in a natural way via "metrics" μ, we may afford a more stratified approach to connection and separation properties. So we propose a notion of a graded connection $C(r,s)$.

6.1 From graded connections to connections

We let

$Bd_r X \varepsilon Kl(\mu_r^+ X)$ where $Z\varepsilon\mu_r^+(X) \Longleftrightarrow Z\varepsilon\mu_r(X) \wedge non(Z\varepsilon\mu_s(X), s > r)$

and then

$X\varepsilon C(r,s)(Y) \Longleftrightarrow \exists W. W\varepsilon el(Bd_r X) \wedge W\varepsilon el(Bd_s Y).$

We have then clearly:

(i) $X\varepsilon C(1,1)(X)$; (ii) $X\varepsilon C(r,s)(Y) \Longrightarrow Y\varepsilon C(s,r)(X)$.

Concerning the property (iii), we may have some partial results However, we adopt here a new approach. It is realistic from both theoretical and applicational points of view to assume that we may have "infinitesimal " parts i.e. objects as "small" with respect to μ as desired.

Infinitesimal parts model We adopt a new axiom of infinitesimal parts

(IP) $non(X\varepsilon el(Y)) \Longrightarrow \forall r > 0. \exists Z\varepsilon el(X), s < r. Z\varepsilon\mu_s^+(Y).$

Our rendering of the property (iii) under (IP) is as follows:

$non(X\varepsilon el(Y)) \Longrightarrow \forall r > 0. \exists Z, s < r. Z\varepsilon\mu_s^+(Y). Z\varepsilon C(1,1)(X) \wedge Z\varepsilon C(1,s)(Y).$

Introducing connections Our notion of a connection will depend on a threshold, α, set according to the needs of a context of reasoning.

Given $0 < \alpha < 1$, we let

(CON) $X\varepsilon C_\alpha(Y) \Longleftrightarrow \exists r, s \geq \alpha. X\varepsilon C(r,s)(Y).$

Then we have

(i) $X\varepsilon C_\alpha(X)$, each α;

(ii) $X\varepsilon C_\alpha(Y) \Longrightarrow Y\varepsilon C_\alpha(X)$;

(iii) $X \neq Y \Longrightarrow \exists Z. (Z\varepsilon C_\alpha(X) \wedge non(Z\varepsilon C_\alpha(Y)) \vee Z\varepsilon C_\alpha(Y) \wedge non(Z\varepsilon C_\alpha(X)))$

i.e. the functor C_α has all the properties of connection in the sense of [5] and [2], [3].

Restoring rough mereology from connections We show now that when we adopt mereological notions as they are defined via connections in Calculus of Individuals, we do not get anything new: we come back to rough mereology we started from. The formula

$X\varepsilon el_C(Y) \Longleftrightarrow \forall Z. (Z\varepsilon C(X) \Longrightarrow Z\varepsilon C(Y))$ is the definition of the notion of an element from a connection C. We claim

Proposition 7. $X\varepsilon el_{C_\alpha}(Y) \Longleftrightarrow X\varepsilon el(Y).$

Clearly, $X\varepsilon el_{C_\alpha}(Y) \Longrightarrow X\varepsilon el(Y)$. Assume that $X\varepsilon el(Y)$; $Z\varepsilon C_\alpha(X)$. There is W with $W\varepsilon\mu_r^+(Z)$ and $W\varepsilon\mu_s^+(X)$, $r, s \geq \alpha$. Then by (RM2), $W\varepsilon\mu_{s'}^+(Y)$ with an $s' \geq s$ and so $Z\varepsilon C_\alpha(Y)$. It follows that $X\varepsilon el(Y) \Longrightarrow X\varepsilon el_{C_\alpha}(Y)$.

Any of connections C_α restores thus the original notion of an element, el.

Therefore in our setting of rough mereology, we may have as well the mereotopological setting of [2], [3], [5].

Let us observe that in general $C_\alpha \neq OV$ where $X \varepsilon OV(Y) \iff \exists Z. Z \varepsilon el(X) \wedge Z \varepsilon el(Y)$ is the functor of *overlapping* (in our context, objects may connect each other without necessarily having a part in common).

7 Geometry via rough mereology

It has been shown that in the mereotopological context of Calculus of Individuals one may introduce a geometry (cf. [3]). We show that in the context of rough mereology geometric structures arise naturally without any resort to the intermediate structure of connection. It is well known that elementary geometry may be developed on the basis of eg. the primitive notion of "being closer to ... than to..". We consider here the axioms for this notion going back to Tarski (cf. eg. [4]) and we introduce a name - forming functor on pairs of individual names $CT(Y,Z)$ ($X \varepsilon CT(Y,Z)$ is read "X is closer to Y than to Z") subject to

(CT1) $X \varepsilon CT(Y,Z) \wedge X \varepsilon CT(Z,W) \implies X \varepsilon CT(Y,W)$;
(CT2) $X \varepsilon CT(Y,Z) \wedge Z \varepsilon CT(X,Y) \implies Y \varepsilon CT(X,Z)$;
(CT3) $non(X \varepsilon CT(Y,Y))$;
(CT4) $X \varepsilon CT(Y,Z) \implies X \varepsilon CT(Y,W) \vee X \varepsilon CT(W,Z)$.

We define this notion in the context of rough mereology: for X,Y, we let $\mu^+(X,Y) = r \iff X \varepsilon \mu_r^+(Y)$ and then
$X \varepsilon CT(Y,Z) \iff \max(\mu^+(X,Y), \mu^+(Y,X)) \geq \max(\mu^+(X,Z), \mu^+(Z,X))$.
Then

Proposition 8. *The functor CT thus defined satisfies (CT1)-(CT4).*

We may now follow e.g. the lines of [4], [3] and give definitions of a other geometric notions; for instance, letting
$T(X,Y,Z) \iff \forall W. X = W \vee CT(Y,X,W) \vee CT(Z,X,W)$
we may render the notion that X is positioned between Y and Z and this may permit to define a straight line segment and further notions as pointed to in e.g. [4]. The details will be presented elsewhere (cf. [16]).

8 Conclusion

We have presented a scheme for developing conceptual spatial reasoning under uncertainty in the framework of rough mereology. In this framework, as it will be presented elsewhere, we may develop various approaches to spatial reasoning, including metric geometry based on predicates μ and metrics derived from them.

Acknowledgement

This work has been prepared under the Research Grant No 8T11C 024 17 from the State Committee for Scientific Research(KBN) of the Republic of Poland and with the help of a Research Grant No 1/99 from the Polish–Japanese Institute of Information Technology.

References

1. R. C. Arkin, *Behaviour Based Robotics*, MIT Press, Cambridge, MA, 1998.
2. N. Asher and L. Vieu, Toward a geometry of commonsense: a semantics and a complete axiomatization of mereotopology, in: *Proceedings IJCAI'95*, Montreal, 1995, 846-852.
3. M. Aurnague and L. Vieu, A theory of space for natural language semantics, in: K. Korta and J.M. Larrazàbal (eds.), *Semantics and Pragmatics of Natural Language: Logical and Computational Aspects*, San Sebastian, 1995, 69-126.
4. J. van Benthem, *The Logic of Time*, Reidel, 1983.
5. B. Clarke, A calculus of individuals based on "connection", *Notre Dame Journal of Formal Logic*, 22(3)(1981), 204-218.
6. R. Clay, Relation of Leśniewski's mereology to boolean algebra, *The Journal of Symbolic Logic*, 39(4)(1974), 638-648.
7. M. Dorigo and M. Colombetti, *Robot Shaping. An Experiment in Behavior Engineering*, MIT Press, Cambridge, MA, 1998.
8. D. Kortenkamp, R. P. Bonasso, R. Murphy, *Artificial Intelligence and Mobile Robotics*, AAAI Press/MIT Press, 1998.
9. C. Lejewski, On Leśniewski's ontology, *Ratio*, 1(2)(1958), 15-176.
10. H. Leonard and N. Goodman, The calculus of individuals and its uses, *The Journal of Symbolic Logic*, 5(1940), 45-55.
11. St. Leśniewski, O podstawach matematyki (On the foundations of Mathematics, in Polish), *Przeglad Filozoficzny*, 30(1927), 164-206; 31(1928), 261-291; 32(1929), 60-101; 33(1930), 77-105; 34(1931), 142-170.
12. Z. Pawlak and A. Skowron, Rough membership functions, in: R.R. Yaeger, M. Fedrizzi and J. Kacprzyk (eds.), *Advances in the Dempster - Shafer Theory of Evidence*, John Wiley and S., New York, 1994, 251-271.
13. L. Polkowski and A. Skowron, Rough mereology, *Lecture Notes in Artificial Intelligence*, vol. 869, Springer Verlag, 1994, 85-94.
14. L. Polkowski and A. Skowron, Rough mereology: a new paradigm for approximate reasoning, *Intern. J. Approx. Reasoning*, 15(4) (1996), 333-365.
15. L. Polkowski and A. Skowron, Grammar systems for distributed synthesis of approximate solutions extracted from experience, in: Gh. Paun and A. Salomaa (eds.), *Grammatical Models of Multi-Agent Systems*, Gordon and Breach Sci. Publ., Amsterdam, 1998, 316-333.
16. L. Polkowski, On synthesis of constructs for spatial reasoning via rough mereology, *Fundamenta Informaticae*, to appear.
17. A. Skowron and L. Polkowski, Rough mereological foundations for design, analysis, synthesis and control in distributed systems, *Information Sciences. An Intern. J.*, 104(1-2) (1998), 129-156.
18. J. Słupecki, St. Leśniewski's calculus of names, *Studia Logica*,3(1955), 7-72.
19. B. Sobociński, Studies in Leśniewski's mereology, *Yearbook for 1954-55 of the Polish Society of Arts and Sciences Abroad*, 591954), 34-48.
20. A. Tarski, Zur Grundlegung der Boole'sche Algebra I, *Fundamenta Mathematicae*, 24(1935), 177-198.
21. E. Čech, Topological spaces, in: J. Novàk (ed.), *Topological Papers of Eduard Čech*, Academia, Prague, 1966, 436-472.

An Algorithm for Finding Equivalence Relations from Tables with Non-deterministic Information

Hiroshi SAKAI and Akimichi OKUMA

Department of Computer Engineering, Kyushu Institute of Technology

Tobata, Kitakyushu 804, Japan

e-mail: sakai@comp.kyutech.ac.jp

Abstract. Rough sets theory depending upon DIS($Deterministic\ Information\ System$) is now becoming a mathematical foundation of soft computing. Here, we pick up NIS($Non\text{-}deterministic\ Information\ System$) which is more general system than DIS and we try to develop the rough sets theory depending upon NIS. We first give a definition of definability for every object set X, then we propose an algorithm for checking it. To find an adequate equivalence relation from NIS for X is the most important part in this algorithm, which is like a resolution. According to this algorithm, we implemented some programs by prolog language on the workstation.

1 Introduction

Rough sets theory is seen as a mathematical foundation of soft computing, which covers some areas of research in AI, i.e., knowledge, imprecision, vagueness, learning, induction[1,2,3,4]. We recently see many applications of this theory to knowledge discovery and data mining[5,6,7,8,9].

In this paper, we deal with rough sets in NIS(Non-deterministic Information System), which will be an advancement from rough sets in DIS(Deterministic Information System). According to [1,2], we define every $DIS = (OB, AT, \{VAL_a| a \in AT\}, f)$, where OB is a set whose element we call $object$, AT is a set whose element we call $attribute$, VAL_a for $a \in AT$ is a set whose element we call $attribute\ value$ and f is a mapping such that $f : OB * AT \rightarrow \cup_{a \in AT} VAL_a$, which we call $classification\ function$. For every $x, y(x \neq y) \in OB$, if $f(x,a) = f(y,a)$ for every $a \in AT$ then we see there is a relation for x and y. This relation becomes an equivalence relation on OB, namely we can always define an equivalence relation EQ on OB. If a set $X(\subset OB)$ is the union of some equivalence classes in EQ, then we call X is $definable$ in DIS. Otherwise we call X is $rough$[1].

Now we go to the NIS. We define every $NIS = (OB, AT, \{VAL_a|a \in AT\}, g)$, where g is a mapping such that $g : OB * AT \rightarrow P(\cup_{a \in AT} VAL_a)$ (Power set for $\cup_{a \in AT} VAL_a$)[3,4]. We need to remark that there are two interpretations for mapping g, namely AND-interpretation and OR-interpretation. For example, we can give the following two interpretations for $g(tom, language) = \{English, Polish, Japanese\}$.

(AND-interpretation) Tom can use three languages, English, Polish and Japanese. Namely, we see $g(tom, language)$ is $English \wedge Polish \wedge Japanese$.

(OR-interpretation) Tom can use either one of language in English, Polish or

Japanese. Namely we see $g(tom, language)$ is *English* $\vee$ *Polish* $\vee$ *Japanese*. The OR-interpretation seems to be more important for g. Because, it is related to incomplete information and uncertain information. Furthermore, knowledge discovery, data mining and machine learning from incomplete information and uncertain information will be important issue. In such situation, we discuss NIS with OR-interpretation. We have already proposed incomplete information and selective information for OR-interpretation[10], where we distinguished them by the existence of unknown real value. In this paper, we extend the contents in [10] and develop the algorithm for finding equivalence relations in NIS.

2 Aim and Purpose in Handling NIS

Now in this section, we show the aim and purpose in handling NIS. Let's consider the following example.

Example 1. Suppose the next NIS_1 such that $OB = \{1, 2, 3, 4\}$, $AT = \{A, B, C\}$, $\cup_{a \in AT} VAL_a = \{1, 2, 3\}$ and g is given by the following table. In this table,

OB	A	B	C
1	1 $\vee$ 2	2	1 $\vee$ 2 $\vee$ 3
2	1	2	1 $\vee$ 2 $\vee$ 3
3	1	1 $\vee$ 2	2
4	1	2	2 $\vee$ 3

Table 1. Non-deterministic Table for NIS_1

if we select an element for every disjunction then we get a DIS. There are $72 (=2*3*3*2*2)$ $DISs$ for this NIS_1. In this case, we have the following issues.

Issue 1: For a set $\{1, 2\} (\subset OB)$, if we select 1 from $g(1, A)$ and 3 from $g(1, C)$, $g(2, C)$ and $g(4, C)$ then $\{1, 2\}$ is not definable. However, if we select 1 from $g(1, C)$ and $g(2, C)$ then $\{1, 2\}$ is definable. How can we check such definability for every subset X of OB ?

Issue 2: How can we get all possible equivalence relations from 72 $DISs$? Do we have to check 72 $DISs$ sequentially ?

Issue 3: Suppose there are following information for attribute D: $g(1, D) = \{1\}$, $g(2, D) = \{1\}$, $g(3, D) = \{2\}$ and $g(4, D) = \{2\}$, respectively. In this case, which DIS from NIS_1 makes $(A, B, C) \rightarrow D$ consistent ? How can we get all $DISs$ which make $(A, B, C) \rightarrow D$ consistent ?

These issues come from the fact such that the equivalence relation in DIS is always unique but there are some possible equivalence relations for NIS.

Now we just a little show the real execution for Issue 2 to clarify how our system works.

```
?-relationall.
[1] [[1,2,3,4]] 1          [10] [[1,4],[2],[3]] 5
[2] [[1,2,3],[4]] 1        [11] [[1],[2,3,4]] 5
[3] [[1,2,4],[3]] 3        [12] [[1],[2,3],[4]] 4
[4] [[1,2],[3,4]] 2        [13] [[1],[2,4],[3]] 14
[5] [[1,2],[3],[4]] 5      [14] [[1],[2],[3,4]] 8
[6] [[1,3,4],[2]] 2        [15] [[1],[2],[3],[4]] 19
[7] [[1,3],[2,4]] 1        POSSIBLE CASES 72
[8] [[1,3],[2],[4]] 1      EXEC_TIME=0.1566100121(sec)
[9] [[1,4],[2,3]] 1        yes
```

In the above execution, we see there are 15 kinds of equivalence relations and there are 19 $DISs$ whose equivalence relation is $\{\{1\},\{2\},\{3\},\{4\}\}$. According to this execution, we can see that 2 cases of $\{\{1,2\},\{3,4\}\}$, 5 cases of $\{\{1,2\},\{3\},\{4\}\}$, 8 cases of $\{\{1\},\{2\},\{3,4\}\}$ and 19 cases of $\{\{1\},\{2\},\{3\}, \{4\}\}$ make $(A,B,C) \rightarrow D$ consistent by Proposition 4.1 in [1]. In the subsequent sections, we discuss the definability of every set in NIS as well as the above issues.

3 An Algorithm for Checking Definability of Set in DIS

In this section, we simply refer to an algorithm to detect the definability of set in DIS. Here, we suppose an equivalence relation EQ in the DIS and we use $[x]$ to express an equivalence class with object x.

An Algorithm in DIS
(1) Make a set $SUP(= \cup_{x \in X}[x])$.
(2) If $SUP = X$ then X is definable in DIS else go to the next step (3).
(3) Make a set $INF(= \cup\{[x] \in EQ|[x] \subset X\})$, then lower and the upper approximation of X are INF and SUP, respectively.

The above algorithm manages the definability of set X, upper and the lower approximation of X. We will propose a new algorithm in NIS depending upon the above one.

4 Some Definitions and Properties in NIS

We first give some definitions then we show a proposition.

Definition 1. For $NIS = (OB, AT, \{VAL_a|a \in AT\}, g)$, we call $NIS' = (OB, AT, \{VAL_a|a \in AT\}, g')$ which satisfies the following (1) and (2) an *extension* from NIS.
(1) $g'(x,a) \subset g(x,a)$ for every $x \in OB$, $a \in AT$.
(2) $g'(x,a)$ is a singleton set for every $x \in OB$, $a \in AT$.
Here, we can see every extension from NIS is a DIS, because every attribute value is fixed uniquely.

Definition 2. For every extension NIS' from NIS, we call the equivalence relation in NIS' a *possible equivalence relation* in NIS. We also call every element in this relation a *possible equivalence class* in NIS.

In every DIS, we know the definability of a set X, so we give the next definition.

Definition 3. A set $X(\subset OB)$ is *definable* in NIS, if X is definable in some extensions from NIS.

We soon remind a way to detect the definability of a set X in NIS, namely we sequentially make every extension from NIS and execute the program by algorithm in DIS. However, we need the same number of files as extensions from NIS. Furthermore, if X is not definable in NIS then we have to execute the same program for all extensions. So we propose another way from now on. We give the following definitions.

Definition 4. Suppose $NIS = (OB, AT, \{VAL_a | a \in AT\}, g)$. If $g(x, a)$ is a singleton set for every $a \in AT$ then we call that object x is *fixed*. Furthermore, $OB_{fixed} = \{x \in OB|$ object x is fixed $\}$.

Definition 5. Suppose $NIS = (OB, AT, \{VAL_a | a \in AT\}, g)$ and $g(x, a)$ is not a singleton set for some $a \in AT$. By picking up an element in such $g(x, a)$, we can make object x fixed. Here, we call a set of pairs $\{[attribute, picked\_element]\}$ *selection* in x. For a selection θ, x_θ expresses the fixed tuple for x.

In Example 1, if we take a selection $\theta = \{[A, 1], [C, 1]\}$, then the 1_θ is $(1, 2, 1)$. For $\theta = \{[B, 2]\}$, the 3_θ is $(1, 2, 2)$.

Definition 6. Suppose $NIS = (OB, AT, \{VAL_a | a \in AT\}, g)$. For every $x(\in OB)$ and selection θ in x, we give the following definitions.

(1) $inf(x, \theta) = \{x\} \cup \{y \in OB_{fixed}|$ x_θ and the tuple for y are the same $\}$.

(2) $sup(x, \theta) = \{y \in OB|$ there is a selection θ' such that $x_\theta = y_{\theta'}\}$.

According to these definitions, we get the following proposition.

Proposition 1.

(1) The $inf(x, \theta)$ is the minimal possible equivalence class including object x for the selection θ.

(2) For every $y \in (sup(x, \theta) - inf(x, \theta))$, there are selections θ' and θ'' such that $x_\theta = y_{\theta'}$ and $x_\theta \neq y_{\theta''}$.

(3) A subset $X(\subset OB)$ which satisfies $inf(x, \theta) \subset X \subset sup(x, \theta)$ for some x and θ can be a possible equivalence class.

(Proof) (1) For x and θ, the tuple for every $y \in inf(x, \theta)$ is the same and fixed. So $inf(x, \theta)$ is a minimal possible equivalence class with x for the selection θ.

(2) For $y \in (sup(x, \theta) - inf(x, \theta))$, we get $y \in sup(x, \theta)$ and $y \notin inf(x, \theta)$. By the definition of sup, there is a selection θ' such that $x_\theta = y_{\theta'}$. If $y \in OB_{fixed}$ then $y \in inf(x, \theta)$, which makes contradiction to $y \notin inf(x, \theta)$. So $y \notin OB_{fixed}$, and there exists at least another selection θ'' such that $y_{\theta''} \neq x_\theta$.

(3) According to (1) and (2), $inf(x, \theta) \cup M$ for $M \subset (sup(x, \theta) - inf(x, \theta))$ can be a possible equivalence class.

In this proposition, the (3) is related to the definability of set in NIS and we use this property. However, we have to remark that $inf(x, \theta)$ and $sup(x, \theta)$ are not independent in every x. The $inf(x, \theta)$ and $sup(x, \theta)$ are mutually related to other $inf(y, \theta')$ and $sup(y, \theta')$. We show it in the next example.

Example 2. Suppose NIS_1 in Example 1. The $OB_{fixed} = \emptyset$ and we get the following subset of all inf and sup.

(A) $inf(1, \{[A,1],[C,3]\}) = \{1\}$, $sup(1, \{[A,1],[C,3]\}) = \{1,2,4\}$.
(B) $inf(3, \{[B,1]\}) = \{3\}$, $sup(3, \{[B,1]\}) = \{3\}$.
(C) $inf(3, \{[B,2]\}) = \{3\}$, $sup(3, \{[B,2]\}) = \{1,2,3,4\}$.
(D) $inf(4, \{[C,2]\}) = \{4\}$, $sup(4, \{[C,2]\}) = \{1,2,3,4\}$.
(E) $inf(4, \{[C,3]\}) = \{4\}$, $sup(4, \{[C,3]\}) = \{1,2,4\}$.

Here in (A), the following sets $\{1\}$, $\{1,2\}$, $\{1,4\}$ and $\{1,2,4\}$ can be a possible equivalence class by (3) in Proposition 1. However, if we make $\{1,2\}$ a possible equivalence class, then we implicitly make object $4 \notin [1](= [2])$. It implies selection $[C,3]$ for object 4 is rejected, because $4_{\{[C,3]\}}$ is (1,2,3) which is the same as $1_{\{[A,1],[C,3]\}}$. Namely, we can not use (E) and we have to revise (C) and (D) as follows:

(C') $inf(3, \{[B,2]\}) = \{3\}$, $sup(3, \{[B,2]\}) = \{3,4\}$.
(D') $inf(4, \{[C,2]\}) = \{4\}$, $sup(4, \{[C,2]\}) = \{3,4\}$.

If we use (B) then $[3] = \{3\}$ and reject the (C'), because either (B) or (C') hold. Here, we have to revise (D') as follows:

(D") $inf(4, \{[C,2]\}) = \{4\}$, $sup(4, \{[C,2]\}) = \{4\}$.

For this (D"), only $\{4\}$ can be a possible equivalence class. Finally we get a possible equivalence relation $\{\{1,2\},\{3\},\{4\}\}$ and the selections are $\{[A,1],[C,3]\}$ for object 1, $\{[C,3]\}$ for 2, $\{[B,1]\}$ for 3 and $\{[C,2]\}$ for 4. These selections specify a DIS from NIS. We also know that sets like $\{1,2,3\}$ and $\{3,4\}$ are definable in NIS but $\{2,3\}$ is not defiable in this DIS.

5   Proposal of An Algorithm in NIS

The following is the overview of proposing algorithm.
An Algorithm for Checking Definability of Set in NIS
Suppose we are given $inf(x,\theta)$ and $sup(x,\theta)$ for every $x(\in OB)$.
Input: A set $X(\subset OB)$.
Output: X is definable in NIS or not.
(1) Set $X^* = X$.
(2) For the first element $x(\in X^*)$, find $X'(\subset X^*)$ such that $inf(x,\theta) \subset X' \subset sup(x,\theta)$ for some θ.
(3) The usable $inf(y,\theta')$ and $sup(y,\theta')$ for $y \in OB$ are restricted by selecting X' in (2). So, check the usable inf and sup, and go to (4).
(4) If there is no contradiction in (3), then set $[x] = X'$, $X^* = X^* - X'$ and go to (2). Especially if $X^* = \emptyset$ then we conclude X is definable. To find other cases, backtrack to (2). If there is contradiction in (3), then backtrack to (2) and try another X'. If there is no branch for backtrack, then we conclude X is not definable.

In this algorithm, if we set $X = OB$ then we can get all possible equivalence relations. This algorithm seems to be simple and natural, but managing the inf and sup is very complicated. We also need to discuss how we get inf and sup information from NIS.

6 Implementation of Proposing Algorithm in NIS

Now in this section, we show the implementation of a prover for NIS. We depend upon prolog language on workstation for implementing this prover. Our prover consists of the following two subsystems:

(1) *File translator from data file to an internal expression.*

(2) *Query interpreter with some subcommands.*

6.1 Data File for NIS

Here, we show the data file in prolog, which is very simple. We use two kinds of atomic formulas:

$object(number\_of\_objects, number\_of\_attributes)$.

$data(object, tuple\_data)$.

The following is the real data file for NIS_1.

```
object(4,3). data(1,[[1,2],2,[1,2,3]]). data(2,[1,2,[1,2,3]]).
data(3,[1,[1,2],2]). data(4,[1,2,[2,3]]).
```

We use a list to express disjunction. This data structure is so easy that we can soon make this file from every non-deterministic table. There is no restrictions for every number of items except prolog and workstation's restriction.

6.2 File Translator from Data File to Internal Expression

This translator creates an internal expression from every data file, which consists of the following three kinds of atomic formulas.

$cond(object, number\_for\_selection, tuple\_for\_this\_selection)$.

$pos(object, number\_of\_all\_selections)$.

$conn([object, number\_for\_selection], [slist, slist1], [mlist, mlist1], maylist)$.

As for the 2nd, 3rd and 4th arguments in $conn$, we will show their contents by using real execution. The following is the translation of data file.

```
?-consult(nkbtf.pl).
yes
?-go.
File Name for Read Open:'nkbda23.pl'.
File Name for Write Open:'out.pl'.
EXEC_TIME=0.05459904671(sec)
yes
```

In this translation, *nkbtf.pl* is the translator and *nkbda23.pl* is a data file for NIS_1. The file *out.pl* keeps the internal expression for NIS_1. The following is a part of internal expression for object 3.

```
cond(3,1,[1,1,2]).
cond(3,2,[1,2,2]).
pos(3,2).
conn([3,1],[[3],[1]],[[],[]],[[3,1]]).
conn([3,2],[[3],[2]],[[1,2,4],[3,2,1]],[[3,2],[1,3],[2,2],[4,1]]).
```

The $pos(3, 2)$ shows there are two selections for object 3 and $cond(3, 1, [1, 1, 2])$

does $[1, 1, 2]$ is the tuple for the first selection $\theta (= \{[B, 1]\})$. In this selection, the 2nd argument in $conn([3, 1], \_, \_, \_)$ shows $inf(3, \theta) = \{3\}$ and 3rd argument does $sup(3, \theta) - inf(3, \theta) = \emptyset$. Similarly for the second selection $\theta'(= \{[B, 2]\})$ which makes tuple $[1, 2, 2]$, we get $inf(3, \theta') = \{3\}$ and $sup(3, \theta') - inf(3, \theta') = \{1, 2, 4\}$. Here, we identify the selections θ with the second argument in *cond*. For example, we identify a selection $\theta = \{[B, 1]\}$ as the second argument 1 in $cond(3, 1, [1, 1, 2])$.

Definition 7. For $cond(object, number\_for\_selection, tuple)$, we call *number_ for_selection* an *index* of selection θ and we do $[object, number\_for\_selection]$ an *index* of the fixed tuple.

6.3 An Algorithm for Translator

Now we simply show the translation algorithm, which consists of two phases. In *Phase*1, we create $cond(object, \_, \_)$ and $pos(object, \_)$ from $data(object, \_)$. For every $data(object, list)$, we first make the cartesian products from *list* then sequentially we assert $cond(object, selection, fixed\_tuple)$, and finally we assert $pos(object, last\_number)$.

In *Phase*2, we make every $conn([object, selection], \_, \_, \_)$ from every *cond*. For every $cond(object, selection, fixed\_tuple)$, we first initialize lists $[slist, slist1]$ and $[mlist, mlist1]$ and we find other $cond(object', selection', fixed\_tuple)$. If $pos(object', 1)$ then we add $[object', selection']$ to $[slist, slist1]$ else we do to $[mlist, mlist1]$. We continue it for all selections. Finally, we assign the union of $[slist, slist1]$ and $[mlist, mlist1]$ to *maylist* and assert $conn([object, selection], [slist, slist1], [mlist, mlist1], maylist)$. We have realized the translator according to this algorithm.

6.4 An Algorithm for Handling Usable inf and sup

In proposing algorithm, the most difficult part is to handle every subset of objects from usable inf and sup. The usable inf and sup are dynamically revised, so we need to manage what are the usable inf and sup. For example in the translated $conn([3, 2], [[3], [2]], [[1, 2, 4], [3, 2, 1]], \_)$, every $\{3\} \cup M (M \subset \{1, 2, 4\})$ can be a possible equivalence class by Proposition 1. To make $\{1, 3\}$ a possible equivalence class, we need to positively use object 1 in $\{1, 2, 4\}$ and negatively use objects 2 and 4 in $\{1, 2, 4\}$.

Definition 8. For $X \subset OB$, suppose $inf(x, \theta) \subset X \subset sup(x, \theta)$ for some $x \in OB$. In this case, we call every element in X *positive use* of index $[x, \theta]$ and every element in $(sup(x, \theta) - X)$ *negative use* of $[x, \theta]$.

To manage such two kinds of usage, we adopt a positive list $PLIST$ and a negative list $NLIST$. The $PLIST$ keeps indexes $[object, selection]$ which have been applied as positive use, and the $NLIST$ keeps indexes which have been applied as negative use. For these two lists and positive and negative use, we have the following remarks.

Remark for Positive Use of $[x, \theta]$

Suppose the index for x_θ is $[x, num]$. The x_θ is applicable as positive use only when $[x, \_] \notin PLIST$ and $[x, num] \notin NLIST$.

Remark for Negative Use of $[x, \theta]$

Suppose the index for x_θ is $[x, num]$. The x_θ is applicable as negative use in the following cases;

(1) $[x, num] \in NLIST$.

(2) $[x, num] \notin NLIST$, $[x, num] \notin PLIST$ and $[x, num'] \in PLIST$ for $num \neq num'$.

(3) $[x, num] \notin NLIST$, $[x, \_] \notin PLIST$ and there is at least $[x, num''] \notin NLIST$ for $num \neq num''$.

The above remarks avoid the contradiction such that $[x, \theta]$ is applied not only positive use but also negative use. The third condition in negative use shows that $[x, num] \in NLIST$ for all num does not hold.

Now we show the algorithm for finding a possible equivalence class.

An Algorithm: candidate

Input: $X = \{x_1, \cdots, x_n\} \subset OB$, inf, sup, current $PLIST$ and $NLIST$.

Output: There is a possible equivalence class $[x_1] \subset X$ such that $inf(x_1, \theta) \subset [x_1] \subset sup(x_1, \theta)$ or not.

(1) Pick up a selection θ such that $inf(x_1, \theta) \subset X$. If we can not pick up such selection then respond there is no possible equivalence class.

(2) If every element in $inf(x_1, \theta)$ is applicable as positive use then go to (3) else go to (1) and try another selection.

(3) Pick up a set $M(\subset (sup(x_1, \theta) - inf(x_1, \theta)))$ and go to (4). If we can not pick up any other M then go to (1) and try another selection.

(4) If $M \subset (X - inf(x_1, \theta))$ and every element in M is applicable as positive use then set $PLIST \leftarrow PLIST \cup \{[y, \theta']|y \in inf(x_1, \theta) \cup M, y_{\theta'} = x_{1, \theta}\}$ and go to (5) else go to (3) and try another M.

(5) If every element in $(sup(x_1, \theta) - (inf(x_1, \theta) \cup M))$ is applicable as negative use then go to (6) else go to (3) and try another M.

(6) Set $NLIST \leftarrow NLIST \cup \{[y, \theta']|y \in (sup(x_1, \theta) - (inf(x_1, \theta) \cup M)), y_{\theta'} = x_{1, \theta}\}$. Respond $[x_1](= inf(x_1, \theta) \cup M)$ can be a possible equivalence class.

According to this algorithm, we realized a program *candidate* which responses a possible equivalence class depending upon the current $PLIST$ and $NLIST$.

6.5 Realization of Query Interpreter and Its Subcommands

Now we show the basic programs *class* depending upon the algorithm *candidate*. This *class* manages the definability of a set in NIS.

```
class(X,Y,EQUIV,Ppre,Pres,Npre,Nres)
  :-X==[],EQUIV=Y,Pres=Ppre,Nres=Npre.
class([X|X1],Y,EQUIV,Ppre,Pres,Npre,Nres)
  :-candidate([X|X1],CAN,Ppre,Pres1,Npre,Nres1),
    minus([X|X1],CAN,REST),
    class(REST,[CAN|Y],EQUIV,Pres1,Pres,Nres1,Nres).
```

In *class*, the second argument Y keeps the temporary set of equivalence classes, the fourth argument $Ppre$ does the temporary $PLIST$ and the sixth argument $Npre$ does the temporary $NLIST$. In the second clause, we first make a set

$CAN(\subset [X|X1])$ which satisfies all conditions, then we execute the *class* for a set $([X|X1]-CAN)$ again. If this $([X|X1]-CAN)$ is empty set, then the first clause is called and the temporary items are unified to response variable $EQUIV$, *Pres* and *Nres*. After finding a refutation for *class*, we get an equivalence relation and *DIS*. We have also prepared some subcommands depending upon *class*, *classex*, *relation*, *relationex* and *relationall*.

Now, we just show the real execution times for some $NISs$.

(CASE1) In NIS_1, we got two $DISs$ for $relationex([[1,2],[3,4]])$ in 0.0697(sec).

(CASE2) The number of object is 20, attribute is 10, DIS from NIS is 648(= $2^3 * 3^4$). It took 0.1646(sec) for translation. For $class([1,2,3,4,5])$, we got no DIS in 0.0018(sec). For $class([1,2,3,4,5,6])$, we got 324 $DISs$ in 0.0481(sec). For *relationall* which is the most heavy query, we got 48 possible equivalence relations in 2.0513(sec).

(CASE3) The number of object is 70, attribute is 4, DIS from NIS is 34992(= $2^4 * 3^7$). It took 0.3875(sec) for translation. For $class([1,2,3,4,5])$, we got no DIS in 0.0053(sec). For *relationall*, we got 4 possible equivalence relations in 215.6433(sec). The relations come from 20736 $DISs$, 2592 $DISs$, 10368 $DISs$ and 1296 $DISs$, respectively.

7 Concluding Remarks

In this paper, we discussed the definability of set in NIS and proposed an algorithm for checking it. The algorithm *candidate* takes the important roll for realizing some programs, which will be a good tool for handling NIS. We will apply our framework to machine learning and knowledge discovery from NIS.

REFERENCES

[1] Z.Pawlak: Rough Sets, Kluwer Academic Publisher, 1991.
[2] Z.Pawlak: Data versus Logic A Rough Set View, Proc. 4th Int'l. Workshop on Rough Set, Fuzzy Sets and Machine Discovery, pp.1-8, 1996.
[3] E.Orlowska and Z.Pawlak: Logical Foundations of Knowledge Representation, Pas Reports, 537, 1984.
[4] A.Nakamura, S.Tsumoto, H.Tanaka and S.Kobayashi: Rough Set Theory and Its Applications, Journal of Japanese Society for AI, Vol.11, No.2, pp.209-215, 1996.
[5] J.Grzymala-Busse: A New Version of the Rule Induction System LERS, Fundamenta Informaticae, Vol.31, pp.27-39, 1997.
[6] J.Komorowski and J.Zytkow(Eds.): Principles of Data Mining and Knowledge Discovery, Lecture Notes in AI, Vol.1263, 1997.
[7] Z.Ras and S.Joshi:Query Approximate Answering System for an Incomplete DKBS, Fundamenta Informaticae, Vol.30, pp.313-324, 1997.
[8] S.Tsumoto: PRIMEROSE, Bulletin of Int'l. Rough Set Society, Vol.2, No.1, pp.42-43, 1998.
[9] N.Zhong, J.Dong, S.Fujitsu and S.Ohsuga: Soft Techniques to Rule Discovery in Data, Transactions of Information Processing Society of Japan, Vol.39, No.9, pp.2581-2592, 1998.
[10] H.Sakai: Some Issues on Nondeterministic Knowledge Bases with Incomplete and Selective Information, Proc. RSCTC'98, Lecture Notes in AI, Vol.1424, pp.424-431, 1998.

On the Extension of Rough Sets under Incomplete Information

Jerzy Stefanowski[1] and Alexis Tsoukiàs[2]

[1] Institute of Computing Science, Poznań University of Technology, 3A Piotrowo,
60-965 Poznań, Poland, e-mail: Jerzy.Stefanowski@cs.put.poznan.pl
[2] LAMSADE - CNRS, Université Paris Dauphine, 75775 Paris Cédex 16, France
e-mail: tsoukias@lamsade.dauphine.fr

Abstract. The rough set theory, based on the conventional indiscernibility relation, is not useful for analysing incomplete information. We introduce two generalizations of this theory. The first proposal is based on non symmetric similarity relations, while the second one uses valued tolerance relation. Both approaches provide more informative results than the previously known approach employing simple tolerance relation.

1 Introduction

Rough set theory has been developed since Pawlak's seminal work [5] (see also [6]) as a tool enabling to classify objects which are only "roughly" described, in the sense that the available information enables only a partial discrimination among them although they are considered as different objects. In other terms, objects considered as "distinct" could happen to have the "same" or "similar" description, at least as far as a set of attributes is considered. Such a set of attributes can be viewed as the possible dimensions under which the surrounding world can be described for a given knowledge. An explicit hypothesis done in the classic rough set theory is that all available objects are completely described by the set of available attributes. Denoting the set of objects as $A = \{a_1, \cdots a_n\}$ and the set of attributes as $C = \{c_1, \cdots c_m\}$ it is considered that $\forall a_j \in A, c_i \in C$, the attribute value always exists, i.e. $c_i(a_j) \neq \emptyset$.

Such a hypothesis, although sound, contrast with several empirical situations where the information concerning the set A is only partial either because it has not been possible to obtain the attribute values (for instance if the set A are patients and the attributes are clinical exams, not all results may be available in a given time) or because it is definitely impossible to get a value for some object on a given attribute.

The problem has been already faced in literature by Grzymala [2], Kryszkiewicz [3, 4], Słowiński and Stefanowski [7]. Our paper enhances such works by distinguishing two different semantics for the incomplete information: the "missing" semantics (unknown values allow any comparison) and the "absent" semantics (unknown values do not allow any comparison) and explores three different formalisms to handle incomplete information tables: tolerance relations, non symmetric similarity relations and valued tolerance relations.

The paper is organized as follows. In section 2 we discuss the tolerance approach introduced by Kryszkiewicz [3]. Moreover, we give an example of incomplete information table which will be used all along the paper in order to help the understanding of the different approaches and allow comparisons. In section 3 an approach based on non symmetric similarity relations is introduced using some results obtained by Słowiński and Vanderpooten [8]. We also demonstrate that the non symmetric similarity approach refines the results obtained using the tolerance relation approach. Finally, in section 4 a valued tolerance approach is introduced and discussed as an intermediate approach among the two previous ones. Conclusions are given in the last section.

2 Tolerance relations

In the following we briefly present the idea introduced by Kryszkiewicz [3]. In our point of view the key concept introduced in this approach is to associate to the unavailable values of the information table a "null" value to be considered as "everything is possible" value. Such an interpretation corresponds to the idea that such values are just "missing", but they do exist. In other words, it is our imperfect knowledge that obliges us to work with a partial information table. Each object potentially has a complete description, but we just miss it for the moment. More formally, given an information table $IT = (A, C)$, a subset of attributes $B \subseteq C$ we denote the missing values by $*$ and we introduce the following binary relation T:

$\forall x, y \in A \times A \ T(x, y) \Leftrightarrow \forall c_j \in B \ c_j(x) = c_j(y)$ or $c_j(x) = *$ or $c_j(y) = *$

Clearly T is a reflexive and symmetric relation, but not necessarily transitive. We call the relation T a "tolerance relation". Further on let us denote by $I_B(x)$ the set of of objects y for which $T(x, y)$ holds taking into account attributes B. We call such a set the "tolerance class of x", thus allowing the definition of a set of tolerance classes of the set A. We can now use the tolerance classes as the basis for redefining the concept of lower and upper approximation of a set Φ using the set of attributes $B \subseteq C$. We have:

$\Phi_B = \{x \in A | I_B(x) \subseteq \Phi\}$ the lower approximation of Φ

$\Phi^B = \{x \in A | I_B(x) \cap \Phi \neq \emptyset\}$ the upper approximation of Φ

It is easy to observe that $\Phi^B = \bigcup \{I(x) | x \in \Phi\}$ also. Let us introduce now an example of incomplete information table which will be further used in the paper. *Example 1.* Suppose the following information table is given

A	a_1	a_2	a_3	a_4	a_5	a_6	a_7	a_8	a_9	a_{10}	a_{11}	a_{12}
c_1	3	2	2	$*$	$*$	2	3	$*$	3	1	$*$	3
c_2	2	3	3	2	2	3	$*$	0	2	$*$	2	2
c_3	1	2	2	$*$	$*$	2	$*$	0	1	$*$	$*$	1
c_4	0	0	0	1	1	1	3	$*$	3	$*$	$*$	$*$
d	Φ	Φ	Ψ	Φ	Ψ	Ψ	Φ	Ψ	Ψ	Φ	Ψ	Φ

where $a_1, ..., a_{12}$ are the available objects, $c_1, ..., c_4$ are four attributes which values (discrete) range from 0 to 3 and d is a decision attribute classifying objects either to the set Φ or to the set Ψ.

Using the tolerance relation approach to analyse the above example we have the following results: $I_C(a_1) = \{a_1, a_{11}, a_{12}\}$, $I_C(a_2) = \{a_2, a_3\}$, $I_C(a_3) = \{a_2, a_3\}$, $I_C(a_4) = \{a_4, a_5, a_{10}, a_{11}, a_{12}\}$, $I_C(a_5) = \{a_4, a_5, a_{10}, a_{11}, a_{12}\}$, $I_C(a_6) = \{a_6\}$, $I_C(a_7) = \{a_7, a_8, a_9, a_{11}, a_{12}\}$, $I_C(a_8) = \{a_7, a_8, a_{10}\}$, $I_C(a_9) = \{a_7, a_9, a_{11}, a_{12}\}$, $I_C(a_{10}) = \{a_4, a_5, a_8, a_{10}, a_{11}\}$, $I_C(a_{11}) = \{a_1, a_4, a_5, a_7, a_9, a_{10}, a_{11}, a_{12}\}$, $I_C(a_{12}) = \{a_1, a_4, a_5, a_7, a_9, a_{11}, a_{12}\}$. From which we can deduce that: $\Phi_C = \emptyset$, $\Phi^C = \{a_1, a_2, a_3, a_4, a_5, a_7, a_8, a_9, a_{10}, a_{11}, a_{12}\}$, $\Psi_C = \{a_6\}$, $\Psi^C = A$

The results are quite poor. Moreover there exist elements which intuitively could be classified in Φ or in Ψ, while they are not. Take for instance a_1. We have complete knowledge about it and intuitively there is no element perceived as similar to it. However, it is not in the lower approximation of Φ. This is due to "missing values" of a_{11} and a_{12} which enables them to be considered as "similar" to a_1. Of course this is "safe" because potentially the two objects could come up with exactly the same values of a_1.

A reduct is defined similarly as in the "classical" rough set the same model, i.e. it is a minimal subset of attributes that preserves lower approximations of object classification as for all attributes . In Example 1, the set of attributes $\{c_1, c_2, c_4\}$ is the only reduct. Kryszkiewicz [3] discussed the generation of decision rules from incomplete information tables. She considered mainly generalized decision rules of the form $\wedge_i(c_i, v) \rightarrow \vee(d, w)$. If the decision part contains one disjunct only, the rule is certain. Let B be a set of condition attributes which occur in a condition part of the rule $s \rightarrow t$. A decision rule is true if for each object x satisfying condition part s, $I_B(x) \subseteq [t]$. It is also required that the rule must have non-redundant condition part. In our example, we can find only one certain decision rule: $(c_1 = 2) \wedge (c_2 = 3) \wedge (c_4 = 1) \rightarrow (d = \Psi)$.

3 Similarity Relations

We introduce now a new approach based on the concept of a not necessarily symmetric similarity relation. Such a concept has been first introduced in general rough set theory by Słowiński and Vanderpooten [8] in order to enhance the concept of indiscernability relation. We first introduce what we call the "absent values semantics" for incomplete information tables. In this approach we consider that objects may be described "incompletely" not only because of our imperfect knowledge, but also because definitely impossible to describe them on all the attributes. Therefore we do not consider the unknown values as uncertain, but as "non existing" and we do not allow to compare unknown values.

Under such a perspective each object may have a more or less complete description, depending on how many attributes has been possible to apply. From this point of view an object x can be considered similar to another object y only if they have the same known values. More formally, denoting as usual the unknown value as $*$, given an information table $IT = (A, C)$ and a subset of attributes

$B \subseteq C$ we introduce a similarity relation S as follows:
$\forall x, y \ S(x, y) \Leftrightarrow \forall c_j \in B : \ c_j(x) \neq *, \ c_j(x) = c_j(y)$

It is easy to observe that such a relation although not symmetric is transitive. The relation S is a partial order on the set A. Actually it can be seen as a representation of the inclusion relation since we can consider that "*x is similar to y*" iff the "*the description of x*" is included in "*the description of y*". We can define for any object $x \in A$ two sets:
$R(x) = \{y \in A | S(y, x)\}$ the set of objects similar to x
$R^{-1}(x) = \{y \in A | S(x, y)\}$ the set of objects to which x is similar

Clearly $R(x)$ and $R^{-1}(x)$ are two different sets. We can now define for the lower and upper approximation of a set Φ as follows:
$\Phi_B = \{x \in A | R^{-1}(x) \subseteq \Phi\}$ the lower approximation of Φ
$\Phi^B = \bigcup \{R(x) | x \in \Phi\}$ the upper approximation of Φ

In other terms we consider as surely belonging to Φ all objects which have objects similar to them belonging to Φ. On the other hand any object which is similar to an object in Φ could potentially belong to Φ. Comparing our approach with the tolerance relation based one we can state the following result.

Theorem 1. *Given an information table $IT = (A, C)$ and a set Φ, the upper and lower approximations of Φ obtained using a non symmetric similarity relation are a refinement of the ones obtained using a tolerance relation.*

Proof. Denote as Φ_B^T the lower approximation of Φ using the tolerance approach and Φ_B^S the lower approximation of Φ using the similarity approach, Φ_T^B and Φ_S^B being the upper approximations respectively. We have to demonstrate that: $\Phi_B^T \subseteq \Phi_B^S$ and $\Phi_S^B \subseteq \Phi_T^B$. Clearly we have that: $\forall x, y \ S(x, y) \rightarrow T(x, y)$ since the conditions for which the relation S holds are a subset of the conditions for which the relation T holds. Then it is easy to observe that: $\forall x \ R(x) \subseteq I(x)$ and $R^{-1}(x) \subseteq I(x)$.

1. $\Phi_B^T \subseteq \Phi_B^S$. By definition $\Phi_B^T = \{x \in A | I(x) \subseteq \Phi\}$ and $\Phi_B^S = \{x \in A | R^{-1}(x) \subseteq \Phi\}$. Therefore if an object x belongs to Φ_B^T we have that $I_B(x) \subseteq \Phi$ and since $R^{-1}(x) \subseteq I(x)$ we have that $R^{-1}(x) \subseteq \Phi$ and therefore the same object x will belong to Φ_B^S. The inverse is not always true. Thus the lower approximation of Φ using the non symmetric similarity relation is at least as rich as the lower approximation of Φ using the tolerance relation.

2. $\Phi_S^B \subseteq \Phi_T^B$. By definition $\Phi_S^B = \bigcup_{x \in \Phi} R(x)$ and $\Phi_T^B = \bigcup_{x \in \Phi} I(x)$ and since $R(x) \subseteq I(x)$ the union of the sets $R(x)$ will be a subset of the union of the sets $I(x)$. The inverse is not always true. Therefore the upper approximation of Φ using the non symmetric similarity relation is at most as rich as the upper approximation of Φ using the tolerance relation.

Continuation of Example 1. Let us come back to the example introduced in section 1. Using all attributes C we have the following results: $R^{-1}(a_1) = \{a_1\}$, $R(a_1) = \{a_1, a_{11}, a_{12}\}$, $R^{-1}(a_2) = \{a_2, a_3\}$, $R(a_2) = \{a_2, a_3\}$, $R^{-1}(a_3) = \{a_2, a_3\}$, $R(a_3) = \{a_2, a_3\}$, $R^{-1}(a_4) = \{a_4, a_5\}$, $R(a_4) = \{a_4, a_5, a_{11}\}$, $R^{-1}(a_5) = \{a_4, a_5\}$, $R(a_5) = \{a_4, a_5, a_{11}\}$, $R^{-1}(a_6) = \{a_6\}$, $R(a_6) = \{a_6\}$, $R^{-1}(a_7) =$

$\{a_7, a_9\}$, $R(a_7) = \{a_7\}$, $R^{-1}(a_8) = \{a_8\}$, $R(a_8) = \{a_8\}$, $R^{-1}(a_9) = \{a_9\}$, $R(a_9) = \{a_7, a_9, a_{11}, a_{12}\}$, $R^{-1}(a_{10}) = \{a_{10}\}$, $R(a_{10}) = \{a_{10}\}$, $R^{-1}(a_{11}) = \{a_1, a_4, a_5, a_9, a_{11}, a_{12}\}$, $R(a_{11}) = \{a_{11}\}$, $R^{-1}(a_{12}) = \{a_1, a_9, a_{12}\}$, $R(a_{12}) = \{a_{11}, a_{12}\}$. From which we can deduce that: $\Phi_C = \{a_1, a_{10}\}$, $\Phi^C = \{a_1, a_2, a_3, a_4, a_5, a_7, a_{10}, a_{11}, a_{12}\}$, $\Psi_C = \{a_6, a_8, a_9\}$, $\Psi^C = \{a_2, a_3, a_4, a_5, a_6, a_7, a_8, a_9, a_{11}, a_{12}\}$.

The new approximations are more informative than the tolerance based ones. Moreover, we find now in the lower approximations of the sets Φ and Ψ some of the objects which intuitively we were expecting to be there. Obviously such an approach is less "safe" than the tolerance based one, since objects can be classified as "surely in Φ" although very little is known about them (e.g. object a_{10}). However, under the "absent values" semantic we do not consider a partially described object as "little known", but as "known" just on few attributes.

The subset C' of C is a reduct with respect to a classification if it is minimal subset of attributes C that keeps the same lower approximation of this classification. We observe that according to definition of the relation an object "totally unknown" (having in all attributes an unknown value) is not similar to any other object. If we eliminate one or more attributes which will make an object to become "totally unknown" on the remaining attributes we lose relevant information for the classification. We can conclude that all such attributes have to be in the reducts. In example 1 there is one reduct $\{c_1, c_2, c_4\}$ - it leads to the same classes $R^{-1}(x)$ and $R(x)$ as using all attributes.

The decision rule is defined as $s \rightarrow t$ (where $s = \wedge_i (c_i, v)$ and $t = (d, w)$). The rule is true if for each object x satisfying s, its class $R(x) \subseteq [t]$. The condition part cannot contain redundant conditions.

In example 1, the following certain decision rules can be generated:
$(c_1 = 1) \rightarrow (d = \Phi)$, $(c_3 = 1) \wedge (c_4 = 0) \rightarrow (d = \Phi)$, $(c_1 = 3) \wedge (c_4 = 0) \rightarrow (d = \Phi)$
$(c_2 = 3) \wedge (c_4 = 1) \rightarrow (d = \Psi)$, $(c_2 = 0) \rightarrow (d = \Psi)$, $(c_3 = 0) \rightarrow (d = \Psi)$
The absent value semantics gives more informative decision rules than tolerance based approach. Nevertheless these two different approaches (the tolerance and the non symmetric similarity) appear to be two extremes, in the middle of which it could be possible to use a more flexible approach.

4 Valued tolerance relations

Going back to the example of section 2, let's consider the elements a_1, a_{11} and a_{12}. Under both the tolerance relation approach and the non symmetric similarity relation approach we have: $T(a_{11}, a_1), T(a_{12}, a_1), S(a_{11}, a_1), S(a_{12}, a_1)$. However we may desire to express the intuitive idea that a_{12} is "more similar" to a_1 than a_{11} or that a_{11} is "less similar" to a_1 than a_{12}. This is due to the fact that in the case of a_{12} only one value is unknown and the rest all are equal, while in the case of a_{11} only one value is equal and the rest are unknown. We may try to capture such a difference using a valued tolerance relation.

The reader may notice that we can define different types of valued tolerance (or similarity) using different comparison rules. Moreover a valued tolerance (or similarity) relation can be defined also for complete information tables. Actually the approach we will present is independent from the specific formula adopted for the valued tolerance and can be extended to any type of valued relation.

Given a valued tolerance relation for each element of A we can define a "tolerance class" that is a fuzzy set with membership function the "tolerance degree" to the reference object. It is easy to observe that if we associate to the non zero tolerance degree the value 1 we obtain the tolerance classes introduced in section 2. The problem is to define the concepts of upper and lower approximation of a set Φ. Given a set Φ to describe and a set $Z \subseteq A$ we will try to define the degree by which Z approximates from the top or from the bottom the set Φ. Under such a perspective, each subset of A may be a lower or upper approximation of Φ, but to different degrees. For this purpose we need to translate in a functional representation the usual logical connectives of negation, conjunction etc..:

1. A negation is a function $N : [0,1] \mapsto [0,1]$, such that $N(0) = 1$ and $N(1) = 0$. An usual representation of the negation is $N(x) = 1 - x$.

2. A T-norm is a continuous, non decreasing function $T : [0,1]^2 \mapsto [0,1]$ such that $T(x,1) = x$. Clearly a T-norm stands for a conjunction. Usual representations of T-norms are: the min: $T(x,y) = \min(x,y)$; the product: $T(x,y) = xy$; the Łukasiewicz T-norm: $T(x,y) = \max(x + y - 1, 0)$.

3. A T-conorm is a continuous, non decreasing function $S : [0,1]^2 \mapsto [0,1]$ such that $S(0,y) = y$. Clearly a T-conorm stands for a disjunction. Usual representations of T-conorms are: the max: $S(x,y) = \max(x,y)$; the product: $S(x,y) = x + y - xy$; the Łukasiewicz T-conorm: $S(x,y) = \min(x + y, 1)$.

If $S(x,y) = N(T(N(x), N(y)))$ we have the equivalent of the De Morgan law and we call the triplet $\langle N, T, S \rangle$ a De Morgan triplet. $I(x,y)$, the degree by which x may imply y is again a function $I : [0,1]^2 \mapsto [0,1]$. However, the definition of the properties that such a function may satisfy do not make the unanimity. Two basic properties may be desired: the first claiming that $I(x,y) = S(N(x),y)$ translating the usual logical equivalence $x \rightarrow y =_{def} \neg x \lor y$; the second claiming that whenever the truth value of x is not greater than the truth value of y, then the implication should be true ($x \leq y \Leftrightarrow I(x,y) = 1$). It is almost impossible to satisfy both the two properties. In the very few cases where this happens other properties are not satisfied (for a discussion see [1]).

Coming back to our lower and upper approximations we know that given a set $Z \subseteq A$, a set Φ and attributes $B \subseteq C$ the usual definitions are:

1. $Z = \Phi_B \Leftrightarrow \forall z \in Z, \ \Theta(z) \subseteq \Phi$, 2. $Z = \Phi^B \Leftrightarrow \forall z \in Z, \ \Theta(z) \cap \Phi \neq \emptyset$ $\Theta(z)$ being the "indiscernability (tolerance, similarity etc.)" class of element z. The functional translation of such definitions is straightforward. Having:

$\forall x \ \phi(x) =_{def} T_x \phi(x); \ \exists x \ \phi(x) =_{def} S_x \phi(x); \ \Phi \subseteq \Psi =_{def} T_x (I(\mu_\Phi(x), \mu_\Psi(x)));$
$\Phi \cap \Psi \neq \emptyset =_{def} \exists x \ \phi(x) \land \psi(x) =_{def} S_x (T(\mu_\Phi(x), \mu_\Psi(x)))$ we get:

1. $\mu_{\Phi_B}(Z) = T_{z \in Z}(T_{x \in \Theta(z)}(I(R(z,x), \hat{x})))$,

2. $\mu_{\Phi^B}(Z) = T_{z \in Z}(S_{x \in \Theta(z)}(T(R(z,x), \hat{x})))$,

where: $\mu_{\Phi_B}(Z)$ is the degree for set Z to be a lower approximation of Φ; $\mu_{\Phi^B}(Z)$ is the degree for set Z to be an upper approximation of Φ; $\Theta(z)$ is the tolerance class of element z; T, S, I are the functions previously defined; $R(z,x)$ is the membership degree of element x in the tolerance class of z; $\hat{x}$ is the membership degree of element x in the set Φ ($\hat{x} \in \{0,1\}$).

Continuation of Example 1. Considering that the set of possible values on each attribute is discrete we make the hypothesis that there exists a uniform probability distribution among such values. More formally, consider c_j an attribute of an information table $IT = (A, C)$ and associate to it the set $E_j = \{e_j^1, \cdots e_j^m\}$ of all its possible values. Given an element $x \in A$ the probability that $c_j(x) = e_j^i$ is $1/|E_j|$. Therefore given any two elements $x, y \in A$ and an attribute c_j, if $c_j(y) = e_j^i$, the probability $R_j(x, y)$ that x is similar to y on the attribute c_j is $1/|E_j|$. On this basis we can compute the probability that two elements are similar on the whole set of attributes as the joint probability that the values of the two elements are the same on all the attributes: $R(x, y) = \prod_{c_j \in C} R_j(x, y)$. Applying this rule to objects we obtain the following table 1 concerning the valued tolerance relation.

	a_1	a_2	a_3	a_4	a_5	a_6	a_7	a_8	a_9	a_{10}	a_{11}	a_{12}
a_1	1	0	0	0	0	0	0	0	0	0	1/64	1/4
a_2	0	1	1	0	0	0	0	0	0	0	0	0
a_3	0	1	1	0	0	0	0	0	0	0	0	0
a_4	0	0	0	1	1/256	0	0	0	0	1/1024	1/1024	1/64
a_5	0	0	0	1/256	1	0	0	0	0	1/1024	1/1024	1/64
a_6	0	0	0	0	0	1	0	0	0	0	0	0
a_7	0	0	0	0	0	0	1	1/256	1/16	0	1/1024	1/64
a_8	0	0	0	0	0	0	1/256	1	0	1/1024	0	0
a_9	0	0	0	0	0	0	1/16	0	1	0	1/64	1/4
a_{10}	0	0	0	1/1024	1/1024	0	0	1/1024	0	1	1/4096	0
a_{11}	1/64	0	0	1/1024	1/1024	0	1/1024	0	1/64	1/4096	1	1/256
a_{12}	1/4	0	0	1/64	1/64	0	1/64	0	1/4	0	1/256	1

Table 1: Valued tolerance relation for Example 1.

If we consider element a_1, the valued tolerance relation $R(a_1, x)$, $x \in A$ will result in the vector $[1, 0, 0, 0, 0, 0, 0, 0, 0, 0, 1/64, 1/4]$ which actually represents the tolerance class $\Theta(a_1)$ of element a_1. The reader may notice that the crisp tolerance class of element a_1 was the set $\{a_1, a_{11}, a_{12}\}$ which corresponds to the vector $[1, 0, 0, 0, 0, 0, 0, 0, 0, 0, 1, 1]$. Following our "probabilistic approach" we may choose for T and S the product representation, while for I we will satisfy the De Morgan property thus obtaining: $T(x, y) = xy$, $S(x, y) = x + y - xy$, $I(x, y) = 1 - x + xy$. Clearly our choice of $I(x, y)$ does not satisfy the second property of implication. However, the reader may notice that in our specific case we have a peculiar implication from a fuzzy set ($\Theta(z)$) to a regular set (Φ), such that $\hat{x} \in \{0, 1\}$. The application of any implication satisfying the second property will reduce the valuation to the set $\{0, 1\}$ and therefore the whole degree $\mu_{\Phi_B}(Z)$ will collapse to $\{0, 1\}$ and thus to the usual lower approximation. With such considerations we obtain:

$$\mu_{\Phi_B}(Z) = \prod_{z \in Z} \prod_{x \in \Theta(z)} (1 - R(z, x) + R(z, x)\hat{x})$$
$$\mu_{\Phi^B}(Z) = \prod_{z \in Z} (1 - \prod_{x \in \Theta(z)} (1 - R(z, x)\hat{x}))$$

Consider now the set Φ and as set Z consider the element a_1, where $R(a_1, x)$

was previously introduced and $\hat{x}$ takes the values $[1,1,0,1,0,0,1,0,0,1,0,1]$. We obtain $\mu_{\Phi_C}(a_1) = 0.98$ and $\mu_{\bar{\Phi}C}(a_1) = 1$. Operationally we could choose a set Z as lower (upper) approximation of set Φ as follows:

1. take all elements for which $\mu(\Theta(z) \to \Phi) = 1$ $(\mu(\Theta(z) \cap \Phi) = 1)$;
2. then add elements in a way such that $\mu(\Theta(z) \to \Phi) > k$ $(\mu(\Theta(z) \cap \Phi) > k)$, (for decreasing values of k, let's say 0.99, 0.98 etc.), thus obtaining a family of lower (upper) approximations with decreasing membership function $\mu_{\Phi_B}(Z)$ $(\mu_{\bar{\Phi}B}(Z))$;
3. fix a minimum level λ enabling to accept a set Z as a lower (upper) approximation of Φ (thus $\mu_{\Phi_B}(Z) \geq \lambda$).

The concept of reduct and decision rules are also generalized in the valued tolerance case. Given the decision table (A, C) and the partition $\mathcal{Y} = \Phi_1, \Phi_2, \dots \Phi_n$, the subset of attributes $C' \subset C$ is a reduct iff it does not decease the degree of lower approximation obtained with C, i.e. if $z_1, z_2, \dots, z_n$ is a family of lower approximations of $\Phi_1, \Phi_2, \dots \Phi_n$ then $\forall_{i=1,\dots,n} z_i \; \mu_{\Phi_{iC}}(z_i) \leq \mu_{\Phi_{iC'}}(z_i)$.

In order to induce classification rules from the decision table on hand we may accept now rules with a "credibility degree" derived from the fact that objects may be similar to the conditional part of the rule only to a certain degree, besides the fact the implication in the decision part is also uncertain. More formally we give the following representation for a rule ρ_i: $\rho_i^J =_{def} \bigwedge_j (c_j(a_i) = v) \to (d = w)$ where: $J \subseteq C$, v is the value of attribute c_j, w is the value of attribute d.

As usual we may use relation $s(x, \rho_i)$ in order to indicate that element x "supports" rule ρ_i or that, x is similar to some extend to the condition part of rule ρ_i. We denote as $S(\rho_i) = \{x : s(x, \rho_i) > 0\}$ and as $W = \{x : d(x) = w\}$. Then ρ_i is a decision rule iff: $\forall x \in S(\rho_i) : \Theta(x) \subseteq W$. We can compute a credibility degree for any rule ρ_i calculating the truth value of the previous formula which can be rewritten as: $\forall x, y \; s(x, \rho_i) \to (R(x, y) \to W(y))$. We get: $\mu(\rho_i) = T_x(I_y(s(x, \rho_i), I(\mu_{\Theta(x)}(y), \mu_W(y))))$. Finally it is necessary to check whether J is a non-redundant set of conditions for rule ρ_i, i.e. to look if it is possible to satisfy the condition: $\exists \hat{J} \subset J : \mu(\rho_i^{\hat{J}}) \geq \mu(\rho_i^J)$ or not.

Continuation of Example 1. Consider again the incomplete table and take as candidate the rule: $\rho_1 : (c_1 = 3) \wedge (c_2 = 2) \wedge (c_3 = 1) \wedge (c_4 = 0) \to (d = \Phi)$. Since in the paper we have chosen for the functional representation of implication the satisfaction of De Morgan law and for T-norms the product, we get:

$$\mu(\rho_i) = \prod_{x \in S(\rho_i)} (1 - s(x, \rho_i) + s(x, \rho_i) \prod_{y \in \Theta(x)} (1 - \mu_{\Theta(x)}(y) + \mu_{\Theta(x)}(y)\mu_W(y)))$$

where $s(x, \rho_i)$ represents the "support" degree of element x to the rule ρ_i. We thus get that $\mu(\rho_1) = 0.905$. However, the condition part of rule ρ_1 is redundant and is transformed to: $\rho_1 : (c_1 = 3) \wedge (c_3 = 1) \wedge (c_4 = 0) \to (d = \Phi)$ with degree $\mu(\rho_1) = 0.905$. This rule is supported by objects $S(\rho_1) = \{a_1, a_{11}, a_{12}\}$. For the set Ψ we have one rule: $\rho_2 : (c_1 = 2) \wedge (c_2 = 3) \wedge (c_4 = 1) \to (d = \Psi)$ with degree $\mu(\rho_2) = 1.0$ and a supporting object a_6.

Operationally a user may first fix a threshold of credibility for the rules to accept and then could operate a sensitivity analysis on the set of rules that is possible to accept in an interval of such threshold.

5 Conclusions

Rough set theory has been conceived under the implicit hypothesis that all objects in a universe can be evaluated under a given set of attributes. However, it can be the case that several values are not available for various reasons. In our paper we introduce two different semantics in order to distinguish such situations. "Missing values" imply that non available information could always become available and that in order to make "safe" classifications and rules induction we might consider that such missing values are equal to everything. Tolerance relations (which are reflexive and symmetric, but not transitive) capture in a formal way such an approach. "Absent values" imply that not available information cannot be used in comparing objects and that classification and rules induction should be performed with the existing information since the absent values could never become available. Similarity relations (which in our case are reflexive and transitive, but not symmetric) are introduced in our paper in order to formalize such an idea. We demonstrate in the paper that our approach always lead to more informative results with respect to the tolerance relation based approach (although less safe).

A third approach is also introduced in the paper, as an intermediate position among the two previously presented. Such an approach is based on the use of a valued tolerance relation. A valued relation could appear for several reasons not only because of the non available information and in fact the approach presented has a more general validity. However in this paper we limit ourselves in discussing the missing values case. A functional extension of the concepts of upper and lower approximation is introduced in this paper so that to any subset of the universe a degree of lower (upper) approximation can be associated. Further on such a functional extension enables to compute a credibility degree for any decision rule induced by the classification.

References

1. Dubois D., Lang J., Prade H., Fuzzy sets in approximate reasoning *Fuzzy Sets and Systems* 40 (1991), 203–244.
2. Grzymala-Busse J. W. On the unknown attribute values in learning from examples. *Proc. of Int. Symp. on Methodologies for Intelligent Systems*, 1991, 368–377.
3. Kryszkiewicz M., Rough set approach to incomplete information system. *Information Sciences* 112 (1998), 39–49.
4. Kryszkiewicz M., Properties of incomplete information systems in the framework of rough sets. In Polkowski L., Skowron A. (eds.) *Rough Sets in Data Mining and Knowledge Discovery*, Physica-Verlag, 1998, 422–450.
5. Pawlak Z., Rough sets. *Int. J. Computer and Information Sci.*, 11, 1982, 341–356.
6. Pawlak Z., *Rough sets. Theoretical aspects of reasoning about data*. Kluwer, 1991.
7. Słowiński R., Stefanowski J., Rough classification in incomplete information systems. *Math. Computing Modelling*, 12 (10/11), 1989, 1347-1357.
8. Słowiński R., Vanderpooten D., A generalized definition of rough approximation based on similarity. *IEEE Transactions on Data and Knowledge Engineering*, 1999 (to apear).

On Rough Relations:
An Alternative Formulation

Y.Y. Yao and Tao Wang

Department of Computer Science, University of Regina
Regina, Saskatchewan, Canada S4S 0A2
E-mail: {yyao,htwang}@cs.uregina.ca

Abstract. Another formulation of the notion of rough relations is presented. Instead of using two equivalence relations on two universes, or a joint equivalence relation on their Cartesian product, we start from specific classes of binary relations obeying certain properties. The chosen class of relations is a subsystem of all binary relations and represents relations we are interested. An arbitrary relation is approximated by a pair of relations in the chosen class.

1 Introduction

The theory of rough sets is built on partitions of the universe defined by equivalence relations [6, 16]. A partition of the uninverse represents a granulated view of the universe, in which equivalence classes are considered to be basic granules. It is assumed that information is available for only the basic granules. One has to consider each equivalence class as a whole instead of individual elements of the universe. For inferring information about an arbitrary subset of the universe, it is necessary to consider its approximations by equivalence classes. More specifically, a set is described by a pair of lower and upper approximations. From existing studies of rough sets, we can identify at least two formulations, the partition based method and subsystem based method [14, 15]. In partition based approach, the lower approximation is the union of equivalence classes contained in the set, and the upper approximation is the union of equivalence classes having a nonempty intersection with the set. In subsystem based approach, one can use equivalence classes as basic building blocks and construct a subsystem of the power set by taking unions of equivalence classes. The constructed subsystem is in fact an σ-algebra of subsets of the universe. That is, it contains both the empty set and the entire set, and is closed under set intersection and union. The lower approximation is the largest subset in the subsystem that is contained in the set to be approximated, and the upper approximation is the smallest subset in the subsystem that contains the set to be approximated. Each of the two formulations captures different aspects of rough set approximations. They can be used to obtain quite distinctive generalizations of rough set theory [15, 17].

A binary relation is a set of pairs, i.e., a subset of the Cartesian product of two universes. It is therefore very natural to generalize rough sets to the notion of rough relations. The majority of existing studies on rough relations is relied on partition based approach. It involves two equivalence relations on two universes, or a joint equivalence relation on their Cartesian product. This straightforward definition of rough relations was proposed by Pawlak [7, 9]. Generalizations of rough relations, along the same line of argument, have been made by Düntsch [3], Stepaniuk [11, 12, 13], and Skowron and Stepaniuk [10]. An implication of the partition based formulation is that the properties of lower and upper approximations depend on the relation to be approximated. Although a binary relation is a set of pair, it is set equipped with additional properties, such as reflexivity, symmetry, and transitivity. The added information provided by binary relations is not fully explored in many studies of rough relations. For some applications, we may only be interested in approximating a relation in terms of relations with special properties [4]. The subsystem based approach may be useful, as one can choose the subsystem so that all relations in the subsystem have some desired properties. Greco et al. [4] implicitly used subsystem based approach for the approximation of preferential information.

The main objective of this paper is to present an alternative formulation of rough relations by extending the subsystem based method. In Section 2, we review two formulations of rough set approximations. In Section 3, a subsystem based formulation of rough relations is introduced. Special types of subsystems are used for defining rough relation approximations. This study is complementary to existing studies, and the results may provide more insights into the understanding and applications of rough relations.

2 Two Formulations of Rough Set Approximations

Let $E \subseteq U \times U$ denote an equivalence relation on a finite and nonempty universe U, where $U \times U = U^2$ is the Cartesian product of U. That is, E is reflexive, symmetric, and transitive. The pair $apr = (U, E)$ is referred to as a Pawlak approximation space. The equivalence relation E partitions U into disjoint subsets known as equivalence classes. That is, E induces a quotient set of the universe U, denoted by U/E. Equivalence classes are called elementary sets. They are interpreted as basic observable, measurable, or definable subsets of U. The empty set $\emptyset$ and a union of one or more elementary sets are interpreted as composite ones. The family of all such subsets is denoted by $\mathrm{Def}(U)$. It defines a topology space $(U, \mathrm{Def}(U))$ in which $\mathrm{Def}(U)$, a subsystem of the power set of U, consists of both closed and open sets. Two formulations of rough sets can be obtained by focusing on the partition U/E and the topology $\mathrm{Def}(U)$, respectively.

An arbitrary subset $X \subseteq U$ is approximated by a pair of subsets of U called lower and upper approximations, or simply a rough set approximation [6]. The lower approximation $\underline{apr}(X)$ is the union of all elementary sets contained in X, and the upper approximation $\overline{apr}(X)$ is the union of all elementary sets which have a nonempty intersection with X. They are given by:

$$(\text{def1}) \quad \underline{apr}(X) = \bigcup \{[x]_E \mid x \in U, [x]_E \subseteq X\},$$
$$\overline{apr}(X) = \bigcup \{[x]_E \mid x \in U, [x]_E \cap X \neq \emptyset\},$$

where $[x]_E$ denotes the equivalence class containing x:

$$[x]_E = \{y \mid xEy, \; x, y \in U\}. \tag{1}$$

For rough set approximations, we have the following properties:

(L1) $\underline{apr}(X \cap Y) = \underline{apr}(X) \cap \underline{apr}(Y),$

(L2) $\underline{apr}(X) \subseteq X,$

(L3) $\underline{apr}(X) = \underline{apr}(\underline{apr}(X)),$

(L4) $\overline{apr}(X) = \underline{apr}(\overline{apr}(X)),$

and

(U1) $\overline{apr}(X \cup Y) = \overline{apr}(X) \cup \overline{apr}(Y),$

(U2) $X \subseteq \overline{apr}(X),$

(U3) $\overline{apr}(X) = \overline{apr}(\overline{apr}(X)),$

(U4) $\underline{apr}(X) = \overline{apr}(\underline{apr}(X)).$

The two approximations are dual to each other in the sense that $\underline{apr}(-X) = -\overline{apr}(X)$ and $\overline{apr}(-X) = -\underline{apr}(X)$. The properties with the same number may be considered as dual properties. It is possible to compute the lower approximation of $X \cap Y$ based on the lower approximations of X and Y. However, it is impossible to compute the upper approximation of $X \cap Y$ based on the upper approximations of X and Y. Similar observation can also be made for the approximations of $X \cup Y$.

By the properties of rough set approximations, $\underline{apr}(X)$ is indeed the greatest definable set contained in X, $\overline{apr}(X)$ is the least definable set containing X. The following equivalent definition can be used [6, 14]:

(def2) $\underline{apr}(X) = \bigcup \{Y \mid Y \in \text{Def}(U), Y \subseteq X\},$

           $\overline{apr}(X) = \bigcap \{Y \mid Y \in \text{Def}(U), X \subseteq Y\}.$

For a subset $X \in \text{Def}(U)$, we have $X = \underline{apr}(X) = \overline{apr}(X)$. Thus, we can say that subsets in $\text{Def}(U)$ have exact representations. For other subsets of U, both lower and upper approximations do not equal to the set itself, which leads to approximate representations of the set. It should be clear by now the reason for calling elements of $\text{Def}(U)$ definable sets. Mathematically speaking, subsets in $\text{Def}(U)$ may be considered as fixed points of approximation operators $\underline{apr}$ and $\overline{apr}$. Every other element is approximated using the fixed points. That is, $\underline{apr}(X)$ is the best approximation of X from below, and $\overline{apr}(X)$ is the best approximation of X from above.

Although both definitions are equivalent, they offer quite different interpretations for rough set approximations. Definition (def1) focuses on equivalence classes, which clearly shows how relationships between elements of U are used. The approximations of an arbitrary subset of the universe stem from the granulation of universe by an equivalence relation. This definition can be extended to define approximation operators based on other types of binary relations [17]. Definition (def2) focuses on a subsystem of U with special properties. With less elements in the subsystem than that in the power set, certain elements of the power set have to be approximated. The formulation can be easily applied to situations where a binary relation is not readily available. It has been used to study approximation in mathematical structures such as topological spaces, closure systems, Boolean algebras, lattices, and posets [1, 14, 15].

In generalizing definition (def2), subsystems of the power set must be properly chosen [15]. The subsystem for defining lower approximations must contain the empty set $\emptyset$ and be closed under union, and the subsystem for defining upper approximations must contain the entire set U and be closed under intersection. In other words, the subsystem for defining upper approximation must be a closure system [2]. In general, the two subsystems are not necessarily the same, nor dual to each other [1, 15]. The subsystem Def(U) induced by an equivalence relation is only a special case.

3 Rough Relation Approximations

This section first reviews a commonly used formulation of rough relations based on definition (def1) and discusses its limitations. By extending definition (def2), we present a new formulation.

3.1 A commonly used formulation

A binary relation R on a universe U is a set of ordered pairs of elements from U, i.e., $R \subseteq U \times U$. The power set of $U \times U$, i.e., $2^{U \times U}$, is the set of all binary relations on U. The empty binary relation is denoted by $\emptyset$, and the whole relation is $U \times U$. One may apply set-theoretic operations to relations and define the complement, intersection, and union of binary relations. By taking $U \times U$ as a new universe, one can immediately study approximations of binary relations. For clarity, we only consider binary relations on the same universe, instead of the general case where relations are defined on more than two distinct universes [7].

Suppose E_1 and E_2 are two equivalence relations on U. They induce two approximation spaces $apr_1 = (U, E_1)$ and $apr_2 = (U, E_2)$. The product relation $E = E_1 \times E_2$:

$$(x, y)E(v, w) \iff xE_1v, yE_2w, \tag{2}$$

is an equivalence relation on $U \times U$. It gives rise to a product approximation space $apr = (U \times U, E_1 \times E_2)$. In the special case, a single approximation space $apr_U = (U, E_U)$ can be used to derive the product approximation space $apr = (U \times U, E_U \times E_U)$. The notion of product approximation space forms a basis for rough relation approximations. For an equivalence relation $E \subseteq (U \times U)^2$, the equivalence class containing (x, y):

$$[(x, y)]_E = \{(v, w) \mid (x, y)E(v, w), (x, y), (v, w) \in U \times U\},$$
$$= [x]_{E_1} \times [y]_{E_2}, \tag{3}$$

is in fact a binary relation on U. It is called an elementary definable relation. The empty relation $\emptyset$ and unions of elementary definable relations are referred to as definable relations. The family of definable relations is denoted by $\mathrm{Def}(U \times U)$. Although definable relations are constructed from an equivalence relation E on $U \times U$, relations in $\mathrm{Def}(U \times U)$ are not necessarily reflexive, symmetric, or transitive. This can be easily seen from the fact that the elementary relations $[(x, y)]_E$ do not necessarily have any of those properties.

Given a binary relation $R \subseteq U \times U$, by definition (def1) we can approximate it by two relations:

$$(\text{def1}) \quad \underline{apr}(R) = \bigcup \{[(x,y)]_E \mid (x,y) \in U \times U, [(x,y)]_E \subseteq R\},$$

$$\overline{apr}(R) = \bigcup \{[(x,y)]_E \mid (x,y) \in U \times U, [(x,y)]_E \cap R \neq \emptyset\}.$$

Equivalently, definition (def2) can be used with respect to the subsystem $\mathrm{Def}(U \times U)$. The rough relation approximations are dual to each other and satisfy properties (L1)-(L4) and (U1)-(U4). Since a binary relation is a set with added information, one can observe the following additional facts [7, 11, 13]:

1. Suppose $E = E_U \times E_U$. If $E_U \neq I_U$, neither $\underline{apr}(I_U)$ nor $\overline{apr}(I_U)$ is the identity relation, where $I_U = \{(x, x) \mid x \in U\}$ denotes the identity relation on U.

2. For a reflexive relation R, $\overline{apr}(R)$ is reflexive, and $\underline{apr}(R)$ is not necessarily reflexive.

3. For a symmetric relation R, both $\underline{apr}(R)$ and $\overline{apr}(R)$ are symmetric.

4. For a transitive relation R, $\underline{apr}(R)$ and $\overline{apr}(R)$ are not necessarily transitive.

5. For an equivalence relation R, $\underline{apr}(R)$ and $\overline{apr}(R)$ are not necessarily equivalence relations.

6. Suppose $E = E_U \times E_U$. For an equivalence relation R, $\overline{apr}(R)$ is an equivalence relation if and only if $\overline{apr}(R) = (R \cup E_U)^*$, and $\underline{apr}(R)$ is an equivalence relation if and only if $E_U \subseteq R$, where R^* denotes the reflexive and transitive closure of a relation R.

One can therefore conclude that the lower and upper approximations of a relation may not have all the properties of the relation to be approximated. If an arbitrary relation is approximated by elements of $\mathrm{Def}(U \times U)$, one cannot expect certain properties of its approximations. However, in some situations, it may be desirable that a relation is approximated by relations having certain specific properties. We clearly cannot achieve this goal with the standard formulation of rough relations.

3.2 A new formulation

If subsystems of $U \times U$ are properly chosen, some of the difficulties identified in the last section can be avoided by generalizing definition (def2). In what follows, a new formulation is presented, with focus on properties, such as reflexivity, symmetry, and transitivity, of binary relations.

Let $P = \{reflexive,\ symmetric,\ transitive\} = \{r, s, t\}$ denote a set of properties of binary relations on U. For $A \subseteq P$, the set of binary relations satisfying all properties in A is denoted by S_A. For instance, $S_{\{r,s\}}$ consists of all reflexive and symmetric relations (i.e., tolerance or compatibility relations). One can verify the following properties:

1. The system $S_{\{r\}}$ is closed under both intersection and union. It does not contain the empty relation, i.e., $\emptyset \notin S_{\{r\}}$, and contains the whole relation, i.e., $U \times U \in S_{\{r\}}$.
2. The system $S_{\{s\}}$ is closed under both intersection and union. It contains both the empty relation and the whole relation.
3. The system $S_{\{t\}}$ is closed under intersection, but not closed under union. It contains both the empty relation and the whole relation.
4. The system of compatibility relations $S_{\{r,s\}}$ is closed under both intersection and union. It contains the whole relation, and does not contain the empty relation.
5. The system $S_{\{r,t\}}$ is closed under intersection and not closed under union. It contains the whole relation, and does not contain the empty relation.
6. The system $S_{\{s,t\}}$ is closed under intersection. It contains both the empty relation and the whole relation.
7. The system of equivalence relations $S_{\{r,s,t\}}$ is closed under intersection, but not closed under union. It contains the whole relation, and does not contain the empty relation.

They represent all possible subsystems with properties in the set P. It is interesting to note that the subsystem $\mathrm{Def}(U \times U)$ induced by an equivalence relation on $U \times U$ does not belong to any of the above classes. Subsystems that can be used for various approximations are summarized as follows:

Lower approximation:
   $S_{\{r\}} \cup \{\emptyset\}$, $S_{\{s\}}$, $S_{\{r,s\}} \cup \{\emptyset\}$.
Upper approximation:
   All subsystems.
Lower and upper approximations:
   $S_{\{r\}} \cup \{\emptyset\}$, $S_{\{s\}}$, $S_{\{r,s\}} \cup \{\emptyset\}$.

Although every subsystem can be used for defining upper approximation, only three subsystems can be used for lower approximation.

Given a subsystem $S_l \subseteq 2^{U \times U}$ containing $\emptyset$ and being closed under union, and a subsystem $S_u \subseteq 2^{U \times U}$ containing $U \times U$ and being closed under intersection, the rough relation approximation of a binary relation R is defined by:

$$(\mathrm{def2}) \quad \underline{apr}(R) = \bigcup \{Q \mid Q \in S_l, Q \subseteq R\},$$
$$\overline{apr}(R) = \bigcap \{Q \mid Q \in S_u, R \subseteq Q\}.$$

In the special case, two subsystems can be the same. For example, one may use the subsystem of compatibility relations. By definition, rough relation approximations satisfy properties (L2), (L3), (U2), (U3), and the following weaker version of (L1) and (U1):

$$(L0) \quad R \subseteq Q \Longrightarrow \underline{apr}(R) \subseteq \underline{apr}(Q),$$
$$(U0) \quad R \subseteq Q \Longrightarrow \overline{apr}(R) \subseteq \overline{apr}(Q).$$

A detailed discussion of such subsystems in the setting of rough set approximations can be found in a recent paper by Yao [15].

Regarding the subsystems characterized by properties in $P = \{r, s, t\}$, we have the following results:

(i). Suppose the pair of subsystems $(S_{\{r\}} \cup \{\emptyset\}, S_{\{r\}})$ is used for defining lower and upper approximations. We have:

$$\underline{apr}(R) = \begin{cases} \emptyset & \text{if } I_U \not\subseteq R, \\ R & \text{if } I_U \subseteq R, \end{cases}$$
$$\overline{apr}(R) = R \cup I_U.$$

(ii). For the subsystem $S_{\{s\}}$, we have:

$$\underline{apr}(R) = R \cap R^{-1},$$
$$\overline{apr}(R) = R \cup R^{-1},$$

where $R^{-1} = \{(y, x) \mid x R y\}$ is the inverse of the relation R.

(iii). For the subsystem $S_{\{t\}}$, we have:

$$\overline{apr}(R) = R^+,$$

where R^+ denotes the transitive closure of the binary relation R.

(iv). For the subsystem $S_{\{r,s\}} \cup \{\emptyset\}$, we have:

$$\underline{apr}(R) = \begin{cases} \emptyset & \text{if } I_U \not\subseteq R, \\ R \cap R^{-1} & \text{if } I_U \subseteq R, \end{cases}$$
$$\overline{apr}(R) = R \cup I_U \cup R^{-1}.$$

(v). For the subsystem $S_{\{r,t\}}$, we have:

$$\overline{apr}(R) = I_U \cup R^+.$$

(vi). For the subsystem $S_{\{s,t\}}$, we have:

$$\overline{apr}(R) = (R \cup R^{-1})^+.$$

(vii). For the subsystem $S_{\{r,s,t\}}$, we have:

$$\overline{apr}(R) = (R \cup I_U \cup R^{-1})^+.$$

One can see that the lower approximation is obtained by removing certain pairs from the relation, while the upper approximation is obtained by adding certain pairs to the relation, so that the required properties hold. This interpretation of approximation is intuitively appealing. The definition (def2) only provides a formal description of rough relation approximations. In practice, one can easily obtain the approximations without actually constructing the subsystems and using definition (def2).

When the subsystem $S_{\{r\}} \cup \{\emptyset\}$ is used for lower and upper approximations, reflexive relations are fixed points. That is, both lower and upper approximations of a reflexive relation equal to the relation itself. Similar observations hold for other subsystems.

Our formulation of rough relation approximations is very flexible. In approximating a relation, two different subsystems may be used, one for lower approximation, and the other for upper approximation. For example, one may approximate an arbitrary binary relation from below by a compatibility relation, and from above by an equivalence relation. If the relation is reflexive, then the lower approximation is obtained by deleting pairs that violate the property of symmetry, while the upper approximation is obtained by adding pairs so that the transitivity holds. Such a pair of lower and upper approximations provides a good characterization of the original relation. The subsystems discussed so far are some examples. In general, one can construct various subsystems for approximation as long as they obey certain properties. The subsystem for lower approximation must contain $\emptyset$ and be closed under union, and the subsystem for upper approximation must contain $U \times U$ and be closed under intersection. For example, for defining both lower and upper approximations one may select a subset of $S_{\{r,s,t\}} \cup \{\emptyset\}$ such that it is closed under both intersection and union.

4 Conclusion

A binary relation is not simply a set of pairs, but a set with additional information and properties. The problem of rough *relation* approximation may therefore be different from rough *set* approximations. In contrast to other related studies, the main purpose of this paper is to investigate possibilities of using such extra information in approximating relations. An alternative formulation of rough relations is proposed based on subsystems of binary relations with certain properties, instead of using equivalence relations. From a quite different point of view, our formulation explicitly addresses some fundamental issues which have been overlooked in existing studies of rough relations.

The results as shown by (i)-(vii) are simple and they could have been obtained easily without the introduction of the new framework. However, the importance of the approach may not be taken lightly. The recognition and utilization of special classes of binary relations for approximating other binary relations may have significant implications on the understanding and applications of rough relation approximations. The results may be applied to rough *function* approximations [8]. In this paper, we only considered three properties of binary relations.

With our formulation, other properties of binary relations can also be considered. Order relations (i.e., preference relations) play a very important role in decision theory [4, 5]. It may be useful to apply the proposed method for approximating order relations.

References

1. Cattaneo, G. Abstract approximation spaces for rough theories, in: *Rough Sets in Knowledge Discovery*, Polkowski, L. and Skowron, A. (Eds.), Physica-Verlag, Heidelberg, 59-98, 1998.
2. Cohn, P.M. *Universal Algebra*, Harper and Row Publishers, New York, 1965.
3. Düntsch, I. Rough sets and algebras of relations, in: *Incomplete Information: Rough Set Analysis*, Orlowska, E. (Ed.), Physica-Verlag, Heidelberg, 95-108, 1998.
4. Greco, S., Matarazzo, B. and Slowinski, R. Rough approximation of preferential information by dominance relations, *Proceedings of Fourth International Workshop on Rough Sets, Fuzzy Sets, and Machine Discovery*, 125-130, 1996.
5. Greco, S., Matarazzo, B. and Slowinski, R. A new rough set approach to multicriteria and multiattribute classification, in: *Rough Sets and Current Trends in Computing*, Polkowski, L. and Skowron, A. (Eds.), Springer-Verlag, Berlin, 60-67, 1998.
6. Pawlak, Z. Rough sets, *International Journal of Computer and Information Sciences*, 11, 341-356, 1982.
7. Pawlak, Z. Rough relations, *Bulletin of Polish Academy of Sciences, Technical Sciences*, 34, 557-590, 1986.
8. Pawlak, Z. Rough functions, *Bulletin of Polish Academy of Sciences, Technical Sciences*, 35, 249-251, 1987.
9. Pawlak, Z. Rough sets, rough relations and rough functions, *Fundamenta Informaticae*, 27, 103-108, 1996.
10. Skowron, A. and Stepaniuk, J. Approximation of relations, in: *Rough Sets, Fuzzy Sets and Knowledge Discovery*, Ziarko, W.P. (Ed.), Springer-Verlag, London, 161-166, 1994.
11. Stepaniuk, J. Properties and applications on rough relations, *Proceedings of the Fifth International Workshop on Intelligent Information Systems*, 136-141, 1996.
12. Stepaniuk, J. Approximation spaces in extensions of rough set theory, in: *Rough Sets and Current Trends in Computing*, Polkowski, L. and Skowron, A. (Eds.), Springer-Verlag, Berlin, 290-297, 1998.
13. Stepaniuk, J. Rough relations and logics, in: *Rough Sets in Knowledge Discovery*, Polkowski, L. and Skowron, A. (Eds.), Physica-Verlag, Heidelberg, 248-260, 1998.
14. Yao, Y.Y. Two views of the theory of rough sets in finite universes, *International Journal of Approximate Reasoning*, 15, 291-317, 1996.
15. Yao, Y.Y. On generalizing Pawlak approximation operators, in: *Rough Sets and Current Trends in Computing*, Polkowski, L. and Skowron, A. (Eds.), Springer-Verlag, Berlin, 298-307, 1998.
16. Yao, Y.Y. A comparative study of fuzzy sets and rough sets, *Information Sciences*, 109, 227-242, 1998.
17. Yao, Y.Y. and Lin, T.Y. Generalization of rough sets using modal logic, *Intelligent Automation and Soft Computing, an International Journal*, 2, 103-120, 1996.

Formal Rough Concept Analysis*

Jamil Saquer   Jitender S. Deogun

deogun(jsaquer)@cse.unl.edu

The Department of Computer Science,
University of Nebraska,
Lincoln, NE 68588, USA

abstract

In this paper, we present a novel approach for approximating concepts in the
framework of formal concept analysis. Two main problems are investigated. The
first, given a set A of objects (or a set B of features), we want to find a formal
concept that approximates A (or B). The second, given a pair (A, B), where
A is a set of objects and B is a set of features, the objective is to find formal
concepts that approximate (A, B). The techniques developed in this paper use
ideas from rough set theory. The approach we present is different and more
general than existing approaches.

1 Introduction

Formal concept analysis (FCA) is a mathematical framework developed by Rudolf
Wille and his colleagues at Darmstadt/Germany that is useful for representation and analysis of data [8]. A pair consisting of a set of objects and a set of
features common to these objects is called a concept. Using the framework of
FCA, concepts are structured in the form of a lattice called the concept lattice.
The concept lattice is a useful tool for knowledge representation and knowledge
discovery [2]. Formal concept analysis has also been applied in the area of conceptual modeling that deals with the acquisition, representation and organization
of knowledge [4]. Several concept learning methods have been implemented in
[1, 2, 3] using ideas from formal concept analysis.

Not every pair of a set of objects and a set of features defines a concept [8].
Furthermore, we might be faced with a situation where we have a set of features
(or a set of objects) and need to find the best concept that approximates these
features (or objects). For example, when a physician diagnosis a patient, he finds
a disease whose symptoms are the closest to the symptoms that the patient has.

---

*This research was supported in part by the Army Research Office, Grant No. DAAH04-
96-1-0325, under DEPSCoR program of Advanced Research Projects Agency, Department of
Defense and by the U.S. Department of Energy, Grant No. DE-FG02-9 7ER1220.

In this case we can think of the symptoms as features and diseases as objects. It is therefore of fundamental importance to be able to find concept approximations regardless how little information is available.

In this paper we present a general approach for approximating concepts. We first show how a set of objects (or features) can be approximated by a concept. We prove that our approximations are the best that can be achieved using rough sets. We then extend our approach to approximate a pair of a set of objects and a set of features.

2   Background

Relationships between objects and features in FCA is given in a context which is defined as a triple (G, M, I), where G and M are sets of objects and features (also called attributes), respectively, and $I \subseteq G \times M$. If object g possesses feature m, then $(g, m) \in I$ which is also written as gIm. The set of all common features to a set of objects A is denoted by $\beta(A)$ and defined as $\{m \in M \mid gIm \; \forall g \in A\}$. Similarly, the maximal set of objects possessing all the features in a set of features B is denoted by $\alpha(B)$ and given by $\{g \in G \mid gIm \; \forall m \in B\}$. A formal concept is defined as a pair (A, B) where $A \subseteq G$, $B \subseteq M$, $\beta(A) = B$ and $\alpha(B) = A$. A is called the extent of the concept and B is called its intent.

Using the above definitions of α and β, it is easy to verify that $A_1 \subseteq A_2$ implies that $\beta(A_1) \supseteq \beta(A_2)$, and $B_1 \subseteq B_2$ implies that $\alpha(B_1) \supseteq \alpha(B_2)$ for every $A_1, A_2 \subseteq G$, and $B_1, B_2 \subseteq M$ [8]. Let $\mathcal{C}(G, M, I)$ denote the set of all concepts of the context (G, M, I) and (A_1, B_1) and (A_2, B_2) be two concepts in $\mathcal{C}(G, M, I)$. (A_1, B_1) is called a subconcept of (A_2, B_2) which is denoted by $(A_1, B_1) \leq (A_2, B_2)$ whenever A_1 is a subset of A_2 (or equivalently B_1 contains B_2). The relation $\leq$ is an order relation on $\mathcal{C}(G, M, I)$.

In the sequel we give an overview of few basic rough set theory terms. Let U be a nonempty finite set of objects called the Universe. Let A be a set of attributes. Associate with each $a \in A$ a set V_a of all possible values of a called its domain. Let $a(x)$ denote the value of the attribute a for element x. Let B be a subset of A (B can be equal to A). A binary relation R^B on U is defined as $xR^By \iff a(x) = a(y) \forall a \in B$. Clearly, R^B is an equivalence relation and thus forms a partition on U. Let $[x]_B$ denote the equivalence class of x with respect to R^B. When B is clear from context, we will write $[x]$ instead of $[x]_B$. Let U/R^B denote the set of all equivalence classes determined by R^B. Equivalence classes of the relation R^B are called B-elementary sets (or just elementary sets). Any finite union of elementary sets is called a definable set.

Given a set $X \subseteq U$, X may not be definable. The relation R^B can be used to characterize X by a pair of definable sets called its lower and upper approximations. The lower and upper approximations of X with respect to R^B (or set of attributes B) are defined as $\underline{B(X)} = \{m \in U \mid [m]_B \subseteq X\}$ and $\overline{B(X)} = \{m \in U \mid [m]_B \cap X \neq \emptyset\}$, respectively. Clearly, the lower approximation of X is the greatest definable set contained in X and the upper approximation of X is the smallest definable set containing X. The difference between the upper and lower approximations of X is known as the boundary region of X and is denoted by $BND(X)$. If $BND(X)$ is an empty set, then X is a definable set with respect to B; on the other hand, if $BND(X)$ is not empty, then X is referred to as a rough set with respect to B [6].

3 Existing Approach

The existing approach for approximating concepts is due to Kent and is called rough concept analysis [5]. It relies on the existence of an equivalence relation, E, on the set of objects, G, that is provided by an expert. A pair (G, E) where E is an equivalence relation on G is called an approximation space. An E-definable formal context of G-objects and M-attributes is a formal context (G, M, I) whose elementary extents $\{Im \mid m \in M\}$ are E-definable subsets of G-objects where $Im = \{g \in G \mid gIm\}$.

The lower and upper E-approximations of I with respect to (G, E) are denoted by $\underline{I}_E$ and $\overline{I}^E$, respectively, and given by

$$\underline{I}_E = \{(g, m) \mid [g]_E \subseteq Im\} \quad \text{and} \quad \overline{I}^E = \{(g, m) \mid [g]_E \cap Im \neq \emptyset\}.$$

The formal context (G, M, I) can be approximated by the lower and upper contexts $(G, M, \underline{I}_E)$ and $(G, M, \overline{I}^E)$. The rough extents of an attribute set $B \subseteq M$ with respect to $\underline{I}_E$ and $\overline{I}^E$ are defined by

$$\alpha(\underline{B}_E) = \underline{\alpha(B)}_E = \bigcap_{m \in B} \underline{Im}_E \quad \text{and} \quad \alpha(\overline{B}^E) = \overline{\alpha(B)}^E = \bigcap_{m \in B} \overline{Im}^E$$

Any formal concept $(A, B) \in \mathcal{C}(G, M, I)$ can be approximated by means of $\underline{I}_E$ and $\overline{I}^E$. The lower and upper E-approximations of (A, B) are given by

$$\underline{(A, B)}_E = (\alpha(\underline{B}_E), \beta(\alpha(\underline{B}_E))) \quad \text{and} \quad \overline{(A, B)}^E = (\alpha(\overline{B}^E), \beta(\alpha(\overline{B}^E)))$$

4 Formal Rough Concept Analysis

In the previous section we presented an overview of the existing approach for approximating concepts. This approach is not direct because upper and lower approximations for the context (G, M, I) have to be found first and then used for approximating a pair (A, B) of objects and features. The resulting upper and lower approximations of (G, M, I) depend on the approximation space (G, E) as described in Section 3. This means that different equivalence relations on G would result in different answers. Furthermore, the set A was not used in the approximation. This means that all pairs that have the same set of features will always have the same lower and upper E-approximations.

In this section we present a different and more general approach for approximating concepts. Our approach is consistent in that the relation or relations we use in the approximation are defined in a way that assures that the same answer is always given. We first show how a set A of objects (or B of features) can be approximated by a concept whose extent (intent) approximates A (B). Then we show how a pair (A, B) can be be approximated by one or two concepts.

First, a few definitions need to be given. Let (G, M, I) be a context, not every subset $A \subseteq G$ is an extent nor every subset $B \subseteq M$ is an intent. Wille [8] has shown that $A \subseteq \alpha(\beta(A))$ for any $A \subseteq G$ and $B \subseteq \beta(\alpha(B))$ for any $B \subseteq M$. Furthermore, $\beta(\bigcup_{i \in J} A_i) = \bigcap_{i \in J} \beta(A_i)$ and $\alpha(\bigcup_{i \in J} B_i) = \bigcap_{i \in J} \alpha(B_i)$ where J is and index set. This later result will be used later. A set of objects A is called feasible if $A = \alpha(\beta(A))$. Similarly a set of features is feasible if $B = \beta(\alpha(B))$. If A is feasible, then clearly $(A, \beta(A))$ is a concept. Similarly, if B is feasible, then $(\alpha(B), B)$ is a concept. Let us also say that a set $A \subseteq G$ is definable if it is the union of feasible extents; otherwise, we say that A is nondefinable. Similarly, $B \subseteq M$ is definable if it is the union of feasible intents; otherwise, B is nondefinable. A pair (A, B) is called a definable concept if both A and B are definable, $\alpha(B) = A$ and $\beta(A) = B$; otherwise, (A, B) is a nondefinable concept.

4.1 Approximating a Set of Objects

Given a set of objects $A \subseteq G$, we are interested in finding a definable concept that approximates A. We have the following cases:

Case 1: A is feasible. Clearly $(A, \beta(A))$ is a definable concept. Therefore, $(A, \beta(A))$ is the best approximation.

Case 2: A is definable. Since A is definable, it can be written as $A = A_1 \cup A_2 \ldots \cup A_n$, where each A_i, $i = 1, \ldots n$, is feasible.
Hence, $\beta(A) = \beta(A_1 \cup A_2 \ldots \cup A_n) = \beta(A_1) \cap \beta(A_2) \cap \ldots \cap \beta(A_n) = \bigcap_{i=1}^{i=n} \beta(A_i)$.
Therefore, when A is definable, the best approximation is obtained by

$$(A, \beta(A)) = (\bigcup_{i=1}^{i=n} A_i, \bigcap_{i=1}^{i=n} \beta(A_i)).$$

Case 3: A is nondefinable. If A is nondefinable, it is not as straightforward to find a definable concept that approximates A. Our approach is to think of A as a rough set. We first find a pair of definable sets $\underline{A}$ and $\overline{A}$ that best approximate A. $\underline{A}$ and $\overline{A}$ are then used in finding two concepts that best approximate A.

Let $gI = \{m \in M \mid gIm\}$ denote the set of all features that are possessed by the object g. Define a relation R on G as follows:

$$g_1 R g_2 \quad \text{iff} \quad g_1 I = g_2 I \quad \text{where} \quad g_1, g_2 \in G.$$

Clearly, R is reflexive, symmetric and transitive. Thus, R is an equivalence relation on G. Let G/R be the set of all equivalence classes induced by R on G.

Lemma 4.1 *Each equivalence class $X \in G/R$ is a feasible extent.*

Proof. Assume not, that is, there is an $X \in G/R$ which is not a feasible extent. Therefore, $X \subset \beta(\alpha(X))$ and $X \neq \beta(\alpha(X))$. So there is an object $g \in \beta(\alpha(X)) - X$ such that $gIm \; \forall m \in \alpha(X)$. But this is a contradiction because by the definition of R, g must be in X because it has all the features in $\alpha(X)$. Therefore, $X = \beta(\alpha(X))$ which means that X is a feasible extent. $\square$

It follows from the previous lemma that each equivalence class $X \in G/R$ is a feasible extent and thus is an elementary extent. Therefore, define the lower and upper approximations of $A \in G$ with respect to R as

$$\underline{A} = \{g \in G \mid [g] \subseteq A\}, \quad \text{and} \quad \overline{A} = \{g \in G \mid [g] \cap A \neq \emptyset\}$$

Now, we can find two concepts that approximate A. The lower approximation is given by $(\underline{A}, \beta(\underline{A}))$ and the upper approximation is given by $(\overline{A}, \beta(\overline{A}))$.

Lemma 4.2 *If A is a nondefinable extent, then the best lower and upper approximations are given by $(\underline{A}, \beta(\underline{A}))$ and $(\overline{A}, \beta(\overline{A}))$, respectively.*

Proof. $\underline{A}$ is a union of feasible extents and thus is a definable extent. Therefore, $(\underline{A}, \beta(\underline{A}))$ is a definable concept. Similarly, we can show that $(\overline{A}, \beta(\overline{A}))$ is a definable concept.
Since $\underline{A} \subseteq A \subseteq \overline{A}$, we have $(\underline{A}, \beta(\underline{A})) \leq (A, \beta(A)) \leq (\overline{A}, \beta(\overline{A}))$. Furthermore, $\underline{A}$ is the greatest definable extent contained in A and $\overline{A}$ is the least definable extent containing A. This implies that $(\underline{A}, \beta(\underline{A}))$ is the greatest definable subconcept of $(A, \beta(A))$ and $(\overline{A}, \beta(\overline{A}))$ is the least definable superconcept of $(A, \beta(A))$. Therefore, $(\underline{A}, \beta(\underline{A}))$ and $(\overline{A}, \beta(\overline{A}))$ are the best lower and upper approximations. $\square$

4.2 Approximating a Set of Features

Because approximating a set of features is similar to approximating a set of objects and because of limitations of space, we will omit some unnecessary details.

Case 1: B is feasible. The concept $(\alpha(B), B)$ best approximates B.

Case 2: B is definable. B can be written as $B = \bigcup_{i=1}^{i=l} B_i$ where each B_i, is feasible. Hence, $\alpha(B) = \alpha(\bigcup_{i=1}^{i=l} B_i) = \bigcap_{i=1}^{i=l} \alpha(B_i)$ Therefore, B can be approximated by the definable concept $(\alpha(B), B) = (\bigcap_{i=1}^{i=l} \alpha(B_i), \bigcup_{i=1}^{i=l} B_i)$

Case 3: B is nondefinable. Let $Im = \{g \in G \mid gIm\}$ be the set of all objects that posses the attribute m. Define a relation R' on M as follows:

$$m_1 R' m_2 \quad \text{iff} \quad Im_1 = Im_2 \quad \text{where} \quad m_1, m_2 \in M.$$

Clearly, R' is an equivalence relation. Let G/R' be the set of all equivalence classes induced by R' on M.

Lemma 4.3 *Each equivalence class $Y \in G/R'$ is a feasible intent and thus an elementary set.*

Using the result of the previous lemma, the lower and upper approximations of $B \in M$ with respect to R are defined by

$$\underline{B} = \{m \in M \mid [m] \subseteq B\}, \quad \text{and} \quad \overline{B} = \{m \in M \mid [m] \cap B \neq \emptyset\}$$

The next lemma, which can be proved similar to lemma 4.2, gives two concepts that best approximate B.

Lemma 4.4 *If B is a nondefinable intent, then the best lower and upper approximations are given by $(\alpha(\underline{B}), \underline{B})$ and $(\alpha(\overline{B}), \overline{B})$, respectively.*

4.3 Approximating A Concept

Given a pair (A, B) where $A \subseteq G$ and $B \subseteq M$, we want to find one or two concepts approximating (A, B). Four different cases need to be considered:
I) Both A and B are definable, II)A is definable and B is not, III)B is definable and A is not, and IV) Both A and B are nondefinable.

4.3.1 Both A and B are Definable

Four subcases need to be considered.

1. Both A and B are feasible. If $\beta(A) = B$, then $\alpha(B)$ must equal to A because both A and B are feasible. Thus the given concept (A, B) is definable and no approximation is needed.
If $\beta(A) \neq B$, (and thus $\alpha(B) \neq A$), let $\beta(A) = A'$ and $\alpha(B) = B'$. Since both A and B are feasible, then both (A, A') and (B', B) are definable concepts in (G, M, I). Consider the two concepts $(A \cup B', \beta(A \cup B')) = (A \cup B', A' \cap B)$ and $(\alpha(A' \cup B), A' \cup B) = (A \cap B', A' \cup B)$. We notice that $\beta(B') = B$ and $\alpha(A') = A$ because B and A are feasible. Furthermore, $(A \cap B', A' \cup B) \leq (A, B) \leq (A \cup B', A' \cap B)$. Therefore, the lower and upper approximations of (A, B) are given by $\underline{(A, B)} = (A \cap B', A' \cup B)$ and $\overline{(A, B)} = (A \cup B', A' \cap B)$.

2. A is feasible and B is not. Since B is definable, it can be written as a union of feasible intents. Let $B = \bigcup_{i=1}^{i=m} B_i$ where B_i is feasible for $i = 1, 2, \ldots, m$. Let $\alpha(B_i) = B_i'$, for $i = 1, 2, \ldots, m$, and $\alpha(B) = B'$.

$$B' = \alpha(B) = \alpha(\bigcup_{i=1}^{i=m} B_i) = \bigcap_{i=1}^{i=m} \alpha(B_i) = \bigcap_{i=1}^{i=m} B_i'.$$

Therefore, the lower and upper approximations of (A, B) are given by

$$\underline{(A, B)} = (A \cap B', A' \cup B) = (\bigcap_{i=1}^{i=m}(A \cap B_i'), \bigcup_{i=1}^{i=m}(A' \cup B_i)),$$

and

$$\overline{(A,B)} = (A \cup B', A' \cap B) = (\bigcap_{i=1}^{i=m}(A \cup B_i'), \bigcup_{i=1}^{i=m}(A' \cap B_i)).$$

3. B is feasible and A is not. This case is similar to the previous case and details are omitted. The lower and upper approximations are given by

$$\underline{(A,B)} = (A \cap B', A' \cup B) = (\bigcup_{i=1}^{i=l}(A_i \cap B'), \bigcap_{i=1}^{i=l}(A_i' \cup B)), \quad \text{and}$$
$$\overline{(A,B)} = (A \cup B', A' \cap B) = (\bigcup_{i=1}^{i=l}(A_i \cup B'), \bigcap_{i=1}^{i=l}(A_i' \cap B)).$$

Where $A = \bigcup_{i=1}^{i=l} A_i$ and each A_i is feasible for $i = 1, 2, \ldots, l$.

4. Both A and B are not feasible. Since A and B are definable, they can be written as unions of feasible extents and intents, respectively.
Let $A = \bigcup_{i=1}^{i=l} A_i$ and $B = \bigcup_{j=1}^{j=k} B_j$, where A_i and B_j are feasible for $i = 1, 2, \ldots, l$ and $j = 1, 2, \ldots, k$. Let A', A_i', B', and B_i' denote $\beta(A)$, $\beta(A_i)$, $\alpha(B)$, and $\alpha(B_i)$, respectively. Then,

$$A' = \beta(A) = \beta(\bigcup_{i=1}^{i=l} A_i) = \bigcap_{i=1}^{i=l} \beta(A_i) = \bigcap_{i=1}^{i=l} A_i', \quad \text{and}$$
$$B' = \alpha(B) = \alpha(\bigcup_{j=1}^{j=k} B_j) = \bigcap_{j=1}^{j=k} \alpha(B_j) = \bigcap_{j=1}^{j=k} B_j'.$$

The lower and upper approximations of (A, B) are given by

$$\underline{(A,B)} = (A \cap B', A' \cup B)$$
$$= ((\bigcup_{i=1}^{i=l} A_i) \cap (\bigcap_{j=1}^{j=k} B_j'), (\bigcap_{i=1}^{i=l} A_i') \cup (\bigcup_{j=1}^{j=k} B_j)), \quad \text{and}$$
$$\overline{(A,B)} = (A \cup B', A' \cap B)$$
$$= ((\bigcup_{i=1}^{i=l} A_i) \cup (\bigcap_{j=1}^{j=k} B_j'), (\bigcap_{i=1}^{i=l} A_i') \cap (\bigcup_{j=1}^{j=k} B_j)).$$

4.3.2 A is Definable and B is not

Since A is definable, it can be written as $A = \bigcup_{i=1}^{i=l} A_i$ where each A_i is feasible. Define a binary relation R' on M such that for $m_1, m_2 \in M$, $m_1 R' m_2$ if $Im_1 = Im_2$. Clearly, R' is an equivalence relation and thus can be used in creating a rough approximation for any subset of M as was done earlier. Let $\underline{B}$ and $\overline{B}$ be the lower and upper approximations of B with respect to R'.

$\underline{B}$ and $\overline{B}$ can be used in creating lower and upper approximations for (A, B) in the context (G, M, I). The lower and upper approximations are given by

$$\underline{(A,B)} = (A \cap \alpha(\overline{B}), \beta(A) \cup \overline{B}), \quad \text{and} \quad \overline{(A,B)} = (A \cup \alpha(\underline{B}), \beta(A) \cap \underline{B}).$$

The concept $(A \cap \alpha(\overline{B}), \beta(A) \cup \overline{B})$ is definable because $\beta(A \cap \alpha(\overline{B})) = \beta(A) \cup \overline{B}$. Similarly, $(A \cup \alpha(\underline{B}), \beta(A) \cap \underline{B})$ is definable because $\beta(A \cup \alpha(\underline{B})) = \beta(A) \cap \underline{B}$.

We can show that the approximations developed above are indeed correct by observing that $\underline{B} \subseteq B \subseteq \overline{B}$ which implies that $\beta(A) \cap \underline{B} \subseteq B \subseteq \beta(A) \cup \overline{B}$ Therefore,

$$(A \cap \alpha(\overline{B}), \beta(A) \cup \overline{B}) \leq (A, B) \leq (A \cup \alpha(\underline{B}), \beta(A) \cap \underline{B})$$

To get the final answer we need to substitute $\bigcup_{i=1}^{i=l} A_i$ in place of A. However, we choose not to do that in this context to make the results easier to read.

4.3.3 B is Definable and A is not

The scenario here is similar to that in the previous subsection and we will just sketch the results. Since B is definable, it can be written as $B = \bigcup_{j=1}^{j=k} B_j$ where B_j is feasible for $j = 1, 2, \ldots, k$. Define a relation R on G by $g_1 R g_2$ iff $g_1 I = g_2 I$. R is an equivalence relation on G and can be used in approximating the nondefinable set A. Let $\underline{A}$ and $\overline{A}$ represent the lower and upper approximations of A with respect to R. The lower and upper approximations of (A, B) are given by

$$\underline{(A, B)} = (\underline{A} \cap \alpha(B), \beta(\underline{A}) \cup B) \quad \text{and} \quad \overline{(A, B)} = (\overline{A} \cup \alpha(B), \beta(\overline{A}) \cap B).$$

4.3.4 Both A and B are Nondefinable

Neither A nor B can be used in approximating the nondefinable concept (A, B). However, we can combine the approaches from the previous two subsections and define R to be an equivalence relation on G and define R' to be an equivalence relation on M using the same definitions from the previous two subsections. Let the lower and upper approximations of A with respect to G be given by $\underline{A}$ and $\overline{A}$, respectively. Similarly, let $\underline{B}$ and $\overline{B}$ denote the lower and upper approximations of B with respect to R'. Now, $\underline{A}, \overline{A}, \underline{B}$ and $\overline{B}$ are definable sets that can be used for approximating the nondefinable concept (A, B). The lower approximations is given by

$$\underline{(A, B)} = (\underline{A} \cap \alpha(\overline{B}), \beta(\underline{A}) \cup \overline{B}),$$

Similarly, the upper approximations is given by

$$\overline{(A, B)} = (\overline{A} \cup \alpha(\underline{B}), \beta(\overline{A}) \cap \underline{B}).$$

Clearly, both $(\underline{A} \cap \alpha(\overline{B}), \beta(\underline{A}) \cup \overline{B})$ and $(\overline{A} \cup \alpha(\underline{B}), \beta(\overline{A}) \cap \underline{B})$ are definable concepts. Furthermore, it is easy to prove that

$$(\underline{A} \cap \alpha(\overline{B}), \beta(\underline{A}) \cup \overline{B}) \leq (A, B) \leq (\overline{A} \cup \alpha(\underline{B}), \beta(\overline{A}) \cap \underline{B}).$$

Hence, our proposed approximations are correct.

5 Conclusion and Future Work

This paper presents a new approach for approximating concepts using rough sets. Using this approach the given context is used directly for finding upper and lower approximations for nondefinable concepts without any context approximation which is a major difference from the approach given in [5]. Another major difference between our approach and the existing approach is that in the existing approach an equivalence relation on the set of objects has to be given by an expert before the approximation process can start. Different equivalence relations would usually result in different approximations for a given nondefinable concept. Our natural choice of an equivalence relation R on the set of objects G is such that objects that share the same set of features are grouped together in the same equivalence class. This choice guarantees the uniqueness of our approximations. Furthermore, such a definition of R can help automate the process of approximating concepts.

Using the new approach, we showed how a set of objects, features, or a nondefinable concept can be approximated by a definable concept. We proved that the approximations found for a set of features or objects are the best one can get. The ideas developed in this paper are useful for information retrieval, knowledge acquisition and conceptual modeling. An implementation of a system that uses the ideas developed in this paper is currently in progress.

References

[1] Carpineto, C. and Romano, G., A lattice Conceptual Clustering System and its Application to Browsing Retrieval, Machine Learning, 10, 95-122, 1996.

[2] Godin, R., and Missaoui, R., An Incremental Concept Formation for Learning from Databases, Theoretical Computer Science, 133, 387-419, 1994.

[3] Ho, T. B., An Approach to Concept Formation Based on Formal Concept Analysis, IEICE Trans. Information and Systems, E78-D, 553-559, 1995.

[4] Kangassalo, H., On the concept of concept for conceptual modeling and concept deduction, in Information Modeling and Knowledge Bases III, Ohsuga et al. (eds.), IOS Press, 17-58, 1992.

[5] Kent, Robert E., Rough Concept Analysis: A Synthesis of Rough Set and Formal Concept Analysis, Fundamenta Informaticae v27, 169-181 1996.

[6] Pawlak, Z., Rough Set Theory and Its Applications to Data Analysis, Cybernetics and Systems, v29, n7, 1998.

[7] Pawlak, Z., Rough sets, International Journal of Information and Computer Science, 11, pp. 341-356, 1982.

[8] Wille, R. , Restructuring Lattice Theory: an Approach Based on Hierarchies of Concepts, In Ivan Rival ed., Ordered sets, Reidel, Dordecht-Boston, pp. 445-470, 1982.

Noise Reduction in Telecommunication Channels Using Rough Sets and Neural Networks

Rafal Krolikowski and Andrzej Czyzewski

Sound Engineering Department, Technical University of Gdansk,
ul. Narutowicza 11/12, 80-952 Gdansk, Poland

{rafal, andrzej}@sound.eti.pg.gda.pl

Abstract. A new concept of reduction of non-stationary noise affecting audio signals transmitted in telecommunication channels is proposed. This concept exploits some features of the human auditory system as well as some methods originated from soft computing domain, i.e. rough set-based reasoning and neural processing. The foundations of the engineered method and a description of applied decision algorithms are presented. A number of experiments have been prepared, and some of them have already been carried out. A brief discussion of these experiments' results and some conclusions are also included.

1 Introduction

The commonly used noise reduction methods do not use some subjective properties of the human auditory system, which have been successfully exploited in audio coding standards. However, as was revealed by the results of experiments carried out by the authors, auditory masking can be also used to the suppression of noise corrupting audio signals. The mathematical foundations of this perceptual approach and relevant algorithms were presented in some recent authors' papers [4][6].

In all noise reduction methods, there is a need to know at least approximated statistics of the noise. This problem becomes more complex in the case of non-stationary noise, since such a method requires choosing a certain noise statistics from among others. Hence, a problem of an efficient decision system occurs. Therefore intelligent algorithms (i.e. rough sets or neural networks) can be very helpful as decision systems in this field of application [2][4].

2 Psychoacoustics Principles

The concept of critical bands is related to propagation and processing of acoustic signals in the human auditory system. Well-proven phenomena reveal that the inner ear behaves as a bank of band-pass filters which analyse a broad spectral range in subbands independently from others. These subbands are called critical bands, and a perceptual unit of frequency has been introduced. It is called Bark and is related to the width of a single subband. Often used transformation to this subjective scale of hearing is the following relation proposed by Zwicker [11]:

$$b = 13 \cdot \text{arctg}(0.76 \cdot 10^{-3} \cdot f) + 3.5 \cdot \text{arctg}\left[(\frac{f}{7500})^2\right], \qquad (1)$$

where b, f denote frequency in Barks and Hz, respectively.

Another psychoacoustic phenomenon is related to masking. Some tones can be inaudible in the presence of others, especially when one of them is louder, and their frequencies are not too distant. These tones which mask others are called maskers, and this phenomenon is fundamental for contemporary audio coding standards [8]. More details on psychoacoustics can be found in abundant literature [11].

3 Description Of The Perceptual Noise Reduction System

The perceptual noise reduction system (Fig. 2) is fed by two inputs: the noisy signal $y(m)$ and the noise patterns $\tilde{n}(m)$. The signal $y(m)$ consists of the original audio signal $x(m)$ corrupted by the noise $n(m)$, and is transformed to the spectral representation $Y(j\omega)$ with the use of the DFT procedure. In turn, the patterns $\tilde{n}(m)$ are assumed to be correlated to the noise $n(m)$, and are taken from empty passages of the signal transmitted in a telecommunication channel. The signal $\tilde{n}(m)$ is delivered to the Noise Estimation Module which task is to collect essential information on the noise $n(m)$. At its output, the time-frequency noise estimation $\rho(t, j\omega)$ is obtained. Both this estimation $\rho(t, j\omega)$ and the spectrum of the corrupted audio $Y(j\omega)$ are supplied to the Decision Systems. Its first task is to select one of the collected spectral estimations $\rho(j\omega) \subset \rho(t, j\omega)$ which is correlated best to the corrupting noise in a given moment. The second task is to qualify the elements of the signal $Y(j\omega)$ for two disjoint sets: the set U of the useful or the set D of the useless elements. It is necessary to know, which spectral components are maskers (useful), and which ones are to be masked (useless).

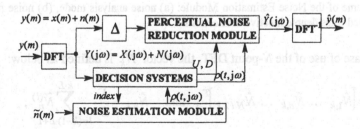

Fig. 1. General lay-out of the noise reduction system

The spectrum of the corrupted signal $Y(j\omega)$ as well as the sets U, D and the chosen noise estimation $\rho(t, j\omega)$ are fed to the Perceptual Noise Reduction Module that executes a perceptual algorithm of noise reduction. Next, the output $\hat{Y}(j\omega)$ is processed by the inverse DFT procedure, and finally the restored signal $\hat{y}(m)$ is obtained, which is subjectively perceived as less noisy than the original one.

4 Implementation of the Noise Reduction System

4.1 Noise Estimation Module

The Noise Estimation Module's run can be divided into two modes: the noise analysis mode and the noise reduction mode (Fig. 2). In the first one (Fig. 2a), the patterns $\tilde{n}(m)$ are analysed in *the Extraction of Noise Parameters* block where they are transformed into the spectral domain, averaged upon subsequent L frames and analysed. As a result, two kinds of output are obtained: the average power spectrum $\hat{N}_k$, and the associated vector $V_k^{\hat{n}}$ of coefficients related to the spectrum $\hat{N}_k$. The index k denotes the time interval, within which elements of these vectors are computed. Subsequent, the both vectors are collected in the *Table of Vectors*. The content of the table is used during the training of decision algorithms in the Decision Systems and the noise reduction mode (Fig. 2b). In this latter mode, due to a query to the table, the appropriate spectrum $\hat{N}_j$ is output, which is expected to be mostly correlated to the noise currently corrupting the audio. The query *index* value is produced by a decision system, and denotes the index of a desired spectrum in the table.

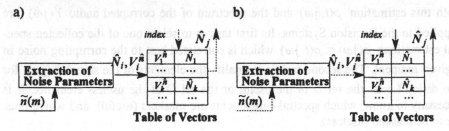

Fig. 2. Scheme of the Noise Estimation Module: (a) noise analysis mode, (b) noise reduction mode. Dashed lines denote inactive connections

In the case of use of the N-point DFT, the vector $\hat{N}_k$ is defined as below:

$$\hat{N}_k = \left[\hat{N}_{1,k} \ \ldots \ \hat{N}_{n,k} \ \ldots \ \hat{N}_{N/2,k}\right]^T, \quad \text{and:} \quad \hat{N}_{n,k} = \frac{1}{L} \cdot \sum_{l=(k-1)\cdot L+1}^{k\cdot L} \tilde{N}_n^{(l)}, \tag{2}$$

where $\hat{N}_{n,k}$ is averaged on the basis of last L values of the spectral power $\tilde{N}_n$.

The associated vector $V_k^{\hat{n}}$ serves as a key vector and is exploited during the noise reduction mode when a spectrum $\hat{N}_j$ is searched for. This vector should be unique, however in practice the condition is hard to be ensured. Its elements are expected to reflect quantitatively a noisy character of the average spectrum $\hat{N}_k$. Therefore two kinds of parameters are considered that turned out to be very robust in contemporary perceptual coding schemes [8]: the *Spectral Flatness Measure* [5] and the *unpredictability measure* [1]. These parameters are computed in each critical band, and their definitions for the *l*-th frame are given further.

Application of the Spectral Flatness Measure. The *SFM* parameter is defined as the ratio of the geometric G_m to the arithmetic A_m mean of the power spectrum [5], and is given in dB. In the *b*-th subband, the parameter can be redefined as follows:

$$SFM_b^{(l)} = 10\log_{10} G_m^{(l)} / A_m^{(l)} \qquad (3)$$

Hence, the vector $V_k^{\hat{n}}$ can be described in the following way:

$$V_k^{\hat{n}} = \begin{bmatrix} SFM_{1,k} & \dots & SFM_{b,k} & \dots & SFM_{B,k} \end{bmatrix}^T \text{ and: } SFM_{b,k} = \frac{1}{L} \cdot \sum_{l=(k-1)\cdot L+1}^{k\cdot L} SFM_b^{(l)}. \qquad (4)$$

Application of the Unpredictability Measure. Introducing denotations of the spectral magnitude prediction $\hat{r}_i^{(l)}$ and the phase prediction $\hat{\phi}_i^{(l)}$ of the *i*-th spectral component on the basis of the their last two real values as below:

$$\hat{r}_i^{(l)} = r_i^{(l-1)} + (r_i^{(l-1)} - r_i^{(l-2)}) \quad \text{and} \quad \hat{\phi}_i^{(l)} = \phi_i^{(l-1)} + (\phi_i^{(l-1)} - \phi_i^{(l-2)}), \qquad (5)$$

the unpredictability measure $c_i^{(l)}$ is defined as the Euclidean distance between the real values of $r_i^{(l)}$, $\phi_i^{(l)}$ and the predicted ones ($\hat{r}_i^{(l)}$, $\hat{\phi}_i^{(l)}$) as in the formula [1]:

$$c_i^{(l)} = \frac{\text{dist}\left((\hat{r}_i^{(l)}, \hat{\phi}_i^{(l)}), (r_i^{(l)}, \phi_i^{(l)})\right)}{r_i^{(l)} + \text{abs}\left(\hat{r}_i^{(l)}\right)}. \qquad (6)$$

In such a case, the *k*-th element of the vector $V_k^{\hat{n}} = \begin{bmatrix} C_{1,k} & \dots & C_{b,k} & \dots & C_{B,k} \end{bmatrix}^T$ is averaged upon last L frames and within the *b*-th critical band in the following way:

$$C_{b,k} = \frac{1}{L} \cdot \sum_{l=(k-1)\cdot L+1}^{k\cdot L} C_b^{(l)}, \quad \text{where: } C_b^{(l)} = \frac{1}{\text{count}(b)} \cdot \sum_{i=lower(b)}^{upper(b)} c_i^{(l)}, \qquad (7)$$

where *upper*(b) and *lower*(b) denote indexes of the first and the last spectral bin in the *b*-th subband which contains *count*(b) components.

4.2 Decision Systems

The Decision Systems module (Fig. 3) is fed by the spectral representation of the noisy signal $Y(j\omega)$. First, the input signal is processed in the *Extraction of Parameters* block which task is to obtain a vector of parameters V_i^y, and these parameters are expected to be mostly related to the noisy character of the input $Y(j\omega)$. Therefore elements of V_i^y are defined by analogy to the key vector $V_k^{\hat{n}}$ elements, and computed as in the formula (4) or (7). The vector V_i^y is next supplied to the *Decision System* I which objective is to give the *index* value of the noise spectrum $\hat{N}_j$ that should be most correlated to the noise present in the noisy signal $Y(j\omega)$.

Having received the desired vector $\hat{N}_j$ from the *Table of Vectors*, this vector is compared with the spectral representation $Y(j\omega)$ in the *Decision System* II which produces two output sets: the set U of useful and the set D of useless components.

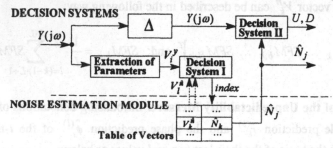

Fig. 3. Scheme of the Decision Module

Implementation of the Decision System I. Decision-making in the *Decision System* I can be based on the rough sets or neural reasoning, and the run of the system can be divided into two modes: the training- and the execution one. In the first case, the content of the *Table of Vectors* is exploited, which is depicted in Fig. 3 with dashed lines. The intelligent decision algorithms are considered further in the paragraph.

Application of Rough Sets. In the training mode, related to rule discovery, a part of the *Tables of Vectors* is treated as a decision table, where elements of the key vector $V_k^{\hat{n}}$ defined by the formulas (4) or (7) are conditional attributes and the vector's index in the *Table of Vectors* is a decision attribute. Therefore the k-th object in the table is described by the following relationship:

$$SFM_{1,k}, \dots, SFM_{b,k}, \dots SFM_{B,k} \Rightarrow k \quad \text{or:} \quad C_{1,k}, \dots, C_{b,k}, \dots C_{B,k} \Rightarrow k \quad (8)$$

where the variables: *SFM, C* are given as in the formulae (4) and (7), respectively.

The rule discovery procedure is based on the rough set principles [7] and algorithm [2]. It can be noticed that only conditional attributes require quantization. So far, only the uniform quantization has been used. In the execution mode, the input vector of noisy audio parameters V_i^y is quantized, and next processed by the set of generated rules. In result, the *index* value of the noisy spectrum $\hat{N}_j$ in the *Tables of Vectors* is obtained.

Application of Neural Networks. The applied neural network is a classic feedforward structure with hidden neurons [10]. In the preliminary experiments, only one hidden layer was considered. The number of input units is equal to the number of elements in the vector $V_k^{\hat{n}}$, given by eq. (4) or (7). There is only one output neuron which produces the entry value (*index* value) to the Table of Vectors. For all input and hidden units, the activation function is sigmoidal. However, since the index value is an integer, this function for the output unit is considered as sigmoidal or linear.

In the training mode, only a part of the *Table of Vectors* is used in the training set: the key vector $V_k^{\hat{n}}$ is an input vector, whereas its index in the table is a desired output, associated with this vector. Thus, the training set is as follows:

$$\left\{ (V_1^{\hat{n}},1),...,(V_k^{\hat{n}},k),...,(V_K^{\hat{n}},K) \right\}. \tag{9}$$

As a training method, the standard Error Backpropagation Training Algorithm [10] is applied, and the error measure is based on the mean squared error. If the neuron's activation function of the output neuron is sigmoidal, the desired index value should be scaled down in order to match the interval [0,1].

In the execution mode, the network's output (related to the *index* value) must be rounded up to the nearest integer. Additionally, if the output unit processes according to sigmoidal function, before the round-up the neuron's output must be scaled up by the same factor by which was scaled down.

Implementation of the Decision System II. In the *Decision System* II, the division into useful and useless elements is executed according to the following simple procedure. All these components which spectral powers Y exceed the double average value of the representative noise estimation $\hat{N}_j$ are assumed to be the useful elements. In turn, the remaining components are regarded as useless ones. Hence, in the case of use of the N-point DFT the sets U and D can defined as follows:

$$\begin{aligned} U &= \left\{ n, Y_n : Y_n \geq 2 \cdot \hat{N}_{n,j} \quad \text{and} \quad n = 1,...,N/2 \right\} \\ D &= \left\{ n, Y_n : Y_n < 2 \cdot \hat{N}_{n,j} \quad \text{and} \quad n = 1,...,N/2 \right\} \end{aligned} \tag{10}$$

In general, there can be various methods of such a division [2], and a choice of one of them has significant influence on the subjective quality of a restored audio.

4.3 Perceptual Noise Reduction Module

The task of the module is to process the spectral representation of the noisy signal as follows. All useful spectral components are reduced according to the spectral subtraction principles [9], whereas the remaining useless components are masked using the psychoacoustic approach. This perceptual approach is a separate complex issue which is not related to the application of intelligent tools, and due to the space limitations it is not described in this paper. However, the applied perceptual models, engineered methods together with appropriate algorithms were presented extensively in details in the recent publications of the authors [4][6].

5 Experiments

There were two objectives of the experiments: verification of the engineered method for non-stationary noise reduction and comparison of different decision algorithms. Some verification tests were carried out first, in order to check, whether application of intelligent tools could improve the quality of restored audio signals. The results were encouraging enough, for the next comparison experiments to be prepared.

In order to assess the quality of a decision algorithm, the result of such an algorithm should be compared with the desired output. In the case of the research, it was necessary to know whether the noise spectrum pointed by the decision system was the best choice, and if not, an error measure was needed.

For the purpose of the comparison experiments, two recordings were made: a male voice (5.81 s) and a non-stationary noise (2.79 s) taken from a radio channel. Next, the original audio was corrupted by the additive noise, and in the same time, elements of the key vectors of the noise were computed and collected. Since the original audio and the noise were given, it was known, which part of the noisy voice was described by which key vector and noise spectrum vector. The parameters of these recordings were as follows. They both were mono, sampled with 16-bit resolution and with the frequency equalled to 8192 Hz, which resulted in $B = 18$ critical bands. During the audio processing, the signals were divided into frames and overlapped. Since the Hamming window function was used, the overlap size was the half of the frame length N. In the experiments three values of N are used: 128, 256, 512, and their influence on the time- and frequency resolution are shown in Tab. 1. Additionally, in Tab. 1 there is a number of the noise key vectors, assuming that signal are averaged on last $L = 4$ frames. It can be notices that the number of these vectors is also the number of objects in a decision table or the number of training vectors.

Table 1. Influence of frame size N on time- and frequency resolution and training set contents

N	Time-resolution	frequency resolution	Number of vectors
128	7.83 ms	64 Hz	88
256	15.63 ms	32 Hz	44
512	31.25 ms	16 Hz	22

The comparison tests were divided with respect to the following variables:

- various frame size: $N = 128$, $N = 256$, $N = 512$
- various key vector types: based on the *SFM* parameters (4) and the unpredictability measure (7)
- various quantization steps: 0.1, 0.5, 1
- various number of hidden neurons: 10, 15, 20
- various neuron's activation function of the last unit: sigmoidal, linear

Hence, a single test attempt can be described by a set of parameters which are valid for a given decision algorithm. Thus, the following denotations are proposed: (*N, vector*, RS, *quantization step*) for application of rough sets, and (*N, vector*, NN, *hidden neurons, output unit*) for application of neural nets. Totally, there are 32 tests, such as the following exemplary ones: (512,*SFM*,NN,10,sigmoidal), (512,*C*,NN,10,sigmoidal), (512,*SFM*,RS,0.5).

In order to assess the quality of a decision system, the error measure E is introduced. This measure is the average value of the errors $E^{(i)}$ produced for all I frames of the noisy signal, and is defined according to the following expression:

$$E = \frac{1}{I} \cdot \sum_{i=1}^{I} E^{(i)} \quad \text{and} \quad E^{(i)} = \sum_{b=1}^{B} \left(V_{b,i}^{y} - V_{b,index}^{\hat{n}} \right)^2, \quad (11)$$

where $V_{b,i}^{y}$ is related to the vector of parameters of the noisy audio, and $V_{b,index}^{\hat{n}}$ is the b-th element of the key vector, placed at position *index* in the *Table of Vectors*.

By analogy, the optimal error measure E_{opt} and the maximum error E_{max} can be introduced. Assuming that the i-th frame of the audio is corrupted by the noise described by the j-th vector in the *Table of Vectors*, these measures is defined as below:

$$E_{opt} = \frac{1}{I} \cdot \sum_{i=1}^{I} E_{opt}^{(i)} \quad \text{and} \quad E_{opt}^{(i)} = \sum_{b=1}^{B} \left(V_{b,i}^{y} - V_{b,j}^{\hat{n}} \right)^2, \quad (12)$$

$$E_{max} = \frac{1}{I} \cdot \sum_{i=1}^{I} E_{max}^{(i)} \quad \text{and} \quad E_{max}^{(i)} = \max_{k=1,\dots,K} \left[\sum_{b=1}^{B} \left(V_{b,i}^{y} - V_{b,k}^{\hat{n}} \right)^2 \right], \quad (13)$$

where K is the number of vectors in the *Table of Vectors*.

Finally, the quality coefficient q is introduced as follows:

$$q = 1 - \frac{|E - E_{opt}|}{E_{max} - E_{opt}}. \quad (14)$$

For the completed surveys this quality coefficient is as follows:

- (512,*SFM*,NN,10,sigmoidal): $q = 81.38\%$
- (512,*C*,NN,10,sigmoidal): $q = 85.21\%$
- (512,*SFM*,RS,0.5): $q = 78.43\%$

It can be assumed that the best result for the test (512,C,NN,10,sigmoidal) is due to the use of the more precise key vector type (based on the unpredictability measure). In turn, the worst result obtained with rough sets can be caused mainly by the inefficient quantization step.

6 Conclusions

In the paper, the engineered system for non-stationary noise reduction has been presented, which exploits reasoning based on neural processing and rough sets. A number of experiments have been carried out in order to assess the quality of particular decision systems. The results of experiments show that computational intelligence and soft computing algorithms are applicable to the control of perceptual coding algorithms applied to noise reduction.

References

1 Brandenburg, K.: Second Generation Perceptual Audio Coding: The Hybrid Coder. Proc. of the 90th Audio Eng. Soc. Conv., Monteux, France, (1990) preprint 2937
2. Czyzewski, A.: Learning Algorithms for Audio Signal Enhancement. J. of Audio Eng. Soc. 45 (1997) 931-943
3. Czyzewski, A., Krolikowski, R.: Application of Intelligent Decision Systems to the Perceptual Noise Reduction of Audio Signals. Proc. of the 5th European Congress on Intelligent Techniques and Soft Computing, Aachen, Germany 1 (1997) 188-192
4. Czyzewski, A., Krolikowski, R.: Noise Reduction in Audio Employing Auditory Masking Approach. Proc. of 106th Audio Eng. Soc. Conv., Munich, Germany, (1999) preprint 4930
5. Johnston, J.: Transform Coding of Audio Signals Using Perceptual Noise Criteria. IEEE J. of Select. Areas in Communication 6 (1988) 314-323
6. Krolikowski, R., Czyzewski, A.: Noise Reduction in Acoustic Signals Using the Perceptual Coding. Proceedings of the 137th Meeting Acoustic Society of America, Berlin, Germany, (1999) CD-Preprint
7. Pawlak, Z.: Rough Sets. International Journal of Computer and Information Sciences 11 (1982) 341-356
8. Shlien, S.: Guide to MPEG-1 Audio Standard. IEEE Trans. Broadcasting 40 (1994) 206-218
9. Vaseghi, S.: Advanced Signal Processing and Digital Noise Reduction. Wiley & Teubner, New York (1997)
10. Zurada, J.: Introduction to Artificial Neural Networks. West Publ. Comp., St. Paul (1992)
11. Zwicker, E., Fastl, H.: Psychoacoustics, Facts and Models. Springer Verlag, Berlin (1990)

Acknowledgements

The research is sponsored by the State Committee for Scientific Research, Warsaw, Poland. Grant No. 8 T11D 021 12.

Rough Set Analysis of Electrostimulation Test Database for the Prediction of Post-Operative Profits in Cochlear Implanted Patients

Andrzej Czyzewski, Henryk Skarzynski, Bozena Kostek and Rafal Krolikowski

Institute of Physiology and Pathology of Hearing,
ul. Pstrowskiego 1, 00-943 Warsaw, Poland

kid@sound.eti.pg.gda.pl

Abstract. A new method of examining the hearing nerve in deaf people has been developed at the Institute of Physiology and Pathology of Hearing in Warsaw. It consists in testing deaf people by speech signal delivered through a ball shaped microelectrode connected to the modulated current source and attached to the promontory area. The electric current delivered to the ball shaped electrode is modulated with real speech signal which is transposed downwards the frequency scale. A computer database of patients' data and electrostimulation test results has been created. This database was analyzed using the rough set method in order to find rules allowing prediction of hearing recovery of cochlear implantation candidates. The Rough Set Class Library (RSCL) has been developed in order to implement data mining procedures to the engineered database of electrostimulation test results. The RSC Library supports symbolic approach to data processing. Additionally, the library is equipped with a set of data quantization methods that may be a part of an interface between external data environment and the rough set-based kernel of the system. The results of studies in the domain of prediction of post-operative profits of deaf patients based on the rough set analysis of electrostimulation test database are presented and discussed in the paper.

1 Introduction

The electrostimulation tests are treated as an important tool in preoperative diagnostics of deaf people who are candidates for the cochlear implantation. The idea of electrical stimulation, which goes back to A. Volta, is to electrically stimulate terminations of the fibers of the auditory nerve in order to evoke an auditory sensation in the central nervous system. The electrical stimulation is the application of electrical current to the audiovestibular nerve in order to assess its integrity [1]. It can be performed applying either invasive or non-invasive technique. Using invasive technique a promontory needle-electrode is applied transtympanically [3],[8] or ball-shaped electrode is placed in the round window niche following a transcanal tympanomeatal incision and removal of the bony overhang overlying the round window membrane

[4]. A non-invasive alternative is extratympanic ear canal test [6][7][9]. During this test the metal electrode is inserted into the ear canal, which has been filled with physiological saline solution [2]. It is assumed that the non-invasive attachment considerably simplifies preoperative electrostimulation. This is especially important when testing children is considered [5].

The results of electrical stimulation tests are highly diversified, depending on a patient's age and condition, thus there was a need to create a database provided with adequate tools allowing to find and to study the correlation between obtained test results and patients' ability to receive and understand sounds. However, in order to achieve the best perspective in cochlear implantation it is necessary not only to diagnose properly the auditory nerve status, but at the same time to evaluate the future benefits of the cochlear implant to the patient. The procedure developed at the Institute of Physiology and Pathology of Hearing in Warsaw allows determining some vital characteristics of the hearing sense that help to make decisions regarding the cochlear implantation [11]. The testing based on the electrical stimulation via the external auditory canal filled with saline is performed using the ball shaped electrode and the spectral transposition of speech signal. Moreover, patients' data such as personal data and health history are included in the database. The rough set algorithm was engineered enabling an analysis of highly diversified database records in order to find some dependencies between data and making possible to predict results of cochlear implantation basing on the results obtained previously in other patients.

2 Method of Speech Processing

The method of spectral transposition of speech signal was engineered earlier for the use in special hearing aids applied in the profound hearing loss [10]. Some essential modifications were introduced to the previously designed algorithm in order to adjust it to the current needs.

The simplified scheme of the algorithm of spectral transposition of speech signals is shown in Fig. 1. It can be noticed that the structure of the algorithm is based on the voice coder (vocoder). The natural speech signal is delivered to the transposer. Since the energy of sounds lowers upwards the frequency scale, the signal is preemphasized by 6 dB/oct. By analogy, the transposed signal is deemphasized at the output by the same ratio, i.e. 6 dB/oct. Additionally, in order to get rid of some disturbances that may occur while manipulating the signal, a low-pass filter is applied. The deemphasized signal is compressed in the Compressor module, because of the serious limitation of the dynamic ratio of signals received by the electrically stimulated auditory nerve.

The detection of voiced speech portions is based on the cepstrum analyzis method. When voiced sounds are pronounced, a Fundamental Frequency Generator is activated. In such a case, the synthesized sound is a result of a convolution of a periodic stimulus and the impulse response of a vocal tract represented by spectral envelopes. The detected vocal tone frequency is then divided by a factor selected from the range of 1.5 to 3, depending on the width of patient's auditory nerve response frequency band. The resynthesis of speech with the lower vocal tone frequency allows to maintain speech formants in the low frequency band related to patients' auditory nerve sensitivity characteristics. The above procedures make possible to resynthesize speech in the lower band of the frequency scale in such a way that the formant frequency ratios are preserved. It helps to maintain the synthetic speech intelligible, though it is received within a narrow frequency band only.

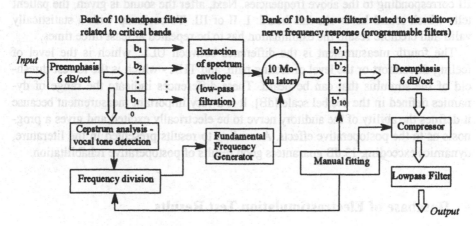

Fig. 1. General lay-out of the noise reduction system

3 Examination of the Patients

A set of charts prepared in order to facilitate co-operation with patients. The charts include responses, such as auditory sensation received by the patient: soft, comfortable, loud, very loud, etc. The charts also include information on the type of received signal. During the examination, three standard tests are performed:

- TMU - dynamics range defined by the auditory threshold (THR) and uncomfortable loudness
- TDL - Time Difference Limen test
- TID - test of frequency differentiation

In the TMU test, the values of intensity in [μA] of the electrical stimuli evoking an auditory response are determined. In the TDL test, the time of subjecting to the stimulus is differentiated. The first step is to determine the level of comfortable hearing (MCL) for the patient for a stimulus of 125 Hz (or 62.5 Hz). If the dynamics in this range of frequencies is as high as 10 dB, this result is recognized as good. Next, the patient listens to three sounds, of which one is longer and two are shorter. The purpose of this test is to find whether the patient is capable of differentiating the sequence in which the sounds are given. The difference in the duration of the long and short sound changes, depending on the patient's response. The result of this test is given in miliseconds [ms].

If the result obtained by the patient is less than 120 ms, this means that the patient can recognize time relations well. It gives a good perspective for a patient's achieving speech comprehension in an acoustic manner. In the next test (TID) - the frequency of the stimulus given is differentiated. This test is done in three frequencies: 31.25 Hz, 62.5 Hz, 125 Hz. For these frequencies the level of the most comfortable hearing is determined. Three different sounds are demonstrated to the patient: I, II, III corresponding to the above frequencies. Next, after the sound is given, the patient tells which sound he or she has heard: I, II or III. To come up with a statistically valid and credible result, the examination has to be repeated at least three times.

The fourth measurement is the difference between ULL, which is the level of feeling discomfort or the level of feeling pain, and THR - which is the lowest threshold of the stimulus that can be heard. These differences indicate the range of dynamics defined in the decibel scale [dB]. It is a very important measurement because it defines the ability of the auditory nerve to be electrically excited and gives a prognosis as to the postoperative effects. According to results presented in the literature, dynamics exceeding 13 dB guarantees good results of postoperative rehabilitation.

4 Database of Electrostimulation Test Results

In order to evaluate the results obtained in preoperative electrostimulation tests a set of examining techniques is used after the cochlear implantation. These are: screening tests and auditory speech perception, recognition and identification tests, the latter consisting in using various speech elements, such as single words, vowels, monosyllable, onomatopoeias, etc. This aimed at assigning a correlation between preoperative and postoperative test results. For this purpose the mentioned database containing results obtained by more than 150 implanted patients has been created at the Institute of Physiology and Pathology of Hearing in Warsaw. It includes also personal data and some additional factors pertaining educational and social skills, a degree of motivation, how early a hearing aid was prescribed that provides constant acoustic stimulation of the auditory system, etc.

The created database has been designed for testing by techniques recognized as data mining or soft computing. These techniques are very valuable in clinical diagnostics because they can trace some hidden relations between data, not visible in the case when patients' data are not complete or there are many records included in the database. The database addresses the following issues: a) personal data; b) cause of deafness; c) kind of deafness (prelingual, perilingual, postlingual); d) time passed since deafness; e) dates of examinations; f) number of tests performed; g) found dynamic range of hearing nerve sensitivity [dB]; h) found frequency band of hearing nerve sensitivity [Hz]; i) TMU measurement; j) TDL test result; k) TID test result; l) some factors which may influence the results of electrostimulation tests (e.g.: progressive hearing loss, acoustic trauma, use of hearing aids, ...); m) use of transposition during the voice communication; n) patient's motivation; o) patient social skills; p) vowels recognition ability; q) monosyllable recognition ability; r) onomatopoeias recognition ability; s) simple commands recognition ability.

The patients are divided to the two ranges of age: 0-18 years; more than 18 years. There are 5 groups of patients distinguished with regard of the time passed since deafness: less than 1 year; 1-5 years; 5-10 years; 10-20 years; more than 20 years.

As is easy to observe in above list, the database contains highly diversified information, namely text strings (a-b), integers (c-h), real numbers (i-k), binary flags (l-m), and grades (n-s). The processing of such of information can be done efficiently by the algorithm based on the rough set method. The results of measurements (g-k) need to be quantized automatically by some adequate algorithms.

5 Rough Set Analysis of Electrostimulation Database

The library of rough set procedures was engineered at the Sound Engineering Department of the Technical University of Gdansk. This library makes possible to include some rough set data analysis procedures in the engineered database software. A description of the rough set class library is given below in this chapter.

5.1 Rough Set Class Library

The *Rough Set Class Library* is an object-oriented library of procedures/functions which goal is to process data according to principles of the rough set theory. The RSCL takes all necessary actions related to data mining and knowledge discovery. The engineered library is designed to run in the DOS/Windows environment and compiled with the use of Borland™ C++ Compiler v. 3.1.

In general, implemented functions in RSCL comprise the following tasks:
- rule induction
- processing of rules
- fundamental operations on rough sets: partition of the universe into equivalence classes (i.e. sets of objects indiscernible with respect to a given set), calculation of lower and upper approximation, calculation of boundary region, calculation of positive and negative region
- supply of auxiliary functions: showing the winning rule and its rough measure, calculating number/percentage of certain and uncertain rules, showing range of the rough measure for all rules, computing cardinality of data sets.

The kernel of the rough-set-based data processing system works on symbolic data, i.e. symbolic representations of attributes and decisions in an information table. In order to facilitate fitting the system's kernel to external, the most frequent non-symbolic, data, some methods of quantization are supplied in the library [12]. These discretization procedures are as follows: Equal Interval Width Method, Statistical Clustering Method and Maximum Distance Method.

5.2 The Rough Set Class Implementation

In this paragraph is given a brief description of some RSCL functions that are related to the principles of rough-set-based data processing.

- *quantization of data in the decision table*
  The function performs discretization of data values in the decision table according to a given quantization method.
  Input: a decision table, a quantization method.
  Output: the decision table with quantized values.

- *quantization of conditional attributes*
  The function performs discretization of values of the conditional attributes according to a given quantization method.
  Input: a set of conditional attributes, a quantization method.
  Output: the set of conditional attributes in an adequate symbolic representation.

- *dequantization of decision attributes*
  The function executes the inverse action to the quantization algorithm -replaces the symbolic value of data with adequate crisp values.
  Input: a symbolic decision (decision attributes), a quantization method.
  Output: crisp values of the decision.

- *rule induction*
  The function induces a rule base. The action is performed on the basis of the principles of the rough set method.
  Input: a decision table.
  Output: a table with discovered knowledge (induced rules).

- *processing of rules*
  The function deduces a decision on the basis of an event, i.e. set of conditional attributes. The rule base has to be induced earlier.
  Input: an event (set of conditional attributes).
  Output: a decision (set of decision attributes).

- *partition of the universe into equivalence classes*
  The procedure yields a set of equivalence classes. The decision table is partitioned into some sets with respect to a certain set of attributes.
  Input: a decision table (the universe), a set of conditional attributes.
  Output: a set of equivalence classes.

- *calculation of lower- and upper approximation and boundary region*
  The functions compute sets, which are lower- and upper approximations and a boundary region of a set of objects with respect to a certain set of attributes.
  Input: a set of objects (in the form of a decision table), a set of attributes.
  Output: a resultant set of objects.

- *calculation of positive and negative region*
  The functions calculate positive or negative regions of classification for a certain set of attributes.
  Input: a decision table, two sets of attributes.
  Output: a resultant set of objects.

115

5.3 Rough Set Processing of Electrostimulation Data

The results of electrostimulation tests are collected in forms, separately for each patient. Then they should be transformed to the decision tables used in the rough set decision systems (Tab. 1) Therefore, objects t_1 to t_n represent patient cases. Attributes A_1 to A_m are to be denoted as tested parameters, introduced in Par. 4 (a-s). They are used as conditional attributes. The data values are defined by a_{11} to a_{nm} as numbers or grades (quantized values). The decision D is understood as a value assigned to the overall grade (*OVERALL GRADE*). This quantity represents the expected post-operative profits express in the descriptive scale as follows:

OVERALL GRADE = 1 - meaning: predicted hearing recovery profits - none
OVERALL GRADE = 2 - meaning: predicted hearing recovery profits – low
OVERALL GRADE = 3 - meaning: predicted hearing recovery profits – fair
OVERALL GRADE = 4 - meaning: predicted hearing recovery profits – well
OVERALL GRADE = 5 - meaning: predicted hearing recovery profits – very good

Table 1. Decision table used in electrostimulation database

Attribute \ Patient	A_1	A_2	...	A_m	D
t_1	a_{11}	a_{12}	...	a_{1m}	d_1
⋮	⋮	⋮	⋱	⋮	⋮
t_n	a_{n1}	a_{n2}	...	a_{nm}	d_n

The engineered decision system employs both learning and testing algorithms [15]. During the first phase rules are derived that from the basis for the second phase performance. The generation of decision rules starts from rules of the length equals 1, then the system generates rules of the length equals 2, etc. The maximum rule length may be defined by the operator. The system induces both certain and possible rules. It is assumed that the rough set measure (μ_{RS}) for possible rules should exceed the value 0.5. Moreover, only such rules are taken into account, that have been preceded by any shorter rule operating on the same parameters. The system produces rules of the following form:

$(attribute\_A_1 = value\_a_{11})$ and ... and $(attribute\_A_m = value\_a_{nm}) \Rightarrow (OverallGrade\_d_i)$

The data were gathered from all subjects during the interviews and electrostimulation test sessions. Some exemplary data records are presented in Tab. 2. Having results of several patients, these data are then processed by the rough set algorithm.

Table 2. Fragments of electrostimulation database records (described in Par. 4)

Values /Grades Patient	(b)	(c)	(d)	(f)	(g)	⋯	(d)	Overall Grade
t_1	otitis	3	1	14	11	⋯	3	2
⋮	⋮	⋮	⋮	⋮	⋮	⋱	⋮	⋮
t_n	acoust. trauma	1	3	14	14	⋯	5	5

For the discussed example (Tab. 2) the following strongest rules were obtained, being in a good accordance with the principles and practice of audiology:

IF (g = high) **THEN** (*OverallGrade* = very good) $\mu_{RS} = 1$

IF (d = high) & (s = low) **THEN** (*OverallGrade* = low) $\mu_{RS} = 0.92$

IF (c = 1) & (g = low) & (s = low) **THEN** (*OverallGrade* = none) $\mu_{RS} = 0.8$

IF (b = trauma) & (c = 3) **THEN** (*OverallGrade* = well) $\mu_{RS} = 0.76$

IF (f = high) & (g = high) **THEN** (*OverallGrade* = well) $\mu_{RS} = 0.72$

Every new patient record can be tested using previously induced rules and on this basis a predictive diagnosis of post-operative profits can be automatically provided by the system. This diagnosis expressed as a grade value may be used as a supportive or a contradictory factor in the process of qualification of the deaf patient to cochlear implantation. The accuracy of decisions produced by the intelligent database analysis algorithm is expected to grow higher as the number of patient records increase.

6 Conclusions

In this paper a new procedure for testing candidates for hearing implantation has been presented. The obtained frequency range during the electrostimulation tests with the modified ball-shaped electrode allows to deliver not only tones to the auditory nerve but using the signal processing device also speech signal can be received by some completely deaf patients. This may be helpful to properly diagnose and qualify patients to cochlear implantation and to give the patients some kind of sound experience before the implantation.

The engineered RSC Library procedures offering a symbolic approach to data processing have been implemented in the constructed electrostimulation test database allowing to analyze data records automatically in order to mine rules showing some hidden dependencies between patients' data and the expected hearing recovery after cochlear implantation.

The proposed method of prediction of post-operative results is presently at the experimental stage and requires some more testing in order to reveal its full potential. Nevertheless, a diagnosis provided by the algorithm may be already used as a supportive or a contradictory factor in the process of qualification of deaf patients to cochlear implantation.

117

References

1. American Speech-Language-Hearing Association: Electrical Stimulation for Cochlear Implant Selection and Rehabilitation. ASHA **34**, Suppl. 7 (1992) 13-16
2. Kerber, M., Klasek, O., Stephan K., Wieser K.: Advantages of the Ear-Canal-Electrode in Preoperative Stimulation. In: Hochmair-Desoyer, I., J., Hochmair, E., S. (eds.): Advances in Cochlear Implants (1994) 85-88
3. House, W., F., Brackmann, D., E.: Electrical Promontory Testing in Differential Diagnosis of Sensori-Neural Hearing Impairment. Presented at the 77th Annual Meeting of the American Laryng. Society, Inc., Palm Beach, Florida, USA (1974)
4. Shipp, D., B., Nedzelski, J., M.: Prognostic Value of Round Window Psychophysical Measurements with Adult Cochlear Implant Candidates. In: Hochmair-Desoyer, I., J., Hochmair, E., S. (eds.): Advances in Cochlear Implants (1994) 79-84
5. Wagner, H., Gerhardt, H., J., Stürzebecher, E., Werbs, M.: Preoperative Assessment of Function of the Auditory Nerve Using Electroaudiometry and a Notched-Noise Auditory Brain Stem Response Technique. Proc. of Speech and Hearing Symposium, Melbourne, Australia (1994) 198-201
6. Nowosielski, J., Redhead, T., Kattula, S.: Extratympanic Electrocochleography with a Conductive Fluid and Flexible Electrode. British Journal of Audiology **11** (1991) 345-349
7. Bochenek, W., Chorzempa, A., Hazell, J., Kiciak, J., Kukwa, A.: Non-Invasive Electrical Stimulation of the Ear Canal as a Communication Aid in Acquired Total Deafness. British Journal of Audiology **23** (1989) 285-291
8. Allen, A., A., Singh, R., S., Sood, R., K., Sik, M., J.: Experience with Promontory Stimulation. In: Hochmair-Desoyer, I., J., Hochmair, E., S. (eds.): Advances in Cochlear Implants (1994) 93-96
9. Muchnik, Ch., Taitelbaum, R., Tene, S., Hildesheimer, M.: Auditory Temporal Resolution and Open Speech Recognition. Cochlear Implant Recip. Scand Audiol. **23** 105-109
10. Czyzewski, A., Krolikowski, R., Kostek, B., Skarzynski, H., Lorens, A.: A Method for Spectral Transposition of Speech Signal Applicable in Profound Hearing Loss. Pres. at the IEEE Acoustic and Speech Signal Processing Worhshop, Mohonk Mountain, USA (1997)
11. Skarzynski, H., Czyzewski, A., Kostek, B.: Prediction of Post-Operative Profits in Cochlear Implanted People Using Electrical Stimulation Test Results. Proc. of the Forum Acusticum '99, Berlin, Germany (1999)
12. Skowron, A., Nguyen Son, H.: Quantization of Real Value Attributes: Rough Set and Bool. Reason. Approach. ICS Report, Warsaw Univ. of Techn., Warsaw, Poland (1995)
13. Chmielewski, M.,R., Grzymala-Busse, J.,W. et al.: The Rule Induction System LERS. Proc. of the 1st Int. Workshop on Rough Sets, Poznan-Kiekrz, Poland (1992)
14. Pawlak, Z.: Rough Sets - Present State and the Future. Foundations of Computing and Decision Sciences **18** (1993)
15. Kostek , B.: Soft Computing in Acoustics. Physica-Verlag, Heidelberg, New York (1999)

Acknowledgements

The research is sponsored by the Committee for Scientific Research (KBN), Warsaw, Poland. Grant No. 4 P05C 082 13.

A Rough Set-Based Approach
to Text Classification

Alexios Chouchoulas[1] and Qiang Shen[1]

Division of Informatics, The University of Edinburgh
{alexios,qiangs}@dai.ed.ac.uk

Abstract. A non-trivial obstacle in good text classification for information filtering and retrieval (IF/IR) is the dimensionality of the data.
This paper proposes a technique using Rough Set Theory to alleviate
this situation. Given corpora of documents and a training set of examples of classified documents, the technique locates a minimal set of
co-ordinate keywords to distinguish between classes of documents, reducing the dimensionality of the keyword vectors. This simplifies the creation of knowledge-based IF/IR systems, speeds up their operation, and
allows easy editing of the rule bases employed. The paper describes the
proposed technique, discusses the integration of a keyword acquisition
algorithm with a rough set-based dimensionality reduction algorithm,
and provides experimental results of a proof-of-concept implementation.

1 Introduction

Information Filtering (IF) and Information Retrieval (IR) is rapidly acquiring
importance as the volume of electronically stored information explodes. Text
classification is an important part of information filtering in that it categorises
documents within text corpora. The user may then handle the various classes
of documents in different ways and devote attention only to those classes that
merit it. For instance, an E-mail classification system could divide incoming mail
into business-related messages, personal messages and useless, unsolicited mail
to be deleted automatically.

However, a non-trivial obstacle in good text classification is the dimensionality of the data. In most IF/IR techniques, each document is described by a
vector of extremely high dimensionality — typically one value per word or pair
of words in the document [1]. The vector ordinates are used as preconditions to a
rule which decides what class the document belongs to. Document vectors commonly comprise tens of thousands of dimensions [2], which renders the problem
all but intractable for even the most powerful computers. The use of the cosine
of the angle between two vectors [3] as a comparison metric further increases the
number of operations to be performed for the classification of one document.

This paper proposes a technique using Rough Set Theory [4] that can help
cope with this situation. Given corpora of documents and a set of examples
of classified documents, the technique can quickly locate a minimal set of coordinate keywords to distinguish between classes of documents. As a result, it

dramatically reduces the dimensionality of the keyword space. The resulting set of keywords (or preconditions) is typically small enough to be understood by a human. This simplifies the creation of knowledge-based IF/IR systems, allowing easy editing of the rule bases.

The background of text classification and Rough Set theory is discussed in Sect. 2. Section 3 provides a description of the proposed system. Section 4 describes experimental results. The paper is concluded in Sect. 5.

2 Background

2.1 The Rough Set Attribute Reduction Method (RSAR)

Rough set theory [4] is a flexible, formal mathematical tool that can be applied to reducing the dimensionality of datasets. RSAR removes redundant input attributes from datasets of nominal values, while ensuring that no information essential for the task at hand is lost. The approach is very efficient, taking advantage of conventional Set Theory operations. It works by maximising a quantity known as *degree of dependency*. The degree of dependency $\gamma_P(X)$ of a set Y of decision attributes on a set of conditional attributes X provides a measure of how important that set of decision attributes is in classifying the dataset examples into the classes in Y. If $\gamma_X(Y) = 0$, then classification Y is independent of the attributes in X, hence the decision attributes are of no use to this classification. If $\gamma = 1$, then Y is completely dependent on X, hence the attributes are indispensable. Values $0 < \gamma_X(Y) < 1$ denote partial dependency, which shows that only some of the attributes in X may be useful, or that the dataset was flawed to begin with.

To calculate $\gamma_X(Y)$, it is necessary to define the *indiscernibility* relation. Given a subset of the set of attributes, $P \subseteq \mathbb{A}$, two objects x and y in a dataset $\mathbb{U}$ are *indiscernible* with respect to P if and only if $f(x, Q) = f(y, Q) \ \forall \ q \subseteq \mathbb{P}$ (where $f(\alpha, B)$ is the classification function represented in the dataset, returning the classification of object α using the conditional attributes contained in the set B). The indiscernibility relation for all $P \in \mathbb{A}$ is written as IND(P). U/IND(P) is used to denote the partition of $\mathbb{U}$ given IND(P):

$$\mathrm{U/IND}(P) = \bigotimes \{q \in P : \mathrm{U/IND}(q)\} \tag{1}$$

where the operator $\otimes$, as applied to two sets of sets A and B, is defined as:

$$A \otimes B = \{X \cap Y : \forall \ X \in A, \ \forall \ Y \in B, \ X \cap Y \neq \emptyset\} \tag{2}$$

Rough Sets approximate traditional sets using a pair of other sets, named the *lower* and *upper approximation* of the set in question. The lower and upper approximation of a set $P \subseteq \mathbb{U}$, given an equivalence relation IND(P), is defined as:

$$\underline{P}Y = \bigcup \{X : X \in U/\mathrm{IND}(P), X \subseteq Y\} \tag{3}$$

$$\overline{P}Y = \bigcup \{X : X \in U/\mathrm{IND}(P), X \cap Y \neq \emptyset\} \tag{4}$$

Suppose that P and Q are equivalence relations in U, the *positive region* $\mathrm{POS}_P(Q)$ contains all objects in U that can be classified in attributes Q using the information in attributes P and is defined as:

$$\mathrm{POS}_P(Q) = \bigcup_{X \in Q} \underline{P}X \tag{5}$$

From this, the degree of dependency $\gamma_P(Q)$ is given by:

$$\gamma_P(Q) = \frac{\| \mathrm{POS}_P(Q) \|}{\| U \|} \tag{6}$$

where $\| Set \|$ is the cardinality of *Set*.

The naïve version of the RSAR algorithm evaluates $\gamma_P(Q)$ for all possible subsets of the dataset's conditional attributes, stopping when it either reaches 1, or there are no more combinations to investigate. The QUICKREDUCT Algorithm (described in [5, 6]) escapes the NP-hard nature of the naïve version by searching the tree of attribute combinations in a depth-first manner. It starts off with an empty subset and adds attributes one by one, each time selecting the attribute whose addition to the current subset will offer the highest increase of $\gamma_P(Q)$. The algorithm stops when a $\gamma_P(Q)$ of 1 is reached, or when all attributes have been added (in which case the dataset could not be correctly classified to begin with).

It is evident from this that the RSAR will not compromise with a set of conditional attributes that contains large part of the information of the initial set — it will *always* attempt to reduce the attribute set without losing *any* information significant to the classification at hand.

Since the QUICKREDUCT algorithm builds conditional attribute subsets incrementally, it is possible to influence its decisions by suitable re-ordering of the conditional attributes in the dataset such that attributes suspected to be of more importance are placed before (to the left of) attributes of lesser importance. This is done by the integrated system described in this paper, in order to suggest to RSAR to utilise the better, more characteristic keywords before others.

2.2 Text Classification

Text classification aims to separate groups (corpora) of documents into categories. Like all classification tasks, it may be tackled either by comparing new documents with previously classified ones (distance-based techniques), or by using rule-based approaches.

Most text classification techniques that do not operate at the semantics level work on a hyperplane whose axes represent the presence of different keywords. Depending on the specific technique, the axes of this keyword space may be discrete (e.g. boolean) or continuous (e.g. frequency of keywords, importance of keyword, et cetera). The dimensionality of the keyword space depends on the cardinality of the universal set of keywords, defined as the union of all possible keywords of all documents examined, as shown in (7). Any document D_j (a set of keywords) in the corpora at hand can then be represented by a multidimensional keyword vector $\underline{x}_{D_j}$:

$$\mathbb{U} = \bigcup_i D_i = \{k_1, k_2, \ldots, k_n\} \tag{7}$$

$$\underline{x}_{D_j} = \langle f(D_j, k_1), f(D_j, k_2), \ldots, f(D_j, k_n)\rangle \tag{8}$$

Where each ordinate $f(D_j, k_i)$ $(1 \leq i \leq n)$ in the vector represents the *weight* of the keyword k_i in $\mathbb{U}$. The weight is the result of some metric $f(D_j, k_i)$, measuring the presence, frequency importance or other quantifiable aspect of the keyword k_i in the document.

The two most commonly used text classification approaches are outlined below.

Distance-Based Text Classification Distance-based classification involves the comparison of high-dimensionality keyword vectors. In cases where the vector describes groups of documents, it identifies the centre of a cluster of documents. Documents are classified by comparing their document vectors. The metric most commonly used is the cosine of the angle between the two vectors, derived in terms of the scalar or inner product, though other metrics are also available [1].

The set of keywords representing one document (or a cluster of documents) may be obtained by numerous different algorithms that scan corpora of documents for keywords, ranking them by perceived importance. Weight calculation metrics range from a simple frequency-proportional weighting technique that naïvely attaches more importance to the common words in a document, to the inverse document frequency that emphasises those keywords that are common in the document in question, yet uncommon in the overall collection of documents.

Datasets for such systems are almost always built automatically and maintained using paradigms like learning by observation, example or imitation that abstract away the actual calculation of weights and formation of vectors. This allows the user to obtain complex intelligent-like text classification behaviour with a minimum of effort. Unfortunately, the dimensionality of the document vectors is typically extremely high (usually in the tens of thousands), a detail that greatly slows down classification tasks and makes storage of document vectors expensive [2].

Rule-Based Text Classification Rule-based text classification has been in use for a relatively long time and is an established method of classifying documents.

Common applications include the *kill-file* article filters used by Usenet client software and van den Berg's autonomous E-mail filter, Procmail.

In this context, keyword vectors are viewed as rule preconditions; the class a document belongs to is used as the rule decision:

$$k_1, k_2, \ldots, k_n \in \mathbb{U}$$
$$r_i(D) = p(D, k_1) \wedge p(D, k_2) \wedge \ldots \wedge p(D, k_n) \Rightarrow D \in \mathbb{D} \tag{9}$$

Where k_i are document keywords, $\mathbb{U}$ is the universal keyword set, D is one document, $\mathbb{D}$ is a document class, $r_i(D)$ is rule i applied to D and $p(D, k_i)$ is a function evaluating to *true* if D contains keyword k_i such that it satisfies some metric (e.g. a minimum frequency or weight). Not all keywords in the universal set need be checked for. This allows rule-based text classifiers to exhibit a notation much terser than that of vector-based classifiers, where a vector must always have the same dimensionality as the keyword space.

In most cases, rules are written by the human user. Most typical rule bases simply test for the presence of specific keywords in the document (so $p(D, k_i) \Leftrightarrow k_i \in D$). For example, a Usenet client may use a kill-file to filter out newsgroup articles by some specific person by looking for the person's name in the article's 'From' field. Such rule-based approaches are inherently simple to understand, which accounts for their popularity among end-users. Unfortunately, complex needs often result in very complex rule bases, ones that users have difficulty maintaining by hand.

3 The Proposed System

This paper proposes a system that builds text classification rule bases, although it is trivial to adapt it to distance-based approaches by using a different keyword acquisition sub-system (as described in sect. 3.1). The modularity of the proposed technique allows this.

The application domain chosen to test the system is E-mail, since real-life corpora of E-mail messages are very easy to obtain. Most users of E-mail keep 'folders' of messages related in some way. This provides training data for the system. Like many documents, E-mail messages are structured: each message comprises a header, itself comprising a number of fields, and a body. Keywords may need to be treated differently, depending on where in the message they are encountered. For example, the Message-ID field is a unique identifier of an E-mail message and, as such, a notorious opportunity for over-training. This rigidly structured nature of E-mail messages makes the domain attractive as a test-case for a text classification system.

The system comprises two main stages, as shown in fig. 1: the keyword acquisition stage reads corpora of documents (folders of similar E-mail messages), locates candidate keywords, estimates their importance and builds an intermediate dataset of high dimensionality. The RSAR part examines the dataset and

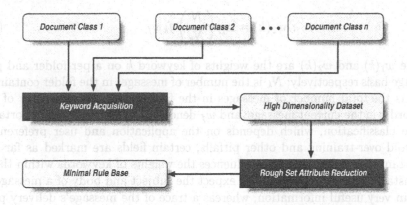

Fig. 1. Data flow through the system.

removes redundancy, insofar as this is possible, leaving a dataset or rule base containing a drastically reduced number of preconditions per rule.

3.1 Keyword Acquisition

This sub-system uses a set of E-mail folders as input. Each folder is expected to contain similar E-mail messages. Folders (in UNIX mailbox format) contain standard E-mail messages as described in the RFC-822 document [7]. Messages are sequentially read; each field (key/value pair) within each message is treated separately. The body of a message is treated as a very long field.

Within each field, words are isolated and pre-filtered to avoid very short or long keywords, or keywords that are not words (e.g. long numbers or random sequences of characters). Every word or pair of consecutive words in the text is considered a candidate keyword. The name of the current field is also added to the keyword, so that information on where in the message the keyword occurred is retained.

At this stage, the keyword acquisition sub-system has two modes of operation: one, dubbed *single cluster* mode, generates a set of keywords characterising each entire folder; the other generates separate sets of keywords for each message in each folder, ultimately used to create one classification rule per message, hence dubbed *one-per-message*. In the former case, shown in (10), keyword weights are calculated such that keywords common to most messages are deemed more important. In the latter case (11), the weighting metric emphasises keywords that show the message's difference from other messages in the collection. This applies pressure to diversify the keyword sets, rather than create multiple copies of the same set of keywords for each message. The weighting functions are as follows:

$$w_1(k) = -\log\left(\frac{N_k}{N}\right) f_k w_f \tag{10}$$

$$w_2(k) = \log\left(\frac{N}{N_k}\right) f_k w_f \tag{11}$$

Where $w_1(k)$ and $w_2(k)$ are the weights of keyword k on a per-folder and per-message basis respectively; N_k is the number of messages in the folder containing k; N is the total number of messages in the folder; f_k is the frequency of the keyword k in the current message; and w_f denotes the current field's importance to the classification, which depends on the application and user preferences. To avoid over-training and other pitfalls, certain fields are marked as far less important than others and this influences the weights of keywords within them. For instance, it is relatively safe to expect the subject and body of a message to contain very useful information, whereas a trace of the message's delivery path (the Received field) is unlikely to provide useful keywords.

Finally, before the weighted keyword is added to the set of keywords, it passes through two filters: one is a low-pass filter removing words so uncommon that are definitely not good keywords; the other is a high-pass filter that removes far too common words such as auxiliary verbs, articles et cetera. This gives the added advantage of language-independence to the keyword acquisition algorithm: most similar methods rely on English thesauri and lists of common English words to perform the same function. Finally, all weights are normalised before the keyword sets are output. This allows for more homogeneous handling of weights in the next stages and avoids counter-intuitive results as identified in [1].

It must be emphasised that any keyword acquisition algorithm may be substituted for the one described above, as long as it outputs weighted keywords.

3.2 Rough Set Attribute Reduction

The RSAR used here is exactly as described previously. It reads the sets of keywords generated by the keyword acquisition algorithm above. A dataset is constructed by evaluating the union of all sets of weighted keywords; the keywords are sorted in order of decreasing weight. Where one keyword has two or more different weights, the maximum one is used. Each keyword maps to one conditional attribute in the dataset. The decision attribute is the name of the folder from where the keyword set was extracted. Missing values in the dataset denote the absence of that particular keyword in the particular keyword set.

For example, the two sets of keywords below, describing two documents D_1 and D_2, may be used to build the dataset shown in table 1.

$$D_1 = \{\langle k_1, 0.19\rangle, \langle k_2, 0.98\rangle, \langle k_3, 0.72\rangle, \langle k_4, 0.87\rangle\} \tag{12}$$

$$D_2 = \{\langle k_4, 0.31\rangle, \langle k_5, 0.42\rangle, \langle k_6, 0.56\rangle\} \tag{13}$$

Since RSAR is better suited to nominal datasets, the dataset is thus quantised. Two different quantisation methods are available: a boolean quantisation, where a value of 1 signifies the keyword's presence and a value of 0 signifies its absence; and a quantisation of the normalised weight space into eleven values,

Table 1. Dataset produced from (12) and (13), assuming they belong to folders α and β respectively.

	k_2	k_4	k_3	k_6	k_5	k_1	$\rightarrow$	Class
D_1	0.98	0.87	0.72			0.19	$\rightarrow$	α
D_2		0.31		0.56	0.42		$\rightarrow$	β

calculated by $\lfloor 10w \rfloor$ (where w is a keyword weight and $\lfloor \cdot \rfloor$ is the *floor* function, evaluating to the greatest integer less than or equal to its argument). The two methods are there to allow better interfacing with various classifiers as well as to evaluate the best technique for quantising weights in the application domain at hand.

Having quantised the intermediate, high-dimensionality dataset, the RSAR can now execute the QUICKREDUCT algorithm to remove all redundant decision attributes. This results in a dataset of radically reduced dimensionality. Since each object in the dataset comprises a set of conditional attributes accompanied by a decision attribute (the document class the object belongs to), it can be viewed as a set of production rules, with conditional attribute values being the rule preconditions and the document class being the rule's decision. The dataset is simply post-processed to remove duplicate rules and output in the form of a rule base.

4 Experimental Results

The use of the above proposed system allows the generation of rule bases of extremely reduced dimensionality from corpora of documents, without reducing the classification accuracy, inasmuch as this is possible. To demonstrate this, seven different corpora of E-mail messages were used. The corpora were chosen so as to provide a wide spectrum of characteristics found in sets of documents: some are homogeneous, containing messages by a single author; others contain text from multiple authors with different writing styles. Small and larger corpora are mixed to ensure that the size of the document collection does not influence the quality of the resultant rule base. Each collection of messages represents one class of documents.

Random groups of two to five such message folders are chosen and fed to the system. All combinations of keyword generation (single cluster or one-per-message) and quantisation method (boolean, weight integer part) are investigated. Table 2 shows the average dimensionality reduction for each combination. Dimensionality reduction is shown in orders of magnitude (base 10). Note that, although the one-per-message dataset generation technique shows greater reduction, the dimensionality is typically much higher than for the single cluster method. Average pre-reduction dimensionality is 26,827 for the one-per-message technique and 338 for the single-cluster technique. The resultant rule bases have

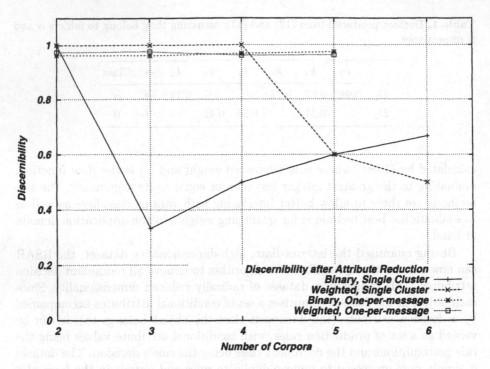

Fig. 2. Discernibility after dimensionality reduction.

rules involving one to six preconditions, with the boolean rule bases exhibiting slightly higher dimensionality due to boolean conditional attributes' lesser information content.

Figure 2 shows the average discernibility in the generated rule bases. A discernibility of 1 indicates no loss of information after the RSAR algorithm has executed. A classifier employing the rule base in question will, assuming the training set is a good statistical sample, achieve a classification accuracy very similar to the discernibility of the rule base. As shown in the figure, generating one rule for an entire folder of E-mail messages does not allow for good classification. Discernibility can drop to unacceptable levels and varies widely depending on the content of the text corpora. By contrast, generating multiple rules allows for much better coverage of the feature space — discernibility is high enough to

Table 2. Average dimensionality reduction in orders of magnitude for all four combinations of operation modes.

	Single cluster	One-per-message
Boolean	2.02	3.62
Weighted	2.29	3.70

127

satisfactory and appears to be almost constant. Binary quantisation seems to offer slightly less satisfactory results than the weighted method.

In terms of the linguistic nature of the rule bases, it is considerably easier for a human to read and understand rules that span entire corpora, rather than rules that describe single documents. It is also possible to judge the quality of the rule base in an intuitive manner, simply by reading it.

5 Conclusion

Text classification relies heavily on the acquisition of sets of co-ordinate keywords describing documents. This paper has summarised a technique that reduces the need to choose a restricted number of good quality keywords by allowing larger collections of keywords to be built. The dimensionality of these sets of keywords is then reduced with the aid of Rough Set Theory, while maintaining intact the information contained in the keyword sets. The technique is efficient, language-independent and particularly flexible due to its modular nature. The decrease in dimensionality is drastic, due to RSAR's reduction of the keyword set to the minimum required for the classification at hand.

The system can be interfaced to a number of text classifiers to produce highly optimised rule-based text classification applications, while still allowing the user to read and intuitively understand the rule bases.

The approach is still in its early states of research, which accounts for the less-than-acceptable results of the single-cluster rule generation method. Further investigation into suitable keyword weighting metrics and rule induction especially designed for text classification is in progress, as is an actual test-bed application of the present technique to the classification of in-coming E-mail messages.

References

1. C. J. van Rijsbergen *Information Retrieval.* Butterworths, United Kingdom (1990)
2. T. Joachims Text Categorization with Support Vector Machines: Learning with Many Relevant Features. *European Conf. Mach. Learning, ECML98."* (1998)
3. A. Moukas and P. Maes Amalthaea: An Evolving Multi-Agent Information Filtering and Discovery System for the WWW. *Journal of Autonomous Agents and Multi-Agent Systems,* **1** (1998) 59–88.
4. Z. Pawlak *Rough Sets: Theoretical Aspects of Reasoning About Data.* Kluwer Academic Publishers, Dordrecht (1991)
5. Q. Shen and A. Chouchoulas Combining Rough Sets and Data-Driven Fuzzy Learning. Accepted for publication in *Pattern Recognition.*
6. A. Chouchoulas and Q. Shen Rough Set-Aided Rule Induction for Plant Monitoring. *Proceedings of the 1998 International Joint Conference on Information Science (JCIS'98),* **2** (1998) 316–319.
7. D. H. Crocker, *RFC 822, Standard for the Format of ARPA Internet Text Messages.* Dept. of Electrical Engineering, Univ. of Delaware (1982)

Modular Rough Fuzzy MLP: Evolutionary Design

Pabitra Mitra, Sushmita Mitra and Sankar K. Pal

Machine Intelligence Unit
Indian Statistical Institute
Calcutta 700035, INDIA
EMail: {pabitra_r,sushmita,sankar}@isical.ac.in}

Abstract. This article describes a way of designing a hybrid system for classification and rule generation, in soft computing paradigm, integrating rough set theory with a fuzzy MLP using an evolutionary algorithm. An l-class classification problem is split into l two-class problems. Crude subnetworks are initially obtained for each of these two-class problems via rough set theory. These subnetworks are then combined and the final network is evolved using a GA with restricted mutation operator which utilizes the knowledge of the modular structure already generated, for faster convergence. The GA tunes the fuzzification parameters, and the network weights and structure simultaneously, by optimizing a single fitness function.

1 Introduction

Soft Computing is a consortium of methodologies which works synergetically (not competitively) and provides, in one form or another, flexible information processing capability for handling real life ambiguous situations [1]. There are ongoing efforts to integrate artificial neural networks (ANNs) with fuzzy set theory, genetic algorithms (GAs), rough set theory and other methodologies in the *soft computing* paradigm [2].

Knowledge-based networks [3,4] constitute a special class of ANNs that consider crude domain knowledge to generate the initial network architecture, which is later refined in the presence of training data. Recently, the theory of rough sets has been used to generate knowledge-based networks.

A recent trend in neural network design for large scale problems is to split the original task into simpler subtasks, and use a subnetwork module for each of the subtasks [5]. The *divide and conquer* strategy leads to super-linear speedup in training. It has been shown that by combining the output of several subnetworks in an ensemble, one can improve the generalization ability over that of a single large network [6].

In the present article an evolutionary strategy is suggested for designing a connectionist system, integrating fuzzy sets and rough sets. The basic building block is a Rough Fuzzy MLP [7]. The evolutionary training algorithm generates the weight values for a parsimonious network and simultaneously tunes the fuzzification parameters by optimizing a single fitness function. Rough set theory is used to obtain a set of probable knowledge-based subnetworks which form the initial population of the GA. These modules are then integrated and evolved with a *restricted* mutation operator that helps preserve extracted localized rule structures as potential solutions. An restricted mutation operator is implemented, which utilizes the knowledge of modular structure evolved to achieve faster convergence.

2 Fuzzy MLP

The fuzzy MLP model [8] incorporates fuzziness at the input and output levels of the MLP, and is capable of handling exact (numerical) and/or inexact (linguistic) forms of input data. Any input feature value is described in terms of some combination of membership values in the linguistic property sets *low* (L), *medium* (M) and *high* (H). Class membership values (μ) of patterns are represented at the output layer of the fuzzy MLP. During training, the weights are updated by backpropagating errors with respect to these membership values such that the contribution of uncertain vectors is automatically reduced.

A single layer feedforward MLP is used. An n-dimensional pattern $F_i = [F_{i1}, F_{i2}, \ldots, F_{in}]$ is represented as a $3n$-dimensional vector

$$F_i = [\mu_{low(F_{i1})}(F_i), \ldots, \mu_{high(F_{in})}(F_i)] = [y_1^0, y_2^0, \ldots, y_{3n}^0] , \quad (1)$$

where the μ values indicate the membership functions of the corresponding linguistic π-sets *low, medium* and *high* along each feature axis and $y_1^0, \ldots, y_{3n}^0$ refer to the activations of the $3n$ neurons in the input layer.

When the input feature is numerical, we use the π–fuzzy sets (in the one dimensional form), with range [0,1], represented as

$$\pi(F_j; c, \lambda) = \begin{cases} 2(1 - \frac{\|F_j - c\|}{\lambda})^2, & \text{for } \frac{\lambda}{2} \le \|F_j - c\| \le \lambda \\ 1 - 2(\frac{\|F_j - c\|}{\lambda})^2, & \text{for } 0 \le \|F_j - c\| \le \frac{\lambda}{2} \\ 0 , & \text{otherwise} , \end{cases} \quad (2)$$

where $\lambda (> 0)$ is the radius of the π–function with c as the central point. Note that features in linguistic and set forms can also be handled in this framework [8].

3 Rough Fuzzy MLP

Here we describe the Rough Fuzzy MLP [7]. Let $S = < U, A >$ be a decision table, with C and $D = \{d_1, \ldots, d_l\}$ its sets of condition and decision attributes respectively. Divide the decision table $S = < U, A >$ into l tables $S_i = < U_i, A_i >$, $i = 1, \ldots, l$, corresponding to the l decision attributes $d_1, \ldots, d_l$, where

$$U = U_1 \cup \ldots \cup U_l \text{ and } A_i = C \cup \{d_i\}.$$

The size of each S_i ($i = 1, \ldots, l$) is first reduced with the help of a threshold on the number of occurrences of the same pattern of attribute values. This will be elicited in the sequel. Let the reduced decision table be denoted by T_i, and $\{x_{i1}, \ldots, x_{ip}\}$ be the set of those objects of U_i that occur in $T_i, i = 1, \ldots, l$.

Now for each d_i-reduct $B = \{b_1, ..., b_k\}$ (say), a discernibility matrix (denoted $\mathbf{M}_{d_i}(B)$) from the d_i-discernibility matrix is defined as follows:

$$c_{ij} = \{a \in B : a(x_i) \neq a(x_j)\}, \tag{3}$$

for $i, j = 1, ..., n$.

For each object $x_j \in x_{i_1}, ..., x_{i_p}$, the discernibility function $f_{d_i}^{x_j}$ is defined as

$$f_{d_i}^{x_j} = \bigwedge \{\bigvee(c_{ij}) : 1 \leq i, j \leq n, \; j < i, \; c_{ij} \neq \emptyset\}, \tag{4}$$

where $\bigvee(c_{ij})$ is the disjunction of all members of c_{ij}. Then $f_{d_i}^{x_j}$ is brought to its conjunctive normal form (c.n.f). One thus obtains a dependency rule r_i, viz. $P_i \leftarrow d_i$, where P_i is the disjunctive normal form (d.n.f) of $f_{d_i}^{x_j}, j \in i_1, ..., i_p$.

The dependency factor df_i for r_i is given by

$$df_i = \frac{card(POS_i(d_i))}{card(U_i)}, \tag{5}$$

where $POS_i(d_i) = \bigcup_{X \in I_{d_i}} l_i(X)$, and $l_i(X)$ is the lower approximation of X with respect to I_i. In this case $df_i = 1$ [7].

Consider the case of feature F_j for class c_k in the l-class problem domain. The inputs for the ith representative sample $\mathbf{F}_i$ are mapped to the corresponding three-dimensional feature space of $\mu_{low(F_{ij})}(\mathbf{F}_1), \mu_{medium(F_{ij})}(\mathbf{F}_1)$ and $\mu_{high(F_{ij})}(\mathbf{F}_1)$, by eqn. (1). Let these be represented by L_j, M_j and H_j respectively. Then consider only those attributes which have a numerical value greater than some threshold Th ($0.5 \leq Th < 1$). This implies clamping only those features demonstrating high membership values with one, while the others are fixed at zero. As the method considers multiple objects in a class a separate $n_k \times 3n$-dimensional attribute-value decision table is generated for each class c_k (where n_k indicates the number of objects in c_k).

Let there be m sets $O_1, ..., O_m$ of objects in the table having identical attribute values, and $card(O_i) = n_{k_i}, i = 1, ..., m$, such that $n_{k_1} \geq ... \geq n_{k_m}$ and $\sum_{i=1}^{m} n_{k_i} = n_k$. The attribute-value table can now be represented as an $m \times 3n$ array. Let $n_{k'_1}, n_{k'_2}, ..., n_{k'_m}$ denote the distinct elements among $n_{k_1}, ..., n_{k_m}$ such that $n_{k'_1} > n_{k'_2} > ... > n_{k'_m}$. Let a heuristic threshold function be defined as

$$Tr = \left\lceil \frac{\sum_{i=1}^{m} \frac{1}{n_{k'_i} - n_{k'_{i+1}}}}{Th} \right\rceil, \tag{6}$$

so that all entries having frequency less than Tr are eliminated from the table, resulting in the reduced attribute-value table. Note that the main motive of introducing this threshold function lies in reducing the size of the resulting network. One attempts to eliminate noisy pattern representatives (having lower values of n_{k_i}) from the reduced attribute-value table.

While designing the initial structure of the rough fuzzy MLP, the union of the rules of the l classes is considered. The input layer consists of $3n$ attribute values while the output layer is represented by l classes. The hidden layer nodes model the first level (innermost) operator in the antecedent part of a rule, which can be either a conjunct or a disjunct. The output layer nodes model the outer level operands, which can again be either a conjunct or a disjunct. For each inner level operator, corresponding to one output class (one dependency rule), one hidden node is dedicated. Only those input attributes that appear in this conjunct/disjunct are connected to the appropriate hidden node, which in turn is connected to the corresponding output node. Each outer level operator is modeled at the output layer by joining the corresponding hidden nodes. Note that a single attribute (involving no inner level operators) is directly connected to the appropriate output node via a hidden node, to maintain uniformity in rule mapping.

Let the dependency factor for a particular dependency rule for class c_k be $df = \alpha = 1$ by eqn. (5). The weight w_{ki}^1 between a hidden node i and output node k is set at $\frac{\alpha}{fac} + \varepsilon$, where fac refers to the number of outer level operands in the antecedent of the rule and ε is a small random number taken to destroy any symmetry among the weights. Note that $fac \geq 1$ and each hidden node is connected to only one output node. Let the initial weight so clamped at a hidden node be denoted as β. The weight $w_{ia_j}^0$ between an attribute a_j (where a corresponds to low (L), $medium$ (M) or $high$ (H)) and hidden node i is set to $\frac{\beta}{facd} + \varepsilon$, such that $facd$ is the number of attributes connected by the corresponding inner level operator. Again $facd \geq 1$. Thus for an l-class problem domain there are at least l hidden nodes. All other possible connections in the resulting fuzzy MLP are set as small random numbers. It is to be mentioned that the number of hidden nodes is determined from the dependency rules.

4 Modular Knowledge-Based Network

It is believed that the use of Modular Neural Network (MNN) enables a wider use of ANNs for large scale systems. Embedding modularity (*i.e.* to perform local and encapsulated computation) into neural networks leads to many advantages compared to the use of a single network. For instance, constraining the network connectivity increases its learning capacity and permits its application to large scale problems [5]. It is easier to encode *a priori* knowledge in modular neural networks. In addition, the number of network parameters can be reduced by using modularity. This feature speeds computation and can improve the generalization capability of the system.

We use two phases. First an l-class classification problem is split into l two-class problems. Let there be l sets of subnetworks, with $3n$ inputs and one output node each. Rough set theoretic concepts are used to encode domain knowledge into each of the subnetworks, using eqns (3)-(6). The number of hidden nodes and connectivity of the knowledge-based subnetworks is automatically determined. A two-class problem leads to the generation of one or more crude subnetworks, each encoding a particular decision rule. Let each of these constitute a pool. So we obtain $m \geq l$ pools of knowledge-based modules. Each pool k is perturbed to

132

generate a total of n_k subnetworks, such that $n_1 = \ldots = n_k = \ldots = n_m$. These pools constitute the initial population of subnetworks, which are then evolved independently using genetic algorithms.

At the end of training, the modules/subnetworks corresponding to each two-class problem are concatenated to form an initial network for the second phase. The inter module links are initialized to small random values as depicted in Fig. 1. A set of such concatenated networks forms the initial population of the GA. The mutation probability for the inter-module links is now set to a high value, while that of intra-module links is set to a relatively lower value. This sort of *restricted* mutation helps preserve some of the localized rule structures, already extracted and evolved, as potential solutions. The initial population for the GA of the entire network is formed from all possible combinations of these individual network modules and random perturbations about them. This ensures that for complex multi-modal pattern distributions all the different representative points remain in the population. The algorithm then searches through the reduced space of possible network topologies.

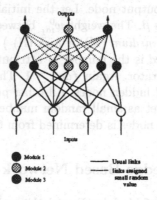

Fig. 1. Intra and Inter module links

5 Evolutionary Design

Genetic algorithms are highly parallel and adaptive search processes based on the principles of natural selection. Here we use GAs for evolving the weight values as well as the structure of the fuzzy MLP used in the framework of modular neural networks. The input and output fuzzification parameters are also tuned. Unlike other theory refinement systems which train only the *best* network approximation obtained from the domain theories, the initial population here consists of all possible networks generated from rough set theoretic rules. This is an advantage because potentially valuable information may be wasted by discarding the contribution of less successful networks at the initial level itself.

Genetic algorithms involve three basic procedures - encoding of the problem parameters in the form of binary strings, application of genetic operators like crossover and mutation, selection of individuals based on some objective function to create a new population. Each of these aspects is discussed below with relevance to our algorithm.

5.1 Chromosomal Representation

The problem variables consist of the weight values and the input/output fuzzification parameters. Each of the weights is encoded into a binary word of 16 bit length, where [000...0] decodes to -128 and [111...1] decodes to 128. An additional bit is assigned to each weight to indicate the presence or absence of the link. If this bit is 0 the remaining bits are unrepresented in the phenotype. The total number of bits in the string is therefore dynamic [9]. Thus a total of 17 bits are assigned for each weight. The fuzzification parameters tuned are the centers (c) and radius (λ) for each of the linguistic attributes *low*, *medium* and *high* of each feature (eqn. 2). These are also coded as 16 bit strings in the range [0, 2].

Initial population is generated by coding the networks obtained by rough set based knowledge encoding, and by random perturbations about them. A population size of 64 was considered.

5.2 Genetic Operators

Crossover It is obvious that due to the large string length, single point crossover would have little effectiveness. Multiple point crossover is adopted, with the distance between two crossover points being a random variable between 8 and 24 bits. This is done to ensure a high probability for only one crossover point occurring within a word encoding a single weight. The crossover probability is fixed at 0.7.

Mutation The search string being very large, the influence of mutation is more on the search. Each of the bits in the string is chosen to have some mutation probability (*pmut*). This mutation probability however has a spatio-temporal variation. The maximum value of *pmut* is chosen to be 0.4 and the minimum value as 0.01. The mutation probabilities also vary along the encoded string, with the bits corresponding to inter-module links being assigned a probability *pmut* (*i.e.*, the value of *pmut* at that iteration) and intra-module links assigned a probability *pmut*/10. This is done to ensure least alterations in the structure of the individual modules already evolved. Hence, the mutation operator indirectly incorporates the domain knowledge extracted through rough set theory.

5.3 Choice of fitness function

In GAs the fitness function is the final arbiter for string creation, and the nature of the solution obtained depends on the objective function. An objective function of the form described below is chosen.

$$F = \alpha_1 f_1 + \alpha_2 f_2 ,$$

where

$$f_1 = \frac{No.\ of\ Correctly\ Classified\ Sample\ in\ Training\ Set}{Total\ No.\ of\ Samples\ in\ Training\ Set}$$

$$f_2 = 1 - \frac{No.\ of\ links\ present}{Total\ No.\ of\ links\ possible}.$$

Here α_1 and α_2 determine the relative weightage of each of the factors. α_1 is taken to be 0.9 and α_2 is taken as 0.1, to give more importance to the classification score compared to the network size in terms of number of links. Note that we optimize the network connectivity, weights and input/output fuzzification parameters simultaneously.

5.4 Selection

Selection is done by the *roulette wheel* method. The probabilities are calculated on the basis of ranking of the individuals in terms of the objective function, instead of the objective function itself. Fitness ranking overcomes two of the biggest problems inherited from traditional fitness scaling : *over compression* and *under expansion.*

Elitism is incorporated in the selection process to prevent oscillation of the fitness function with generation. The fitness of the best individual of a new generation is compared with that of the current generation. If the latter has a higher value - the corresponding individual replaces a randomly selected individual in the new population.

6 Implementation and Results

The genetic-rough-neuro-fuzzy algorithm has been implemented on speech data.

Let the proposed methodology be termed Model S. Other models compared include:

Model O: An ordinary MLP trained using backpropagation (BP) with weight decay. Model F: A fuzzy multilayer perceptron trained using BP [8] (with weight decay).

Model R: A fuzzy multilayer perceptron trained using BP (with weight decay), with initial knowledge encoding using rough sets [7].

Model FM: A modular fuzzy multilayer perceptron trained with GAs along with tuning of the fuzzification parameters. Here the term modular refers to the use of subnetworks corresponding to each class, that are later concatenated using GAs.

135

A threshold value of 0.5 is applied on the fuzzified inputs to generate the attribute value table used in rough set encoding, such that $y_i^0 = 1$ if $y_i^0 \geq 0.5$ and 0 otherwise. Here, 50% of the samples are used as training set and the network is tested on the remaining samples.

The speech data **Vowel** deals with 871 Indian Telegu vowel sounds. These were uttered in a consonant-vowel-consonant context by three male speakers in the age group of 30 to 35 years. The data set has three features: F_1, F_2 and F_3 corresponding to the first, second and third vowel formant frequencies obtained through spectrum analysis of the speech data. There are six classes δ, a, i, u, e, o. These overlapping classes will be denoted by $c_1, c_2, \ldots, c_6$.

The rough set theoretic technique is applied on the vowel data to extract some knowledge which is initially encoded among the connection weights of the subnetworks. The data is first transformed into a nine dimensional linguistic space.

The dependency rules obtained are :

$$c_1 \leftarrow M_1 \vee L_3, \ c_1 \leftarrow M_1 \vee M_3, \ c_2 \leftarrow M_2 \vee M_3, \ c_3 \leftarrow (L_1 \wedge M_3) \vee (L_1 \wedge H_3),$$
$$c_4 \leftarrow (L_2 \wedge M_3) \vee (L_1 \wedge L_2) \vee (L_1 \wedge L_3) \vee (L_2 \wedge L_3)$$
$$c_4 \leftarrow (L_2 \wedge H_3) \vee (L_1 \wedge L_2) \vee (L_1 \wedge L_3) \vee (L_2 \wedge L_3), \ c_5 \leftarrow (M_1 \wedge M_3) \vee (H_1 \wedge M_3),$$
$$c_5 \leftarrow (H_1 \wedge M_3) \vee (H_1 \wedge H_3) \vee (M_3 \wedge H_3) c_6 \leftarrow L_1 \vee M_3,$$
$$c_6 \leftarrow M_1 \vee M_3, \ c_6 \leftarrow L_1 \vee H_3, \ c_6 \leftarrow M_1 \vee H_3.$$

The above rules are used to get initial subnetwork modules using the scheme outlined in Section 3. The integrated network contains a single hidden layer with 15 nodes. In all, 32 such networks are obtained. The remaining 32 networks are obtained by small (20%) random perturbations about them, to generate an initial population of 64 individuals.

The performance of Model S along with its comparison with other models using the same number of hidden nodes is presented in Table 1. In the first phase of the GA (for models FM and S), each of the subnetworks are partially trained for 10 sweeps each. It is observed that Model S performs the best with the least network size after being trained for only 90 sweeps in the final phase. Comparing Models F and R, we observe that the incorporation of domain knowledge in the latter through rough sets boosts its performance. Similarly, using the modular approach with GA in Model FM improves its efficiency over that of Model F. Note that Model S encompasses both models R and FM. Hence it results in the least redundant yet effective model.

7 Conclusions

A methodology for integrating rough sets with fuzzy MLP using genetic algorithms for designing a knowledge-based network for pattern classification and rule generation is presented. The proposed algorithm involves synthesis of several MLP modules, each encoding the rough set rules for a particular class. These

Table 1. Comparative performance of different models for **Vowel** data

Models	Model O		Model F		Model R		Model FM		Model S	
	Train	Test	Train	Test	Train	Test	Train	Test	Train	Test
c_1(%)	11.20	8.10	15.70	14.21	44.12	42.41	42.40	32.50	62.20	58.42
c_2(%)	75.71	76.40	82.58	88.41	88.81	87.53	95.02	88.89	100.0	88.89
c_3(%)	80.00	85.48	90.91	92.42	88.41	88.71	90.91	89.50	94.24	92.42
c_4(%)	71.43	65.20	93.22	87.21	88.23	87.44	90.91	90.00	90.20	90.25
c_5(%)	68.57	59.10	80.00	78.57	94.22	93.45	82.21	80.42	85.84	82.42
c_6(%)	76.47	71.10	96.21	93.90	94.45	94.21	100.0	100.0	95.10	94.90
Net(%)	65.23	64.20	84.36	81.82	86.82	85.81	85.48	82.45	87.21	85.81
# links	131		210		152		124		84	
Iterations	5600		5600		2000		200		90	

knowledge-based modules are refined using a GA. The genetic operators are implemented in such a way that they help preserve the modular structure already evolved. It is seen that this methodology along with modular network decomposition results in superior performance in terms of classification score, training time, and network sparseness (thereby enabling easier extraction of rules) as compared to earlier hybridizations.

References

1. L. A. Zadeh, "Fuzzy logic, neural networks, and soft computing," *Communications of the ACM*, vol. 37, pp. 77–84, 1994.
2. *Proc. of Fifth International Conference on Soft Computing (IIZUKA98)*, Iizuka, Fukuoka, Japan, October 1998.
3. L. M. Fu, "Knowledge-based connectionism for revising domain theories," *IEEE Transactions on Systems, Man, and Cybernetics*, vol. 23, pp. 173–182, 1993.
4. G. G. Towell and J. W. Shavlik, "Knowledge-based artificial neural networks," *Artificial Intelligence*, vol. 70, pp. 119–165, 1994.
5. B. M. Happel and J. J. Murre, "Design and Evolution of Modular Neural Network Architectures," *Neural Networks*, vol. 7, pp. 985–1004, 1994.
6. L. Hansen and P. Salamon, "Neural network ensembles," *IEEE Transactions on Pattern Analysis and Machine Intelligence*, vol. 12, pp. 993–1001, 1990.
7. M. Banerjee, S. Mitra, and S. K. Pal, "Rough fuzzy MLP: Knowledge encoding and classification," *IEEE Transactions on Neural Networks*, vol. 9, no. 6, pp. 1203–1216, 1998.
8. S. K. Pal and S. Mitra, "Multi-layer perceptron, fuzzy sets and classification," *IEEE Transactions on Neural Networks*, vol. 3, pp. 683–697, 1992.
9. V. Maniezzo, "Genetic evolution of the topology and weight distribution of neural networks," *IEEE Transactions on Neural Networks*, vol. 5, pp. 39–53, 1994.

Approximate Reducts and Association Rules
– Correspondence and Complexity Results –

Hung Son Nguyen, Dominik Ślęzak

Institute of Mathematics
Warsaw University
Banacha 2, 02-097 Warsaw, Poland
email: son@mimuw.edu.pl, slezak@alfa.mimuw.edu.pl

Abstract. We consider approximate versions of fundamental notions of theories of rough sets and association rules. We analyze the complexity of searching for α-reducts, understood as subsets discerning "α-almost" objects from different decision classes, in decision tables. We present how optimal approximate association rules can be derived from data by using heuristics for searching for minimal α-reducts. NP-hardness of the problem of finding optimal approximate association rules is shown as well. It makes the results enabling the usage of rough sets algorithms to the search of association rules extremely important in view of applications.

1 Introduction

Theory of rough sets ([5]) provides efficient tools for dealing with fundamental data mining challenges, like data representation and classification, or knowledge description (see e.g. [2], [3], [4], [8]). Basing on the notions of information system and decision table, the language of reducts and rules was proposed for expressing dependencies between considered features, in view of gathered information.

Given a distinguished feature, called decision, the notion of decision reduct is constructed over, so called, discernibility matrix ([7]), where information about all pairs of objects with different decision values is stored. A reduct is any minimal (in sense of inclusion) subset of non-decision features (conditions) which discern all such pairs, necessary to be considered, e.g., with respect to proper decision classification of new cases.

In real applications, basing on such deterministic reducts, understood as above, is often too restrictive with respect to discerning all necessary pairs. Indeed, deterministic dependencies may require too many conditions to be involved to. Several approaches to uncertainty representation of decision rules and reducts were proposed to weaken the above conditions (see e.g. [6], [8], [9]). In the language of reducts and their discernibility characteristics, we can say that such uncertainty or imprecision can be connected with a ratio of pairs from different decision classes which remain not discerned by such an approximate reduct.

Applications of rough sets theory to the generation of rules, for classification of new cases or representation of data information, are usually restricted to searching for decision rules with a fixed decision feature related to a rule's consequence. Recently, however, more and more attention is paid on, so called, associative mechanism of rules' generation, where all attributes can occur as involved to conditions or consequences of particular rules (compare with, e.g., [1], [10]). Relationship between techniques of searching for optimal association rules and rough sets optimization tasks, like, e.g., the templates generation, were studied in [3]. In this paper we would like to focus on approximate association rules, analyzing both complexity of related search tasks and their correspondence to approximate reducts.

The reader may pay attention on similarities between construction of proofs of complexity results concerning approximate reducts and association rules. We believe that presented techniques can be regarded as even more universal for simple and intuitive characteristics of related optimization tasks. What even more important, however, is the correspondence between the optimization problems concerning the above mentioned notions - Although the problems of finding both minimal approximate reducts and all approximate reducts are NP-hard, the existence of very efficient and fast heuristics for solving them (compare, e.g., with [4]) makes such a correspondence very important tool for development of appropriate methods of finding optimal approximate association rules.

The paper is organized as follows: In Section 2 we introduce basic notions of rough sets theory and consider the complexity of searching for minimal approximate (in sense of discernibility) reducts in decision tables. In Section 3 we introduce the notion of association rule as strongly related to the notion of template, known from rough sets theory. Similarly as in case of approximate reducts, we show the NP-hardness of the problem of finding optimal approximate (in sense of a confidence threshold) association rule corresponding to a given template. In Section 4 we show how optimal approximate association rules can be searched for as minimal approximate reducts, by using an appropriate decision table representation. In Section 5 we conclude the paper with pointing the directions of further research.

2 Approximate reducts

An *information system* is a pair $\mathbb{S} = (U, A)$, where U is a non-empty, finite set called the *universe* and A is a non-empty, finite set of *attributes*. Each $a \in A$ corresponds to function $a : U \to V_a$, where V_a is called the *value set* of a. Elements of U are called *situations*, *objects* or *rows*, interpreted as, e.g., cases, states, patients, observations. We also consider a special case of information system: *decision table* $\mathbb{S} = (U, A \cup \{d\})$, where $d \notin A$ is a distinguished attribute called *decision* and the elements of A are called *conditional attributes* (conditions).

In a given information system, in general, we are not able to distinguish all pairs of situations (objects) by using attributes of the system. Namely, different situations can have the same values on considered attributes. Hence, any set of attributes divides the universe U onto some classes which establish a partition of U ([5]). With any subset of attributes $B \subseteq A$ we associate a bi-

nary relation $ind(B)$, called a *B-indiscernibility relation*, which is defined by
$ind(B) = \{(u, u') \in U \times U : \text{for every } a \in B,\ a(u) = a(u')\}$.

Let $\mathbb{S} = (U, A)$ be an information system. Assume that $U = \{u_1, ..., u_N\}$,
and $A = \{a_1, ..., a_n\}$. By $\mathbb{M}(\mathbb{S})$ we denote an $N \times N$ matrix $(c_{i,j})$, called the
discernibility matrix of $\mathbb{S}$, such that $c_{i,j} = \{a \in A : a(u_i) \neq a(u_j)\}$ for $i, j =
1, ..., N$. Discernibility matrices are useful for deriving possibly small subsets
of attributes, still keeping the knowledge encoded within a system. Given $\mathbb{S} =
(U, A)$, we call as a *reduct* each subset $B \subseteq A$ being minimal in sense of inclusion,
intersecting with each non-empty $c_{i,j}$, i.e., such that

$$\forall_{i,j}(c_{i,j} \neq \emptyset) \Rightarrow (B \cap c_{i,j} \neq \emptyset)$$

The above condition states the reducts as minimal subsets of attributes which
discern all pairs of objects possible to be discerned within an information system. In a special case, for decision table $\mathbb{S} = (U, A \cup \{d\})$, we may weaken this
condition, because not all pairs are *necessary* to be discerned, to keep knowledge concerning decision d - we modify elements of corresponding discernibility
matrix with respect to formula

$$[c_{i,j} = \{a \in A : a(u_i) \neq a(u_j)\} \Leftrightarrow d(u_i) \neq d(u_j)] \wedge [c_{i,j} = \emptyset \Leftrightarrow d(u_i) = d(u_j)]$$

In this paper we are going to focus on decision tables, so, from now, we will
understand reducts as corresponding to such modified matrices.

Extracting reducts from data is a crucial task in view of tending to possibly
clear description of decision in terms of conditional attributes. In view of the
above formulas, such a description can be regarded as deterministic, relatively
to gathered information (one can show that the above definition of reduct in
a decision table is equivalent to that based on *generalized decision functions*,
considered, e.g., in [8]). Still, according to real life applications, we often cannot
afford to handle subsets of conditions defining d even in such a relative way.
Thus, in some applications (see e.g. [6]), we prefer to use α-approximations of
reducts, where $\alpha \in (0, 1]$ is a real parameter.

We consider two versions of such approximations. The first of them is related
to the task of discerning almost all pairs of objects with different decision classes,
regardless of information provided by conditional attributes: The set of attributes
$B \subseteq A$ is called an α-reduct iff it is minimal in sense of inclusion, intersecting at
least $\alpha \cdot 100\%$ of pairs necessary to be discerned with respect to decision, what
means that

$$\frac{|\{c_{i,j} : B \cap c_{i,j} \neq \emptyset\}|}{|\{(u_i, u_j) : d(u_i) \neq d(u_j)\}|} \geq \alpha$$

Appropriate tuning of parameter α in the above inequality provides representation of inconsistent information, alternative to approaches based on generalized
or other decision functions, proposed, e.g., in [8] or [9]. What similar, however, is
the complexity characteristics, well known for $\alpha = 1$, of the following problem:

Theorem 1. *For a given $\alpha \in (0, 1)$, the problem of finding the minimal (in sense
of cardinality) α-reduct is NP-hard with respect to the number of conditional
attributes.*

Because of the lack of space, let us just mention that the proof of Theorem 1 can be obtained by deriving (in polynomial time) the problem of minimal graph covering (i.e. the problem of finding minimal set of vertices which cover all edges in a given graph) to that considered above. Let us illustrate this derivation with the following example:

Example 1

Let us consider the Minimal α-Reduct Problem for $c = 0.8$. We illustrate the proof of Theorem 1 by the graph $\mathbf{G} = (V, E)$ with five vertices $V = \{v_1, v_2, v_3, v_4, v_5\}$ and six edges $E = \{e_1, e_2, e_3, e_4, e_5, e_6\}$. First we compute $k = \left\lfloor \frac{\alpha}{1-\alpha} \right\rfloor = 4$. Hence, decision table $\mathbf{S}(G)$ consists of five conditional attributes $\{a_{v_1}, a_{v_2}, a_{v_3}, a_{v_4}, a_{v_5}\}$, decision a^* and $(4+1)+6 = 11$ objects $\{x_1, x_2, x_3, x_4, x^*, u_{e_1}, u_{e_2}, u_{e_3}, u_{e_4}, u_{e_5}, u_{e_6}\}$. Decision table $\mathbf{S}(G)$ constructed from the graph G is presented below:

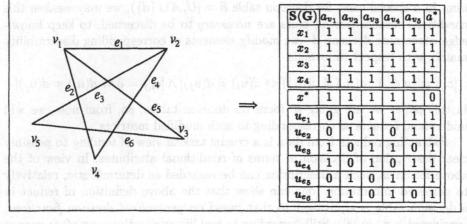

$\mathbf{S}(G)$	a_{v_1}	a_{v_2}	a_{v_3}	a_{v_4}	a_{v_5}	a^*
x_1	1	1	1	1	1	1
x_2	1	1	1	1	1	1
x_3	1	1	1	1	1	1
x_4	1	1	1	1	1	1
x^*	1	1	1	1	1	0
u_{e_1}	0	0	1	1	1	1
u_{e_2}	0	1	1	0	1	1
u_{e_3}	1	0	1	1	0	1
u_{e_4}	1	0	1	0	1	1
u_{e_5}	0	1	0	1	1	1
u_{e_6}	1	1	0	1	0	1

Analogous result can be obtained for the Minimal Relative α-Reduct Problem, where relative α-reducts are understood as subsets $B \subseteq A$ being minimal in sense of inclusion, satisfying inequality

$$\frac{|\{c_{i,j} : B \cap c_{i,j} \neq \emptyset\}|}{|\{c_{i,j} : c_{i,j} \neq \emptyset\}|} \geq \alpha$$

In such a case, since procedure illustrated by **Example 1** is not appropriate any more, we have to use more sophisticated representation of a graph by a decision table. Instead of the formal proof, again, let us just modify the previous illustration. Appropriate modification can be seen in **Example 2** below.

Although the above results may seem to reduce the possibilities of dealing with rough set tools in an effective way, a number of random algorithms for finding approximately optimal solutions to mentioned problems can be proposed. The power of heuristics possible to be implemented by using rough set algorithmic techniques (see e.g. [4]) is worth remembering because the majority of interesting data mining problems is known to be NP-hard as well. Thus, the analysis of correspondence between such problems and the search for (approximate) reducts can turn out to be very fruitful in view of many applications.

Example 2

In case of the Minimal Relative α-Reduct Problem, $\alpha = 0.8$, graph $G = (V, E)$ from the above Example can be translated to decision table $S'(G)$, where, comparing to $S(G)$, we add four new conditions a'_1, a'_2, a'_3, a'_4. One can show that from a given minimal relative α-reduct in $S'(G)$ we can derive (in a polynomial time with respect to the number of conditions) minimal graph covering for G.

$S'(G)$	a_{v_1}	a_{v_2}	a_{v_3}	a_{v_4}	a_{v_5}	a'_1	a'_2	a'_3	a'_4	a^*
x_1	1	1	1	1	1	1	1	1	0	1
x_2	1	1	1	1	1	1	1	0	1	1
x_3	1	1	1	1	1	1	0	1	1	1
x_4	1	1	1	1	1	0	1	1	1	1
x^*	1	1	1	1	1	1	1	1	1	0
u_{e_1}	0	0	1	1	1	1	1	1	1	1
u_{e_2}	0	1	1	0	1	1	1	1	1	1
u_{e_3}	1	0	1	1	0	1	1	1	1	1
u_{e_4}	1	0	1	0	1	1	1	1	1	1
u_{e_5}	0	1	0	1	1	1	1	1	1	1
u_{e_6}	1	1	0	1	0	1	1	1	1	1

3 Approximate association rules

Association rules and their generation can be defined in many ways (see [1]). As we mentioned in Introduction, we are going to introduce them as related to so called *templates*.

Given an information table $S = (U, A)$, by *descriptors* we mean the terms of the form $(a = v)$, where $a \in A$ is an attribute and $v \in V_a$ is a value in the domain of a (see [4]). The notion of descriptor can be generalized by using terms of the form $(a \in S)$, where $S \subseteq V_a$ is a set of values. By a template we mean the conjunction of descriptors, i.e. $T = D_1 \wedge D_2 \wedge ... \wedge D_m$, where $D_1, ... D_m$ are either simple or generalized descriptors. We denote by $length(T)$ the number of descriptors in T.

An object $u \in U$ satisfies template $T = (a_{i_1} = v_1) \wedge ... \wedge (a_{i_m} = v_m)$ if and only if $\forall_j a_{i_j}(u) = v_j$. Hence, template T describes the set of objects having the common property: "*the values of attributes $a_{j_1}, ..., a_{j_m}$ on these objects are equal to $v_1, ..., v_m$, respectively*". The *support* of T is defined by $support(T) = |\{u \in U : u$ satisfies $T\}|$.

Long templates with large support are preferred in many Data Mining tasks. Regarding on a concrete optimization function, problems of finding optimal large templates are known as being NP-hard with respect to the number of attributes involved into descriptors, or remain open problems (see e.g. [3]). Nevertheless, the large templates can be found quite efficiently by *Apriori* and *AprioriTid* algorithms (see [1], [10]). A number of other methods for large template generation has been proposed e.g. in [4].

According to the presented notation, association rules can be defined as implications of the form $(P \Rightarrow Q)$, where P and Q are different simple templates. Thus, they take the form

$$(a_{i_1} = v_{i_1}) \wedge ... \wedge (a_{i_k} = v_{i_k}) \Rightarrow (a_{j_1} = v_{j_1}) \wedge ... \wedge (a_{j_l} = v_{j_l}) \qquad (1)$$

Usually, for a given information system S, the quality of association rule $\mathcal{R} = P \Rightarrow Q$ is evaluated by two measures called *support and confidence* with respect to S. The support of rule $\mathcal{R}$ is defined by the number of objects from S satisfying condition $(\mathbf{P} \wedge \mathbf{Q})$, i.e.

$$support(\mathcal{R}) = support(\mathbf{P} \wedge \mathbf{Q})$$

The second measure, confidence of $\mathcal{R}$, is the ratio of support of $(\mathbf{P} \wedge \mathbf{Q})$ and support of $\mathbf{P}$, i.e.

$$confidence(\mathcal{R}) = \frac{support(\mathbf{P} \wedge \mathbf{Q})}{support(\mathbf{P})}$$

The following problem has been investigated by many authors (see e.g. [1], [10]):

> FOR A GIVEN INFORMATION TABLE S, AN INTEGER s, AND A REAL NUMBER $c \in (0,1)$, FIND AS MANY AS POSSIBLE ASSOCIATION RULES $\mathcal{R} = (\mathbf{P} \Rightarrow \mathbf{Q})$ SUCH THAT $support(\mathcal{R}) \geq s$ AND $confidence(\mathcal{R}) \geq c$.

All existing association rule generation methods consist of two main steps:

1. Generate as many as possible templates $\mathbf{T} = D_1 \wedge D_2 ... \wedge D_k$ such that $support(\mathbf{T}) \geq s$ and $support(\mathbf{T} \wedge D) < s$ for any descriptor D (i.e. maximal templates among these which are supported by not less than s objects).
2. For any template $\mathbf{T}$, search for decomposition $\mathbf{T} = \mathbf{P} \wedge \mathbf{Q}$ such that:
   (a) $support(\mathbf{P}) \leq \frac{support(\mathbf{T})}{c}$,
   (b) $\mathbf{P}$ is the smallest template satisfying (a).

In this paper we show that the second above step can be solved using rough set methods. Let us assume that template $\mathbf{T} = D_1 \wedge D_2 \wedge ... \wedge D_m$, which is supported by at least s objects, has been found. For a given confidence threshold $c \in (0,1]$ decomposition $\mathbf{T} = \mathbf{P} \wedge \mathbf{Q}$ is called c-irreducible if $confidence(\mathbf{P} \Rightarrow \mathbf{Q}) \geq c$ and for any decomposition $\mathbf{T} = \mathbf{P}' \wedge \mathbf{Q}'$ such that $\mathbf{P}'$ is a sub-template of $\mathbf{P}$, $confidence(\mathbf{P}' \Rightarrow \mathbf{Q}') < c$.

We are especially interested in approximate association rules, corresponding to $c < 1$. The following gives analogy of this case to well known result concerning the search for deterministic association rules.

Theorem 2. *For a fixed $c \in (0,1)$, the problem of searching for the shortest association rule from the template $\mathbf{T}$ for a given table S with confidence limited by c (Optimal c-Association Rule Problem) is NP-hard, with respect to the length of $\mathbf{T}$.*

The proof of this theorem is similar to that of Theorem 1. We illustrate it by example:

> **Example 3**
> Let us consider the Optimal c-Association Rules Problem for $c = 0.8$. We illustrate the proof of Theorem 2 analogously to the illustration from **Example 1**, related to the proof of Theorem 1. For graph $\mathbf{G} = (V, E)$, we compute $k = \left\lceil \frac{c}{1-c} \right\rceil = 4$ like previously. The only difference is that instead of decision table $\mathbf{S}(\mathbf{G})$ we begin to consider information system $\mathbf{S}''(\mathbf{G})$, where a^* is a non-decision attribute, like the others.

4 Application of approximate reducts to the search of approximate association rules

In this section we are going to show that the problem of searching for optimal approximate association rules from the given template is equivalent to the problem of searching for minimal α-reducts in an appropriately modified decision table. We construct new decision table $\mathbb{S}|_{\mathbf{T}} = (U, A|_{\mathbf{T}} \cup d)$ from original information table $\mathbb{S}$ and template $\mathbf{T}$ as follows:

1. $A|_{\mathbf{T}} = \{a_{D_1}, a_{D_2}, ..., a_{D_m}\}$ is the set of attributes corresponding to the descriptors of $\mathbf{T}$, such that $a_{D_i}(u) = \begin{cases} 1 \text{ if object } u \text{ satisfies } D_i, \\ 0 \text{ otherwise.} \end{cases}$

2. Decision attribute d determines whether a given object satisfies template $\mathbf{T}$, i.e. $d(u) = \begin{cases} 1 \text{ if object } u \text{ satisfies } \mathbf{T}, \\ 0 \text{ otherwise.} \end{cases}$

The following theorem describes the relationship between the optimal association rule and the minimal reduct search problems.

Theorem 3. *For information table* $\mathbb{S} = (U, A)$, *template* $\mathbf{T}$, *set of descriptors* $\mathbf{P}$ *and parameter* $c \in (0, 1]$, *implication* $\bigwedge_{D_i \in \mathbf{P}} D_i \Rightarrow \bigwedge_{D_j \notin \mathbf{P}} D_j$ *is a c-irreducible association rule from* $\mathbf{T}$ *iff* $\mathbf{P}$ *is an α-reduct in* $\mathbb{S}|_{\mathbf{T}}$, *for* $\alpha = 1 - (\frac{1}{c} - 1)/(\frac{N}{s} - 1)$, *where* N *is the total number of objects in* U *and* $s = support(\mathbf{T})$. *In particular, the above implication is a deterministic association rule iff* $\mathbf{P}$ *is a reduct in* $\mathbb{S}|_{\mathbf{T}}$.

The following example illustrates the main idea of the method based on the above characteristics. Let us consider the following information table $\mathbb{S}$ with 18 objects and 9 attributes:

S	a_1	a_2	a_3	a_4	a_5	a_6	a_7	a_8	a_9
u_1	0	1	1	1	80	2	2	2	3
u_2	0	1	2	1	81	0	aa	1	aa
u_3	0	2	2	1	82	0	aa	1	aa
u_4	0	1	2	1	80	0	aa	1	aa
u_5	1	1	2	2	81	1	aa	1	aa
u_6	0	2	1	2	81	1	aa	1	aa
u_7	1	2	1	2	83	1	aa	1	aa
u_8	0	2	2	1	81	0	aa	1	aa
u_9	0	1	2	1	82	0	aa	1	aa
u_{10}	0	3	2	1	84	0	aa	1	aa
u_{11}	1	1	2	2	80	0	aa	2	aa
u_{12}	0	2	3	2	82	0	aa	2	aa
u_{13}	0	2	2	1	81	0	aa	1	aa
u_{14}	0	3	2	2	81	2	aa	2	aa
u_{15}	0	4	2	1	82	0	aa	1	aa
u_{16}	0	3	2	1	83	0	aa	1	aa
u_{17}	0	1	2	1	84	0	aa	1	aa
u_{18}	1	2	2	1	82	0	aa	2	aa

$\mathbb{S}|_{\mathbf{T}}$	D_1 $a_1 = 0$	D_2 $a_3 = 2$	D_3 $a_4 = 1$	D_4 $a_6 = 0$	D_5 $a_8 = 1$	d
u_1	1	0	1	0	0	0
u_2	1	1	1	1	1	1
u_3	1	1	1	1	1	1
u_4	1	1	1	1	1	1
u_5	0	1	0	0	1	0
u_6	1	0	0	0	1	0
u_7	0	0	0	0	1	0
u_8	1	1	1	1	1	1
u_9	1	1	1	1	1	1
u_{10}	1	1	1	1	1	1
u_{11}	0	1	0	1	0	0
u_{12}	1	0	0	1	0	0
u_{13}	1	1	1	1	1	1
u_{14}	1	1	0	0	0	0
u_{15}	1	1	1	1	1	1
u_{16}	1	1	1	1	1	1
u_{17}	1	1	1	1	1	1
u_{18}	0	1	1	1	0	0

Assume that template

$$\mathbf{T} = (a_1 = 0) \wedge (a_3 = 2) \wedge (a_4 = 1) \wedge (a_6 = 0) \wedge (a_8 = 1)$$

has been extracted from information table $\mathbb{S}$. One can see that $support(\mathbf{T}) = 10$ and $length(\mathbf{T}) = 5$. Decision table $\mathbb{S}|_{\mathbf{T}}$ is presented below:

$M(\mathbb{S}\|_{\mathbf{T}})$	u_2, u_3, u_4, u_8, u_9 $u_{10}, u_{13}, u_{15}, u_{16}, u_{17}$		
u_1	$D_2 \vee D_4 \vee D_5$		$D_3 \wedge D_5 \Rightarrow D_1 \wedge D_2 \wedge D_4$
u_5	$D_1 \vee D_3 \vee D_4$		$D_4 \wedge D_5 \Rightarrow D_1 \wedge D_2 \wedge D_3$
u_6	$D_2 \vee D_3 \vee D_4$	$= 100\% \Longrightarrow$	$D_1 \wedge D_2 \wedge D_3 \Rightarrow D_4 \wedge D_5$
u_7	$D_1 \vee D_2 \vee D_3 \vee D_4$		$D_1 \wedge D_2 \wedge D_4 \Rightarrow D_3 \wedge D_5$
u_{11}	$D_1 \vee D_3 \vee D_5$		$D_1 \wedge D_2 \wedge D_5 \Rightarrow D_3 \wedge D_4$
u_{12}	$D_2 \vee D_3 \vee D_5$		$D_1 \wedge D_3 \wedge D_4 \Rightarrow D_2 \wedge D_5$
u_{14}	$D_3 \vee D_4 \vee D_5$		$D_1 \wedge D_2 \Rightarrow D_3 \wedge D_4 \wedge D_5$
u_{18}	$D_1 \vee D_5$		$D_1 \wedge D_3 \Rightarrow D_3 \wedge D_4 \wedge D_5$

Matrix rules, $= 100\%$ group:
$D_3 \wedge D_5 \Rightarrow D_1 \wedge D_2 \wedge D_4$
$D_4 \wedge D_5 \Rightarrow D_1 \wedge D_2 \wedge D_3$
$D_1 \wedge D_2 \wedge D_3 \Rightarrow D_4 \wedge D_5$
$D_1 \wedge D_2 \wedge D_4 \Rightarrow D_3 \wedge D_5$
$D_1 \wedge D_2 \wedge D_5 \Rightarrow D_3 \wedge D_4$
$D_1 \wedge D_3 \wedge D_4 \Rightarrow D_2 \wedge D_5$

$= 90\%$ group:
$D_1 \wedge D_2 \Rightarrow D_3 \wedge D_4 \wedge D_5$
$D_1 \wedge D_3 \Rightarrow D_3 \wedge D_4 \wedge D_5$
$D_1 \wedge D_4 \Rightarrow D_2 \wedge D_3 \wedge D_5$
$D_1 \wedge D_5 \Rightarrow D_2 \wedge D_3 \wedge D_4$
$D_2 \wedge D_3 \Rightarrow D_1 \wedge D_4 \wedge D_5$
$D_2 \wedge D_5 \Rightarrow D_1 \wedge D_3 \wedge D_4$
$D_3 \wedge D_4 \Rightarrow D_1 \wedge D_2 \wedge D_5$

The discernibility function f corresponding to matrix $M(\mathbb{S}|_{\mathbf{T}})$ is the following:

$$f = (D_2 \vee D_4 \vee D_5) \wedge (D_1 \vee D_3 \vee D_4) \wedge (D_2 \vee D_3 \vee D_4) \wedge (D_1 \vee D_2 \vee D_3 \vee D_4)$$
$$\wedge (D_1 \vee D_3 \vee D_5) \wedge (D_2 \vee D_3 \vee D_5) \wedge (D_3 \vee D_4 \vee D_5) \wedge (D_1 \vee D_5)$$

After simplification we obtain six reducts: $f = (D_3 \wedge D_5) \vee (D_4 \wedge D_5) \vee (D_1 \wedge D_2 \wedge D_3) \vee (D_1 \wedge D_2 \wedge D_4) \vee (D_1 \wedge D_2 \wedge D_5) \vee (D_1 \wedge D_3 \wedge D_4)$ for decision table $\mathbb{S}|_{\mathbf{T}}$. Thus, we have found from $\mathbf{T}$ six deterministic association rules with full confidence.

For $c = 0.9$, we would like to find α-reducts for decision table $\mathbb{S}|_{\mathbf{T}}$, where $\alpha = 1 - \frac{\frac{1}{c} - 1}{\frac{N}{s} - 1} = 0.86$. Hence we search for a set of descriptors covering at least $\lceil (n - s)(\alpha) \rceil = \lceil 8 \cdot 0.86 \rceil = 7$ elements of discernibility matrix $M(\mathbb{S}|_{\mathbf{T}})$. One can see that each of the following sets $\{D_1, D_2\}$, $\{D_1, D_3\}$, $\{D_1, D_4\}$, $\{D_1, D_5\}$, $\{D_2, D_3\}$, $\{D_2, D_5\}$, $\{D_3, D_4\}$ intersects with exactly 7 members of discernibility matrix $M(\mathbb{S}|_{\mathbf{T}})$. In the above table we present all association rules achieved from these sets.

The problems of finding both minimal α-reducts and all α-reducts are NP-hard, so we usually cannot afford for such exhaustive computations like these presented above. However, one should remember that the above is just an illustrative example and in real life applications we can use very efficient and fast heuristics for solving α-reduct problems (see e.g. [4] for further references). In particular, it makes presented derivation very important tool for development of appropriate methods of finding optimal approximate association rules.

5 Conclusions

Searching for minimal α-reducts is well known problem in Rough Sets theory. A great effort has been involved to solve these problems. One can find numerous applications of α-reducts in the knowledge discovery domain. In this paper we have shown, that the problem of searching for the shortest α-reducts is NP-hard. We also investigated the application of α-reducts to association rule generation. Still, further development of the language of association rules is needed for applications. In the next papers we are going to present such a development, together with new, rough set based algorithms for the association rule generation.

Acknowledgements Supported by ESPRIT project 20288 CRIT-2 and KBN Research Grant 8T11C02412.

References

1. Agrawal R., Mannila H., Srikant R., Toivonen H., Verkamo A.I., 1996. Fast discovery of assocation rules. In V.M. Fayad, G.Piatetsky Shapiro, P. Smyth, R. Uthurusamy (eds): *Advanced in Knowledge Discovery and Data Mining*, AAAI/MIT Press, pp.307-328.
2. J. Bazan. A comparison of dynamic non-dynamic rough set methods for extracting laws from decision tables. In: L. Polkowski and A. Skowron (Eds.), *Rough Sets in Knowledge Discovery 1: Methodology and Applications*, Physica-Verlag, Heidelberg, 1998, 321–365.
3. H.S. Nguyen and S.H. Nguyen. Pattern extraction from data, *Fundamenta Informaticae* 34 (1998) 129–144.
4. Nguyen S. Hoa, A. Skowron, P. Synak. Discovery of data pattern with applications to decomposition and classification problems. In L. Polkowski, A. Skowron (eds.): Rough Sets in Knowledge Discovery 2. Physica-Verlag, Heidelberg 1998, pp. 55–97.
5. Z. Pawlak. *Rough Sets - Theoretical Aspects of Reasoning about Data*. Kluwer Academic Publishers, Dordrecht, 1991.
6. Skowron A. Synthesis of adaptive decision systems from experimental data. In A. Aamodt, J. Komorowski (eds), Proc. of the Fifth Scandinavian Conference on Artificial Intelligence (SCAI'95), May 1995, Trondheim, Norway, IOS Press, Amsterdam, 220–238.
7. A. Skowron and C. Rauszer. The discernibility matrices and functions in information systems, in: R. Słowiński (Ed.), *Intelligent decision support: Handbook of applications and advances of the rough sets theory*, Kluwer Academic Publishers, Dordrecht, 1992, 331-362.
8. D. Ślęzak. Decision information functions for inconsistent decision tables analysis. Accepted to International Conference EUFIT'99.
9. D. Ślęzak. Reasoning in decision tables with frequency based implicants. In preparation.
10. Mohammed Javeed Zaki, Srinivasan Parthasarathy, Mitsunori Ogihara, Wei Li, 1997. New Parallel Algorithms for Fast Discovery of Association Rules. In Data Mining and Knowledge Discovery : An International Journal, special issue on Scalable High-Performance Computing for KDD, Vol. 1, No. 4, December 1997, pp 343-373.

Handling Missing Values in Rough Set Analysis of Multi-attribute and Multi-criteria Decision Problems

Salvatore Greco[1], Benedetto Matarazzo[1], Roman Slowinski[2]

[1]Faculty of Economics, University of Catania, Corso Italia, 55,
95129 Catania, Italy, {salgreco, matarazz}@mbox.unict.it
[2]Institute of Computing Science, Poznan University of Technology,
60965 Poznan, Poland, slowinsk@sol.put.poznan.pl

Abstract. Rough sets proved to be very useful for analysis of decision problems concerning objects described in a data table by a set of condition attributes and by a set of decision attributes. In practical applications, however, the data table is often not complete because some data are missing. To deal with this case, we propose an extension of the rough set methodology. The adaptation concerns both the classical rough set approach based on indiscernibility relations and the new rough set approach based on dominance relations. While the first approach deals with multi-attribute classification problems, the second approach deals with multi-criteria sorting problems. The adapted relations of indiscernibility or dominance between two objects are considered as directional statements where a subject is compared to a referent object having no missing values. The two rough set approaches handling the missing values boil down to the original approaches when the data table is complete. The rules induced from the rough approximations are robust in a sense that each rule is supported by at least one object with no missing values on condition attributes or criteria used by the rule.

1 Introduction

The rough sets philosophy introduced by Pawlak [5, 6] is based on the assumption that with every object of the universe there is associated a certain amount of information (data, knowledge), expressed by means of some attributes used for object description. It proved to be an excellent tool for analysis of decision problems [7, 10] where the set of attributes is divided into disjoint sets of condition attributes and decision attributes describing objects in a data table.

The key idea of rough sets is approximation of knowledge expressed by decision attributes using knowledge expressed by condition attributes. The rough set approach answers several questions related to the approximation: (a) is the information contained in the data table consistent? (b) what are the non-redundant subsets of condition attributes ensuring the same quality of approximation as the whole set of condition attributes? (c) what are the condition attributes which cannot be eliminated from the approximation without decreasing the quality of approximation? (d) what minimal "*if ..., then ...*" decision rules can be induced from the approximations?

The original rough set approach is not able, however, to discover and process inconsistencies coming from consideration of *criteria*, i.e. condition attributes with preference-ordered scales. For this reason, Greco, Matarazzo and Slowinski [1, 2] have proposed a new rough set approach that is able to deal with inconsistencies typical to consideration of criteria and preference-ordered decision classes. This innovation is mainly based on substitution of the indiscernibility relation by a dominance relation in the rough approximation of decision classes. An important consequence of this fact is a possibility of inferring from exemplary decisions the preference model in terms of decision rules being logical statements of the type *"if...,* *then..."*. The separation of certain and doubtful knowledge about the decision maker's preferences is done by distinction of different kinds of decision rules, depending whether they are induced from lower approximations of decision classes or from the boundaries of these classes composed of inconsistent examples that do not observe the dominance principle. Such preference model is more general than the classical functional or relational model in multi-criteria decision making and it is more understandable for the users because of its natural syntax.

Both the classical rough set approach based on the use of indiscernibility relations and the new rough set approach based on the use of dominance relations suffer, however, from another deficiency: they require the data table to be complete, i.e. without *missing values* on condition attributes or criteria describing the objects.

To deal with the case of missing values in the data table, we propose an adaptation of the rough set methodology. The adaptation concerns both the classical rough set approach and the dominance-based rough set approach. While the first approach deals with *multi-attribute classification*, the second approach deals with *multi-criteria sorting*. Multi-attribute classification concerns an assignment of a set of objects to a set of pre-defined classes. The objects are described by a set of (regular) attributes and the classes are not necessarily ordered. Multi-criteria sorting concerns a set of objects evaluated by criteria, i.e. attributes with preference-ordered scales. In this problem, the classes are also preference-ordered.

The adapted relations of indiscernibility or dominance between two objects are considered as directional statements where a subject is compared to a referent object. We require that the referent object has no missing values. The two adapted rough set approaches maintain all good characteristics of their original approaches. They also boil down to the original approaches when there are no missing values. The rules induced from the rough approximations defined according to the adapted relations verify some suitable properties: they are either exact or approximate, depending whether they are supported by consistent objects or not, and they are robust in a sense that each rule is supported by at least one object with no missing value on the condition attributes or criteria represented in the rule.

The paper is organized in the following way. In section 2, we present the extended rough sets methodology handling the missing values. It is composed of four sub-sections – first two are devoted to adaptation of the classical rough set approach based on the use of indiscernibility relations; the other two undertake the adaptation of the new rough set approach based on the use of dominance relations. In order to illustrate the concepts introduced in section 2, we present an illustrative example in section 3. Section 4 groups conclusions.

2 Rough approximations defined on data tables with missing values

For algorithmic reasons, the data set about objects is represented in the form of a data table. The rows of the table are labelled by *objects*, whereas columns are labelled by *attributes* and entries of the table are *attribute-values*, called *descriptors*.

Formally, by a *data table* we understand the 4-tuple S=<U,Q,V,f>, where U is a finite set of objects, Q is a finite set of attributes, $V = \bigcup_{q \in Q} V_q$ and V_q is a domain of the attribute q, and f:U×Q→V is a total function such that $f(x,q) \in V_q \cup \{*\}$ for every q∈Q, x∈U, called an *information function*. The symbol "*" indicates that the value of an attribute for a given object is unknown (*missing*).

If set Q is divided into set C of *condition attributes* and set D of *decision attributes*, then such a data table is called *decision table*. If the domain (scale) of a condition attribute is ordered according to a decreasing or increasing preference, then it is a *criterion*. For condition attribute q∈C being a criterion, S_q is an *outranking relation* [8] on U such that xS_qy means "x is at least as good as y with respect to criterion q". We suppose that S_q is a total preorder, i.e. a strongly complete and transitive binary relation, defined on U on the basis of evaluations $f(\cdot,q)$. The domains of "regular" condition attributes are not ordered.

We assume that the set D of decision attributes is a singleton {d}. Decision attribute d makes a partition of U into a finite number of classes Cl={Cl_t, t∈T}, T={1,...,n}, such that each x∈U belongs to one and only one $Cl_t \in Cl$. The domain of d can be preference-ordered or not. In the former case, we suppose that the classes are ordered such that the higher is the class number the better is the class, i.e. for all r,s∈T, such that r>s, the objects from Cl_r are preferred (strictly or weakly) to the objects from Cl_s. More formally, if S is a *comprehensive outranking relation* on U, i.e. if for all x,y∈U, xSy means "x is at least as good as y", we suppose: [x∈Cl_r, y∈Cl_s, r>s] ⇒ [xSy and *not* ySx]. These assumptions are typical for consideration of a *multi-criteria sorting problem*.

In the following sub-sections of this section we are considering separately the multi-attribute classification and the multi-criteria sorting with respect to the problem of missing values. The first idea of dealing with missing values in the rough set approach to the multi-attribute classification problem in the way described below has been given in [3].

2.1 Multi-attribute classification problem with missing values

For any two objects x,y∈U, we are considering a directional comparison of y to x; object y is called subject and object x, referent. We say that subject y is *indiscernible* with referent x with respect to condition attributes P⊆C (denotation yI_Px) if for every q∈P the following conditions are met:

- f(x,q)≠* ,
- f(x,q)=f(y,q) or f(y,q)=*.

The above means that the referent object considered for indiscernibility with respect to P should have no missing values on attributes from set P.

The binary relation I_P is not necessarily reflexive because for some $x \in U$ there may exist $q \in P$ for which $f(x,q)=*$ and, therefore, we cannot state xI_Px. Moreover, I_P is also not necessarily symmetric because the statement yI_Px cannot be inverted if there exist $q \in P$ for which $f(y,q)=*$. However, I_P is transitive because for each $x,y,z \in U$, the statements xI_Py and yI_Pz imply xI_Pz. This is justified by the observations that object z can substitute object y in the statement xI_Py because yI_Pz and both y and z, as referent objects, have no missing values.

For each $P \subseteq C$ let us define a set of objects having no missing values on $q \in P$:

$$U_P=\{x \in U: f(x,q) \neq * \text{ for each } q \in P\}.$$

It is easy to see that the restriction of I_P to U_P (in other words, the binary relation $I_P \cap U_P \times U_P$ defined on U_P) is reflexive, symmetric and transitive, i.e. it is an equivalence binary relation.

For each $x \in U$ and for each $P \subseteq Q$ let $I_P(x)=\{y \in U: yI_Px\}$ denote the class of objects indiscernible with x. Given $X \subseteq U$ and $P \subseteq Q$, we define lower approximation of X with respect to P as

$$\underline{I}_P(X)=\{x \in U_P: I_P(x) \subseteq X\}. \tag{1}$$

Analogously, we define the upper approximation of X with respect to P as

$$\bar{I}_P(X)=\{x \in U_P: I_P(x) \cap X \neq \varnothing\}. \tag{2}$$

Let us observe that if $x \notin U_P$ then $I_P(x)=\varnothing$ and, therefore, we can also write

$$\bar{I}_P(X)=\{x \in U: I_P(x) \cap X \neq \varnothing\}.$$

Let $X_P=X \cap U_P$. For each $X \in U$ and for each $P \subseteq C$: $\underline{I}_P(X) \subseteq X_P \subseteq \bar{I}_P(X)$ (*rough inclusion*) and $\underline{I}_P(X)=U_P - \bar{I}_P(U-X)$ (*complementarity*).

The P-*boundary* of X in S, denoted by $Bn_P(X)$, is equal to $Bn_P(X)= \bar{I}_P(X) - \underline{I}_P(X)$.

$Bn_P(X)$ constitutes the "doubtful region" of X: according to knowledge expressed by P nothing can be said with certainty about membership of its elements in the set X.

The following concept will also be useful [9]. Given a partition $Cl=\{Cl_t, t \in T\}$, $T=\{1,...,n\}$, of U, the P-boundary with respect to $k>1$ classes $\{Cl_{t1},...,Cl_{tk}\} \subseteq \{Cl_1,...,Cl_n\}$ is defined as

$$Bd_P(\{Cl_{t1},...,Cl_{tk}\}) = \left(\bigcap_{t=t1,...,tk} Bn_P(Cl_t) \right) \cap \left(\bigcap_{t \neq t1,...,tk} (U - Bn_P(Cl_t)) \right).$$

The objects from $Bd_P(\{Cl_{t1},...,Cl_{tk}\})$ can be assigned to one of the classes $Cl_{t1},...,Cl_{tk}$ but P provides not enough information to know exactly to what class.

Let us observe that a very useful property of lower approximation within classical rough sets theory is that if an object $x \in U$ belongs to the lower approximation with respect to $P \subseteq C$, then x belongs also to the lower approximation with respect to $R \subseteq C$ when $P \subseteq R$ (this is a kind of monotonicity property). However, definition (1) does not satisfy this property of lower approximation, because it is possible that $f(x,q) \neq *$ for all $q \in P$ but $f(x,q)=*$ for some $q \in R-P$. This is quite problematic with respect to definition of some key concepts of the rough sets theory, like quality of approximation, reduct and core.

Therefore, another definition of lower approximation should be considered to restore the concepts of quality of approximation, reduct and core in the case of missing values. Given $X \subseteq U$ and $P \subseteq Q$, this definition is the following:

$$I_P^*(X)= \bigcup_{R\subseteq P} I_R(X). \qquad (3)$$

$I_P^*(X)$ will be called *cumulative* P-lower approximation of X because it includes all the objects belonging to all R-lower approximations of X, where $R\subseteq P$.

It can be shown that another type of indiscernibility relation, denoted by I_P^*, permits a direct definition of the cumulative P-lower approximation in a classic way. For each $x,y\in U$ and for each $P\subseteq Q$, $y\ I_P^*\ x$ means that $f(x,q)=f(y,q)$ or $f(x,q)=*$ and/or $f(y,q)=*$, for every $q\in P$. Let $I_P^*(x)=\{y\in U: y\ I_P^*\ x\}$ for each $x\in U$ and for each $P\subseteq Q$. I_P^* is a reflexive and symmetric but not transitive [4]. Let us observe that the restriction of I_P^* to U_P^* is reflexive, symmetric and transitive when $U_P^*=\{x\in U: f(x,q)\neq*$ for at least one $q\in P\}$.

Theorem 1. (*Definition (3) expressed in terms of* I_P^*) $\ I_P^*(X)=\{x\in U_P^*: I_P^*(x)\subseteq X\}$.

Using I_P^* we can give definition of the P-upper approximation of X:

$$\bar{I}_P^*(X)=\{x\in U_P^*: I_P^*(x)\cap X\neq\varnothing\}. \qquad (4)$$

For each $X\subseteq U$, let $X_P^*=X\cap U_P^*$. Let us remark that $x\in U_P^*$ if and only if there exists $R\neq\varnothing$ such that $R\subseteq P$ and $x\in U_R$. For each $X\subseteq U$ and for each $P\subseteq C$: $I_P^*(X)\subseteq X_P^*\subseteq \bar{I}_P^*(X)$ (*rough inclusion*) and $I_P^*(X)=U_P^* - \bar{I}_P^*(U-X)$ (*complementarity*).

The P-boundary of X approximated with I_P^* is equal to $Bn_P^*(X)=\bar{I}_P^*(X) - I_P^*(X)$.

Given a partition $Cl=\{Cl_t, t\in T\}$, $T=\{1,...,n\}$, of U, the P-boundary with respect to $k>1$ classes $\{Cl_{t1},...,Cl_{tk}\}\subseteq \{Cl_1,...,Cl_n\}$ is defined as

$$Bd_P^*(\{Cl_{t1},...,Cl_{tk}\}) = \left(\bigcap_{t=t1,...,tk} Bn_P^*(Cl_t)\right)\cap\left(\bigcap_{t\neq t1,...,tk}\left(U - Bn_P^*(Cl_t)\right)\right).$$

The objects from $Bd_P^*(\{Cl_{t1},...,Cl_{tk}\})$ can be assigned to one of the classes $Cl_{t1},...,Cl_{tk}$, however, P and all its subsets provide not enough information to know exactly to what class.

Theorem 2. (*Monotonicity of the accuracy of approximation*) For each $X\subseteq U$ and for each $P,T\subseteq C$, such that $P\subseteq T$, the following inclusion holds: i) $I_P^*(X)\subseteq I_T^*(X)$.

Furthermore, if $U_P^*=U_T^*$, the following inclusion is also true: ii) $\bar{I}_P^*(X)\supseteq \bar{I}_T^*(X)$.

Due to Theorem 2, when augmenting a set of attributes P, we get a lower approximation of X that is at least of the same cardinality. Thus, we can restore for the case of missing values the key concepts of the rough sets theory: accuracy and quality of approximation, reduct and core.

2.2 Decision rules for multi-attribute classification with missing values

Using the rough approximations (1), (2) and (3), (4), it is possible to induce a generalized description of the information contained in the decision table in terms of *decision rules*. These are logical statements (implications) of the type "*if ..., then...*", where the antecedent (condition part) is a conjunction of elementary conditions concerning particular condition attributes and the consequence (decision part) is a disjunction of possible assignments to particular classes of a partition of U induced by decision attributes. Given a partition $Cl=\{Cl_t, t \in T\}$, $T=\{1,...,n\}$, of U, the syntax of a rule is the following:

"*if* $f(x,q_1) = r_{q1}$ *and* $f(x,q_2) = r_{q2}$ *and* ... $f(x,q_p) = r_{qp}$, *then* x is assigned to Cl_{t1} *or*

$$Cl_{t2} \ or \ ... \ Cl_{tk}",$$

where $\{q_1,q_2,...,q_p\} \subseteq C$, $(r_{q1},r_{q2},...,r_{qp}) \in V_{q1} \times V_{q2} \times ... \times V_{qp}$ and $\{Cl_{t1},Cl_{t2},...,Cl_{tk}\} \subseteq \{Cl_1, Cl_2,...,Cl_n\}$. If the consequence is univocal, i.e. k=1, then the rule is *exact*, otherwise it is *approximate* or *uncertain*.

Let us observe that for any $Cl_t \in \{Cl_1,Cl_2,...,Cl_n\}$ and $P \subseteq Q$, the definition (1) of P-lower approximation of Cl_t can be rewritten as:

$$\underline{I}_P(Cl_t)=\{x \in U_P: \text{ for each } y \in U, \ if \ yI_Px, \ then \ y \in Cl_t\}. \tag{1'}$$

Thus the objects belonging to the lower approximation $\underline{I}_P(Cl_t)$ can be considered as a basis for induction of exact decision rules.

Therefore, the statement "*if* $f(x,q_1) = r_{q1}$ *and* $f(x,q_2) = r_{q2}$ *and* ... $f(x,q_p) = r_{qp}$, *then* x is assigned to Cl_t", is accepted as an exact decision rule iff there exists at least one $y \in \underline{I}_p(Cl_t)$, $P=\{q_1,...,q_p\}$, such that $f(y,q_1) = r_{q1}$ and $f(y,q_2)=r_{q2}$ and ... $f(y,q_p)=r_{qp}$.

Given $\{Cl_{t1},...,Cl_{tk}\} \subseteq \{Cl_1,Cl_2,...,Cl_n\}$ we can write:

$Bd_P(\{Cl_{t1},...,Cl_{tk}\}) = \{x \in U_P: \text{ for each } y \in U, \ if \ yI_Px, \ then \ y \in Cl_{t1} \ or \ ... \ Cl_{tk}\}.$ (2')

Thus, the objects belonging to the boundary $Bd_P(\{Cl_{t1},...,Cl_{tk}\})$ can be considered as a basis for induction of approximate decision rules.

Since each decision rule is an implication, a *minimal* decision rule represents such an implication that there is no other implication with an antecedent of at least the same weakness and a consequent of at least the same strength.

We say that $y \in U$ *supports* the exact decision rule if $[f(y,q_1)=r_{q1}$ and/or $f(y,q_1)=*]$ and $[f(y,q_2)=r_{q2}$ and/or $f(y,q_2)=*]$ ... and $[f(y,q_p)=r_{qp}$ and/or $f(y,q_p)=*$] and $y \in Cl_t$. Similarly, we say that $y \in U$ *supports* the approximate decision rule if $[f(y,q_1)=r_{q1}$ and/or $f(y,q_1)=*]$ and $[f(y,q_2)=r_{q2}$ and/or $f(y,q_2)=*]$ ... and $[f(y,q_p)=r_{qp}$ and/or $f(y,q_p)=*]$ and $y \in Bd_C^*(\{Cl_{t1},...,Cl_{tk}\})$.

A set of decision rules is *complete* if it fulfils the following conditions:

- each $x \in \underline{I}_C^*(Cl_t)$ supports at least one exact decision rule suggesting an assignment to Cl_t, for each $Cl_t \in Cl$,

- each $x \in Bd_C^*(\{Cl_{t1},...,Cl_{tk}\})$ supports at least one approximate decision rule suggesting an assignment to Cl_{t1} *or* Cl_{t2} *or* ... Cl_{tk}, for each $\{Cl_{t1},Cl_{t2},...,Cl_{tk}\} \subseteq \{Cl_1,Cl_2,...,Cl_n\}$.

We call *minimal* each set of minimal decision rules that is complete and non-redundant, i.e. exclusion of any rule from this set makes it non-complete.

2.3 Multi-criteria sorting problem with missing values

Formally, for each $q \in C$ being a criterion there exists an outranking relation [8] S_q on the set of objects U such that xS_qy means "x is at least as good as y with respect to criterion q". We suppose that S_q is a total preorder, i.e. a strongly complete and transitive binary relation defined on U on the basis of evaluations $f(\cdot,q)$. Precisely, we assume that xS_qy iff $f(x,q) \geq f(y,q)$.

Also in this case, we are considering a directional comparison of subject y to referent x, for any two objects $x,y \in U$. We say that subject y *dominates* referent x with respect to criteria $P \subseteq C$ (denotation $y D_P^+ x$) if for every criterion $q \in P$ the following conditions are met:

- $f(x,q) \neq *$,
- $f(y,q) \geq f(x,q)$ or $f(y,q) = *$.

We say that subject y *is dominated by* referent x with respect to criteria $P \subseteq C$ (denotation $x D_P^- y$) if for every criterion $q \in P$ the following conditions are met:

- $f(x,q) \neq *$,
- $f(x,q) \geq f(y,q)$ or $f(y,q) = *$.

The above means that the referent object considered for dominance D_P^+ and D_P^- should have no missing values on criteria from set P.

The binary relations D_P^+ and D_P^- are not necessarily reflexive because for some $x \in U$ there may exist $q \in P$ for which $f(x,q) = *$ and, therefore, we cannot state neither $x D_P^+ x$ nor $x D_P^- x$. However, D_P^+ and D_P^- are transitive because for each $x,y,z \in U$,
(i) $x D_P^+ y$ and $y D_P^+ z$ imply $x D_P^+ z$, and (ii) $x D_P^- y$ and $y D_P^- z$ imply $x D_P^- z$. Implication (i) is justified by the observation that object z can substitute object y in the statement $x D_P^+ y$ because $y D_P^+ z$ and both y and z, as referent objects, have no missing values. As to implication (ii), object x can substitute object y in the statement $y D_P^- z$ because $x D_P^- y$ and both x and y, as referent objects, have no missing values.

For each $P \subseteq C$ we restore the definition of set U_P from sub-section 2.1. It is easy to see that the restrictions of D_P^+ and D_P^- to U_P (in other words, the binary relations $D_P^+ \cap U_P \times U_P$ and $D_P^- \cap U_P \times U_P$ defined on U_P) are reflexive and transitive, i.e. they are partial preorders.

The sets to be approximated are called *upward union* and *downward union* of preference-ordered classes, respectively:

$$Cl_t^{\geq} = \bigcup_{s \geq t} Cl_s, \quad Cl_t^{\leq} = \bigcup_{s \leq t} Cl_s, \quad t=1,...,n.$$

The statement $Cl_t^{\geq} = \bigcup_{s \geq t} Cl_s$ means "x belongs at least to class Cl_t", while $Cl_t^{\leq} = \bigcup_{s \leq t} Cl_s$ means "x belongs at most to class Cl_t".

Let us remark that $Cl_1^{\geq} = Cl_n^{\leq} = U$, $Cl_n^{\geq} = Cl_n$ and $Cl_1^{\leq} = Cl_1$. Furthermore, for t=2,...,n, we have $Cl_t^{\geq} = U - Cl_{t-1}^{\leq}$ and $Cl_{t-1}^{\leq} = U - Cl_t^{\geq}$.

Given $P \subseteq C$ and $x \in U$, the "granules of knowledge" used for approximation are:

- a set of objects dominating x, called P-*dominating set*, $D_P^+(x) = \{y \in U: y D_P^+ x\}$,

- a set of objects dominated by x, called P-*dominated set,* $D_P^-(x) = \{y \in U: x \, D_P^- \, y\}$.

For any $P \subseteq C$ we say that $x \in U$ belongs to $Cl_t^{\geq}$ *without any ambiguity* if $x \in Cl_t^{\geq}$ and for all the objects $y \in U$ dominating x with respect to P, we have $y \in Cl_t^{\geq}$, i.e. $D_P^+(x) \subseteq Cl_t^{\geq}$. Furthermore, we say that $x \in U$ *could belong* to $Cl_t^{\geq}$ if there would exist at least one object $y \in Cl_t^{\geq}$ dominated by x with respect to P, i.e. $y \in D_P^-(x)$.

Thus, with respect to $P \subseteq C$, the set of all objects belonging to $Cl_t^{\geq}$ without any ambiguity constitutes the P-*lower approximation* of $Cl_t^{\geq}$, denoted by $\underline{P}(Cl_t^{\geq})$, and the set of all objects that could belong to $Cl_t^{\geq}$ constitutes the P-*upper approximation* of $Cl_t^{\geq}$, denoted by $\overline{P}(Cl_t^{\geq})$, for t=1,...,n:

$$\underline{P}(Cl_t^{\geq}) = \{x \in U_P: D_P^+(x) \subseteq Cl_t^{\geq}\}, \qquad (5.1)$$

$$\overline{P}(Cl_t^{\geq}) = \{x \in U_P: D_P^-(x) \cap Cl_t^{\geq} \neq \varnothing\}. \qquad (5.2)$$

Analogously, one can define P-*lower approximation* and P-*upper approximation* of $Cl_t^{\leq}$, for t=1,...,n:

$$\underline{P}(Cl_t^{\leq}) = \{x \in U_P: D_P^-(x) \subseteq Cl_t^{\leq}\}, \qquad (6.1)$$

$$\overline{P}(Cl_t^{\leq}) = \{x \in U_P: D_P^+(x) \cap Cl_t^{\leq} \neq \varnothing\}. \qquad (6.2)$$

Let $(Cl_t^{\geq})_P = Cl_t^{\geq} \cap U_P$ and $(Cl_t^{\leq})_P = Cl_t^{\leq} \cap U_P$, t=1,...,n. For each $Cl_t^{\geq}$ and $Cl_t^{\leq}$, t=1,...,n, and for each $P \subseteq C$: $\underline{P}(Cl_t^{\geq}) \subseteq (Cl_t^{\geq})_P \subseteq \overline{P}(Cl_t^{\geq})$, $\underline{P}(Cl_t^{\leq}) \subseteq (Cl_t^{\leq})_P \subseteq \overline{P}(Cl_t^{\leq})$ (*rough inclusion*). Moreover, for each $Cl_t^{\geq}$, t=2,...,n, and $Cl_t^{\leq}$, t=1,...,n-1, and for each $P \subseteq C$: $\underline{P}(Cl_t^{\geq}) = U_P - \overline{P}(Cl_{t-1}^{\leq})$, $\underline{P}(Cl_t^{\leq}) = U_P - \overline{P}(Cl_{t+1}^{\geq})$ (*complementarity*).

The P-*boundaries* (P-doubtful regions) of $Cl_t^{\geq}$ and $Cl_t^{\leq}$ are defined as:

$$Bn_P(Cl_t^{\geq}) = \overline{P}(Cl_t^{\geq}) - \underline{P}(Cl_t^{\geq}), \quad Bn_P(Cl_t^{\leq}) = \overline{P}(Cl_t^{\leq}) - \underline{P}(Cl_t^{\leq}), \quad \text{for t=1,...,n.}$$

Due to complementarity of the rough approximations [1], the following property holds: $Bn_P(Cl_t^{\geq}) = Bn_P(Cl_{t-1}^{\leq})$, for t=2,...,n, and $Bn_P(Cl_t^{\leq}) = Bn_P(Cl_{t+1}^{\geq})$, for t=1,...,n-1.

To preserve the monotonicity property of the lower approximation (see subsection 2.1) it is necessary to use another definition of the approximation for a given $Cl_t^{\geq}$ and $Cl_t^{\leq}$, t=1,...,n, and for each $P \subseteq C$:

$$\underline{P}(Cl_t^{\geq})^* = \bigcup_{R \subseteq P} \underline{R}(Cl_t^{\geq}), \qquad (7.1)$$

$$\underline{P}(Cl_t^{\leq})^* = \bigcup_{R \subseteq P} \underline{R}(Cl_t^{\leq}). \qquad (7.2)$$

$\underline{P}(Cl_t^{\geq})^*$ and $\underline{P}(Cl_t^{\leq})^*$ will be called *cumulative* P-lower approximations of unions $Cl_t^{\geq}$ and $Cl_t^{\leq}$, t=1,...,n, because they include all the objects belonging to all R-lower approximations of $Cl_t^{\geq}$ and $Cl_t^{\leq}$, respectively, where $R \subseteq P$.

It can be shown that another type of dominance relation, denoted by D_P^*, permits a direct definition of the cumulative P-lower approximations in a classical way. For each $x,y \in U$ and for each $P \subseteq Q$, $y \, D_P^* \, x$ means that $f(y,q) \geq f(x,q)$ or $f(x,q) = *$ and/or $f(y,q) = *$, for every $q \in P$.

Given $P \subseteq C$ and $x \in U$, the "granules of knowledge" used for approximation are:

- a set of objects dominating x, called P-*dominating set*, $D_P^{+*}(x) = \{y \in U : y \, D_P^* \, x\}$,
- a set of objects dominated by x, called P-*dominated set*, $D_P^{-*}(x) = \{y \in U : x \, D_P^* \, y\}$.

D_P^* is reflexive but not transitive. Let us observe that the restriction of D_P^* to U_P^* is reflexive and transitive when $U_P^* = \{x \in U : f(x,q) \neq *$ for at least one $q \in P\}$.

Theorem 3. (*Definitions (7.1) and (7.2) expressed in terms of* D_P^*)

$$\underline{P}(Cl_t^{\geq})^* = \{x \in U_P^* : D_P^{+*}(x) \subseteq Cl_t^{\geq}\}, \quad \underline{P}(Cl_t^{\leq})^* = \{x \in U_P^* : D_P^{-*}(x) \subseteq Cl_t^{\leq}\}.$$

Using D_P^* we can give definition of the P-upper approximations of $Cl_t^{\geq}$ and $Cl_t^{\leq}$, complementary to $\underline{P}(Cl_t^{\geq})^*$ and $\underline{P}(Cl_t^{\leq})^*$, respectively:

$$\overline{P}(Cl_t^{\geq})^* = \{x \in U_P^* : D_P^{-*}(x) \cap Cl_t^{\geq} \neq \varnothing\}, \tag{8.1}$$

$$\overline{P}(Cl_t^{\leq})^* = \{x \in U_P^* : D_P^{+*}(x) \cap Cl_t^{\leq} \neq \varnothing\}. \tag{8.2}$$

For each $Cl_t^{\geq} \subseteq U$ and $Cl_t^{\leq} \subseteq U$, let $(Cl_t^{\geq})^* = Cl_t^{\geq} \cap U_P^*$ and $(Cl_t^{\leq})^* = Cl_t^{\leq} \cap U_P^*$. Let us remark that $x \in U_P^*$ if and only if there exists $R \neq \varnothing$ such that $R \subseteq P$ and $x \in U_R$. For each $Cl_t^{\geq}$ and $Cl_t^{\leq}$, t=1,...,n, and for each $P \subseteq C$: $\underline{P}(Cl_t^{\geq})^* \subseteq (Cl_t^{\geq})^* \subseteq \overline{P}(Cl_t^{\geq})^*$, $\underline{P}(Cl_t^{\leq})^* \subseteq (Cl_t^{\leq})^* \subseteq \overline{P}(Cl_t^{\leq})^*$ (*rough inclusion*). Moreover, for each $Cl_t^{\geq}$, t=2,...,n, and $Cl_t^{\leq}$, t=1,...,n-1, and for each $P \subseteq C$: $\underline{P}(Cl_t^{\geq})^* = U_P^* - \overline{P}(Cl_{t-1}^{\leq})^*$, $\underline{P}(Cl_t^{\leq})^* = U_P^* - \overline{P}(Cl_{t+1}^{\geq})^*$ (*complementarity*).

The P-boundary of and $Cl_t^{\leq}$, t=1,...,n, approximated with D_P^* are equal, respectively, to $Bn_P^*(Cl_t^{\geq}) = \overline{P}(Cl_t^{\geq})^* - \underline{P}(Cl_t^{\geq})^*$, $Bn_P^*(Cl_t^{\leq}) = \overline{P}(Cl_t^{\leq})^* - \underline{P}(Cl_t^{\leq})^*$.

Theorem 4. (*Monotonicity of the accuracy of approximation*) For each $Cl_t^{\geq}$ and $Cl_t^{\leq}$, t=1,...,n, and for each $P,R \subseteq C$, such that $P \subseteq R$, the following inclusions hold:

$$\underline{P}(Cl_t^{\geq})^* \subseteq \underline{R}(Cl_t^{\geq})^*, \quad \underline{P}(Cl_t^{\leq})^* \subseteq \underline{R}(Cl_t^{\leq})^*.$$

Furthermore, if $U_P^* = U_T^*$, the following inclusions are also true:

$$\overline{P}(Cl_t^{\geq})^* \supseteq \overline{R}(Cl_t^{\geq})^*, \quad \overline{P}(Cl_t^{\leq})^* \supseteq \overline{R}(Cl_t^{\leq})^*.$$

Due to Theorem 4, when augmenting a set of attributes P, we get lower approximations of $Cl_t^{\geq}$ and $Cl_t^{\leq}$, t=1,...,n, that are at least of the same cardinality. Thus, we can restore for the case of missing values the key concepts of the rough sets theory: accuracy and quality of approximation, reduct and core.

For every $t \in T$ and for every $P \subseteq C$ we define the *quality of approximation of partition Cl* by set of attributes P, or in short, *quality of sorting*:

$$\gamma_P(Cl) = \frac{card\left(U - \left(\bigcup_{t\in T} Bn_P^*\left(Cl_t^{\leq}\right)\right)\right)}{card(U)} = \frac{card\left(U - \left(\bigcup_{t\in T} Bn_P^*\left(Cl_t^{\geq}\right)\right)\right)}{card(U)}.$$

The quality expresses the ratio of all P-correctly sorted objects to all objects in the decision table.

Each minimal subset $P\subseteq C$ such that $\gamma_P(Cl) = \gamma_C(Cl)$ is called a *reduct* of **Cl** and denoted by $RED_{Cl}(C)$. Let us remark that a decision table can have more than one reduct. The intersection of all reducts is called the *core* and denoted by $CORE_{Cl}(C)$.

2.4 Decision rules for multi-criteria sorting with missing values

Using the rough approximations (5.1), (5.2), (6.1), (6.2) and (7.1), (7.2), (8.1), (8.2), it is possible to induce a generalized description of the information contained in the decision table in terms of "*if ..., then...*" *decision rules*.

Given the preference-ordered classes of partition $Cl=\{Cl_t, t\in T\}$, $T=\{1,...,n\}$, of U, the following three types of decision rules can be considered:

1) $D_{\geq}$-*decision rules* with the following syntax:

"*if* $f(x,q_1)\geq r_{q1}$ *and* $f(x,q_2)\geq r_{q2}$ *and* $...f(x,q_p)\geq r_{qp}$, *then* $x\in Cl_t^{\geq}$",

where $P=\{q_1,...,q_p\}\subseteq C$, $(r_{q1},...,r_{qp})\in V_{q1}\times V_{q2}\times...\times V_{qp}$ and $t\in T$;

2) $D_{\leq}$-*decision rules* with the following syntax:

"*if* $f(x,q_1)\leq r_{q1}$ *and* $f(x,q_2)\leq r_{q2}$ *and* $...f(x,q_p)\leq r_{qp}$, *then* $x\in Cl_t^{\leq}$",

where $P=\{q_1,...,q_p\}\subseteq C$, $(r_{q1},...,r_{qp})\in V_{q1}\times V_{q2}\times...\times V_{qp}$ and $t\in T$;

3) $D_{\geq\leq}$-*decision rules* with the following syntax:

"*if* $f(x,q_1)\geq r_{q1}$ *and* $f(x,q_2)\geq r_{q2}$ *and* $...$ $f(x,q_k)\geq r_{qk}$ *and* $f(x,q_{k+1})\leq r_{qk+1}$ *and* $...$ $f(x,q_p)\leq r_{qp}$, *then* $x\in Cl_s\cup Cl_{s+1}\cup...\cup Cl_t$",

where $O'=\{q_1,...,q_k\}\subseteq C$, $O''=\{q_{k+1},...,q_p\}\subseteq C$, $P=O'\cup O''$, O' and O'' not necessarily disjoint, $(r_{q1},...,r_{qp})\in V_{q1}\times V_{q2}\times...\times V_{qp}$, $s,t\in T$ such that $s<t$.

As it is possible that $\{q_1,...,q_k\}\cap\{q_{k+1},...,q_p\}\neq\emptyset$, in the condition part of a $D_{\geq\leq}$-decision rule we can have "$f(x,q)\geq r_q$" and "$f(x,q)\leq r'_q$", where $r_q\leq r'_q$, for some $q\in C$. Moreover, if $r_q=r'_q$, the two conditions boil down to "$f(x,q)=r_q$".

Since each decision rule is an implication, by a *minimal* decision rule we understand such an implication that there is no other implication with an antecedent of at least the same weakness and a consequent of at least the same strength.

We say that an object *supports* a rule if its evaluation by set C of criteria matches the condition part of the rule.

A set of decision rules is *complete* if it fulfils the following conditions:

- each $y\in \underline{C}(Cl_t^{\geq})^*$ supports at least one $D_{\geq}$-decision rule of the type "*if* $f(x,q_1)\geq r_{q1}$ *and* $f(x,q_2)\geq r_{q2}$ *and* $...$ $f(x,q_p)\geq r_{qp}$, *then* $x\in Cl_r^{\geq}$", with $r,t\in\{2,...,n\}$ and $r\geq t$,

- each $y\in \underline{C}(Cl_t^{\leq})^*$ supports at least one $D_{\leq}$-decision rule of the type "*if* $f(x,q_1)\leq r_{q1}$ *and* $f(x,q_2)\leq r_{q2}$ *and* $...$ $f(x,q_p)\leq r_{qp}$, *then* $x\in Cl_u^{\leq}$", with $u,t\in\{1,...,n-1\}$ and $u\leq t$,

- each $y \in \overline{C}(Cl_s^{\le})^* \cap \overline{C}(Cl_t^{\ge})^*$ supports at least one $D_{\ge\le}$-decision rule of the type "if $f(x,q_1) \ge r_{q1}$ and $f(x,q_2) \ge r_{q2}$ and ... $f(x,q_k) \ge r_{qk}$ and $f(x,q_{k+1}) \le r_{qk+1}$ and ... $f(x,q_p) \le r_{qp}$, $then$ $x \in Cl_v \cup Cl_{v+1} \cup ... \cup Cl_z$", with $s,t,v,z \in T$ and $v \le s < t \le z$.

Let us remark that application of any complete set of decision rules on the objects from the data table results in either exact or approximate reassignment of these objects to the classes Cl_t, $t \in T$. Let us explain this reassignment in more detail.

Given a complete set of rules, and an object $y \in U$, such that $y \notin Bn_C(Cl_s^{\le})$ and $y \notin Bn_C(Cl_s^{\ge})$ for any $s \in T$, the following situations may occur:
- $y \in Cl_t$, $t = 2,...,n-1$; then there exists at least one $D_{\ge}$-decision rule with consequent $x \in Cl_t^{\ge}$, and at least one $D_{\le}$-decision rule with consequent $x \in Cl_t^{\le}$;
- $y \in Cl_1$; then there exists at least one $D_{\le}$-decision rule with consequent $x \in Cl_1^{\le}$;
- $y \in Cl_n$; then there exists at least one $D_{\ge}$-decision rule with consequent $x \in Cl_n^{\ge}$.

In all above situations, intersection of all unions (upward and downward) of classes suggested for assignment in the consequent of rules matching object y will result in (exact) reassignment of y to class Cl_t, $t \in T$.

Similarly, for each object $y \in \overline{C}(Cl_s^{\le})^* \cap \overline{C}(Cl_t^{\ge})^*$, $s < t$, such that $y \notin \overline{C}(Cl_{s1}^{\le})^* \cap \overline{C}(Cl_{t1}^{\ge})^*$, $s1 < [\le]s$ and $t \le [<]t1$, which means that y belongs exclusively to boundaries $Bn_C^*(Cl_v^{\ge})$, $v = s+1,...,t$, and $Bn_C^*(Cl_z^{\le})$, $z = s,...,t-1$, there exists at least one $D_{\ge\le}$-decision rule whose consequent is $x \in Cl_s \cup Cl_{s+1} \cup ... \cup Cl_t$. Thus, in result of application of the complete set of rules to object y, it will be reassigned (approximately) to classes $Cl_s \cup Cl_{s+1} \cup ... \cup Cl_t$.

We call *minimal* each set of minimal decision rules that is complete and non-redundant, i.e. exclusion of any rule from this set makes it non-complete.

3 Conclusions

We adapted the rough sets methodology to the analysis of data sets with missing values. The adaptation concerns both the classical rough set approach based on the use of indiscernibility relations and the new rough set approach based on the use of dominance relations. While the first approach deals with multi-attribute classification problems, the second approach deals with multi-criteria sorting problems. The two adapted rough set approaches maintain all good characteristics of their original approaches. They also boil down to the original approaches when there are no missing values.

The case of missing values is very often met in practice and not many methods can deal satisfactorily with this problem. The way of handling the missing values in our approach seems faithful with respect to available data because the decision rules are *robust* in the sense of being founded on objects existing in the data set and not on hypothetical objects created by putting some possible values instead of the missing

157

ones. This is a distinctive feature of our extension in comparison with the extension proposed by Kryszkiewicz [4].

Acknowledgement

The research of the first two authors has been supported by the Italian Ministry of University and Scientific Research (MURST). The third author wishes to acknowledge financial support from State Committee for Scientific Research, KBN research grant no. 8 T11C 013 13, and from CRIT 2 Esprit Project no. 20288.

References

1. Greco, S., Matarazzo, B., Slowinski, R.: A new rough set approach to multicriteria and multiattribute classification. In: L. Polkowski, A. Skowron (eds.): *Rough sets and Current Trends in Computing (RSTCTC '98)*, Lecture Notes in Artificial Intelligence, vol.1424, Springer-Verlag, Berlin, 1998, pp.60-67
2. Greco, S., Matarazzo, B., Slowinski, R.: The use of rough sets and fuzzy sets in MCDM. In: T. Gal, T. Hanne, T. Stewart (eds.), *Advances in Multiple-Criteria Decision Making*, Kluwer Academic Publishers, Boston, 1999, pp. 14.1-14.59
3. Greco, S., Matarazzo, B., Slowinski, R., Zanakis, S.: Rough set analysis of information tables with missing values. In: *Proc. 5th Internaional Conference of the Decision Sciences Institute*, Athens, Greece, 4-7 July 1999, pp. 1359-1362
4. Kryszkiewicz, M.: Properties of incomplete information systems in the framework of rough sets. In: L. Polkowski, A. Skowron (eds.): *Rough Sets in Knowledge Discovery*. Vol. 1 *Methodology and Applications*. Physica-Verlag, Heidelberg, 1998, pp. 422-450
5. Pawlak, Z.: Rough sets. *International Journal of Information & Computer Sciences* 11 (1982) 341-356
6. Pawlak, Z.: *Rough Sets. Theoretical Aspects of Reasoning about Data*. Kluwer, Dordrecht, 1991
7. Pawlak, Z., Slowinski, R.: Rough set approach to multi-attribute decision analysis. *European Journal of Operational Research* 72 (1994) 443-459
8. Roy, B.: *Méthodologie Multicritère d'Aide à la Décision*. Economica, Paris, 1985
9. Skowron, A., Grzymala-Busse, J.W.: From rough set theory to evidence theory. In R.R Yaeger et al. (eds.): *Advances in the Dempster-Shafer Theory of Evidence*. John Wiley & Sons, Inc. New York, 1994, pp. 193-236
10. Slowinski, R. (ed.): *Intelligent Decision Support. Handbook of Applications and Advances of the Rough Sets Theory*. Kluwer, Dordrecht, 1992

The Generic Rough Set Inductive Logic Programming Model and Motifs in Strings

Arul Siromoney[1] and K. Inoue[2]

[1] School of Computer Science and Engineering,
Anna University, Chennai – 600 025, India
asiro@vsnl.com
[2] Department of Computer Science and Systems Engineering,
Yamaguchi University, Ube 755–8611, Japan
inoue@csse.yamaguchi-u.ac.jp

Abstract. The gRS–ILP model (generic Rough Set Inductive Logic Programming model) provides a framework for Inductive Logic Programming when the setting is imprecise and any induced logic program will not be able to distinguish between certain positive and negative examples. However, in this rough setting, where it is inherently not possible to describe the entire data with 100% accuracy, it is possible to definitively describe part of the data with 100% accuracy. The gRS–ILP model is extended in this paper to motifs in strings. An illustrative experiment is presented using the ILP system Progol on transmembrane domains in amino acid sequences.

Keywords: Rough Set Theory; Inductive Logic Programming; Machine Learning; Knowledge Discovery from Data; Molecular biology;

1 Introduction

Inductive Logic Programming [1] is the research area formed at the intersection of logic programming and machine learning. Inductive Logic Programming (in the example setting) uses background knowledge definite clauses, and positive and negative example ground facts to induce a logic program that describes the examples, where the induced logic program consists of the original background knowledge along with an induced hypothesis (as definite clauses).

Rough set theory [2, 3] defines an indiscernibility relation, where certain subsets of examples cannot be distinguished. A concept is rough when it contains at least one such indistinguishable subset that contains both positive and negative examples. It is inherently not possible to describe the examples accurately, since certain positive and negative examples cannot be distinguished.

The gRS–ILP model [4] introduces a rough setting in Inductive Logic Programming and describes the situation where the background knowledge, declarative bias and evidence are such that it is not possible to induce any logic program from them that is able to distinguish between certain positive and negative examples. Any induced logic program will either cover both the positive and the negative examples in the group, or not cover the group at all, with both the positive and the negative examples in this group being left out.

The gRS–ILP model has useful applications in the definitive description of large data. Knowledge discovery in databases is the non–trivial process of identifying valid, novel, potentially useful, and ultimately understandable patterns in data([5]). This usually involves one of two different goals: prediction and description. Prediction involves using some variables or fields in the database to predict unknown or future values of other variables of interest. Description focuses on finding human–interpretable patterns describing the data. Definitive description is the description of the data with full accuracy. In a rough setting, it is not possible to definitively describe the entire data, since some of the positive examples and negative examples (of the concept being described) inherently cannot be distinguished from each other.

Conventional systems handle a rough setting by using various techniques to induce a hypothesis that describes the evidence as well as possible. They aim to maximize the correct cover of the induced hypothesis by maximizing the number of positive examples covered and negative examples not covered. This means that most of the positive evidence would be described, along with some of the negative evidence. The induced hypothesis cannot say with certainty whether an example definitely belongs to the evidence or not. However, the gRS–ILP model lays a firm theoretical foundation for the definitive description of data in a rough setting. A part of the data is definitively described. The rest of the data can then be described using conventional methods, but not definitively.

This paper extends the gRS–ILP model to motifs in strings, and reports an illustrative experiment using Progol on transmembrane domains in amino acid sequences.

2 Formal definitions of the gRS–ILP model

The generic Rough Set Inductive Logic Programming (gRS–ILP) model introduces the basic definition of elementary sets and a rough setting in ILP [6, 4]. The essential feature of an elementary set is that it consists of examples that cannot be distinguished from each other by any induced logic program in that ILP system. The essential feature of a rough setting is that it is inherently not possible for the consistency and completeness criteria to be fulfilled together, since both positive and negative examples are in the same elementary set. The basic definitions formalised in [7] follow.

2.1 The RSILP system

We first formally define the ILP system in the example setting of [8] as follows.

Definition 2.1. An *ILP system in the example setting* is a tuple $S_{es} = (E_{es}, B)$, where
(1) $E_{es} = E_{es}^+ \cup E_{es}^-$ is the *universe*, where E_{es}^+ is the set of positive examples

(true ground facts), and E_{es}^- is the set of negative examples (false ground facts), and

(2) B is a background knowledge given as definite clauses such that (i) for all $e^- \in E_{es}^-$, $B \not\vdash e^-$, and (ii) for some $e^+ \in E_{es}^+$, $B \not\vdash e^+$. □

Let $S_{es} = (E_{es}, B)$ be an ILP system in the example setting. Then let $\mathcal{H}(S_{es})$ (also written as $\mathcal{H}(E_{es}, B)$) denote the set of all possible definite clause hypotheses that can be induced from E_{es} and B, and be called the *hypothesis space* induced from S_{es} (or from E_{es} and B). Further, let $\mathcal{P}(S_{es})$ (also written as $\mathcal{P}(E_{es}, B) = \{P = B \wedge H \mid H \in \mathcal{H}(E_{es}, B)\}$) denote the set of all the programs induced from E_{es} and B, and be called the *program space* induced from S_{es} (or from E_{es} and B).

Our aim is to find a program $P \in \mathcal{P}(S_{es})$ such that the next two conditions hold: (iii) for all $e^- \in E_{es}^-$, $P \not\vdash e^-$, (iv) for all $e^+ \in E_{es}^+$, $P \vdash e^+$.

The following definitions of Rough Set ILP systems in the gRS–ILP model (abbreviated as *RSILP systems*) use the terminology of [8].

Definition 2.2. An *RSILP system in the example setting* (abbreviated as RSILP–E system) is an ILP system in the example setting, $S_{es} = (E_{es}, B)$, such that there does not exist a program $P \in \mathcal{P}(S_{es})$ satisfying both the conditions (iii) and (iv) above. □

Definition 2.3. An *RSILP–E system in the single–predicate learning context* (abbreviated as RSILP–ES system) is an RSILP–E system, whose *universe E* is such that all examples (ground facts) in E use only one predicate, also known as the *target predicate*. □

A *declarative bias* [8] biases or restricts the set of acceptable hypotheses, and is of two kinds: *syntactic bias* (also called *language bias*) that imposes restrictions on the form (syntax) of clauses allowed in the hypothesis, and *semantic bias* that imposes restrictions on the meaning, or the behaviour of hypotheses.

Definition 2.4. An *RSILP–ES system with declarative bias* (abbreviated as RSILP–ESD system) is a tuple $S = (S', L)$, where
(i) $S' = (E, B)$ is an RSILP–ES system, and
(ii) L is a declarative bias, which is any restriction imposed on the hypothesis space $\mathcal{H}(E, B)$.
We also write $S = (E, B, L)$ instead of $S = (S', L)$. □

For any RSILP–ESD system $S = (E, B, L)$, let
$\mathcal{H}(S) = \{H \in \mathcal{H}(E, B) \mid H \text{ is allowed by } L\}$, and
$\mathcal{P}(S) = \{P = B \wedge H \mid H \in \mathcal{H}(S)\}$.
$\mathcal{H}(S)$ (also written as $\mathcal{H}(E, B, L)$) is called the *hypothesis space* induced from S (or from E, B, and L). $\mathcal{P}(S)$ (also written as $\mathcal{P}(E, B, L)$) denotes the set of all the programs induced by S, and is called the *program space* induced from S (or from E, B, and L).

2.2 Equivalence relation, elementary sets and composed sets

We now define an equivalence relation on the universe of an RSILP–ESD system.

Definition 2.5. Let $S = (E, B, L)$ be an RSILP–ESD system. An indiscernibility relation of S, denoted by $R(S)$, is a relation on E defined as follows:
$\forall x, y \in E$, $(x, y) \in R(S)$ iff
$(P \vdash x \Leftrightarrow P \vdash y)$ for any $P \in \mathcal{P}(S)$ (i.e. iff x and y are inherently indistinguishable by any induced logic program P in $\mathcal{P}(S)$). $\square$

The following fact follows directly from the definition of $R(S)$.

Fact 1 *For any RSILP–ESD system S, $R(S)$ is an equivalence relation.*

Definition 2.6. Let $S = (E, B, L)$ be an RSILP–ESD system. An *elementary set* of $R(S)$ is an equivalence class of the relation $R(S)$. For each $x \in E$, let $[x]_{R(S)}$ denote the elementary set of $R(S)$ containing x. Formally,
$[x]_{R(S)} = \{y \in E \mid (x, y) \in R(S)\}$.
A *composed set* of $R(S)$ is any finite union of elementary sets of $R(S)$. $\square$

Definition 2.7. An RSILP–ESD system $S = (E, B, L)$ is said to be in a *rough setting* iff
$\exists e^+ \in E^+ \; \exists e^- \in E^- \; ((e^+, e^-) \in R(S))$. $\square$

We now define a combination of declarative biases.

Let $S = (E, B)$ be an RSILP–ES system. Let L_1, L_2 and L_3 be declarative biases. $L_1 \wedge L_2$ (resp., $L_1 \vee L_2$) denotes the declarative bias such that $\mathcal{H}(S') = \mathcal{H}(S_1) \cap \mathcal{H}(S_2)$ (resp., $\mathcal{H}(S'') = \mathcal{H}(S_1) \cup \mathcal{H}(S_2)$), where $S' = (E, B, L_1 \wedge L_2)$, $S'' = (E, B, L_1 \vee L_2)$, $S_1 = (E, B, L_1)$ and $S_2 = (E, B, L_2)$ are RSILP ESD systems.

$L_1 \wedge L_2 \wedge L_3$ (resp., $(L_1 \wedge L_2) \vee L_3$) denotes the declarative bias such that $\mathcal{H}(S''') = \mathcal{H}(S_1) \cap \mathcal{H}(S_2) \cap \mathcal{H}(S_3)$ (resp., $\mathcal{H}(S'''') = (\mathcal{H}(S_1) \cap \mathcal{H}(S_2)) \cup \mathcal{H}(S_3)$), where $S''' = (E, B, L_1 \wedge L_2 \wedge L_3)$, $S'''' = (E, B, (L_1 \wedge L_2) \vee L_3)$, $S_1 = (E, B, L_1)$, $S_2 = (E, B, L_2)$ and $S_3 = (E, B, L_3)$ are RSILP–ESD systems. $L_1 \vee L_2 \vee L_3$, $(L_1 \vee L_2) \wedge L_3$, ..., etc. are defined similarly.

2.3 Consistency and completeness in the gRS–ILP model

Let $S = (E, B, L)$ be an RSILP–ESD system, and $\mathcal{P}(S)$ the program space induced by S.

Definition 2.8. The *upper approximation* of S, $Upap(S)$, is defined as the least composed set of $R(S)$ such that $E^+ \subseteq Upap(S)$. $\square$

Definition 2.9. The *lower approximation* of S, $Loap(S)$, is defined as the greatest composed set of $R(S)$ such that $Loap(S) \subseteq E^+$. $\square$

The set $Bndr(S) = Upap(S) - Loap(S)$ is known as the *boundary region* of S (or the *borderline region* of S). The lower approximation of S, $Loap(S)$, is also known as $Pos(S)$, the *positive region* of S. The set $Neg(S) = E - Upap(S)$ is known as the *negative region* of S.

Definition 2.10. The *consistent program space* $\mathcal{P}_{cons}(S)$ of S is defined as
$\mathcal{P}_{cons}(S) = \{P \in \mathcal{P}(S) \mid P \nvdash e^-, \forall e^- \in E^-\}$.
A program $P \in \mathcal{P}(S)$ is *consistent* with respect to S iff $P \in \mathcal{P}_{cons}(S)$.
The *reverse-consistent program space* $\mathcal{P}_{rev-cons}(S)$ of S is defined as
$\mathcal{P}_{rev-cons}(S) = \{P \in \mathcal{P}(S) \mid P \nvdash e^+, \forall e^+ \in E^+\}$.
A program $P \in \mathcal{P}(S)$ is *reverse-consistent* w.r.t. S iff $P \in \mathcal{P}_{rev-cons}(S)$. □

Definition 2.11. The *complete program space* $\mathcal{P}_{comp}(S)$ of S is defined as
$\mathcal{P}_{comp}(S) = \{P \in \mathcal{P}(S) \mid P \vdash e^+, \forall e^+ \in E^+\}$.
A program $P \in \mathcal{P}(S)$ is *complete* with respect to S iff $P \in \mathcal{P}_{comp}(S)$. □

Definition 2.12. The *cover* of a program $P \in \mathcal{P}(S)$ in S is defined as
$cover(S, P) = \{e \in E \mid P \vdash e\}$. □

The following facts follow directly from the definitions.

Fact 2 $\forall P \in \mathcal{P}_{cons}(S)$, $cover(S, P) \subseteq Loap(S)$.

Fact 3 $\forall P \in \mathcal{P}_{comp}(S)$, $cover(S, P) \supseteq Upap(S)$.

Fact 4 $\forall P \in \mathcal{P}_{comp}(S)$, $(E - cover(S, P)) \subseteq (E - Upap(S))$.

Fact 5 $\forall P \in \mathcal{P}_{rev-cons}(S)$, $cover(S, P) \subseteq (E - Upap(S))$.

Fact 6 $\forall P \in \mathcal{P}_{cons}(S)$, $P \vdash e \Rightarrow e \in E^+$.

Fact 7 $\forall P \in \mathcal{P}_{comp}(S)$, $P \nvdash e \Rightarrow e \in E^-$.

Fact 8 $\forall P \in \mathcal{P}_{rev-cons}(S)$, $P \vdash e \Rightarrow e \in E^-$.

These facts are used in a rough setting for the definitive description of data. Definitive description involves the description of the data with 100% accuracy. In a rough setting, it is not possible to definitively describe the entire data, since some of the positive examples and negative examples (of the concept being described) inherently cannot be distinguished from each other. These facts show that definitive description is possible in a rough setting when an example is covered by a consistent program (the example is then definitely positive), covered by a reverse-consistent program (the example is then definitely negative), or is not covered by a complete program (the example is then definitely negative).

2.4 Some useful declarative biases

Let L_{pi} be the declarative bias such that for any RSILP-ESD system $S = (E, B, L_{pi})$, $H \in \mathcal{H}(S) \Rightarrow$ head predicate of C is the target predicate, for any $C \in H$ (predicate invention is not allowed), let L_{rd} be the declarative bias such that for any RSILP–ESD system $S = (E, B, L_{rd})$, $H \in \mathcal{H}(S) \Rightarrow$ head predicate of C is not in the body of C, for any $C \in H$ (directly recursive definition is not allowed), and let L_{eu} be a declarative bias such that for any RSILP–ESD system $S = (E, B, L_{eu})$, $H \in \mathcal{H}(S) \Rightarrow e \notin C$ for any $e \in E$ and any $C \in H$.

Let V be any set of ground atoms. Let $pred(V)$ denote the set of predicate symbols used in V. For each $A \subseteq pred(V)$, let $V_A = \{q(\ldots) \in V \mid q \in A\}$, and $placelist(A) = \{(q, i) \mid q \in A$, and $1 \leq i \leq n_q$ where n_q is the arity of $q\}$. Let B be any background knowledge of an RSILP–ES system. For each $Z \subseteq placelist(A)$, where $A \subseteq pred(B)$, let L_Z be the declarative bias such that, for any RSILP–ES (E, B) with B as the background knowledge:
$\forall H \in \mathcal{H}(E, B, L_Z), \ \forall C \in H$
$[q(t_1, \ldots, t_n) \in C \Rightarrow [q \in A \wedge \forall i \in \{1, \ldots, n\} \ [(q, i) \in Z \Rightarrow t_i$ is a variable]]].

3 The gRS–ILP model and motifs in strings

3.1 Definition of a motif–RSILP–ESD system

Let Σ be a finite alphabet of letters. A *string* over Σ is any sequence of finite length composed of letters from Σ. We use Σ^+ to denote the set of all the strings over Σ. Let the term *substring of a string* have its usual meaning. (Note that the characters in a substring of a string x must occur contiguously in x.) If r ($\in \Sigma^+$) is a substring of a string s ($\in \Sigma^+$), then r is called a *positive motif* of s. If r is not a substring of the string s, then r is called a *negative motif* of s.

Definition 3.13. We define a *motif–RSILP–ESD system* as a 2-tuple $S = (S', \{\Sigma_1, \Sigma_2, \ldots, \Sigma_n\})$, for some finite $n \geq 1$, where:
(1) each Σ_i, $1 \leq i \leq n$, is a finite alphabet of *letters*, and
(2) $S' = (E, B, L)$ is an RSILP–ESD system such that
(i) E is the universe of examples consisting of a unary predicate, say p,
(ii) B is the background knowledge consisting of ground unit clauses, using the following three predicates: *strings* (of arity, say m), *contains* and *abstains* (both of arity 2), where for any $p(x) \in E$:
    (a) $strings(x, s_1, s_2, \ldots, s_{m-1}) \in B \Rightarrow s_1, s_2, \ldots, s_{m-1}$ are attribute strings of the example $p(x)$,
where for each $1 \leq j \leq m - 1$, $s_j \in \Sigma_i^+$ for some $1 \leq i \leq n$,
    (b) $contains(r, s) \in B \Rightarrow r$ ($\in \Sigma_i^+$) is a positive motif of attribute string s ($\in \Sigma_i^+$), $1 \leq i \leq n$, and
    (c) $abstains(r, s) \in B \Rightarrow r$ ($\in \Sigma_i^+$) is a negative motif of attribute string s ($\in \Sigma_i^+$), $1 \leq i \leq n$, and
(iii) L is the declarative bias $L = L_Z \wedge L_{pi} \wedge L_{rd} \wedge L_{eu}$,
where $A = \{strings, contains, abstains\} = pred(B)$ and $Z = \{(strings, 1), (strings, 2), \ldots, (strings, m), (contains, 2), (abstains, 2)\}$. □

It is seen that the motif–RSILP–ESD system is an R–RSILP–ESD system studied in [7].

It is to be noted that the model can be expressed in an alternate manner, with B using 2–arity predicates of the form $d_1(x, s_1)$, $d_2(x, s_2)$, $\ldots$, $d_{m-1}(x, s_{m-1})$, rather than the single m–arity predicate $strings(x, s_1, s_2, \ldots, s_{m-1})$. L_Z is then suitably modified with $A = \{d_1, d_2, \ldots, d_{m-1}, contains, abstains\}$ and $Z = \{(d_1, 1), (d_1, 2), \ldots, (d_{m-1}, 1), (d_{m-1}, 2), (contains, 2), (abstains, 2)\}$.

3.2 An example

We now consider an illustration of a motif–RSILP–ESD system.

The Protein Identification Resource database [9] contains amino acid sequences, with the FEATURE field for each sequence indicating where the transmembrane domains are located within the sequence. The amino acid sequences are cut into substrings in such a manner that positive example attribute strings are entirely within transmembrane domains, and negative example attribute strings are entirely outside transmembrane domains.

The identification of transmembrane domains from amino acid sequences is described in [10]. A decision tree is learnt that can classify any new attribute string as a transmembrane domain. The simple form 'xAy' of a regular pattern language [10] is used in the nodes of a decision tree. 'x' and 'y' are variable substrings and 'A' is a given fixed substring (the motif). The decision tree consists of leaf nodes (labelled with the resulting class) and internal nodes (labelled with a regular pattern of the form 'xAy'). At an internal node, the decision tree tests if the attribute string matches the regular pattern. The 'Y' path of the node of the decision tree is taken when the attribute string is of the form 'xAy', that is, the motif 'A' is contained in the attribute string; and the 'N' path taken otherwise.

The simple form 'xAy' of a regular pattern language determines whether a given motif 'A' is 'contained' in the attribute string. The 'contains' operator has been studied in detail in [11]. The 'contains' and 'abstains' operators are used in [12, 13] to learn transmembrane domains from amino acid sequences. The 'contains' operator is *true* when the motif is contained in the attribute string and *false* otherwise. The operator 'abstains' is the opposite of contains, and is *true* when the motif is not contained in the attribute string.

The Kyte and Doolittle hydropathy index [14] of an amino acid is used to distinguish the amino acids into three distinct categories. The twenty symbol amino acid sequences are transformed into three symbol strings by assigning each amino acid symbol to one of the following three distinct categories: amino acids with positive hydropathy index (1.8 to 4.5) (*), with slightly negative hydropathy index (-1.6 to -0.4) (+), and those with very negative hydropathy index (-4.5 to -3.2) (-). Σ_1 is an alphabet of the 3 letters +, - and *. Σ_2 is an alphabet of 3 letters a, b, n according to whether the amino acid is acidic, basic or neutral.

That is, let $\Sigma_1 = \{+, -, *\}$ and $\Sigma_2 = \{a, b, n\}$.
Let $E = \{p(e1), p(e2), p(e3)\}$.
Let $B = \{strings(e1, +++-++, aaaaaa), strings(e2, ++-, aab), strings(e3, ++-, aba),$
$contains(++, +++-++), contains(+-, +++-++), contains(-+, +++-++), \ldots,$

$abstains(+*, +++-++), \ldots, contains(\mathsf{a}, \mathsf{aaaaaa}), contains(\mathsf{a}, \mathsf{aab}), \ldots,$
$abstains(\mathsf{b}, \mathsf{aaaaaa}), \ldots\}.$
Let $L = L_Z \wedge L_{pi} \wedge L_{rd} \wedge L_{eu}$, where $A = pred(B)$ and $Z = \{(strings, 1),$
$(strings, 2), (strings, 3), (contains, 2), (abstains, 2)\}.$

$S = (S', \Sigma_1, \Sigma_2)$, where $S' = (E, B, L)$, is a motif–RSILP–ESD system.

3.3 Experimental illustration

583 positive and 583 negative examples of the transmembrane data from PIR
[9] are used in this experimental illustration. The amino acid sequences are converted into three symbol strings based on the Kyte and Doolittle hydropathy
index of the amino acid as described above. This is the same translation mechanism used initially in [10]. The symbols 0, 1 and 2 are used instead of the
symbols *, + and -, respectively, that are used in [10]. Motif length of 2 is used.

Progol is an Inductive Logic Programming system written in C by Dr. Muggleton [15]. The syntax for examples, background knowledge and hypothesis is
Dec-10 Prolog. Headless Horn clauses are used to represent negative examples
and constraints. Progol source code and example files are freely available (for
academic research) from ftp.cs.york.ac.uk under the appropriate directory
in pub/ML_GROUP. Progol version 4.4 dated 25.08.98 is used in this study.

Since only one type of string is used in this experimental illustration, a simplified form of the motif–RSILP–ESD system is used. The background knowledge B consists of predicates such as c(p1,s22) or a(p1,s22), when the motif
'22' (equivalent to '--') is present or not present, respectively, in the attribute
string of example tm(p1). The positive examples (E^+) are given as 'tm(p1).'
to 'tm(p583).' and the negative examples (E^-) are given as ':- tm(n1).' to
':- tm(n583).' Appropriate mode declarations are used in Progol to incorporate
the required declarative bias L. Let $S = (E, B, L)$.

The *first step* is any conventional Progol experiment using the data set. Conventionally, the aim is to maximise the correct cover of both positive and negative
examples (in other words, try to increase the number of positive examples covered and decrease the number of negative examples covered). Let this induced
program be known as P for the purpose of this outline.

The *second step* uses Progol with the *default* noise setting of zero, where
any induced hypothesis is consistent and *cannot* cover *any* negative example.
Let this induced consistent program be P_{cons}.

The induced hypothesis of P_{cons} follows.

```
tm(A) :- a(A,s11), a(A,s12), a(A,s22).
tm(A) :- a(A,s20), a(A,s21), c(A,s12).
tm(A) :- a(A,s12), a(A,s22), c(A,s21).
tm(A) :- a(A,s12), a(A,s20), c(A,s22).
tm(A) :- a(A,s02), a(A,s11).
tm(A) :- a(A,s11), a(A,s21), a(A,s22).
```

```
tm(A) :- a(A,s11), a(A,s20).
tm(A) :- a(A,s12), a(A,s20), c(A,s02).
tm(A) :- a(A,s02), a(A,s12), c(A,s20).
tm(A) :- a(A,s10), a(A,s21).
```

The *third step* is to determine a reverse–consistent program denoted by $P_{rev-cons}$, by exchanging the roles of E^+, E^-, and then repeating step 2. The induced hypothesis of $P_{rev-cons}$ follows.

```
tm(A) :- a(A,s00).
tm(A) :- a(A,s10), c(A,s21).
tm(A) :- a(A,s01), c(A,s22).
```

The results are tabulated below.

$|E^+|$	$|E^-|$	$|E|$	$|cover(S, P_{cons})|$	$|cover(S, P_{rev-cons})|$
583	583	1166	249	55

Using Facts 6 and 8 we have the following.
If $P_{cons} \vdash e$, then $e \in E^+$.
If $P_{rev-cons} \vdash e$, then $e \in E^-$.
Otherwise P is used:
If $P \vdash e$, then it is very likely that $e \in E^+$;
else if $P \nvdash e$, then it is very likely that $e \in E^-$.

249 out of 583 positive examples are definitively described by P_{cons} and 55 out of 583 negative examples are definitively described by $P_{rev-cons}$.

Earlier systems conventionally do not use P_{cons} and $P_{rev-cons}$. They handle the rough setting by inducing P to maximize correct cover by maximizing the number of positive examples covered and negative examples not covered. However, this does not definitively describe the data, since P cannot say with certainty whether an example definitely belongs to the evidence or not. When the gRS–ILP model is used, P_{cons} and $P_{rev-cons}$ are induced to definitively describe part of the data. The rest of the data can be described by P, but not definitively.

4 Conclusions

The gRS–ILP model is extended to motifs in strings. An illustrative experiment is presented regarding the definitive description of transmembrane domains from amino acid sequences using Progol. Possibilities for further work include extensions of the gRS–ILP model to areas other than definitive description, such as prediction, and to other application areas.

Acknowledgements

The authors thank Professors S. Miyano, K. Morita, V. Ganapathy, K. M. Mehata, and R. Siromoney for their valuable comments and support; Dr. N. Zhong for help in providing rough set material; and the Japan Society for Promotion of Science for the Ronpaku Fellowship for the first author.

References

1. S. Muggleton. Inductive logic programming. *New Generation Computing*, 8(4):295–318, 1991.
2. Z. Pawlak. Rough sets. *International Journal of Computer and Information Sciences*, 11(5):341–356, 1982.
3. Z. Pawlak. *Rough Sets — Theoretical Aspects of Reasoning about Data*. Kluwer Academic Publishers, Dordrecht, The Netherlands, 1991.
4. A. Siromoney. The generic Rough Set Inductive Logic Programming (gRS-ILP) model. 1999. Submitted for publication.
5. Usama M. Fayyad, Gregory Piatetsky-Shapiro, and Padhraic Smyth. From data mining to knowledge discovery: An overview. In Usama M. Fayyad, Gregory Piatetsky-Shapiro, Padhraic Smyth, and Ramasamy Uthurusamy, editors, *Advances in Knowledge Discovery and Data Mining*, pages 1–36. AAAI Press / The MIT Press, 1997.
6. A. Siromoney. A rough set perspective of Inductive Logic Programming. In Luc De Raedt and Stephen Muggleton, editors, *Proceedings of the IJCAI-97 Workshop on Frontiers of Inductive Logic Programming*, pages 111–113, Nagoya, Japan, 1997.
7. A. Siromoney and K. Inoue. A framework for Rough Set Inductive Logic Programming — the gRS-ILP model. In *Pacific Rim Knowledge Acquisition Workshop*, pages 201–217, Singapore, November 1998.
8. S. Muggleton and L. De Raedt. Inductive logic programming: Theory and Methods. *Journal of Logic Programming*, 19/20:629–679, 1994.
9. PIR. *Protein Identification Resource*. National Biomedical Research Foundation.
10. S. Arikawa, S. Miyano, A. Shinohara, S. Kuhara, Y. Mukouchi, and T. Shinohara. A machine discovery from amino acid sequences by decision trees over regular patterns. *New Generation Computing*, 11:361–375, 1993.
11. Y. Sakakibara and R. Siromoney. A noise model on learning sets of strings. In *Proceedings of the Fifth Annual ACM Workshop on Computational Learning Theory*, pages 295–302. ACM Press, 1992.
12. A. Siromoney and R. Siromoney. Variations and local exceptions in Inductive Logic Programming. In K. Furukawa, D. Michie, and S. Muggleton, editors, *Machine Intelligence*, volume 14, pages 211–232. Oxford University Press, 1995.
13. A. Siromoney and R. Siromoney. A machine learning system for identifying transmembrane domains from amino acid sequences. *Sādhanā — Indian Academy of Sciences Proceedings in Engineering Sciences — Special Issue on Intelligent Systems*, 21:3:317–325, June 1996.
14. J. Kyte and R.F. Doolittle. A simple method for displaying the hydropathic character of protein. *Journal of Molecular Biology*, 157:105–132, 1982.
15. S. Muggleton. Inverse entailment and Progol. *New Generation Computing*, 13:245–286, 1995.

Rough Problem Settings for Inductive Logic Programming

Chunnian Liu[1] and Ning Zhong[2]

[1] Dept. of Computer Science, Beijing Polytechnic University
[2] Dept. of Computer Science and Sys. Eng., Yamaguchi University

Abstract. Inductive Logic Programming (ILP) is a relatively new method in machine learning. Compared with the traditional attribute-value learning methods, it has some advantages (the stronger expressive power and the ease of using background knowledge), but also some weak points. One particular issue is that the theory, techniques and experiences are much less mature for ILP to deal with imperfect data than in the traditional attribute-value learning methods. This paper applies the Rough Set theory to ILP to deal with imperfect data which occur in large real-world applications. We investigate various kinds of imperfect data in ILP and identify a subset of them to tackle. Namely, we concentrate on incomplete background knowledge (where essential predicates/clauses are missing) and on indiscernible data (where some examples belong to both sets of positive and negative training examples), proposing rough problem settings for these cases. The rough settings relax the strict requirements in the standard normal problem setting for ILP, so that rough but useful hypotheses can be induced from imperfect data.

1 Introduction

Inductive Logic Programming (ILP, see [2, 6, 7, 8]) is a relatively new method in machine learning. ILP is concerned with learning from examples within the framework of predicate logic. ILP is relevant to Knowledge Discovery and Data Mining (KDD), and compared with the traditional attribute-value learning methods (the main stream in KDD community up to date), it possesses the following advantages:

- ILP can learn knowledge which is more expressive than that by the attribute-value learning methods, because the former is in predicate logic while the latter is usually in propositional logic.
- ILP can utilize background knowledge more naturally and effectively, because in ILP the examples, the background knowledge, as well as the learned knowledge are all expressed within the same logic framework.

However, when applying ILP to KDD, we can identify some weak points compared with the traditional attribute-value learning methods, such as:

- It is more difficult to handle numbers (especially continuous values) prevailing in real-world databases, because predicate logic lacks effective means for this.

– The theory, techniques and experiences are much less mature for ILP to deal with imperfect data (uncertainty, incompleteness, vagueness, impreciseness, etc. in examples, background knowledge as well as the learned rules) than in the traditional attribute-value learning methods (see [3, 13, 15], for instance).

In [4], we suggested Constraint Inductive Logic Programming (CILP), an integration of ILP and CLP (Constraint Logic Programming), as a solution for the first problem mentioned in the above.

This paper addresses the second problem, applying the Rough Set theory to ILP to deal with some kinds of imperfect data which occur in large real-world applications. Namely, we concentrate on incomplete background knowledge (where essential predicates/clauses are missing) and on indiscernible data (where some examples belong to both sets of positive and negative training examples), proposing rough problem settings for these cases. The rough settings relax the strict requirements in the standard normal problem setting for ILP, so that rough but useful hypotheses can be induced from imperfect data.

This paper is organized as follows: First in Section 2 we give the standard problem setting for ILP which assumes that everything is correct and perfect. Section 3 lists various kinds of imperfect data in ILP and identifies a subset of them to tackle in this paper. Section 4 is a brief review of some part of the Rough Set theory, which is relevant to our purpose in this paper. Section 5 proposes rough problem settings for incomplete background knowledge and for indiscernible data, and discusses the related work. Finally in Section 6 we summarize our work and point out the future research directions.

2   The Normal Problem Setting for ILP

We follow the notations of [8]. Especially, supposing C is a set of clauses $\{c_1, c_2, \ldots\}$, we use $\overline{C}$ to denote the set $\{\sim c_1, \sim c_2, \ldots\}$. The normal problem setting for ILP can be stated as follows:

Given the positive examples E^+ and the negative examples E^- (both are sets of clauses) and the background knowledge B (a finite set of clauses), ILP is to find a theory H (a finite set of clauses) which is correct with respect to E^+ and E^-. That demands:

– $\forall_{e \in E^+} H \cup B \models e$ (completeness wrt. E^+);
– $H \cup B \cup \overline{E^-}$ is satisfiable (consistency wrt. E^-).

The above ILP problem setting is somewhat too general. In most of the ILP literature, the following simplifications are assumed:

– Single predicate learning. The concept to be learned is represented by a single predicate p (called the Target predicate). Examples are instances of the target predicate p and the induced theory is the defining clauses of p. Only the background knowledge B may contain definitions of other predicates which can be used in the defining clauses of the target predicate.

− Restricted within definite clauses. All clauses contained in B and H are definite clauses, and the examples are ground atoms of the target predicate. We can prove that in this case the condition of consistency has an equivalent form: supposing that Σ is a set of definite clauses, E^- is a set of ground atoms, then Σ is consistent with respect to E^- if and only if $\forall_{e \in E^-} \Sigma \not\models e$. This form is more operational than the general condition (i.e., $\Sigma \cup \overline{E^-}$ is satisfiable).

This paper also takes these simplifications. For the convenience of later reference, here we restate the (simplified) normal problem setting for ILP in a more formal way:

Given:

− The target predicate p.
− The positive examples E^+ and the negative examples E^- (two sets of ground atoms of p).
− Background knowledge B (a finite set of definite clauses).

To find:

− Hypothesis H (the defining clauses of p) which is correct with respect to E^+ and E^-, that is,

1. $H \cup B$ is complete with respect to E^+ (i.e. $\forall_{e \in E^+} H \cup B \models e$). We also say that $H \cup B$ covers all positive examples.
2. $H \cup B$ is consistent with respect to E^- (i.e. $\forall_{e \in E^-} H \cup B \not\models e$). We also say that $H \cup B$ rejects any negative examples.

To make the ILP problem meaningful, we assume the following prior conditions:

− B is not complete with respect to E^+. (Otherwise there will be no learning task at all, because the background knowledge itself is the solution).
− $B \cup E^+$ is consistent with respect to E^- (Otherwise there will be no solution to the learning task).

In the above normal problem setting for ILP, everything is assumed correct and perfect. But in large, real-world empirical learning, data are not always perfect. In contrary, uncertainty, incompleteness, vagueness, impreciseness, etc. are frequently observed in training examples, in background knowledge, as well as in the induced hypothesis. Thus ILP has to deal with imperfect data. In this aspect, the theory, measurement, techniques and experiences are much less mature for ILP than in the traditional attribute-value learning methods. This paper addresses this problem, focusing on the potential role of the Rough Set theory in it.

3 Imperfect Data in ILP

We distinguish five kinds of imperfect data encountered in ILP:

1. Imperfect output
   In ILP, even the input data (examples and background knowledge) are "perfect", there are usually more than one hypotheses which can be induced and the preferential order among hypotheses is an important issue. If the input data is imperfect (see below), the situation is more serious: we have imperfect hypotheses. At present, quantitative measures associated with hypotheses in ILP are not as rich as those of the attribute-value learning [15].

2. Noise data
   This includes erroneous argument values in examples, and/or erroneous classification of examples as belonging to E^+ or E^-. The ILP community has made some advances in noise-handling, using heuristics to avoid overly specific hypotheses which will have low prediction accuracy (see [2] for details). The ideas come from the similar techniques developed within the attribute-value learning framework.

3. Too sparse data
   This means that the training examples are too sparse to induce reliable hypothesis H. The noise-handling mechanisms mentioned above usually also take care of too sparse data. Zhong[16, 17] proposes a mechanism considering unseen instances, which can be also extended to ILP.

4. Missing data
   (a) Missing values
       This means that some arguments of some examples have unknown values. A simple way to deal with this problem is to induce a missing value from other examples (e.g. the value occurring in the same argument place of the majority of other examples).
   (b) Missing predicates
       This means that the background knowledge B lacks essential predicates (or essential clauses of some predicates) so that no non-trivial hypothesis H can be induced. (Note that E^+ itself can be always regarded as a hypothesis, but it is trivial). Especially, even though a large amount of positive examples are given, some examples are not generalized by hypotheses if some background knowledge is missing. This is a big topic in the research area of ILP. In recent study of Muggleton, has taken some important steps in the field of ILP[5].

5. Indiscernible data
   This means that some examples belong to both E^+ and E^-. In this case, the prior condition 2' ($B \cup E^+$ is consistent with respect to E^-) in the normal setting is not satisfied, so there will be no solution to the learning task.

As the above list clearly shows, imperfect data handling is a too vast task to tackle in one paper. In the following of this paper, we will concentrate on item 4(b) (Missing predicates) and item 5 (Indiscernible data). In both cases, the requirement in the normal problem setting of ILP that H should be "correct with respect to E^+ and E^-" needs to be relaxed, otherwise there will be no (meaningful) solutions to the ILP problem. We will give rough problem settings in the cases of missing predicates and indiscernible data, using concepts from the Rough Set theory [9, 10].

4 Rough Set Theory

The Rough Set theory is a powerful mathematical model of imprecise information. For reader's convenience, here we review some concepts in the theory [9, 10, 14] which are relevant to our rough problem settings of ILP presented in the next section.

Approximation space $A = (U, R)$. Here U is a set (called the universe) and R is an equivalence relation on U (called an indiscernibility relation). In fact, U is partitioned by R into equivalence classes, elements within an equivalent class are indistinguishable in the approximation space A.

Lower and *upper approximations.* For an equivalence relation R, the *lower* and *upper approximations* of $X \subseteq U$ are defined by

$$\underline{Apr}_A(X) = \bigcup_{[x]_R \subseteq X} [x]_R = \{x \in U \mid [x]_R \subseteq X\} \tag{1}$$

$$\overline{Apr}_A(X) = \bigcup_{[x]_R \cap X \neq 0} [x]_R = \{x \in U \mid [x]_R \cap X \neq 0\} \tag{2}$$

where $[x]_R$ denotes the equivalence class containing x. Furthermore, $\overline{Apr}_A(X)$ can be simply denoted as $\overline{Apr}(X)$ when A is implicit.

Boundary. $Bnd_A(X) = \overline{Apr}_A(X) - \underline{Apr}_A(X)$ is called the *boundary* of X in A.

Rough membership.

- element x surely belongs to X in A if $x \in \underline{Apr}_A(X)$;
- element x possibly belongs to X in A if $x \in \overline{Apr}_A(X)$;
- element x surely does not belong to X in A if $x \notin \overline{Apr}_A(X)$.

5 Rough Problem Settings for ILP

5.1 Rough Problem Setting for Missing Predicates/Clauses

Considering the following ILP problem (adapted from [1]):

Given:

- The target predicate "customer(Name, Age, Sex, Income)"

- The positive examples E^+ :

  customer(a, 30, female, 1).
  customer(b, 53, female, 100).
  customer(d, 50, female, 2).
  customer(e, 32, male, 10).
  customer(f, 55, male, 10).

- The negative examples E^- :

  customer(c, 50, female, 2).
  customer(g, 20, male, 2).

- Background knowledge B defining predicate "married_to(Husband, Wife)" by

  married_to(e, a).
  married_to(f, d).

To find:

- Hypothesis H (the definition of customer/4) which is correct with respect to E^+ and E^-.

The normal problem setting (see Section 2) is perfectly suitable for this problem, and an ILP system can induce the following hypothesis H (a Prolog program defining *customer/4*):

  customer(N, A, S, I) :- I $\geq$ 10.
  customer(N, A, S, I) :- married_to(N', N), customer(N', A', S', I').

However, if predicate *married_to/2* (or its second clause "married_to(f, d)") is missing in the background knowledge B (and no other predicates/clauses in B that tell any essential difference between persons c and d), no meaningful hypothesis will be induced, because no Prolog program defining *customer/4* can explain why person d is a customer while person c is not, given the fact that except their *Names*, all descriptions of the two persons are the same.

This illustrates that even a learning task can be expressed in the normal problem setting for ILP, it is possible that no meaningful hypothesis can be induced due to the lack of essential predicates/clauses in the background knowledge. In order to learn something useful in these cases, the requirement in the normal problem setting of ILP that H should be "correct with respect to E^+ and E^-" has to be relaxed. We propose the following rough problem setting for incomplete background knowledge.

Rough Problem Setting 1 (for missing predicate/clauses)

Given:

- The target predicate p (the set of all ground atoms of p is U)
- An equivalent relation R on U (we have the approximation space $A = (U, R)$)
- $E^+ \subseteq U$ and $E^- \subseteq U$ satisfying the prior condition: $B \cup E^+$ is consistent with respect to E^-
- Background knowledge B (may lack essential predicates/clauses)

Considering the following rough sets:

- $E^{++} = \overline{Apr}_A(E^+)$, containing all positive examples, and those negative examples $E^{-+} = \{e' \in E^- | \exists_{e \in E^+} eRe'\}$
- $E^{--} = E^- - E^{-+}$, containing the "pure" (remaining) negative examples
- $E_{++} = \underline{Apr}_A(E^+)$, containing the "pure" positive examples. That is, $E_{++} = E^+ - E^{+-}$, where $E^{+-} = \{e \in E^+ | \exists_{e' \in E^-} eRe'\}$
- $E_{--} = E^- + E^{+-}$ containing all negative examples and "non-pure" positive examples

To find:

- Hypothesis H^+ (the defining clauses of p) which is *correct* with respect to E^{++} and E^{--}, that is,
  - $H^+ \cup B$ *covers* all examples of E^{++}.
  - $H^+ \cup B$ *rejects* any examples of E^{--}.
- Hypothesis H^- (the defining clauses of p) which is *correct* with respect to E_{++} and E_{--}, that is,
  - $H^- \cup B$ *covers* all examples of E_{++}.
  - $H^- \cup B$ *rejects* any examples of E_{--}.

Returning to our illustrating example, where predicate *Married_to/2* is missing in the background knowledge B. Let R be defined as

"customer(N, A, S, I) R customer(N', A, S, I)",

with the rough problem setting 1, we may induce H^+ as:

customer(N, A, S, I) :- I $\geq$ 10.
customer(N, A, S, I) :- S = female.

which covers all positive examples and the negative example "customer(c, 50, female, 2)", rejecting other negative examples. We may also induce H^- as:

customer(N, A, S, I) :- I $\geq$ 10.
customer(N, A, S, I) :- S = female, A < 50.

which covers all positive examples except "customer(d, 50, female, 2)", rejecting all negative examples.

These hypotheses are rough (because the problem itself is rough), but still useful. On the other hand, if we insist in the normal problem setting for ILP, these hypotheses are not considered as "solutions".

5.2 Rough Problem Setting for Indiscernible Examples

In the KDD community, people have to deal with situations such as: two patients showing the same symptoms have got different diagnostic results; two person records in a database having the same values for condition attributes have different decision attribute values; etc. In ILP, the similar situation is that example e belongs to E^+ as well as to E^-.

In the illustration given in Section 5.1, if we ignore the person names, the target predicate will be "customer(Age, Sex, Income)" and we will encounter indiscernible examples: "customer(50, female, 2)" belongs to E^+ as well as to E^-. Then the problem cannot be expressed in the normal problem setting at all, because the prior condition 2' ($B \cup E^+$ is consistent with respect to E^-) is violated. In order to learn something useful in these cases, the requirement in the normal problem setting of ILP that H should be "correct with respect to E^+ and E^-" has also to be relaxed. We propose the following rough problem setting for indiscernible examples, which essentially is a special case of the above rough setting 1.

Rough Problem Setting 2 (for indiscernible examples)

Given:

- The target predicate p (the set of all ground atoms of p is U)
- $E^+ \subseteq U$ and $E^- \subseteq U$ where $E^+ \cap E^- \neq \emptyset$.
- Background knowledge B

The rough sets to consider and the hypotheses to find:

- Taking the identity relation I as a special equivalent relation R, the remaining description of rough setting 2 is the same as in rough setting 1.

5.3 Related Work

Siromoney[12] also tries to apply the Rough Set theory to ILP. It considers $A = (U, R)$ where U is the set of all possible positive and negative examples, the equivalent relation R on U is defined as eRe' iff for any H which can be induced, either $H \models e, H \models e'$ or $H \not\models e, H \not\models e'$. Then it considers the concept to be learn as a subset C of U, and points out that

$$E^+ \subseteq \underline{Apr}_A(C) \text{ and } E^- \subseteq U - \overline{Apr}_A(C).$$

However it did not distinguish different kinds of imperfect data, nor give any problem settings for ILP. We think this is not surprising, because R and C used in [12] are posterior in the sense that the user does not know them prior. In contrary, we use user-defined R, and E^+, E^- etc. (all known to the user) as the start point, so we can give rough problem settings for ILP, and it is possible to develop ILP systems allowing the user to specify rough problem settings, as well as the normal problem setting. These ILP systems will be able to induce useful hypotheses even when the input data are imperfect.

6 Conclusions and Future Work

This paper addresses the problem of imperfect data handling in Inductive Logic Programming (ILP). We discuss various kinds of imperfect data in ILP, and apply the Rough Set theory to incomplete background knowledge (where essential predicates/clauses are missing) and to indiscernible data (where some examples belong to both sets of positive and negative training examples), proposing rough problem settings for these cases. The rough settings relax the strict requirements in the standard normal problem setting for ILP, so that rough but useful hypotheses can be induced from imperfect data.

Future work in this direction includes:

- Trying to apply the Rough Set theory to other kinds of imperfect data (noise data, too sparse data, missing data, etc.) in ILP.
- Giving quantitative measures associated with hypotheses induced within the rough problem settings of ILP, using appropriate concepts and techniques from the Rough Set theory.
- Developing a new ILP system (or extend an existing ILP system) which allows the user to specify rough problem settings, as well as the normal problem setting. The new ILP system will be able to induce useful hypotheses even when the input data are imperfect.

Acknowledgements

The first author's work of this paper is supported by the Natural Science Foundation of China (NSFC), Beijing Municipal Natural Science Foundation (BMNSF) and Chinese 863 High-Tech Program.

References

1. S. Dzeroski, "Inductive Logic Programming and Knowledge Discovery in Databases ", U.M. Fayyad et al (eds) *Advances in Knowledge Discovery and Data Mining*, MIT Press (1996) 117-151.
2. N. Lavrac, S. Dzeroski, and I. Bratko, "Handling Imperfect Data in Inductive Logic Programming", L. de Raedt (Ed), *Advances in Inductive Logic Programming*, IOS Press (1996) 48-64.
3. T.Y. Lin and N. Cercone (eds), *Rough Sets and Data Mining: Analysis for Imprecise Data*, Kluwer (1997)

4. C. Liu, N. Zhong, S. Ohsuga, "Constraint ILP and its Application to KDD", Proc. of IJCAI-97 Workshop on Frontiers of ILP (1997) 103-104
5. S. Moyle and S. Muggleton, "Learning Programs in the Event Calculus", Proc. 7th Inter. Workshop on ILP (1997) 205-212.
6. S. Muggleton, "Inductive Logic Programming", New Generation Computing, Vol. 8, No 4 (1991) 295-317.
7. S. Muggleton (Ed), *Inductive Logic Programming*, Academic Press (1992).
8. S-H. Nienhuys-Cheng and R. de Wolf, *Foundations of Inductive Logic Programming*, Springer LNAI 1228 (1997).
9. Z. Pawlak, "Rough Sets", International Journal of Computer and Information Science, Vol. 11 (1982) 341-356.
10. Z. Pawlak, *Rough Sets: Theoretical Aspects of Reasoning about Data*, Kluwer (1991).
11. L. de Raedt (Ed), *Advances in Inductive Logic Programming*, IOS Press (1996).
12. A. Siromoney, "A Rough Set Perspective of Inductive Logic Programming", Proc. of the IJCAI'97 Workshop on Frontiers of ILP (1997) 111-113.
13. Y.Y. Yao and T.Y. Lin, "Generalization of Rough Sets Using Modal Logic", Intelligent Automation and Soft Computing, An International Journal, Vol. 2 (1996) 103-120.
14. Y.Y. Yao, S.K.M. Wong, and T.Y. Lin, "A Review of Rough Set Models", T.Y. Lin and N. Cercone (eds.) *Rough Sets and Data Mining* Kluwer (1997) 47-76.
15. Y.Y. Yao and N. Zhong, "An Analysis of Quantitative Measures Associated with Rules", N. Zhong and L. Zhou (Eds), *Methodologies for Knowledge Discovery and Data Mining*, LNAI 1574, Springer-Verlag (1999) 479-488.
16. N. Zhong, J.Z. Dong, and S. Ohsuga, "Data Mining: A Probabilistic Rough Set Approach", L. Polkowski and A. Skowron (Eds), *Rough Sets in Knowledge Discovery*, Physica-Verlag (1998) 127-146.
17. N. Zhong, J.Z. Dong, and S. Ohsuga, "Data Mining Based on GDT and Rough Sets", X. Wu et al (Eds), *Research and Development in Knowledge Discovery and Data Mining*, LNAI 1394, Springer-Verlag (1998) 360-373.

Using Rough Sets with Heuristics for Feature Selection

Juzhen Dong[1], Ning Zhong[1], and Setsuo Ohsuga[2]

[1] Dept. of Computer Science and Sys. Eng., Yamaguchi University
[2] Dept. of Information and Computer Science, Waseda University

Abstract. Practical machine learning algorithms are known to degrade in performance when faced with many features that are not necessary for rule discovery. To cope with this problem, many methods for selecting a subset of features with similar-enough behaviors to merit focused analysis have been proposed. In such methods, the *filter* approach that selects a feature subset using a preprocessing step, and the *wrapper* approach that selects an optimal feature subset from the space of possible subsets of features using the induction algorithm itself as a part of the evaluation function, are two typical ones. Although the filter approach is a faster one, it has some blindness and the performance of induction is not considered. On the other hand, the optimal feature subsets can be obtained by using the wrapper approach, but it is not easy to use because the complexity of time and space. In this paper, we propose an algorithm of using the rough set methodology with greedy heuristics for feature selection. In our approach, selecting features is similar as the filter approach, but the performance of induction is considered in the evaluation criterion for feature selection. That is, we select the features that damage the performance of induction as little as possible.

1   Introduction

Generally speaking, the purpose of building databases in most organizations is that of managing information sources effectively. In other words, data are rarely specially collected/stored in a database for the purpose of mining knowledge in most organizations. Hence, a database always contains a lot of attributes that are redundant and not necessary for rule discovery. If these redundant attributes do not be removed, not only the time complexity of rule discovery will increases, but also the quality of the discovered rules may be much degraded.

Which attribute should be deleted is very difficult to decide for non-experts and even for experts. Clearly, we need additional methods for selecting the feature subset. The problem of feature subset selection is that of finding an optimal subset of features of a database according to some criterion, so that a classifier with the highest possible accuracy can be generated by an inductive learning algorithm that is run on data containing only the subset of features.

Many researchers have investigated this field and several methods have been proposed [1, 3, 11, 6, 5]. A kind of these methods is that of ranking features first according to evaluation measures such as consistency, information, distance, and dependence, and then selecting the features with a higher rank. This kind of methods considers only data but not the classifying properties. The *filter* approach belongs to this type. Another kind of methods such as *wrapper* approach is that of using the induction algorithm itself as an evaluation function for selecting the optimal subsets from the space of all possible subsets. Furthermore, the *rough set* theory provides a mathematical tool to find out all of possible feature subsets (called *reducts*)[1]. Unfortunately, the number of possible subsets are always very large when N is large because there are 2^{N-1} subsets for N features. Therefore examining exhaustively all subsets of features for selecting the optimal one is NP-hard. Most practical algorithms attempt to be fit for the data by solving the NP-hard optimization problem [4].

In this paper, we propose an algorithm of using the rough set theory with greedy heuristics for feature selection. We attempt to find an approach that is not heavy but effective. In our approach, features are selected from the space of features but no the space of reducts, and using the evaluation criterion in which the performance of induction is considered. That is, we select the features that damage the performance of induction as little as possible.

2 Dispensable and Indispensable Features

In the rough set theory, a decision table is denoted $T = (U, A, C, D)$, where U is universe of discourse, A is a family of equivalence relations over U, and $C, D \subset A$ are two subsets of features that are called condition and decision features, respectively[1].

Before describing what are the dispensable and indispensable features, some basic terms and notations on the rough set theory must be explained first.

Lower Approximations:

The lower approximations of a set, $\underline{R}X$, is the set of all elements of U which can be with certainty classified as elements of X, in the knowledge R, where $X \subseteq U$. It can be presented formally:

$$\underline{R}X = \bigcup \{Y \in U/R : Y \subseteq X)$$

The Positive Region:

We also using a positive region to denote the lower approximations of a set. Let P and Q be equivalence relation over U, $P \subset U$ and $Q \subset U$. The P-positive region of Q, $POS_P(Q)$, is the set of all objects of universe U which can be properly classified to classes of U/Q employing knowledge expressed by the classification U/P.

$$POS_P(Q) = \bigcup_{X \in U/Q} \underline{R}X$$

Dispensable and Indispensable Features:

The dispensable and indispensable features are defined as follows:

Let $c \in C$. A feature c is *dispensable* in T, if $POS_{(C-c)}(D) = POS_C(D)$; otherwise feature c is *indispensable* in T.

If c is an indispensable feature, deleting it from T will cause T to be inconsistent. Otherwise, c can be deleted from T. $T = (U, A, C, D)$ is independent if all $c \in C$ are indispensable in T.

Reduct:

The set of features $R \subseteq C$ will be called a reduct of C, if $T = (U, A, R, D)$ is independent and $POS_R(D) = POS_C(D)$.

CORE:

The set of all indispensable features in (C, D) will be denoted by $CORE(C, D)$.

$$CORE(C, D) = \bigcap RED(C, D)$$

where $RED(C, D)$ is the set of all reducts of (C, D).

3 Searching Indispensable Features

All of indispensable features should be contained in an optimal feature subset, because removing any of them will cause inconsistent in a decision table. As defined in Section 2, $CORE$ is the set of all indispensable features. Hence the process of searching indispensable features is that of finding $CORE$.

The discernibility matrix proposed by Skowron [2, 1] can be used for $CORE$ searching. The basic idea of the discernibility matrix can be briefly presented as follows:

Let $T = (U, A, C, D)$ be a decision table, with $U = \{u_1, u_2, \ldots, u_n\}$. By a *discernibility matrix* of T, denoted $M(T)$, we will mean $n \times n$ matrix defined thus:

$$m_{ij} = \{a \in C : a(u_i) \neq a(u_j) \ \land \ D(u_i) \neq D(u_j)\} \text{ for } i, j = 1, 2, \ldots, n.$$

Thus entry m_{ij} is the set of all attributes that discern objects u_i and u_j.

The CORE can be defined now as the set of all single element entries of the discernibility matrix, that is,

$$CORE(C) = \{a \in C : m_{ij} = (a), \quad \text{for some i, j }\}.$$

Searching $CORE$ is to search such a set of features in which each feature is unique to discern some objects.

Example

The discernibility matrix corresponding to the sample database (the decision table) shown in Table 1 with $U = \{u_1, u_2, \ldots, u_7\}$, $C = \{a, b, c\}$, $D = \{d\}$ is as follows:

Table 1. A Sample Database

	a	b	c	d	E
u1	1	0	2	1	1
u2	1	0	2	0	1
u3	1	2	0	0	2
u4	1	2	2	1	0
u5	2	1	0	0	2
u6	2	1	1	0	2
u7	2	1	2	1	1

	u1	u2	u3	u4	u5	u6
u2	-					
u3	b,c,d	b,c				
u4	b	b,d	c,d			
u5	a,b,c,d	a,b,c	-	a,b,c,d		
u6	a,b,c,d	a,b,c	-	a,b,c,d	-	
u7	-	-	a,b,c,d	a,b	c,d	c,d

The CORE is the feature b. We can see that b is the unique feature for discerning $u1$ and $u4$. Furthermore, two reducts are {b, c} and {b, d}. Since the feature a is not contained in any reduct, it could be deleted.

4 Feature Subset Selection

An optimal feature subset selection based on the rough set theory can be viewed as finding such a reduct R, $R \subset C$ with the best classifying properties. R will be used to instead of C in a rule discovery algorithm.

Selecting an optimal reduct R from all subsets of features is not an easy work. Considering the combinations among N features, the number of possible reducts is 2^{N-1}. Hence, selecting the optimal reduct from all of possible reducts is NP-hard. For this reason, many methods for finding approximate results have been proposed [1, 3, 11, 6, 5]. However the features in $CORE$ must be included whatever in an optimal result or in an approximate result. It is obvious that all of indispensable features in $CORE(C, D)$ cannot be deleted from C if the accuracy of a decision table is not changed (dropped). The feature(s) in $CORE$ must be the member of feature subsets. Note that not all of the features in an optimal feature subset must be indispensable. Therefore, The problem of feature subset selection will become how to select the features from dispensable features for forming the best reduct with $CORE$. We use $CORE(C, D)$ as the core of feature subsets. If $CORE$ is not a reduct of (C, D), some of dispensable features must be selected to make up a reduct.

Basically, the feature selection approaches can be divided into two types: the *filter* approach and the *wrapper* approach [11].

4.1 The Filter Approach

The *filter* approach selects the features using a preprocessing step. The main disadvantage of the filter approach is that it totally ignores the effect of the selected feature subset on the performance of the induction algorithm.

The main feature selection strategies of the filter approach are as follows:

1. The minimal subset of features (the MINIMAL_FEATURES bias).
   This bias has severe implications when applied blindly without regard for the resulting induced concept. For example, the ID number of the patient in a medical diagnosis data may be picked as the only feature needed. Given only the ID number, any induction algorithm is expected to generalize very poorly.
2. Selecting the features with a higher rank.
   Ranking a list of features according to some measures. A measure can be based on any of accuracy, consistency, information, distance, and dependence. However, this bias does not help with a redundant feature. Moreover, it may not be wise to use this bias on the data in which some irrelevant feature is strongly correlated to the class feature.

4.2 The Wrapper Approach

In the *wrapper* approach, the features subset selection is done using the induction algorithm as a black box. A search for a good subset is done using the induction algorithm itself as a part of the evaluation function.

The wrapper approach conducts a search in the space of possible subsets of feature. For example, the space of reducts. There are several search methods that can be used for the *wrapper* approach,

- Exhaustive/Complete search
- Heuristic search
- Nondeterministic search

and so on.

When the number of features N is small, the search space may be not so large, but it grows exponentially when N increases. In general, given a search space, the more you search it, the better the subset you can find. But the resource is not unlimited, we have to sacrifice optimality of selected subsets. The sacrifice has also a limit, we must keep the optimality of a feature subset as much as possible while spending as little search time as possible.

Exhaustive/Complete search exhausts all possible subsets and find the optimal ones. It is obvious that no optimal subset can possibly be missed. The number of possible subsets is 2^{N-1}, so that the time complexity of searching all of them is $O(2^{N-1})$. Using heuristics in search avoids brute-force search, but at the same time risks losing optimal subsets. Heuristic search is obviously much faster than exhaustive search since it only searches a part of subsets and finds a near-optimal subset. Nondeterministic search is also called random search strategies. Searching for the next set at random, that is, a current set dose not directly grow or shrink from any previous set following a deterministic rule. There are two characteristics: (1) do not need to wait until the search end; (2) do not know when the optimal set shows up, although we know a better one appears when it is there.

5 Heuristics for Feature Subset Selection

In this section, we describe our approach for feature subset selection. The data we faced are almost very large and the number of features are a quite many, so we select the heuristic search as our search strategy, because exhaustive/complete search is too time consuming and nondeterministic search is difficult to know when the optimal subset appears. Although the heuristic search cannot guarantee that the result must be the best one, it is a better way for solving very-large, complex problems [8].

A search is invalid if it totally ignores the effect of the selected feature subset on the performance of the induction algorithm. Using an induction algorithm itself as a part of the evaluation function like the wrapper approach, no doubt, a good subset can be searched. However, evaluating all subsets of features, even evaluating just a part of feature subsets selected by some strategy, are also time consuming.

For selecting feature subsets from a large database with a lot of features, we select the best features one by one by using the evaluation criterion in an induction algorithm, until a reduct is found. However, unlike the wrapper approach, we do not select the best feature subset from all of possible subsets of features.

The evaluation criterion used in our feature selection approach is that of the rule selection used in the rule discovery system, GDT-RS, developed by us [9, 10]:

1. Selecting the rules that cover as many instances as possible;
2. Selecting the rules that contain as little features as possible, if they cover the same number of instances;
3. Selecting the rules with larger strengths, if they are in the same generalization (condition features) level and cover the same number of instances.

Where the strength of a generalization is related to the number of values in each feature in the generalization. The more the number of values the stronger the generalization.

A feature subset, is good or not, depends on the strengths of the rules discovered by using this subset. The strong the strength, the better the subset. To select the feature that is of benefit to acquire the rules with a larger cover rate and a strong strength, the following selection strategies are used:

- To obtain the subset of features as small as possible, selecting the features that cause the number of consistent instances increases faster.

To avoid the features with a lot of attribute values such as the ID number or continuous attributes are considered first, not only the preprocess of deleting unnecessary attribute and the discretization of continuous attributes is important, but also the features that can generate the strong rules must be selected first. The size of the maximal subset in $POS_R(D)/IND(\{R,D\})$ should be considered since it affects the strengths of rules. In general, the more the number of attribute values in a feature in R, the more the number of subsets, and the smaller the size of the maximal subset. Selecting a feature, by which a bigger subset can be acquired, is a way for our purpose.

Let a cardinality of the lower approximation of a set, $card\ POS_R(D)$, denote the number of consistent instances, $max\_size(POS_R(D)/IND(\{R,D\}))$ denote the size of the maximal subset of the lower approximation of the set $POS_R(D)$. The feature selection can be regarded as selecting such features: if adding them into the subset of features, R, the $card\ POS_R(D)$ increases faster and the $max\_size(POS_R(D)/IND(\{R,D\}))$ is bigger than adding other features.

- When two features have the same performance described above, the one that contains a littler number of different values will be selected. This is for guaranteeing that the number of instances covered by a rule is as many as possible.

Based on the preparations stated above, a heuristic algorithm is described below. At first, we use the features in $CORE$ as the initial feature subset, and then choose the features from dispensable features one by one by using the strategies stated above, and add them into the feature subset, until a reduct is achieved.

A Heuristic Algorithm

Let R be a subset of the selected features, P the set of unselected features, U all of instances, X the contradictory instances, and $EXPECT_k$ the threshold of the accuracy.

In the initial state, $R = CORE(C, D)$, $P = C - CORE(C, D)$,
$$X = U - POS_R(D).$$

Step 1. Calculate the dependent degree, k,
$$k = \gamma_R(D) = \frac{card\ (U - X)}{card\ U}$$

where $card$ denotes the cardinality of the set.
If $k \geq EXPECT\_k$, then stop.

Step 2. For each p in P, calculate
$$v = card\ POS_{R+\{p\}}(D).$$
$$m = card\ max\_set(POS_{R+\{p\}}(D)/IND(R+\{p\}, D)).$$

Step 3. Choose the best feature p with the largest $v * m$, do

$$R = R \cap p,$$
$$P = P - p;$$

Step 4. Remove all of the consistent instances x that are contained in the set of $POS_R(D)$ from X.

Step 5. Goto back to *Step 1.*

Example

We would like to use the sample database shown in Table 1 as an example to explain how to get the feature subset using this algorithm. In Table 1, a, b, c, and d are condition features, E is a decision feature, and $U = \{u1, u2, u3, u4, u5, u6, u7\}$. $\{b\}$ is a unique indispensable feature, because it will cause inconsistent, $\{a_1 c_2 d_1\} \rightarrow E_1$ and $\{a_1 c_2 d_1\} \rightarrow E_0$, if deleting $\{b\}$.

From the following equivalence classes,

$$U/\{b\} = \{\{u1, u2\}, \{u5, u6, u7\}, \{u3, u4\}\}$$
$$U/\{E\} = \{\{u4\}, \{u1, u2, u7\}, \{u3, u5, u6\}\},$$

we know *b-positive region* of E, $POS_b(E)$, is $\{u1, u2\}$. Hence, in the initial state, $R = \{b\}$, $P = \{a, c, d\}$, and $X = \{u3, u4, u5, u6, u7\}$. The initial state is shown as follows:

U	b	E
u3	2	2
u4	2	0
u5	1	2
u6	1	2
u7	1	1

Let $EXPECT\_K = 1$, the termination condition will be $k \geq 1$.

Since $k = 2/7 < 1$, R is not a reduct, we must continue to select. The next candidate is a, c or d. Table 2 gives the results of adding $\{a\}$, $\{c\}$, or $\{d\}$ into R, respectively.

From Table 2 or the following equivalence classes,

$$U/E = \{\{u4\}, \{u7\}, \{u3, u5, u6\}\};$$
$$U/\{a, b\} = \{\{u3, u4\}, \{u5, u6, u7\}\};$$
$$U/\{b, c\} = \{\{u5\}, \{u6\}, \{u7\}, \{u3\}, \{u4\}\};$$
$$U/\{b, d\} = \{\{u5, u6\}, \{u7\}, \{u3\}, \{u4\}\};$$
$$POS_{\{a,b\}}(E) = \emptyset;$$
$$POS_{\{b,c\}}(E) = POS_{\{b,d\}}(E) = \{u3, u4, u5, u6, u7\};$$
$$max\_size(POS_{\{b,c\}}(E)/\{b, c, E\}) = 1;$$
$$max\_size(POS_{\{b,d\}}(E)/\{b, d, E\}) = |\{u5, u6\}| = 2,$$

Table 2. Selecting the second feature from P = {a, c, d}.

U	a	b	E
u3	1	2	2
u4	1	2	0
u5	2	1	2
u6	2	1	2
u7	2	1	1

1. selecting {a}

U	b	c	E
u3	2	0	2
u4	2	2	0
u5	1	0	2
u6	1	1	2
u7	1	2	1

2. selecting {c}

U	b	d	E
u3	2	0	2
u4	2	1	0
u5	1	0	2
u6	1	0	2
u7	1	1	1

3. selecting {d}

we can see that selecting the feature a cannot reduce the number of contradictory instances, but if selecting either c or d, all of instances become consistent. Since the maximal set is in the $U/\{b, d, E\}$, d should be selected first.

After adding d into R, all of instances are consistent and must be removed from U. Hence U is empty, $k = 1$, the process finished. The selected feature subset is {b, d}.

6 Experiment Results

Using the algorithm stated in Section 5, we have tested several databases. Some of them are artificial: Monk1, Monk3; some of them are well made: Mushroom, breast cancer, earthquake; and some of them are real world databases: meningitis, medical treatment, land-slide. Table 3 shows the results of feature selection on these datasets. In Table 3, #attr_n, #inst_n, #CORE, and #attt_n(sel) denote the number of features in a dataset, the number of instances, the number of features in $CORE$, and the number of features selected, respectively.

Table 3. Results of feature selection

Dataset	#attr_n	#inst_n	#CORE	#attr_n(sel)
Monk1	6	124	3	3
Monk3	6	122	4	4
Mushroom	22	8124	0	4
breast_cancer	10	699	1	4
earthquake	16	155	0	3
meningitis	30	140	1	4
medical treatment	57	20920	2	9
land-slide	23	3436	6	8

7 Conclusions

In this paper, we presented an approach for feature selection. It is based on the rough set theory and greedy heuristics. The main advantages of our approach are that it can select a better subset of features quickly and effectively from a large database with a lot of features; the selected features do not damage the performance of induction so much since the performance of induction is considered in the evaluation criterion for feature selection.

References

1. Z. Pawlak. *ROUGH SETS, Theoretical Aspects of Reasoning about Data*, Kluwer (1991).
2. A. Skowron and C. Rauszer. "The Discernibility Matrics and Functions in Information Systems", in R. Slowinski (ed.) *Intelligent Decision Support*, Kluwer (1992) 331-362.
3. A. Skowron and L. Polkowski. "A Synthesis of Decision Systems from Data Tables", in T.Y. Lin and N. Cercone (eds.) *Rough Sets and Data Mining*, Kluwer (1997) 259-299.
4. M. Boussouf. "A Hybrid Approach to Feature Selection", J. Zytkow and M. Quafafou (eds.) *Principles of Data Mining and Knowledge Discovery*, LNAI 1510, Springer (1998) 231-238.
5. Y.Y. Yao, S.K.M. Wong, and C.J. Butz. "On Information-Theoretic Measures of Attribute Importance", Zhong, N. and Zhou, L. (eds.) *Methodologies for Knowledge Discovery and Data Mining*, LNAI 1574, Springer (1999) 231-238.
6. H. Liu and H. Motoda. *Feature Selection*, Kluwer (1998).
7. B. Zupan, M. Bohanec, J. Demsar, and I. Bratko. "Feature Transformation by Function Decomposition", IEEE Intelligent Systems, 13(2) (1998) 38-43.
8. A.V. Aho, J.E. Hopcroft, J.D. Ullman. *Data Structures and Algorithms*, Addison-Wesley (1983)
9. N. Zhong, J.Z. Dong, and S. Ohsuga, "Data Mining based on the Generalization Distribution Table and Rough Sets", X. Wu et al. (eds.) *Research and Development in Knowledge Discovery and Data Mining*, LNAI 1394, Springer (1998) 360-373.
10. J.Z. Dong, N. Zhong, and S. Ohsuga, "Probabilistic Rough Induction: The GDT-RS Methodology and Algorithms", Z.W. Ras and A. Skowron (eds.) *Foundations of Intelligent Systems*, LNAI 1609, Springer (1999) 621-629.
11. R. Kohavi. "Useful Feature Subsets and Rough Set Reducts", *Proc. Third Inter. Workshop on Rough Set and Soft Computing* (1994) 310-317.

The Discretization of Continuous Attributes Based on Compatibility Rough Set and Genetic Algorithm

Lixin Sun and Wen Gao
Dept. of Computer Science and Engineering
Harbin Institute of Technology, Harbin 150001,China
slxyjs@public.hr.hl.cn   wgao@ict.ac.cn

Abstract. Most of the existing discretization approaches discrete each continuous attribute independently, without considering the discretization results of other continuous attributes. Therefore, some unreasonable and superfluous discretization split points are usually created. Based on compatibility rough set model and genetic algorithm, a global discretization approach has been provided. The experimental results indicate that the global discretization approach proposed can significantly decrease the number of discretization split points and the number of rules, but increase the predictive accuracy of the classifier.

1 Introduction

In the practical application of machine learning and data mining, there are many continuous attributes where some symbolic inductive learning algorithm could not be applied unless the continuous attributes are first discretized. There are currently several discretization algorithms[2,5,9,14,15]. Most are independent methods, meaning that they discrete each continuous attribute independently, without considering the discretization results of other continuous attributes.

The final discretization result of a set of continuous attributes mainly depends on the locations and the number of the selected discretization split points that come from different continuous attributes. Usually, the discretization approaches that consider the dependency among multiple continuous attributes are called global discretization. By means of the binarization of continuous attributes[8], the problem of the global discretization of continuous attributes was converted into the problem of selecting the simplest subset of binary attributes[9]. However, the problem with the global discretization approach is that the number of the initial binary attributes is extremely large, because all potential split points are taken into account.

Based on the new compatibility rough set model introduced in section 2, we proposed a novel approach that can generate reasonable sized initial split points. Genetic algorithm is also adopted in order to obtain optimal discretization results.

2 Compatibility Rough Set

The standard rough set model introduced by Pawlak, Z. [10] is based on the equivalence relation on the instances. Many authors have proposed interesting extensions of the initial rough set model[6,7,13,16]. A common feature of these extended models is the induction the unequivalence relation on the instances based on the unequivalence relation on the attribute values. In this study, we intent to introduce a compatibility rough set model based on the compatibility relation directly on the instances instead of the attribute values.

Let $B = \{u_1, \cdots, u_l\}$ is a subset of the full set of the instances, $B \subseteq U$, $C = \{c_1, \cdots, c_m\}$ is the set of the conditional attributes (assume C contains both continuous and symbolic attributes). Each instance u_i corresponds to an attribute value vector $\{v_{i,1}, \cdots, v_{i,m}\}$. We can construct a hyper-region in feature space based on a group of instances by using the instances merging operation $u_1 \oplus \cdots \oplus u_l$, where $\oplus$ denotes the merging operation. Actually, the instances merging operation is realized by the attribute values merging operation $u_1 \oplus \cdots \oplus u_l = \{\{v_{1,1} \oplus \cdots \oplus v_{l,1}\}, \cdots, \{v_{1,m} \oplus \cdots \oplus v_{l,m}\}\}$. For a continuous attribute, $v_{1,1} \oplus \cdots \oplus v_{l,1} = [\min(v_{1,1}, \cdots, v_{l,1}), \max(v_{1,1}, \cdots, v_{l,1})]$, and for a symbolic attribute, $v_{1,1} \oplus \cdots \oplus v_{l,1} = \{v_{1,1} \cup \cdots \cup v_{l,1}\}$.

If the instances in a hyper-region have identical decision label, then the hyper-region is called pure hyper-region. Any pair of instances u_i, u_j in a pure hyper-region is called compatibility. The compatibility on u_i, u_j defines a type of binary relation. Clearly, this binary relation is reflexive and symmetric, therefore the compatibility on the instances is the compatibility relation.

Basing on the compatibility relation, we can only get some compatibility classes on the instances. The union of all the compatibility classes can cover the full set of instances, but usually not every compatibility class is required to cover the set. A set of compatibility classes is called the simplest closed cover, if it can completely cover a full set of instances and contain the least compatibility classes. Because each compatibility class can only cover a proportion of a full set of instances, the problem of searching the simplest closed cover equals to the problem of minimal set cover, it is NP-hard. Next, we will give an greedy algorithm that is able to search for an approximate solution of the simplest closed cover.

Let $U / IND(\{d\}) = \{D_1, \cdots, D_r\}$ is the set of decision equivalence classes of a decision table $A = (U, C \cup \{d\})$, CP is a compatibility class, CPS is the set of compatibility classes, TM is a temporary set.

Step1. $CPS = \{\}$;

Step2. for i=1 to r do { $TM = D_i$;

  While(Card(TM) $\neq 0$)

  { CP = $u_k \in TM$; TM = TM \ u_k ;

  for j=1 to Card(TM) do

  { if(CP is compatibility with $u_j \in D_i$ against $\{D_{i+1}, \cdots, D_r\} \cup CPS$)

   then { CP = CP $\cup u_j$; TM = TM \ u_j ;}

  }

  CPS = CPS $\cup CP$;

  } }

3   Obtaining the Initial Set of Split Points

The major objective of introducing this compatibility rough set model is to generate a moderate sized initial set of split points, from which the simplest set of the discretization split points will be searched. The valued domain of each attribute is divided into some overlapping valued intervals or valued set. For a continuous attribute, each valued interval has two boundaries. The boundaries with smaller value and larger value are called low boundary and high boundary, respectively. The initial split points can be determinate based on the relationship among these boundaries of the valued intervals. The actual procedure is described as follows:

Let $LBS = \{lb_1^i, \cdots, lb_p^i\}$ and $UBS = \{ub_1^i, \cdots, ub_q^i\}$ are two sets of low boundaries and up boundaries of a continuous attribute $c_i \in C$. SPS is the set of split points, TM is a temporary set.

    Step1. $SPS = \{\}$; $TM = \{\}$; $N = 0$;

    Step2. for j=1 to q do

        {if (there is a $lb_{min}^i = \min\{lb_k^i : ub_j^i < lb_k^i \in LBS\}$)

        then look for a $ub_{min}^i = \max\{ub_k^i : lb_{min}^i > ub_k^i \in UBS\}$;

        $TM = TM \cup \{(ub_{max}^i, lb_{min}^i)\}$; $N + +$;}

    Step3. for j=1 to N do

        $\{ sp = (lb_j^i + ub_j^i)/2$;   /* $(lb_j^i, ub_j^i) \in TM$ */

        $SPS = SPS \cup \{sp\}$;}

    Step4. Repeat Step2 and Step3 for each $c_i \in C$.

4   Selecting the Optimal Set of Split Points

The three main issues in applying genetic algorithm to any optimization problem are the choice of an appropriate representation scheme, a fitness function, and the initialization the chromosome population. The natural representation for optimal feature subset selection (OFSS) is exactly the same as the bit string of length N representing the presence or absence of the N possible binary attributes.

The problem of OFSS is usually regarded as the problem of searching state space. Obviously, the closer the starting search states are to the final optimal state, the higher the search efficiency is. H. S. Nguyen [9] had provided a greedy quick discretization method by which we can greedily construct some relative reducts of the binary attributes (the population of the initial chromosomes).

Shan [14] had given an entropy function that can measure the discretization complexity.

$$H_C(D) = -\sum_{j=1}^{m} p(D_j) \sum_{i=1}^{n} p(C_i/D_j) \log(C_i/D_j)$$

Where $p(D_j)$ is the probability that an instance belongs to decision equivalence class D_j, $p(C_i/D_j)$ is the probability that an instance belonging to decision equivalence class D_j is matched by conditional equivalence class C_i.

In order to measure the simplification (the number of the binary attributes) of a relative reduct while the discretization complexity is measured, we slightly modify above entropy function as the fitness function:

$$H(C_{Rem}^b) = Card(C_{Rem}^b) - \sum_{j=1}^{m} p(D_j) \sum_{i=1}^{n} p(C_i / D_j) \log p(C_i / D_j)$$

Where $Card(C_{Rem}^b)$ is the number of the binary attributes in current chromosome.

After certain specific genetic operations, such as cross-over or mutation, the offspring of two relative reducts may not be a new relative reduct. However, being a relative reduct is the essential condition of the optimal subset of the binary attributes. In order to efficiently detect whether a new offspring is a relative reduct, we take the *simplified discernibility factor set* (SDFS) Φ_C modulo decision information, which was proposed by T. Mollestad[11], as a filter. Let C_B is the full set of the binary attributes and C_B' is a subset of C_B, if $\exists \varphi \in \Phi_C$ makes $\varphi \subseteq C_B - C_B'$ is true, then C_B' is a relative reduct, otherwise, C_B' is not a relative reduct. Usually $Card(\Phi_C) \ll Card(U^2)$, so the efficiency of the above set containing inquery is quit high. If only the relative reducts are forwarded to the fitness function, then we can omit a lot of unnessesary fitness function calculation.

5 Experiments and Conclusions

In order to test the effectiveness of the discretization approach based on the compatibility rough set model, an experiment is conducted. We selected nine data sets that are suitable to evaluate the discretizaiton methods from the UCI repository[12]. A general-purpose genetic algorithm program GENESIS [3] was used as searching engine. Parameters for the GA were set using the default values given in GENESIS. In the experiment procedure, each data set is divided into two groups, sixty percent as training set and forty percent as testing set. AE5 rule induction algorithm [4], which is based on the theory of extension matrix, is used to evaluate the predictive accuracy. Table 4 shows the experiment results of the nine selected data sets.

Table 4. The comparison of the number of split points –N. of SP, the number of rules –N. of R and the predictive accuracy of classifier between the discretization of entropy and the discretization of compatibility rough set.

Data Set	Discretization of Entropy			Discretization of Comp. Rough Set		
	N. of SP	N. of R	Accuracy	N. of SP	N. of R	Accuracy
Breast	12	58	81.2%	7	42	92.2%
Diabetes	49	44	64.8%	19	39	69.4%
Echo	23	63	32.5%	8	43	70.0%
Glass	229	93	48.4%	14	20	57.8%
Heart	41	60	63.2%	12	16	73.8%
Hepatiti	57	27	74.2%	13	23	74.2%
Iris	27	19	90.1%	6	8	96.1%
Thyroid	44	78	62.8%	8	39	98.9%
Wine	137	123	47.2%	6	19	92.5%

Comparison of the number of discretization split points, the number of rules and the predictive accuracy proves that the discretization method based on compatibility rough set model can significantly decrease the number of discretization split points and the number of rules, and universally improve the predictive accuracy.

References

1. De Jong, K.: Analysis of the Behavior of a Class of Genetic Adaptive System, Ph.D. Thesis, Department of Computer and Communications Sciences, University of Michigan, Ann Arbor, ML, 1975.
2. Fayyad, U. and Irani, K B.: Multi-Interval Discretization of Continuous Attributes as Preprocessing for Classification Learning. Proceedings of the 13th International Join Conference on Artificial Intelligence. Morgan Kaufmann Publishers,1993, 1022~1027
3. Grefenstette, J.J.: Technical Report CS-83-11, Computer Science Department, Vanderbilt University, 1984.
4. Hong, J.R.: Learning from Examples and a Multi-Purpose Learning System AE5. Chinese Journal of Computers, Vol.12, 1989,98~105.
5. Kerber, R.: ChiMerge: Discretization of Numeric Attributes. Proceedings 10th National Conference on Artificial Intelligence. MIT Press, 1992, 123~127.
6. Krawiec, K., Slowinski, R. and Vanderpooten, D.: Construction of Rough Classifiers Based on Application of a Similarity Relation, Proceedings of the 4th International Workshop on Rough Sets, Fuzzy Sets, and Machine Discovery, Tokyo, Japan, 1996.
7. Kretowski, M. and Stepaniuk, J.: Decision System Based on Tolerance Rough Sets, Proceedings of the 4th International Workshop on Intelligent Information Systems, Augustow, Poland, 1995.
8. Lenarcik, A. and Piasta, Z.: Minimizing the Number of Rules in Deterministic Rough Classifiers, Soft Computing, Lin,Y.Y. and Wildberge, A.M.(Eds.),San Diego,1995,32~35
9. Nguyen,H.S.: Discretization of Real Value Attribute: Boolean Reasoning Approach, Ph.D. Thesis, Warsaw University, 1997.
10. Pawlak, Z.: Rough Sets: Theoretical Aspects of Reasoning about Data, Kluwer Academic Publishers, Boston, 1991.
11. Mollestad, T. and Skowron, A., A Rough Set Framework for Data Mining of Propositional Default Rules, ISMIS'96, Zakopane, Poland, June 1996.
12. Murphy, P.M.: UCI Repository of Machine Learning Database and Domain Theories. At http://www.ics.uci.edu/~mlearn/ML/Repository.html.
13. Slowinski, R. and Vanderpooten, D.: Similarity Relation as a Basis for Rough Approximations, Proceedings of the Second Annual Joint Conference on Information Sciences, Wrightsville Beach, N. Carolina, USA, 1995, 249~250.
14. Shan, N., Hamilton, H.J., Ziarko, W. and Cercone, N.: Discretization of Continuous Valued Attributes in Attribute-Value Systems. Proceedings of the fourth International Workshop on Rough Sets, Fuzzy Sets, and Machine Discovery, Tokyo, Japan, 1996, 74~81.
15. Wu, X.D.: A Bayesian Discretizer for Real-Valued Attributes, The Computer Journal, Vol.39, 1996, 688~691.
16. Yao, Y.Y., and Wong, S.K.M.: Generalization of Rough Sets Using Relationships Between Attribute Values. Proceedings of the Second Annual Joint Conference on Information Sciences, Wrightsville Beach, N. Carolina, USA, 1995, 30~33.

Level Cut Conditioning Approach to the Necessity Measure Specification

Masahiro Inuiguchi and Tetsuzo Tanino

Department of Electronics and Information Systems
Graduate School of Engineering, Osaka University
2-1, Yamada-Oka, Suita, Osaka 565-0871, Japan
{inuiguti, tanino}@eie.eng.osaka-u.ac.jp
http://vanilla.eie.eng.osaka-u.ac.jp

Abstract. A necessity measure N is defined by an implication function. However, specification of an implication function is difficult. Necessity measures are closely related to inclusion relations. In this paper, we propose an approach to necessity measure specification by giving an equivalent parametric inclusion relation between fuzzy sets A and B to $N_A(B) \geq h$. It is shown that, by such a way, we can specify a necessity measure, i.e., an implication function. Moreover, given an implication function, an associated inclusion relation is discussed.

1 Introduction

Possibility theory [2][8] has been applied to many fields such as approximate reasoning, data base theory, decision making, optimization and so forth. In possibility theory, possibility and necessity measures play key roles to handle uncertain information, ambiguous knowledge and vague concepts. There exist quite a lot of possibility and necessity measures and the selection of those measures qualifies the properties of fuzzy reasoning, decision principles and so on. Possibility and necessity measures should reflect the expert's knowledge and/or decision maker's preference. Between possibility and necessity measures, the selection of a necessity measure is much more important since (1) it directly qualifies fuzzy rules and possibility distributions in approximate reasoning (see [1]) and (2) it is used for the measure of safety or robustness.

A necessity measure N under a possibility distribution μ_A (i.e., fuzzy information that x is in A) is defined as

$$N_A(B) = \inf_{x \in X} I(\mu_A(x), \mu_B(x)) \tag{1}$$

by an implication function $I : [0,1] \times [0,1] \to [0,1]$ such that $I(0,0) = I(0,1) = I(1,1) = 1$ and $I(1,0) = 0$, where μ_A and μ_B are membership functions of fuzzy sets A and B of a universal set X. The selection of a necessity measure means that of an implication function. In real world situations, it is not easy for us to select an implication function directly since we are not aware of what kind of implication function is used to evaluate the certainty degree of the conclusion the implication function itself is far from our imagination. On the other hand, the necessity measure is closely related to the inclusion relation between A and B as is defined in the crisp case. Indeed, for several implication functions, equivalent conditions of $N_A(B) \geq h$ are known to be inclusion relations between fuzzy sets A and B with a parameter h. An inclusion relation is much more imaginable for us than an implication function. From this point of view, we will be able to specify an inclusion relation with a parameter h as an equivalent condition to $N_A(B) \geq h$. Such a parametric inclusion relation specification is at least easier than the implication function specification. In this way, the authors [5] succeeded to construct a necessity measure, in other words, an implication function, from a given inclusion relation with respect to $N_A(B) \geq h$ and proposed nine kinds of necessity measures defined by distinct inclusion relations with h. In this paper, an inclusion relation with a parameter h with respect to $N_A(B) \geq h$ is called a 'level cut condition' and the approach to specify a necessity measure by giving a level cut condition is called a 'level cut conditioning approach'.

Since the level cut conditioning approach has not yet studied considerably, there still remain open problems: (Q1) Can we unify the level cut conditions without loss of rationality ?, (Q2) Is there any level cut condition of the necessity measure associated with an arbitrarily given implication function satisfies ?, (Q3) Can any novel necessity measure be derived by this approach ?, (Q4) How utilize the results of this approach to real world problems ? and so on. In this paper, we answer the questions (Q1) and (Q2). To (Q1), we give a generalized level cut condition and show the existence of the necessity measure satisfies the condition. To (Q2), we show that a level cut condition can be obtained when a given implication function satisfies certain properties. On account of limited space, (Q3) and (Q4) are not answered in this paper but in our future papers.

2   Necessity Measures defined by Level Cut Conditions

When A and B are crisp sets, the necessity measure N is uniquely defined by

$$N_A(B) = \begin{cases} 1, \text{ if } A \subseteq B, \\ 0, \text{ otherwise.} \end{cases} \tag{2}$$

The traditional and most well-known necessity measure N^D is the one defined by (1) with Dienes implication function $I^D(a, b) = \max(1 - a, b)$. To this necessity measure, we have (see [5])

$$N_A^D(B) \geq h \Leftrightarrow (A)_{1-h} \subseteq [B]_h, \tag{3}$$

where $(A)_h$ and $[A]_h$ are strong and weak h-level sets of A defined by

$$(A)_h = \{x \mid \mu_A(x) > h\}, \qquad [A]_h = \{x \mid \mu_A(x) \geq h\}. \tag{4}$$

For necessity measures N^G and N^{r-G} defined by (1) with Gödel implication function I^G and reciprocal Gödel implication function I^{r-G} satisfy (see [5])

$$N_A^G(B) \geq h \Leftrightarrow (A)_k \subseteq (B)_k, \ \forall k < h, \tag{5}$$

$$N_A^{r-G}(B) \geq h \Leftrightarrow [A]_{1-h} \subseteq [B]_{1-h}, \ \forall k < h, \tag{6}$$

where I^G and I^{r-G} are defined by

$$I^G(a,b) = \begin{cases} 1, \text{if } a \leq b, \\ b, \text{if } a > b, \end{cases} \qquad I^{r-G}(a,b) = \begin{cases} 1, & \text{if } a \leq b, \\ 1-a, \text{if } a > b. \end{cases} \qquad (7)$$

As shown in (3)–(6), necessity measures are closely related to set inclusion relations. Moreover, those necessity measures are uniquely specified by the right-hand side conditions of (3)–(6) since we have

$$N_A(B) = \sup_h \{h \mid N_A(B) \geq h\}. \qquad (8)$$

From this fact, it is conceivable to specify a necessity measure by giving a necessary and sufficient condition of $N_A(B) \geq h$. From the practical point of view, giving such a condition must be easier than giving an implication function directly to define a necessity measure, since inclusion relations are more imaginable in our mind than implication functions. From this point of view, the authors [5] proposed level cut conditioning approach to define a necessity measure. We succeeded to construct nine kinds of necessity measures giving nine different level cut conditions. In this paper, generalizing our previous results, we discuss measures N^L which satisfy the following condition:

$$N_A^L(B) \geq h \Leftrightarrow m_h(A) \subseteq M_h(B), \qquad (9)$$

where $m_h(A)$ and $M_h(A)$ are fuzzy sets obtained from a fuzzy set A by applying a suitable parametric transformation as will formally be defined later. We assume that the inclusion relation between fuzzy sets is defined normally, i.e.,

$$A \subseteq B \Leftrightarrow \mu_A(x) \leq \mu_B(x), \ \forall x \in X \qquad (10)$$

Let a fuzzy set A have a linguistic label α. Then, roughly speaking, $m_h(A)$ is a fuzzy set corresponding to a linguistic label, "very α", "extremely α" or "typically α" and $M_h(A)$ a fuzzy set corresponding to a linguistic label, "roughly α", "more or less α" or "weakly α". Thus, (9) tries to capture that an event 'x is β' expressed by a fuzzy set B is necessary to a certain extent under information that 'x is α' expressed by a fuzzy set A if and only if the fact 'x is very α' entails the fact 'x is roughly β'. Degrees of stress and relaxation by modifiers 'very' and 'roughly' decrease as the necessity degree h increases, i.e., m_h and M_h satisfy

$$h_1 > h_2 \Rightarrow m_{h_1}(A) \supseteq m_{h_2}(A), \ M_{h_1}(A) \subseteq M_{h_2}(A). \qquad (11)$$

Now let us define m_h and M_h mathematically. $m_h(A)$ and $M_h(A)$ are defined by the following membership functions:

$$\mu_{m_h(A)}(x) = g^m(\mu_A(x), h), \qquad \mu_{M_h(A)}(x) = g^M(\mu_A(x), h), \qquad (12)$$

where functions $g^m : [0,1] \times [0,1] \to [0,1]$ and $g^M : [0,1] \times [0,1] \to [0,1]$ are assumed to satisfy (g1) $g^m(a, \cdot)$ is lower semi-continuous and $g^M(a, \cdot)$ upper semi-continuous for all $a \in [0, 1]$,

(g2) $g^m(1, h) = g^M(1, h) = 1$ and $g^m(0, h) = g^M(0, h) = 0$ for all $h > 0$,

(g3) $g^m(a, 0) = 0$ and $g^M(a, 0) = 1$ for all $a \in [0, 1]$,

(g4) $h_1 \geq h_2$ implies $g^m(a, h_1) \geq g^m(a, h_2)$ and $g^M(a, h_1) \leq g^M(a, h_2)$ for all $a \in [0, 1]$,

(g5) $a \geq b$ implies $g^m(a, h) \geq g^m(b, h)$ and $g^M(a, h) \geq g^M(b, h)$ for all $h \leq 1$,

(g6) $g^m(a, 1) > 0$ and $g^M(a, 1) < 1$ for all $a \in (0, 1)$.

(g1) is required in order to guarantee the existence of a measure satisfies (9) (see Theorem 1). (g2) means that complete members of a fuzzy set A are also complete members of the fuzzy sets $m_h(A)$ and $M_h(A)$ and complete non-members of A are also complete non-members of the fuzzy sets $m_h(A)$ and $M_h(A)$. This implies that $[m_h(A)]_1 = [M_h(A)]_1 = [A]_1$ and $(m_h(A))_0 = (M_h(A))_0 = (A)_0$ for any $h > 0$. (g3) is required so that the left-hand sides of (9) is satisfied when $h = 0$. (g4) coincides with the requirement (11). (g5) means that the membership degrees of $m_h(A)$ and $M_h(A)$ increase as that of A increases. (g6) means that all possible members of A cannot be complete non-members of $m_h(A)$ at the lowest stress level, i.e., $h = 1$ and that all possible non-members of A cannot be complete members of $M_h(A)$ at the lowest relaxation level, i.e., $h = 1$. As described above, those requirements, (g1)–(g6) can be considered as natural.

It should be noted that g^m and g^M are defined by

$$g^m(a, h) = T(a, h), \quad g^M(a, h) = Cnv[I](a, h) = I(h, a), \tag{13}$$

where T and I are a conjunction function and an implication function satisfy

(I1) $I(a, 1) = 1$ and $I(0, a) = 1$, for all $a \in [0, 1]$,

(I2) $I(a, 0) = 0$, for all $a \in (0, 1]$,

(I3) $I(\cdot, a)$ is upper semi-continuous for all $a \in [0, 1]$,

(I4) $a \leq c$ and $b \geq d$ imply $I(a, b) \geq I(c, d)$.

(I5) $I(1, a) < 1$, for all $a \in [0, 1)$.

(T1) $T(0, a) = 0$ and $T(a, 0) = 0$, for all $a \in [0, 1]$,

(T2) $T(1, a) = 1$, for all $a \in (0, 1]$,

(T3) $T(a, \cdot)$ is lower semi-continuous for all $a \in [0, 1]$,

(T4) $a \geq c$ and $b \geq d$ imply $T(a, b) \geq T(c, d)$.

(T5) $T(a, 1) > 0$, for all $a \in (0, 1]$.

Conversely, $I(a, b) = g^M(b, a)$ is an implication function satisfies (I1)–(I5) and $T(a, b) = g^m(a, b)$ is a conjunction function satisfies (T1)–(T5).

Remark 1. In order to express fuzzy sets corresponding to a linguistic labels 'very α' and 'roughly α', $m_h(A)$ and $M_h(A)$ should be satisfy $m_h(A) \subseteq A \subseteq M_h(A)$, $\forall h \in [0, 1]$. In [5], we required so. However, to generalize the results obtained in [5], we dropped this requirement. By this generalization, we can treat conditions including h-level sets. For example, (3) can be expressed by (9) with definitions,

$$g^m(a, h) = \begin{cases} 1, & \text{if } a > 1 - h, \\ 0, & \text{otherwise,} \end{cases} \qquad g^M(a, h) = \begin{cases} 1, & \text{if } a \geq h, \\ 0, & \text{otherwise.} \end{cases}$$

The following proposition guarantees the existence and uniqueness of N^L.

Theorem 1. N^L *exists and is defined by*

$$N_A^L(B) = \sup_{0 \le h \le 1} \{h \mid m_h(A) \subseteq M_h(B)\}. \tag{14}$$

Proof. Suppose $m_h(A) \not\subseteq M_h(B)$ when $\sup_{0 < k \le 1}\{k \mid m_k(A) \subseteq M_k(B)\} \ge h$. From (11) and (12), there exists $x \in X$ such that $g^m(\mu_A(x), k) = \mu_{m_k(A)}(x) \le \mu_{M_k(B)}(x) = g^M(\mu_B(x), k)$, $\forall k < h$ but $g^m(\mu_A(x), h) = \mu_{m_h(A)}(x) > \mu_{M_h(B)}(x) = g^M(\mu_B(x), h)$. This fact implies

$$\liminf_{k \to h} g^m(\mu_A(x), k) < g^m(\mu_A(x), h) \text{ or } \limsup_{k \to h} g^M(\mu_B(x), k) > g^M(\mu_B(x), h).$$

This contradicts the lower semi-continuity of $g^m(a, \cdot)$, $\forall a \in [0, 1]$ and the upper semi-continuity of $g^M(a, \cdot)$, $\forall a \in [0, 1]$. Therefore, we have

$$\sup_{k \le 1}\{k \mid m_k(A) \subseteq M_k(B)\} \ge h \Rightarrow A \subseteq M_h(B)$$

The converse is obvious. Hence, we have (14). $\qquad\square$

The following theorem shows that N^L defined by (14) is a necessity measure.

Theorem 2. N^L *is a necessity measure and the associated implication function* I^L *is defined by*

$$I^L(a, b) = \sup_{0 \le h \le 1} \{h \mid g^m(a, h) \le g^M(b, h)\}. \tag{15}$$

Proof. Let us consider I^L defined by (15). From (g2) and (g3), we obtain $I^L(0, 0) = I^L(0, 1) = I^L(1, 1) = 1$ and $I^L(1, 0) = 0$. Thus, I^L is an implication function. From (g1), we have

$$I^L(a, b) \ge h \Leftrightarrow g^m(a, h) \le g^M(b, h). \tag{$*$}$$

Consider a measure $\Phi_A(B) = \inf_{x \in X} I^L(\mu_A(x), \mu_B(x))$. By ($*$), it is easy to show

$$\Phi_A(B) \ge h \Rightarrow m_h(A) \subseteq M_h(B).$$

Thus, $\Phi_A(B) \le N_A^L(B)$. Suppose $\Phi_A(B) < N_A^L(B) = h^*$. Then there exists an $x^* \in X$ such that $I^L(\mu_A(x^*), \mu_B(x^*)) < h^*$. From ($*$), we have $g^m(\mu_A(x^*), h^*) > g^M(\mu_B(x^*), h^*)$. From (12), $m_{h^*}(A) \not\subseteq M_{h^*}(B)$. From (9), we obtain $N_A^L(B) < h^*$. However, this contradicts $N_A^L(B) \ge h^*$. Hence, $N_A^L(B) = \Phi_A(B)$. $\qquad\square$

It is worth knowing the properties of I^L in order to see the range of implication functions defined by level cut conditions.

Proposition 1. I^L *defined by (15) satisfies (I1), (I4), (I5) and*

(I6) $I(a, 0) < 1$, *for all* $a \in (0, 1]$.

Moreover, I^L satisfies (I2) if and only if $g^m(a,h) > 0$ for all $(a,h) > (0,0)$, and (I7) if and only if $g^M(a,h) < 1$ for all $a < 1$ and $h > 0$, where

(I7) $I(1,a) = 0$, for all $a \in [0,1)$.

Proof. Except for (I5) and (I6), all the properties are straightforward from (g1)–(g6). Form (g2), we obtain

$$I^L(1,a) = \sup_{0\le h\le1} \{h \mid g^M(a,h) \ge 1\}, \quad I^L(a,0) = \sup_{0\le h\le1} \{h \mid g^m(a,h) \le 0\}.$$

From (g1) and (g6), we have (I5) and (I6). □

It should be noted that infinitely many different pairs of (g^m, g^M) produce the same necessity measure as far as the level cut condition, or simply, the condition $g^m(a,h) \le g^M(b,h)$ is equivalent.

Example 1. Let $V(\mathcal{A})$ be the truth value (1 for true and 0 for false) of a statement $\mathcal{A}$. When $g^m(a,h) = \min(a, V(h > 0))$ and $g^M(a,h) = \max(a, V(a \ge h))$, I^L is Gödel implication I^G. When $g^m(a,h) = \min(a, V(a > h))$ and $g^M(a,h) = \max(a, V(h = 0))$, I^L is reciprocal Gödel implication I^{r-G}.

3 Level Cut Conditions Derived from Necessity Measures

In this section, given a necessity measure, or equivalently, given an implication, we discuss how we can obtain the functions g^m and g^M. To do this, requirements (g1)–(g6) are not sufficient to obtain some results. We add a requirement,

(g7) $g^m(\cdot,h)$ is lower semi-continuous and $g^M(\cdot,h)$ upper semi-continuous for all $h \in [0,1]$.

and assume that g^m and g^M satisfy (g1)–(g7). Together with (g1), this additional requirement guarantees that I^L satisfies (I3) and

(I8) $I(a,\cdot)$ is upper semi-continuous.

First, we look into the properties of pseudo-inverses of $g^m(\cdot,h)$ and $g^M(\cdot,h)$ defined by

$$g^{m*}(a,h) = \sup_{0\le b\le1} \{b \mid g^m(b,h) \le a\}, \quad g^{M*}(a,h) = \inf_{0\le b\le1} \{b \mid g^M(b,h) \ge a\}. \tag{16}$$

We have the following propositions.

Proposition 2. $g^{m*}(\cdot,h)$ *is upper semi-continuous and* $g^{M*}(\cdot,h)$ *lower semi-continuous, for all* $h \in [0,1]$.

Proof. When $h = 0$, it is obvious. When $h > 0$, from the definition and (g5), we have

$$\{a \mid g^{m*}(a,h) \ge b^*\} = \bigcap_{0\le b<b^*} \{a \mid a \ge g^m(b,h)\}.$$

This set is closed and hence, $g^{m*}(\cdot,h)$ is upper semi-continuous. The lower semi-continuity of $g^{M*}(\cdot,h)$ can be proved similarly. □

Moreover, from (g1), we can prove the upper semi-continuity of $g^{m*}(a, \cdot)$ and the lower semi-continuity of $g^{M*}(a, \cdot)$.

Proposition 3. *We have*

$$g^m(a, h) \leq g^M(b, h) \Leftrightarrow a \leq g^{m*}(g^M(b, h), h), \qquad (17)$$
$$g^m(a, h) \leq g^M(b, h) \Leftrightarrow g^{M*}(g^m(a, h), h) \leq b. \qquad (18)$$

Proof. From the definition of g^{m*}, we have $a \leq g^{m*}(g^m(a, h), h)$ and $g^{m*}(\cdot, h)$ is non-decreasing. Thus, we have

$$g^m(a, h) \leq g^M(b, h) \Rightarrow a \leq g^{m*}(g^m(a, h), h) \leq g^{m*}(g^M(b, h), h).$$

On the other hand, from the upper semi-continuity of $g^{m*}(\cdot, h)$, we can easily prove $g^m(g^{m*}(a, h), h) \leq a$. Thus, from (g6), we obtain

$$a \leq g^{m*}(g^M(b, h), h) \Rightarrow g^m(a, h) \leq g^m(g^{m*}(g^M(b, h), h), h) \leq g^M(b, h).$$

Hence, (17) is valid. (18) can be proved in the same way. $\qquad \square$

Proposition 3 gives an expression of I^L other than (15), i.e.,

$$I^L(a, b) = \sup_{0 \leq h \leq 1} \{h \mid a \leq g^{m*}(g^M(b, h), h)\}, \quad \text{if } g^m(\cdot, h) \text{ is l. s. c.,} \qquad (19)$$

$$I^L(a, b) = \sup_{0 \leq h \leq 1} \{h \mid g^{M*}(g^m(a, h), h) \leq b\}, \quad \text{if } g^M(\cdot, h) \text{ is u. s. c.} \qquad (20)$$

Moreover, it should be noted that $I^{m*}(a, b) = g^{m*}(g^M(b, a), a)$ is an implication function which satisfies (I1), (I3) (I4), (I5), (I6) and (I8) and $T^{M*}(a, b) = g^{M*}(g^m(a, b), b)$ is a conjunctive function which satisfies (T1), (T3), (T4), (T5), (T6) and (T7):

(T6) $T(1, a) > 0$, for all $a \in (0, 1]$,

(T7) $T(\cdot, a)$ is lower semi-continuous.

This fact together with (19) and (20) remind us functionals σ and ζ both of which yield an implication function from an implication function I and a conjunction function T, respectively (see [4][6][7]), i.e.,

$$\sigma[I](a, b) = \sup_{0 \leq h \leq 1} \{h \mid I(h, b) \geq a\}, \quad \zeta[T](a, b) = \sup_{0 \leq h \leq 1} \{h \mid T(a, h) \leq b\}. \qquad (21)$$

Under assumptions that I satisfies $I(1, 0) = 0$, (I1), (I3), (I6) and (I4-a): $I(a, c) \geq I(b, c)$ whenever $a \leq b$, we can prove $\sigma[\sigma[I]] = I$ and $\sigma[I]$ preserves those properties (see [4][7]). Under the assumptions that T satisfies $T(1, 1) = 1$, (T1), (T3), (T6) and (T4-b): $T(a, b) \leq T(a, c)$ whenever $b \leq c$, we have $\xi[\zeta[T]] = T$ and $\zeta[T]$ satisfies $I(1, 0) = 0$, (I1), (I5), (I8) and (I4-b): $I(a, b) \leq I(a, c)$ whenever $b \leq c$ (see [6][7]), where

$$\xi[I](a, b) = \inf_{0 \leq h \leq 1} \{h \mid I(a, h) \geq b\}. \qquad (22)$$

Moreover, under the assumptions that I satisfies $I(1,0) = 0$, (I1), (I5), (I8) and (I4-b), we have $\zeta[\xi[I]] = I$ and $\xi[I]$ satisfies $T(1,1) = 1$, (T1), (T3), (T6) and (T4-b).

Under the assumptions (g1)–(g7), pairs (I^L, I^{m*}) and (I^L, T^{M*}) satisfy the requirements for $(\sigma[I^L] = I^{m*}$ and $\sigma[I^{m*}] = I^L)$ and $(\xi[I^L] = T^{M*}$ and $\zeta[T^{M*}] = I^L)$, respectively. Hence, given an arbitrary implication function I which satisfies (I1), (I3)–(I6) and (I8), we obtain $I^{m*} = \sigma[I]$ and $T^{M*} = \xi[I]$. One may think that, defining $g^m(a,h) = a$ or $g^M(a,h) = a$ when $h > 0$ so that we have $g^{m*}(a,h) = a$ or $g^{M*}(a,h) = a$ for all $h > 0$, we can obtain g^m and g^M via (13) with substitution of I^{m*} or T^{M*} for I and T. However, unfortunately, there is no guarantee that such g^m and g^M satisfy (g3). The other properties, (g1), (g2), (g4)–(g7), are satisfied as is shown in the following proposition.

Proposition 4. *σ preserves (I2), (I4) and (I8). Moreover, σ preserves (I5) under (I3). On the other hand, when I satisfies (I4-a), $\xi[I]$ satisfies (T4). When I satisfies (I6) and (I8), $\xi[I]$ satisfies (T5). When I satisfies (I7), $\xi[I]$ satisfies (T2). When I satisfies (I3), $\xi[I]$ satisfies (T7).*

From Propositions 1 and 4 together with (13), the following theorem is straightforward.

Theorem 3. *Let I satisfy (I1), (I3)–(I6) and (I8). When I satisfies (I2),*

$$g^m(a,h) = \begin{cases} a, \text{ if } h > 0, \\ 0, \text{ if } h = 0, \end{cases} \qquad g^M(a,h) = \sigma[I](h,a), \qquad (23)$$

satisfy (g1)–(g7). On the other hand, when I satisfies (I7),

$$g^m(a,h) = \xi[I](a,h), \qquad g^M(a,h) = \begin{cases} a, \text{ if } h > 0, \\ 1, \text{ if } h = 0, \end{cases} \qquad (24)$$

satisfy (g1)–(g7).

Theorem 3 gives an answer to (Q2) when I satisfies (I2) or (I7) as well as (I1), (I3)–(I6) and (I8). The complete answer to (Q2) is rather difficult since decompositions of I^{m*} and T^{M*} to g^m and g^M satisfying (g1)–(g7) are not easy. In what follows, we give other answers under certain conditions. The following proposition plays a key role.

Proposition 5. *When g^m is continuous, we have $g^m(g^{m*}(a,h),h) = a$, for all $h \in (0,1]$. Similarly, when g^M is continuous, we have $g^M(g^{M*}(a,h),h) = a$.*

Proof. Because of (g2), the continuity of $g^m(\cdot,h)$ and $g^M(\cdot,h)$ implies that $g^m(\cdot,h)$ and $g^M(\cdot,h)$ are surjective for all $h \in (0,1]$, respectively. Hence, we have $g^{m*}(a,h) = \sup_{0 \le b \le 1}\{b \mid g^m(b,h) = a\}$ and $g^{M*}(a,h) = \inf_{0 \le b \le 1}\{b \mid g^M(b,h) = a\}$. Because of continuity, sup and inf can be replaced with max and min, respectively. Hence, we have the theorem. □

Moreover, the following proposition is straightforward.

Proposition 6. *If a given function $g^m : [0,1] \times [0,1] \to [0,1]$ satisfies (g1)–(g6),*

> *(g8) $g^m(\cdot, h)$ is continuous for all $h \in (0,1]$,*
> *(g9) $g^m(\sigma[I](a,0), a) = 0$ for all $a \in (0,1]$,*
> *(g10) $g^m(a,1) < 1$ for all $a \in [0,1)$,*

then $I^*(a,b) = g^m(\sigma[I](a,b), b)$ satisfies (I1)–(I3), (I5) and (I8). Similarly, if a given function $g^M : [0,1] \times [0,1] \to [0,1]$, such that g^M satisfies (g1)–(g6),

> *(g11) $g^M(\cdot, h)$ is continuous for all $h \in (0,1]$,*
> *(g12) $g^M(\xi[I](1,a), a) = 1$ for all $a \in (0,1]$,*
> *(g13) $g^M(a,1) > 0$ for all $\in (0,1]$,*

then $T^*(a,b) = g^M(\xi[I](a,b), b)$ satisfies (T1)–(T3), (T5) and (T7).

From Propositions 5 and 6, if we find a function g^m (resp. g^M) satisfies (g1)–(g6) and (g8)–(g10) (resp. (g11)–(g13)) and $I^*(a,b) = g^m(\sigma[I](a,b), b)$ (resp. $T^*(a,b) = g^M(\xi[I](a,b), b))$ satisfies (I4) (resp. (T4)), then the pair $(g^m, Cnv[I^*])$ (resp. (T^*, g^M)) is an answer to (Q2) with respect to a given implication function I, where $Cnv[I^*]$ is the converse of an implication function I^*, i.e., $Cnv[I^*](a,b) = I^*(b,a)$.

From Propositions 3 and 4, the necessary and sufficient condition of $N_A(B) \geq h$ becomes (a) $\mu_A(x) \leq \sigma[I](h, \mu_B(x))$, $\forall x \in X$, or (b) $\xi[I](\mu_A(x), h) \leq \mu_B(x)$, $\forall x \in X$. As can be seen easily, (a) if and only if $\max(\mu_A(x), \sigma[I](h,0)) \leq \max(\sigma[I](h, \mu_B(x)), \sigma[I](h,0))$, $\forall x \in X$ and (b) if and only if $\min(\xi[I](\mu_A(x), h), \xi[I](1,h)) \leq \min(\mu_B(x), \xi[I](1,h))$. From this fact, giving bijective and strictly increasing functions $p(\cdot, h) : [\sigma[I](h,0), 1] \to [0,1]$ and $q(\cdot, h) : [0, \xi[I](1,h)] \to [0,1]$, we may define $g^m(a,h) = p(\max(a, \sigma[I](h,0)), h)$ for case (a) and $g^M(a,h) = q(\min(a, \xi[I](1,h)), h)$ for case (b) under certain conditions. From this point of view, we have the following theorem.

Theorem 4. *We have the following assertions:*

1. *Let $p(\cdot, h) : [\sigma[I](h,0), 1] \to [0,1]$ be a bijective and strictly increasing function such that*
   > *(p1) $h_1 \geq h_2$ implies $p(a, h_1) \geq p(a, h_2)$ for all $a \in [\sigma[I](h_2, 0), 1]$,*
   > *(p2) $p(\max(\sigma[I](h,a), \sigma[I](h,0)), h)$ is non-decreasing in h,*
   > *(p3) $\sigma[I](\cdot, 0)$ is continuous,*
   > *then $g^m(a,h) = p(\max(a, \sigma[I](h,0)), h)$ and $g^M(a,h) = p(\max(\sigma[I](h,a), \sigma[I](h,0)), h)$ satisfy (g1)–(g7) and define the level cut condition.*
2. *Let $q(\cdot, h) : [0, \xi[I](1,h)] \to [0,1]$ be a bijective and strictly increasing function such that*
   > *(q1) $h_1 \geq h_2$ implies $q(a, h_1) \leq q(a, h_2)$ for all $a \in [0, \xi[I](1, h_2)]$,*
   > *(q2) $q(\min(\xi[I](a,h), \xi[I](1,h), h)$ is non-increasing in h,*
   > *(q3) $\xi[I](1, \cdot)$ is continuous,*
   > *then $g^m(a,h) = q(\min(\xi[I](a,h), \xi[I](1,h)), h)$ and $g^M(a,h) = q(\min(a, \xi[I](1,h)), h)$ satisfy (g1)–(g7) and define level cut condition.*

Proof. From Proposition 6, it is obvious. $\square$

Table 1. g^m and g^M for I^S, I^R and I^{r-R}

I	$f(0)$	$g^m(a,h)$ $(h>0)$	$g^M(a,h)$ $(h>0)$
I^S	—	$\max(0, 1 - f(a)/f(n(h)))$	$\min(1, f(n(a))/f(n(h)))$
I^R	$< +\infty$	$\max(0, 1 - f(a)/(f(0) - f(h)))$	$\min(1, (f(0) - f(a))/(f(0) - f(h)))$
	$= +\infty$	a	$f^{-1}(\max(0, f(a) - f(h)))$
I^{r-R}	$< +\infty$	$(f(n(a)) - f(h))/(f(0) - f(h))$	$\min(1, f(n(a))/(f(0) - f(h)))$
	$= +\infty$	$n(f^{-1}(\max(0, f(n(a)) - f(h))))$	a

From Theorem 4, we can obtain the level cut condition of a given necessity measure when we find suitable functions p and q. Table 1 shows the level cut condition for S-, R- and reciprocal R-implication functions of a continuous Archimedean t-norm t and strong negation n. A continuous Archimedean t-norm is a conjunction function which is defined by $t(a,b) = f^*(f(a) + f(b))$ with a continuous and strictly decreasing function $f : [0,1] \to [0,+\infty)$ such that $f(1) = 0$, where $f^* : [0,+\infty) \to [0,1]$ is a pseudo-inverse defined by $f^*(r) = \sup\{h \mid f(h) \geq r\}$. A strong negation is a bijective strictly decreasing function $n : [0,1] \to [0,1]$ such that $n(n(a)) = a$. Given a t-norm t and a strong negation n, the associated S-implication function I^S, R-implication function I^R and reciprocal R-implication I^{r-R} are defined as follows (see [3]):

$$I^S(a,b) = n(t(a,n(b))), \quad I^R(a,b) = \zeta[t](a,b), \quad I^{r-R}(a,b) = \zeta[t](n(b),n(a)). \tag{25}$$

References

1. Dubois, D., Prade, H.: Fuzzy Sets in Approximate Reasoning, Part I: Inference with Possibility Distributions. Fuzzy Sets and Syst. **40** (1991) 143–202
2. Dubois, D., Prade, H.: Possibility Theory: An Approach to Computerized Processing of Uncertainty. Plenum Press, New York (1988)
3. Fodor, J., Roubens, M.: Fuzzy Preference Modeling and Multicriteria Decision Making. Kluwer Academic Publishers, Dordrecht (1994)
4. Inuiguchi, M., Kume, Y.: Fuzzy Reasoning Based on Uncertainty Generation Rules Using Necessity Measures. Japanese J. Fuzzy Theory and Syst. 4(3) 329–344
5. Inuiguchi, M., Kume, Y.: Necessity Measures Defined by Level Set Inclusions: Nine Kinds of Necessity Measures and Their Properties. Int. J. General Syst. **22** (1994) 245–275
6. Inuiguchi, M., Sakawa, M.: Interpretation of Fuzzy Reasoning Based on Conjunction Functions in View of Uncertainty Generation Rules Using Necessity Measures. Japanese J. Fuzzy Theory and Syst. 5(3) 323–344
7. Inuiguchi, M., Sakawa, M.: On the Closure of Generation Processes of Implication Functions from a Conjunction Function. In: Yamakawa, T., Matsumoto, G. (eds.): Methodologies for the Conception, Design, and Application of Intelligent Systems – Proceedings of the 4th International Conference on Soft Computing –, Vol.1, World Scientific, Singapore (1996) 327–330
8. Zadeh, L. A.: Fuzzy Sets as a Basis for a Theory of Possibility. Fuzzy Sets and Syst. **1** (1978) 3–28

Four c-Regression Methods and Classification Functions

S. Miyamoto[1], K. Umayahara[2], and T. Nemoto[3]

[1] Institute of Engineering Mechanics and Systems, University of Tsukuba,
Tsukuba-shi, Ibaraki 305-8573, Japan
miyamoto@esys.tsukuba.ac.jp
[2] Institute of Engineering Mechanics and Systems, University of Tsukuba,
Tsukuba-shi, Ibaraki 305-8573, Japan
uma@is.tsukuba.ac.jp
[3] Master's Program in Science and Engineering, University of Tsukuba,
Tsukuba-shi, Ibaraki 305-8573, Japan
nemoto@odin.esys.tsukuba.ac.jp

Abstract. Four methods of c-regression are compared. Two of them are methods of fuzzy clustering: (a) the fuzzy c-regression methods, and (b) an entropy method proposed by the authors. Two others are probabilistic methods of (c) the deterministic annealing, and (d) the mixture distribution method using the EM algorithm. It is shown that the entropy method yields the same formula as that of the deterministic annealing. Clustering results as well as classification functions are compared. The classification functions for fuzzy clustering are fuzzy rules interpolating cluster memberships, while those for the latter two are probabilistic rules. Theoretical properties of the classification functions are studied. A numerical example is shown.

1 Introduction

Recent studies on fuzzy clustering revealed that there are new methods [5, 7–9] based on the idea of regularization. These methods are comparable with the fuzzy c-means [1, 3] and their variations. The c-regression model is well-known among the variations, namely, Hathaway and Bezdek have developed the method of fuzzy c-regression [4]. It is not difficult to show, as we will see below, that the new methods have variations that are applied to the c-regression model.

Another class of methods that may compete with the fuzzy c-means is the mixture distribution model [10] with the EM algorithm [2] for the calculation of solutions. This method is based on the statistical model and hence the two frameworks of fuzziness and statistics are different. Hathaway and Bezdek [4] mention a simple type of the mixture model for the c-regression.

Moreover a method of deterministic annealing has been proposed [11] that is also based on probability theory. This method uses the Gibbs distribution for determining probabilistic allocation of clusters with the heuristic method of using the means for centers. (See also Masulli et. al. [6].)

In addition to the new methods of fuzzy c-means, we have introduced classification functions that interpolate the membership values of the individuals to each cluster. Global characters of clusters are accordingly made clear by using the classification functions [7–9]. The concept of classification functions is applicable to the c-regression model. In contrast, classification functions in probabilistic models are derived from probabilistic rules such as the Bayes rule.

We have thus four methods for the clustering with regression: the two fuzzy models and the other two probabilistic models. These methods should now be compared in theoretical, methodological, and applicational features. In doing this, we can use classification functions.

In the following we first review the four methods briefly, and develop the algorithms for calculating solutions that are not shown in foregoing works. The algorithm for calculating solutions for the entropy method developed by the authors is shown to be equivalent to the method of deterministic annealing, although the two models are different. Theoretical properties of the classification functions are compared. A numerical example is shown to see differences in clustering results and classification functions.

2 Fuzzy methods and probabilistic methods

2.1 Two methods of fuzzy c-regression

A family of methods on the basis of fuzzy c-means have been developed; there are common features in the methods. First, an alternative optimization algorithm is used to find solutions. Second, objective functions for clustering have a common form.

Let $x_i = (x_i^1, \ldots, x_i^p)^T$, $i = 1, \ldots, n$ be individuals to be clustered. They are points in p-dimensional Euclidean space. We consider two types of objective functions:

$$J_1 = \sum_{i=1}^{c} \sum_{k=1}^{n} (u_{ik})^m D_{ik}$$

$$J_2 = \sum_{i=1}^{c} \sum_{k=1}^{n} u_{ik} D_{ik} + \lambda^{-1} \sum_{i=1}^{c} \sum_{k=1}^{n} u_{ik} \log u_{ik}.$$

where u_{ik} is the element of cluster membership matrix $U = (u_{ik})$. The constraint of the fuzzy partition $M = \{ U \mid \sum_{j=1}^{c} u_{jk} = 1, \ 0 \leq u_{ik} \leq 1, i = 1, \ldots, c, k = 1, \ldots, n \}$ is assumed as usual.

The term D_{ik} varies in accordance with the types of clustering problems. In the standard fuzzy c-means, J_1 is used with $D_{ik} = \|x_k - v_i\|^2$, the square of the Euclidean distance between the individual x_k and the center of the cluster i, while J_2 is used with the same D_{ik} in the method of entropy proposed by the authors [7].

Since we consider c-regression models, $D_{ik} = (y_k - f_i(x_k; \beta_i))^2$ is used instead of $\|x_k - v_i\|^2$. Remark that the set of data has the form of (x_k, y_k), $i = 1, \ldots, n$, where x is the p-dimensional independent variable, while y is a scalar dependent variable. We wish to find a function $f_i(x; \beta_i)$ of regression by choosing parameters β_i so that the objective functions are minimized. Among possible choices for f_i, the linear regression is assumed:

$$y = f_i(x; \beta_i) = \sum_{j=1}^{p} \beta_i^j x^j + \beta_i^{p+1} \tag{1}$$

whence $\beta_i = (\beta_i^1, \ldots, \beta_i^{p+1})$ is a $p+1$ dimensional vector parameter.

The term D_{ik} is thus a function of β_i: $D_{ik}(\beta_i) = |y - \sum_{j=1}^{p} \beta_i^j x^j - \beta_i^{p+1}|^2$. Let $B = (\beta_1, \ldots, \beta_c)$, we can express the objective functions as the function of U and B:

$$J_1(U, B) = \sum_{i=1}^{c} \sum_{k=1}^{n} (u_{ik})^m D_{ik}(\beta_i) \tag{2}$$

$$J_2(U, B) = \sum_{i=1}^{c} \sum_{k=1}^{n} u_{ik} D_{ik}(\beta_i) + \lambda^{-1} \sum_{i=1}^{c} \sum_{k=1}^{n} u_{ik} \log u_{ik}. \tag{3}$$

Finally, we note that the following alternative optimization algorithm is used for finding optimal U and B, in which either $J = J_1$ or $J = J_2$.

Algorithm of fuzzy c-regression

R1 Set initial value for $\bar{B}$.
R2 Find optimal solution $\bar{U}$: $J(\bar{U}, \bar{B}) = \min_{U \in M} J(U, \bar{B})$
R3 Find optimal solution $\bar{B}$: $J(\bar{U}, \bar{B}) = \min_{B \in \mathbb{R}^{(p+1)c}} J(\bar{U}, B)$
R4 Check stopping criterion and if convergent, stop. Otherwise go to **R2**.

Assume $J = J_1$ (the standard fuzzy c-regression method is used). It is then easy to see that the solution in **R2** is

$$\bar{u}_{ik} = \left[\sum_{j=1}^{c} \left(\frac{D_{ik}}{D_{jk}} \right)^{\frac{1}{m-1}} \right]^{-1}. \tag{4}$$

while the solution $\bar{\beta}_i$ in **R3** is obtained by solving

$$\left(\sum_{k=1}^{n} (u_{ik})^m z_k z_k^T \right) \beta_i = \sum_{k=1}^{n} (u_{ik})^m y_k z_k \tag{5}$$

with respect to β_i, where $z_k = (x_k^1, \ldots, x_k^p, 1)^T$.

If we use $J = J_2$, we have

$$\bar{u}_{ik} = \frac{e^{-\lambda D_{ik}}}{\sum_{j=1}^{c} e^{-\lambda D_{jk}}}, \qquad \bar{\beta}_i = \left(\sum_{k=1}^{n} u_{ik} z_k z_k^T \right)^{-1} \sum_{k=1}^{n} u_{ik} y_k z_k \tag{6}$$

in **R2** and **R3**, respectively.

2.2 Mixture distribution model for c-regression

The model of the mixture of normal distribution is another useful method for clustering with the EM algorithm [2, 10]. It is based on the statistical concept of maximum likelihood but the results are comparable with those by fuzzy c-means. Application of this model to the c-regression has been mentioned by Hathaway and Bezdek [4].

We simply describe an outline of the algorithm. Notice first that the model assumes that the distribution of the error term $e_i = y - \sum_{j=1}^{p} \beta_i^j x^j - \beta_i^{p+1}$ is Gaussian with the mean 0 and the standard deviation σ_i which is to be estimated in the algorithm.

The distribution is hence assumed to be

$$p(x,y) = \sum_{i=1}^{c} \alpha_i p_i(x,y|C_i), \qquad (\sum_{i=1}^{c} \alpha_i = 1) \tag{7}$$

$$p_i(x,y|C_i) = \frac{1}{\sqrt{2\pi}\sigma_i} \exp\left(-\frac{1}{2\sigma_i^2}(y - \sum_{j=1}^{p} \beta_i^j x^j - \beta_i^{p+1})^2\right) \tag{8}$$

The parameters $\phi_i = (\alpha_i, \sigma_i, \beta_i)$ $(i = 1,\ldots,c)$ should be estimated by the EM algorithm. For simplicity, let $\Phi = (\phi_1,\ldots,\phi_c)$ and assume that $p(x,y|\Phi)$ is the density with the parameter Φ.

Let an estimate of the parameters be ϕ' $(i = 1,\ldots,c)$, then the next estimate α_i by the EM algorithm is given as follows.

$$\alpha_i = \frac{\Psi_i}{n} = \frac{1}{n} \sum_{k=1}^{n} \frac{\alpha_i' p_i(x_k, y_k | \phi_i')}{p(x_k, y_k | \Phi')}, \quad i = 1,\ldots,c,$$

where

$$\psi_{ik} = \frac{\alpha_i' p_i(x_k, y_k | \phi_i')}{p(x_k, y_k | \Phi')}, \quad \Psi_i = \sum_{k=1}^{n} \psi_{ik}.$$

β_i is obtained from solving the following equation:

$$\sum_{j=1}^{p} (\sum_{k=1}^{n} \psi_{ik} x_k^j x_k^\ell)\beta_i^j + (\sum_{k=1}^{n} \psi_{ik} x_k^\ell)\beta_i^{p+1} = \sum_{k=1}^{n} \psi_{ik} y_k x_k^\ell \quad (\ell = 1,\ldots,p)$$

$$\sum_{j=1}^{p} (\sum_{k=1}^{n} \psi_{ik} x_k^j)\beta_i^j + (\sum_{k=1}^{n} \psi_{ik})\beta_i^{p+1} = \sum_{k=1}^{n} \psi_{ik} y_k$$

and finally we have

$$\sigma_i^2 = \frac{1}{\Psi_i} \sum_{k=1}^{n} \psi_{ik} (y_k - \sum_{j=1}^{p} \beta_i^j x_k^j - \beta_i^{p+1})^2, \quad (i = 1,\ldots,c).$$

Remark moreover that the individual (x_k, y_k) is allocated to the cluster C_i using the Bayes formula. Namely, the probability of the allocation is given by

$$p(C_i|x_k, y_k) = \frac{\alpha_i p(x_k, y_k|\phi_i)}{\sum\limits_{j=1}^{c} \alpha_j p(x_k, y_k|\phi_j)} \tag{9}$$

2.3 Deterministic annealing

Gibbs distribution is used for probabilistic rule of allocating individuals to clusters [11]. Namely, the following rule is used:

$$Pr(x \in C_i) = \frac{e^{-\rho\|x-y_i\|^2}}{\sum\limits_{j=1}^{c} e^{-\rho\|x-y_j\|^2}} \tag{10}$$

in which ρ is a parameter and y_i is the cluster representative. For given y_i $(i = 1, \ldots, c)$, $Pr(x \in C_i)$ is determined as above, then the cluster representative is calculated again by the average:

$$y_i = \frac{\sum\limits_{x} x \cdot Pr(x \in C_i)}{\sum\limits_{x} Pr(x \in C_i)}. \tag{11}$$

Iterations of (10) and (11) until convergence provide clusters by this method.

2.4 Deterministic annealing and entropy method

It is now straightforward to use the deterministic annealing for the c-regression. We have

$$\pi_{ik} = Pr((x_k, y_k) \in C_i) = \frac{e^{-\rho D_{ik}(\beta_i)}}{\sum\limits_{j=1}^{c} e^{-\rho D_{jk}(\beta_i)}} \tag{12}$$

while

$$\beta_i = \left(\sum_{k=1}^{n} \pi_{ik} z_k z_k^T \right)^{-1} \sum_{k=1}^{n} \pi_{ik} y_k z_k \tag{13}$$

(Remark: Detailed proof is omitted here to save the space.)

Now, readers can see that the method of entropy and the deterministic annealing provide the equivalent solutions by putting $\rho = \lambda$, although the models are different; the entropy method is a fuzzy model and is based on the alternative optimization, while the deterministic annealing is a probabilistic model and an objective function to be optimized is not assumed. Although we have shown this equivalence in the case of c-regression, the same argument is applicable to the c-means and to other variations of the c-means.

3 Classification Functions

Classification functions in fuzzy c-means and the entropy method have been proposed and studied in Miyamoto and Mukaidono [7]. This idea can be applied to the c-regression.

The classification function in fuzzy clustering means that a new generic observation should be allocated to each cluster with the membership defined by that function, hence the function should have x and y as independent variables in this case of c-regression.

In analogy to the fuzzy c-means, classification functions in fuzzy c-regression is defined by replacing x_k, y_k by the corresponding variables $x = (x^1, \ldots, x^p)$ and y:

$$U_i^1(x, y) = \cfrac{1}{\displaystyle\sum_{\ell=1}^{c} \left(\cfrac{|y - \sum_{j=1}^{p} \beta_i^j x^j - \beta_i^{p+1}|^2}{|y - \sum_{j=1}^{p} \beta_\ell^j x^j - \beta_\ell^{p+1}|^2} \right)^{\frac{1}{m-1}}} \tag{14}$$

and when

$$y = \sum_{j=1}^{p} \beta_i^j x^j + \beta_i^{p+1} \tag{15}$$

for a particular i, the corresponding $U_i^1(x, y) = 1$ and $U_\ell^1(x, y) = 0$ for $\ell \neq i$.

In the case of the entropy method, we have

$$U_i^2(x, y) = \frac{e^{-\lambda |y - \sum_{j=1}^{p} \beta_i^j x^j - \beta_i^{p+1}|^2}}{\displaystyle\sum_{\ell=1}^{c} e^{-\lambda |y - \sum_{j=1}^{p} \beta_\ell^j x^j - \beta_\ell^{p+1}|^2}} \tag{16}$$

Remark that the values of the parameters β_i, $(i = 1, \ldots, c)$ are obtained when the corresponding algorithm of clustering terminates.

As shown above, the classification function for the deterministic annealing is equivalent to that by the entropy method. Thus, $Pr((x, y) \in C_i) = U_i^2(x, y)$ by putting $\lambda = \rho$.

For the mixture distribution model, the same idea is applied. Namely, we can define a classification function, or a discrimination function by replacing the symbols x_k, y_k by the variables x and y. We thus have

$$U_i^m(x, y) = p(C_i | x, y) = \frac{\alpha_i p(x, y | \phi_i)}{\displaystyle\sum_{\ell=1}^{c} \alpha_\ell p(x, y | \phi_\ell)} \tag{17}$$

in which the parameter Φ is obtained from the application of the EM algorithm.

Comparison of classification functions

Some of the theoretical properties of classification function are easily proved. First, notice that the maximum value $U_i^1(x, y) = 1$ is attained at the points where (15) is satisfied. On the other hand, when

$$|y - \sum_{j=1}^{p} \beta_\ell^j x^j - \beta_\ell^{p+1}| \to +\infty$$

for all $1 \le \ell \le c$, we have $U_i^1(x, y) \to \dfrac{1}{c}$.

We next examine the classification function of the entropy method (and equivalently, we are examining the function for the deterministic annealing). It should be remarked that the maximum value of $U_i^2(x, y)$ is not necessarily at the point satisfying (15).

It is easily seen that the value $U_i^2(x, y) = 1$ cannot be attained for a particular (x, y) in contrast to U_i^1. Instead, we have $\lim_{y \to \infty} U_i^2(\tilde{x}, y) = 1$ for some i and an appropriately chosen $x = \tilde{x}$. Whether this property holds or not depends upon the relative positions of the vectors $(\beta_i^1, \ldots, \beta_i^p) \in R^p$, $i = 1, \ldots, c$. Notice that β_i^{p+1} is not included in the discussion below. Hence another vector $\hat{\beta}_i = (\beta_i^1, \ldots, \beta_i^p)$ is used instaed of β_i.

For the following two propositions, the proofs are not difficult and are omitted here. The first proposition formally states the above result.

Proposition 1. If there exists an open half space $S \subset R^p$ such that

$$\{\hat{\beta}_\ell - \hat{\beta}_i : 1 \le \ell \le c, \ell \ne i\} \subset S \tag{18}$$

then there exists $\tilde{x} \in R^p$ such that $\lim_{y \to \infty} U_i^2(\tilde{x}, y) = 1$. If such a half space does not exist, then for all $x \in R^p$, $\lim_{y \to \infty} U_i^2(x, y) = 0$.

A condition for the existence of S in the Proposition 1 is the following.

Proposition 2. Let $T = span\{\hat{\beta}_1, \ldots, \hat{\beta}_c\}$ and $CO\{\hat{\beta}_1, \ldots, \hat{\beta}_c\}$ be the convex hull in T generated by $\{\hat{\beta}_1, \ldots, \hat{\beta}_c\}$. Assume that $\{\hat{\beta}_1, \ldots, \hat{\beta}_c\}$ is independent. Then a condition for the existence of S for a particular i such that (18) holds is that

$$\hat{\beta}_i \notin int(CO\{\hat{\beta}_1, \ldots, \hat{\beta}_c\}).$$

In other words, such a S exists if and only if the vertex of $\hat{\beta}_i$ is not in the interior of the above convex hull.

We thus observe that the maximum value is not at the points of (15) in the entropy method. Analogous results hold for the classification function U_i^m of the mixture distribution model, but propositions of the above type cannot be derived since the function is too complicated. Nevertheless, it is easily seen that the form of the classification function U_i^m becomes equivalent as U_i^2 in a particular case of $\sigma_1 = \cdots = \sigma_c$ and $\alpha_1 = \cdots = \alpha_c$. We therefore expect that the maximum value is not at the points of (15) in the mixture model also.

210

4 A Numerical Example

Figure 1 shows an artificial example of a set of points with two regression lines. The lines in the figure have been obtained from the fuzzy c-regression, but no remarkable differences have been observed concerning the regression lines derived from the four methods.

Figure 2 shows the three-dimensional plot of the classification function for one cluster by the fuzzy c-regression, whereas Figure 3 depicts the plot of $p(C_1|x, y)$ by the mixture distribution model. Readers can see remarkable difference between these two classification functions. The classification function by the entropy method (and the deterministic annealing) in Figure 4 is similar to that in Figure 3.

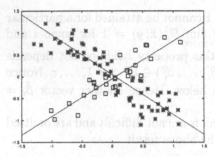

Fig. 1. Two regression lines

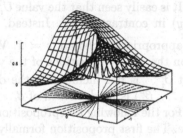

Fig. 2. Classification function by FCR

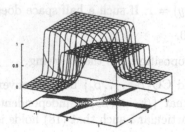

Fig. 3. Classification function by mixture distributions

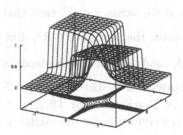

Fig. 4. Classification function by entropy method

5 Conclusion

Four methods of c-regression have been considered and classification functions have been studied. It should be remarked that the classifications function in the standard fuzzy c-regression reveals the shape of regression hyperplane by its maximum values, whereas the entropy method, the deterministic annealing, and the mixture model do not express those shapes of the regressions. Hence to observe outlines and global characteristics of the clusters by the latter class of methods require other types of functions, which we will study from now.

The importance of the entropy method is that it stands between the fuzzy c-means and the mixture model. Moreover the deterministic annealing is equivalent to the entropy method. It thus is based on fuzzy sets and at the same time a probabilistic interpretation is possible.

Future studies include theoretical investigations of the classification functions in the mixture of normal distributions and discussion of other variations of fuzzy c-means using the classification functions.

This study has partly been supported by TARA (Tsukuba Advanced Research Alliance), University of Tsukuba.

References

1. Bezdek,J.C.: Pattern Recognition with Fuzzy Objective Function Algorithms. Plenum, New York (1981)
2. Dempster,A.P., Laird,N.M., Rubin,D.B.: Maximum likelihood from incomplete data via the EM algorithm. J. of the Royal Statistical Society, B., Vol.39 (1977) 1-38
3. Dunn,J.C.: A fuzzy relative of the ISODATA process and its use in detecting compact well-separated clusters. J. of Cybernetics, Vol.3 (1974) 32-57
4. Hathaway,R.J., Bezdek,J.C.: Switching regression models and fuzzy clustering. IEEE Trans. on Fuzzy Syst., Vol.1 (1993) 195-204
5. Li,R.-P., Mukaidono,M.: A maximum entropy approach to fuzzy clustering. Proc. of the 4th IEEE Intern. Conf. on Fuzzy Systems, Yokohama, Japan, March 20-24 (1995) 2227-2232
6. Masulli,F., Artuso,M., Bogus,P., Schenone,A.: Fuzzy clustering methods for the segmentation of multivariate medical images, Proc. of the 7th International Fuzzy Systems Association World Congress, June 25-30, Prague, Chech (1997) 123-128
7. Miyamoto,S., Mukaidono,M.: Fuzzy c - means as a regularization and maximum entropy approach. Proc. of the 7th International Fuzzy Systems Association World Congress, Vol.II, June 25-30, Prague, Chech (1997) 86-92
8. Miyamoto,S., Umayahara,K.: Fuzzy clustering by quadratic regularization. Proc. of FUZZ-IEEE'98, May 4-9, Anchorage, Alaska (1998) 791-795
9. Miyamoto,S., Umayahara,K.: Two methods of fuzzy c-means and classification functions. Proc. of 6th Conf. of the International Federation of Classification Societies, July 21-24, Roma, Italy (1998) 105-110
10. Redner,R.A., Walker,H.F.: Mixture densities, maximum likelihood and the EM algorithm. SIAM Review, Vol.26, No.2 (1984) 195-239
11. Rose,K., Gurewitz,E., Fox,G.: A deterministic annealing approach to clustering. Pattern Recognition Letters, Vol.11 (1990) 589-594

Context-Free Fuzzy Sets
in Data Mining Context

Shusaku Tsumoto[*1] and T.Y. Lin[*2]

*1 Department of Medicine Informatics,
Shimane Medical University, School of Medicine,
89-1 Enya-cho Izumo City, Shimane 693-8501 Japan
*2 Department of Mathematics and Computer Science
San Jose State University
129S 10th St, San Jose, California 95192-0103 USA
E-mail: tsumoto@computer.org, tylin@mathcs.sjsu.edu

Abstract. Conventional studies on rule discovery and rough set methods mainly focus on acquisition of rules, the targets of which have mutually exclusive supporting sets. However, mutual exclusiveness does not always hold in real-world databases, where conventional probabilstic approaches cannot be applied. In this paper, first, we show that these phenomena are easily found in data mining contexts: when we apply attribute-oriented generalization to attributes in databases, generalized attributes will have fuzziness for classification. Secondly, we show that real-world databases may have fuzzy contexts. Then, finally, these contexts should be analyzed by using fuzzy techniques, where context-free fuzzy sets will be a key idea.

1 Introduction

Conventional studies on machine learning[10], rule discovery[2] and rough set methods[5, 12, 13] mainly focus on acquisition of rules, the targets of which have mutually exclusive supporting sets. Supporting sets of target concepts form a partition of the universe, and each method search for sets which covers this partition. Especially, Pawlak's rough set theory shows the family of sets can form an approximation of the partition of the universe. These ideas can easily extend into probabilistic contexts, such as shown in Ziarko's variable precision rough set model[15]. However, mutual exclusiveness of the target does not always hold in real-world databases, where conventional probabilstic approaches cannot be applied.

In this paper, first, we show that these phenomena are easily found in data mining contexts: when we apply attribute-oriented generalization to attributes in databases, generalized attributes will have fuzziness for classification. In this case, we have to take care about the conflicts between each attributes, which can be viewed as a problem with multiple membership functions. Secondly, we will see that real-world databases may have fuzzy contexts. Usually, some kind of experts use multi-valued attributes, corresponding to a list. Especially, in medical context, people may have several diseases during the same period. These cases also violate the assumption of mutual exclusiveness. Then, finally, these contexts should be analyzed by using fuzzy techniques, where context-free fuzzy sets will be a key idea to solve this problem.

2 Attribute-Oriented Generalization and Fuzziness

In this section, first, a probabilistic rule is defined by using two probabilistic measures. Then, attribute-oriented generalization is introduced as tranforming rules.

2.1 Probabilistic Rules

Accuracy and Coverage In the subsequent sections, we adopt the following notations, which is introduced in [9].

Let U denote a nonempty, finite set called the universe and A denote a nonempty, finite set of attributes, i.e., $a : U \to V_a$ for $a \in A$, where V_a is called the domain of a, respectively. Then, a decision table is defined as an information system, $A = (U, A \cup \{d\})$.

The atomic formulas over $B \subseteq A \cup \{d\}$ and V are expressions of the form $[a = v]$, called descriptors over B, where $a \in B$ and $v \in V_a$. The set $F(B, V)$ of formulas over B is the least set containing all atomic formulas over B and closed with respect to disjunction, conjunction and negation.

For each $f \in F(B, V)$, f_A denote the meaning of f in A, i.e., the set of all objects in U with property f, defined inductively as follows.

1. If f is of the form $[a = v]$ then, $f_A = \{s \in U | a(s) = v\}$
2. $(f \wedge g)_A = f_A \cap g_A$; $(f \vee g)_A = f_A \vee g_A$; $(\neg f)_A = U - f_a$

By the use of this framework, classification accuracy and coverage, or true positive rate is defined as follows.

Definition 1.
Let R and D denote a formula in $F(B, V)$ and a set of objects which belong to a decision d. Classification accuracy and coverage(true positive rate) for $R \to d$ is defined as:

$$\alpha_R(D) = \frac{|R_A \cap D|}{|R_A|} (= P(D|R)), \text{ and}$$

$$\kappa_R(D) = \frac{|R_A \cap D|}{|D|} (= P(R|D)),$$

where $|A|$ denotes the cardinality of a set A, $\alpha_R(D)$ denotes a classification accuracy of R as to classification of D, and $\kappa_R(D)$ denotes a coverage, or a true positive rate of R to D, respectively.

1

---

[1] Pawlak recently reports a Bayesian relation between accuracy and coverage[8]:

$$\alpha_R(D)P(D) = P(R|D)P(D) = P(R, D)$$

Definition of Rules

By the use of accuracy and coverage, a probabilistic rule is defined as:

$$R \overset{\alpha,\kappa}{\to} d \quad s.t. \quad R = \wedge_j \vee_k [a_j = v_k], \ \alpha_R(D) \geq \delta_\alpha, \ \kappa_R(D) \geq \delta_\kappa.$$

This rule is a kind of probabilistic proposition with two statistical measures, which is an extension of Ziarko's variable precision model(VPRS) [15].[2]

It is also notable that both a positive rule and a negative rule are defined as special cases of this rule, as shown in the next subsections.

2.2 Attribute-Oriented Generalization

Rule induction methods regard a database as a decision table[5] and induce rules, which can be viewed as reduced decision tables. However, those rules extracted from tables do not include information about attributes and they are too simple. In practical situation, domain knowledge of attributes is very important to gain the comprehensability of induced knowledge, which is one of the reasons why databases are implemented as relational-databases[1]. Thus, reinterpretation of induced rules by using information about attributes is needed to acquire comprehensive rules. For example, terolism, cornea, antimongoloid slanting of palpebral fissures, iris defects and long eyelashes are symptoms around eyes. Thus, those symptoms can be gathered into a category "eye symptoms" when the location of symptoms should be focused on. symptoms should be focused on. The relations among those attributes are hierarchical as shown in Figure 1. This process, grouping of attributes, is called attribute-oriented generalization[1].

Attribute-oriented generalization can be viewed as transformation of variables in the context of rule induction. For example, an attribute "iris defects" should be transformed into an attribute "eye symptoms=yes".It is notable that the transformation of attributes in rules correspond to that of a database because a set of rules is equivalent to a reduced decision table. In this case, the case when eyes are normal is defined as "eye symptoms=no". Thus, the tranformation rule for iris defects is defined as:

$$[iris\text{-}defects = yes] \to [eye\text{-}symptoms = yes] \tag{1}$$

In general, when $[A_k = V_l]$ is a upper-level concept of $[a_i = v_j]$, a transforming rule is defined as:

$$[a_i = v_j] \to [A_k = V_l],$$

and the supporting set of $[A_k = V_l]$ is:

$$[A_i = V_l]_A = \bigcup_{i,j} [a_i = v_j]_a,$$

$$= P(R)P(D|R) = \kappa_R(D)P(R)$$

This relation also suggests that *a priori* and *a posteriori* probabilities should be easily and automatically calculated from database.

[2] This probabilistic rule is also a kind of *Rough Modus Ponens*[7].

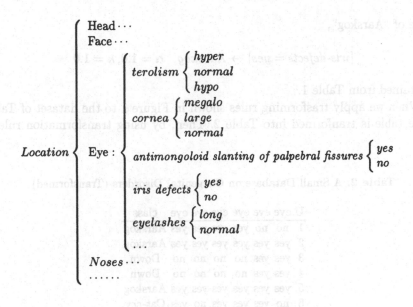

Fig. 1. An Example of Attribute Hierarchy

where A and a is a set of attributes for upper-level and lower level concepts, respectively.

2.3 Examples

Let us illustrate how fuzzy contexts is observed when attribute-oriented generalization is applied by using a small table (Table 1). Then, it is easy to see that

Table 1. A Small Database on Congenital Disorders

U	round	telorism	cornea	slanting	iris-defects	eyelashes	class
1	no	normal	megalo	yes	yes	long	Aarskog
2	yes	hyper	megalo	yes	yes	long	Aarskog
3	yes	hypo	normal	no	no	normal	Down
4	yes	hyper	normal	no	no	normal	Down
5	yes	hyper	large	yes	yes	long	Aarskog
6	no	hyper	megalo	yes	no	long	Cat-cry

DEFINITIONS: round: round face, slanting: antimongoloid slanting of palpebral fissures, Aarskog: Aarskog Syndrome, Down: Down Syndrome, Cat-cry: Cat Cry Syndrome.

a rule of "Aarskog",

$$[iris\text{-}defects = yes] \rightarrow Aarskog \quad \alpha = 1.0, \kappa = 1.0$$

is obtained from Table 1.

When we apply trasforming rules shown in Figure 1 to the dataset of Table 1, the table is tranformed into Table 2. Then, by using transformation rule 1,

Table 2. A Small Database on Congenital Disorders (Transformed)

U	eye	eye	eye	eye	eye	eye	class
1	no	no	yes	yes	yes	yes	Aarskog
2	yes	yes	yes	yes	yes	yes	Aarskog
3	yes	yes	no	no	no	no	Down
4	yes	yes	no	no	no	no	Down
5	yes	yes	yes	yes	yes	yes	Aarskog
6	no	yes	yes	yes	no	yes	Cat-cry

DEFINITIONS: eye: eye-symptoms

the above rule is trasformed into:

$$[eye\text{-}symptoms = yes] \rightarrow Aarskog.$$

It is notable that mutual exclusiveness of attributes has been lost by tranformation. Since five attributes (telorism, cornea, slanting, iris-defects and eyelashes) are generalized into *eye-symptoms*, the candiates for accuracy and coverage will be (5/6, 2/3), (3/4, 3/3), (3/4, 3/3), (3/3, 3/3), and (3/4, 3/3), respectively. Then, we have to select which value is suitable for the context of this analysis.

In [11], one of the authors selected the mimimum value in medical context: accuracy is equal to 3/4 and coverage is equal to 2/3.

Thus, the rewritten rule becomes the following probabilistic rule:

$$[eye\text{-}symptoms = yes] \rightarrow Aarskog \quad \alpha = 3/4 = 0.75, \kappa = 2/3 = 0.67.$$

This examples show that the loss of mutual exclusiveness is directly connected to the emergence of fuziness in a dataset. It it notable that the rule used for transformation is a deterministic one. When this kind of transformation is applied, whether applied rule is deterministic or not, fuzziness will be observed. However, no researchers has pointed out this problem with combination of rule induction and tranformation.

It is also notable that the conflicts between attributes with respect to accuarcy and coverage correponds to the vector representation of membership functions shown in Lin's context-free fuzzy sets[4].

3 Multi-valued Attributes and Fuziness

Another case of the violation of mutual exclusiness is when experts use multi-valued attributes, or a list to describe some attributes in a database. It is a very usual way when we cannot expect the number of inputs for attributes.

For example, in medical context, traffic accidents may injure several parts of bodies. Some patients have the damage only on hands and other ones suffer from multiple injuries, which makes us difficult to fix the number of attributes. Even if we enumerate all the possibilities of injuries and fix the number of columns corresponding to the worst case, most of the patients may have only a small number of them to be input. Usually, medical experts are not good at estimation of possibile inputs, and they are tend to make a list for data storage for the worst cases, although the probability for such cases is very low. For example, if medical experts empirically knows that the number of injuries is at most 20, they will set up 20 columns for input. However, if the averaged number of injuries is 4 or 5, all the remaining attributes will be stored as blank. Table 3 illustrates this observation. Although these attributes look like missing values,they should not be dealt with as missing values and have to be preprocessed: such large columns should be tranformed into binary ones. For the above example, each location of injury will be appended as a column, and if that location is not described in a list, then the value of that column should be set to 0.

Table 3. A Small Database on Fracture

U	f	f	f	f	f	f	f	f	f	ffffffffff
1.	arm	finger	shoulder	-	-	-	-	-	-	----------
2.	foot	-	-	-	-	-	-	-	-	----------
3.	arm	-	-	-	-	-	-	-	-	----------
4.	rib	-	-	-	-	-	-	-	-	----------
5.	head	neck	shoulder	radius	ulnaris	finger	rib	pelvis	femoral	----------
6.	femoral	tibia	calneus	-	-	-	-	-	-	----------

DEFINITIONS: f: fracture.

It is easy to see that mutual exclusiveness of attributes is violated in this case. Readers may say that if data is tranformed into binary attributes then mutual exclusiveness will be recovered. For example, if one of the above attribute-value pairs $[fracture = neck]$ is tranformed into $[neck_f racture = yes]$ and others are tranformed in the same way, then the datatable will be tranformed into a regular information table with binary attributes. It is a very good approach when this attribute is a conditional one. But when a decision attribute is described as a list, then it may be more difficult to deal with. For example, let us consider the case shown in Table 3. Mutual exclusiveness of decision attributes does not hold in this table. One solution is to construct new attributes represented by the conjunciton of several diseases for construction of a new partition of the universe.[3] However, when the number of attribute-value pairs is large, this solution may be quite complicated. Also, the conjunction may not be applicable to some domains.

Table 4. A Small Database on Bacterial Tests

U	Diseases	Diseases	Diseases
1.	Heart Failure	SLE	Renal Failure
2.	Pneumonia	-	-
3.	Pulmonary Emboli	-	-
4.	SLE	PSS	Renal Failure
5.	Liver Cirrohsis	Heart Failure	Hypertension

4 Functional Representation of Context-Free Fuzzy Sets

Lin has pointed out problems with multiple membership functions and introduced relations between context-free fuzzy sets and information tables[4]. The main contribution of context-free fuzzy sets to data mining is that information tables can be used to represent multiple fuzzy membership functions. Usually when we meet multiple membership functions, we have to resolve the conflicts between functions. Lin discusses that this resolution is bounded by the context: min, maximum and other fuzzy operators can be viewed as a *context*. The discussion in Section 2 illustrates Lin's assertion. Especially, when we analyze relational-database, tranformation will be indispensable to data mining of multi-tables. However, tranformation may violate mutual exclusiveness of the target information table. Then, multiple fuzzy membership functions will be observed.

Lin's context-free fuzzy sets shows such analyzing procedures as a simple function as shown in Figure 4. The important parts in this algorithm are the way to construct a list of membership functions and the way to determine whether this algorithm outputs a metalist of a list of membership functions or a list of numerical values obtained by application of fuzzy operators to a list of membership functions.

5 Conclusions

This paper shows that mutual exclusiveness of conditional and decision attributes does not always hold in real-world databases, where conventional probabilstic approaches cannot be applied.

---

[3] This idea is closely related with granular computation[3, 14].

```
procedure Resolution of Multiple Memberships;
  var
    i : integer; L_a, L_i : List;
    A: a list of Attribute-value pairs (multisets:bag);
    F: a list of fuzzy operators;
  begin
    L_i := A;
    while (A ≠ {}) do
      begin
        [a_i = v_j](k) = first(A);
        Applend μ([a_i = v_j](k)) to L_[a_i=v_j]
        /* L_[a_i=v_j]: a list of membership function for attribute-value pairs */
        A := A − [a_i = v_j](k);
      end.
    if (F = {}) then
      /* Context- Free */
      return all of the lists L_[a_i=v_j];
    else
      /* Resolution with Contexts*/
      while (F ≠ {}) do
        begin
          f = first(F);
          Apply f to each L_[a_i=v_j]; μ_f([a_i = v_j]) = f(L_[a_i=v_j])
          Output all of the membership functions μ_f([a_i = v_j])
          F := F − f;
        end.
  end {Resolution of Multiple Memberships};
```

Fig. 2. Resolution of Multiple Fuzzy Memberships

It is surprising that tranformation will easily generate this situation in data mining from relation databases: when we apply attribute-oriented generalization to attributes in databases, generalized attributes will have fuzziness for classification. In this case, we have to take care about the conflicts between each attributes, which can be viewed as a problem with multiple membership functions. Also, real-world databases may have fuzzy contexts when we store multiple-values for each attribute. It is notable that this phenomenon is quite natural at least in medical doamin. Finally, the authors pointed out that these contexts should be analyzed by using fuzzy techniques, where context-free fuzzy sets will be a key idea to solve this problem. It will be our future work to induce fuzzy $if - then$ rules from this database and to compare these fuzzy rules with other conventional approaches.

References

1. Y.D. Cai, N. Cercone and J. Han, Attribute-oriented induction in relational databases. in: Shapiro, G. P. and Frawley, W. J. (eds), *Knowledge Discovery in Databases*, AAAI press, Palo Alto, CA, pp.213-228 (1991).
2. Fayyad, U.M., et al.(eds.): *Advances in Knoweledge Discovery and Data Mining*, AAAI Press (1996).
3. Lin, T.Y. Fuzzy Partitions: Rough Set Theory, in *Proceedings of IPMU'98*, Paris, pp. 1167-1174, 1998.
4. Lin, T.Y. Context Free Fuzzy Sets and Information Tables *Proceedings of EUFIT'98*, Aachen, pp.76-80, 1998.
5. Pawlak, Z., *Rough Sets*. Kluwer Academic Publishers, Dordrecht, 1991.
6. Pawlak, Z. Conflict analysis. In: *Proceedings of the Fifth European Congress on Intelligent Techniques and Soft Computing (EUFIT'97)*, pp.1589–1591, Verlag Mainz, Aachen, 1997.
7. Pawlak, Z. Rough Modus Ponens. *Proceedings of IPMU'98* , Paris, 1998.
8. Pawlak, Z. Rough Sets and Decision Analysis, *Fifth IIASA workshop on Decision Analysis and Support*, Laxenburg, 1998.
9. Skowron, A. and Grzymala-Busse, J. From rough set theory to evidence theory. In: Yager, R., Fedrizzi, M. and Kacprzyk, J.(eds.) *Advances in the Dempster-Shafer Theory of Evidence*, pp.193-236, John Wiley & Sons, New York, 1994.
10. *Readings in Machine Learning*, (Shavlik, J. W. and Dietterich, T.G., eds.) Morgan Kaufmann, Palo Alto, 1990.
11. Tsumoto, S. Knowledge Discovery in Medical Databases based on Rough Sets and Attribute-Oriented Generalization. *Proceedings of IEEE-FUZZ99*, IEEE Press, Anchorage, 1998.
12. Tsumoto, S. Automated Induction of Medical Expert System Rules from Clinical Databases based on Rough Set Theory *Information Sciences* **112**, 67-84, 1998.
13. Tsumoto, S., Automated Discovery of Plausible Rules based on Rough Sets and Rough Inclusion, *Proceedings of PAKDD'99*, (in press), LNAI, Springer-Verlag.
14. Zadeh, L.A., Toward a theory of fuzzy information granulation and its certainty in human reasoning and fuzzy logic. *Fuzzy Sets and Systems* **90**, 111-127, 1997.
15. Ziarko, W., Variable Precision Rough Set Model. *Journal of Computer and System Sciences*. 46, 39-59, 1993.

Applying Fuzzy Hypothesis Testing to Medical Data

Mark Last, Adam Schenker, and Abraham Kandel

Department of Computer Science and Engineering
University of South Florida
4202 E. Fowler Avenue, ENB 118
Tampa, FL 33620, USA
{mlast, aschenke, kandel}@csee.usf.edu

Abstract. Classical statistics and many data mining methods rely on "statistical significance" as a sole criterion for evaluating alternative hypotheses. In this paper, we use a novel, fuzzy logic approach to perform hypothesis testing. The method involves four major steps: hypothesis formulation, data selection (sampling), hypothesis testing (data mining), and decision (results). In the hypothesis formulation step, a null hypothesis and set of alternative hypotheses are created using conjunctive antecedents and consequent functions. In the data selection step, a subset D of the set of all data in the database is chosen as a sample set. This sample should contain enough objects to be representative of the data to a certain degree of satisfaction. In the third step, the fuzzy implication is performed for the data in D for each hypothesis and the results are combined using some aggregation function. These results are used in the final step to determine if the null hypothesis should be accepted or rejected. The method is applied to a real-world data set of medical diagnoses. The automated perception approach is used for comparing the mapping functions of fuzzy hypotheses, tested on different age groups ("young" and "old"). The results are compared to the "crisp" hypothesis testing.

Keywords. Hypothesis testing, fuzzy set theory, data mining, knowledge discovery in databases, approximate reasoning.

1 Introduction

The analysis of medical data has always been a subject of considerable interest for governmental institutions, health care providers, and insurance companies. In this study, we have analyzed a data set, generously provided by the Computing Division of the Israeli Ministry of Health. It includes the demographic data and medical diagnoses (death causes) of 33,134 Israeli citizens who passed away in the year 1993. The file does not contain any identifying information (like names or personal IDs).

In the original database, the medical diagnosis is encoded by an international, 6-digit code (ICD-9-CM). The code provides highly detailed information on the diseases: the 1993 file includes 1,248 distinct codes. Health Ministry officials have grouped these codes into 36 sets of the most common death causes, based on the first three digits of the code.

It is a well-known fact that there is an association between a person's age and the likelihood of having certain diagnoses (e.g., heart diseases are more frequent among older people). Though this association is present in most types of human diseases (and even some unnatural causes of death), it is not necessarily significant, in the practical sense, for any diagnosis. Thus, if a certain disease is more likely by only 2% among people over the age of 40 than among younger people, this can hardly have any impact on the Medicare system. Nevertheless, if the last fact is based on a sufficiently large sample, its *statistical significance* may be very high.

Our purpose here is to find the types of medical diagnoses where the difference between young people and elderly people is *practically significant*. Once these diagnoses are detected, the Ministry of Health (like any other health care organization) can invest a larger part of its budget in preventing the related death causes in certain age groups of the population. Thus, for every possible cause (e.g., cancer, heart disease, or traffic accident) we are testing a single hypothesis saying, "The elderly people are more (less) likely to die from this cause than the young people." Since the number of available hypotheses is strongly limited (the ministry officials have identified 36 sets of major causes), each hypothesis will be tested by a *verification-oriented* approach. For a concise comparison between verification-oriented and discovery-oriented methods of data mining, see Fayyad *et al* [1].

This paper is organized as follows. In the next section, we describe a "crisp" approach to hypothesis testing, aimed at measuring the *statistical significance* of each hypothesis. The limitations of applying this "classical" statistical approach to real-world problems of data analysis are clearly emphasized. In Section 3, we proceed with representing a novel methodology of fuzzy hypothesis testing for verification-based data mining. The analysis of the medical data set by using the "crisp" approach and the fuzzy approach to hypothesis testing is performed in Section 4. Section 5 concludes the paper by comparing the results of the two methods and outlining other potential applications of the Fuzzy Set Theory to the area of data mining.

2 "Crisp" Hypothesis Testing

Statistical hypothesis testing is a process of *indirect proof* [6]. This is because the data analyst assumes a single hypothesis (usually called the null hypothesis) about the underlying phenomenon *to be true*. In the case of medical data, the simplest null hypothesis may be that the likelihood of people under 40 having heart disease is equal to the likelihood of people over 40. The objective of a statistical test is to *verify* the null hypothesis. The test has a "crisp" outcome: the null hypothesis is either rejected or retained (see [6]). According to the statistical theory, *retaining* the null hypothesis should not be interpreted as *accepting* that hypothesis. Retaining just means that we do not have sufficient statistical evidence that the null hypothesis is not true. On the other hand, rejecting the null hypothesis implies that there are an infinite number of alternative hypotheses, one of them being true. In our example, the set of alternative hypotheses includes all non-zero differences between the probabilities of the same disease in the two distinct population groups.

The statistical theory of hypothesis testing deals with a major problem of any data analysis: the limited availability of target data. In many cases, it is either impossible or too expensive to collect information about all the relevant data items. Hence, a *random sample*, selected from the entire population, is frequently used for testing the null hypothesis. In the random sample, like in the entire population, we may find some evidence contradicting the statement of the null hypothesis. This does not necessarily mean that the null hypothesis is wrong: the real data is usually affected by many random factors, known as *noise*. Representing the distribution of noise in the sample cases is an integral part of the null hypothesis. Thus, for comparing means of continuous variables derived from large samples, the assumption of a Normal distribution (based on the Central Limit Theorem) is frequently used.

To compare between the probabilities of a diagnosis in two distinct age groups, we need to perform *the comparison between proportions test* (see [5]). This test is based on two independent random samples, extracted from two populations. The sizes of the samples do not have to be equal, but to apply the Central Limit Theorem, each sample should include at least 30 cases. Furthermore, we assume that each person in the same age group has exactly the same probability of having a certain disease. The last assumption enables us to describe the actual number of "positive" and "negative" cases in each group by using the Binomial distribution.

The massive use of the "crisp" hypothesis testing by many generations of statisticians has not eliminated the confusion associated with its practical application. Retaining a hypothesis is supposed to increase our belief in it – but how much greater should our belief be now? Statistics gives no clear answer. Rejecting a hypothesis leaves us even more confused: we are not supposed to believe in the null hypothesis anymore. However, which alternative hypothesis should be considered true?

Apparently, the significance level may be used as a continuous measure of evaluating hypotheses. However, as indicated by [6], "significant" is a purely technical term and it should not be confused with the practical terms "important," "substantial," "meaningful," etc. Very large samples may lead us to statistically significant conclusions, based on negligible differences between estimates. In other words, statistical significance does not imply practical significance. In the next section, we describe a novel, fuzzy method for determining the validity of a hypothesis on a continuous scale.

3 Fuzzy Hypothesis Testing

The concept of *fuzzy testing*, or more specifically, *fuzzy hypothesis testing* [7] is a verification-based method of data mining. A fuzzy hypothesis test is used to determine the truth (or falsity) of a proposed hypothesis. The hypothesis may involve either crisp or fuzzy data; however, a fuzzy hypothesis test should produce a value on [0,1], which indicates the degree to which the hypothesis is valid for given sample data. This is an extension of the classical hypothesis test, which yields a crisp value in {0,1} (see above). The fuzzy hypothesis test will accept the null hypothesis H_0 to some degree μ and the alternative hypothesis H_1 to some degree 1-μ.

3.1 The Formal Notation

A set of collected data, i.e. a database, is defined:

$$X = \{x_1, x_2, x_3, ..., x_m\}$$

where m is the number of cases (records) in the database and x_i is an n-dimensional vector in an n-dimensional feature space:

$$x_i = <x_{i,1}, x_{i,2}, ..., x_{i,n}>$$

A set $D \subseteq X$ is chosen, called a sample set, which will be used to test the hypothesis. Next, choose a set of hypotheses $H = \{H_0, H_1, ..., H_f\}$ where H_0 is the null hypothesis to accept or reject and H_1 through H_f are the alternate hypotheses we must accept if we reject H_0. A hypothesis can be thought of as an implication of the form:

    if   *condition*$_1$ and *condition*$_2$ and ... *condition*$_k$

    then   x is a member of F with membership $\mu(x_i)$

In other words, a hypothesis is composed of a set C of k conjunctive antecedent conditions and a consequent classification (e.g. cluster, fuzzy set) F. A condition is a comparison of one of the components of x_i and a constant (possibly fuzzy) value. μ is defined as a mapping: $\mu(x_i, H) \rightarrow [0,1]$.

In the medical dataset, examples of conditions include:
- "A person lives in the city of Haifa" (a crisp condition)
- "A person is old" (a fuzzy condition)

The value of μ determines whether the data collected agrees with the hypothesis. A value of $\mu_0 = 1$ means the data is in total agreement with the null hypothesis; a value of $\mu_0 = 0$ means the data totally contradicts the null hypothesis. Additionally, the value of μ for the alternative hypotheses should be the inverse of that of H_0, i.e. $\mu_1 + \mu_2 + ... \mu_f = 1 - \mu_0$.

3.2 Calculating the Sample Size

Since it may not always be practical or possible to use all collected data (i.e. the entire database), a sampling of data, called a sample set, is used to verify the hypotheses.

The sample set D is usually chosen at random from among the set X (the entire database). This random sampling must be large enough to make sure that the set D is "good"; i.e. that D reflects the contents of X. If D = X it must be accepted; the sample is the entire database. If D = Ø, it must be rejected; the sample contains no data. Otherwise, the number of data in D, denoted $d=|D|$, will determine if it is "good."

The following function, called the degree of satisfaction (DoS), is chosen to represent the belief that D is a good sample of X based on d (the sample size) and m (the size of the entire data set):

$$f(d,m) = \begin{cases} \dfrac{\log(\dfrac{d}{m})}{\log(b)} + 1 & \text{when } d > \dfrac{m}{b} \\ 0 & \text{otherwise} \end{cases} \tag{1}$$

where b is a constant that controls the x-intercept of the function (the sample size of zero satisfaction). Larger values of b make the intercept closer to 0. For example, when $b=10$, the x-intercept is at 10% of m (10% of the items are guaranteed to be selected); for $b=100$, the x-intercept is at 1% of m (the minimal sample size is 1%). Figure 1 shows the graph of the function f for $b=10$. In the graph, the x-axis is the percentage of the total data, m, selected for D. In other words, the x-axis is d/m, where 0 is 0% and 1.0 is 100%. The function f is used to select the number of items for D: as $f(d,m) \rightarrow 1$, $d \rightarrow m$. Thus, the sample becomes closer to 100% for higher degrees of satisfaction required.

The function is chosen as it meets the criteria given above for selecting the size of a "good" sample set. If $d=m$ (i.e. the entire database), then $f=1.0$. If $d=0$ (i.e. no data), then $f=0.0$. The introduction of variable b allows us to set a stronger condition of $f=0.0$ when $d < m/b$, if we have a preference that there should be some lower limit on the number of items selected for the sample. We chose the logarithm function because of its shape. From the figure we see that as we add items to the sample, the function f increases faster at the beginning than later, when the sample set is larger. This agrees intuitively with our notion of how a sample works: more items are generally better, but once we have a certain amount of items in our sample the additional information provided by adding more items is less than that of adding the same number of items to a smaller sample.

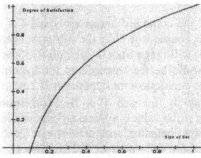

Fig. 1. Plot of function f ($b=10$)

As shown above, the fuzzy method of calculating the sample size does not depend on the hypotheses we are going to test on the data. This approach agrees with the common process of knowledge discovery in databases (see [1]), where the target data is selected before the data mining stage. The procedure for selecting an appropriate sample size, suggested by the statisticians (see [6]), is more complicated and it assumes knowing in advance both the hypotheses to be tested and the underlying distributions. According to [6], the first step is specifying the minimum effect that is "important" to be detected by the hypothesis testing. The linguistic concept of importance is certainly beyond the scope of the statistical inference. However, it is directly related to the process of *approximate reasoning*, easily represented by the Fuzzy Set Theory (see [2]).

3.3 Creating the Mapping Function for Each Hypothesis

The mapping function μ_i maps each vector in D for a given hypothesis H_i to a value in [0,1]. This number represents the degree to which each sample agrees with the hypothesis. In order to determine the agreement, the membership function of the consequent F_i must be known. If the data described by the vector x lies within F_i, then μ_i should equal the degree of membership of x in F_i. Usually F_i will be some geometric function on [0,1], such as a triangular or trapezoidal shaped function.

The vectors in D are compared with the conjunctive conditions in the antecedent of the hypothesis. For crisp conditions, any condition(s), which are false, cause x to be excluded from consideration since they do not lend any support to the null hypothesis or alternative hypotheses. For fuzzy conditions, it may be necessary to use some threshold value to determine if the vector x should be excluded. For example, for a fuzzy value of 0.5 or less, the vector x may be closer to some other fuzzy set. Each fuzzy condition in the antecedent will have a value on [0,1] for each x, and these values must be combined using a t-norm operation, such as *min*. The resulting value indicates the degree to which x supports the antecedent conditions of H. The Dienes-Rescher fuzzy implication [8] is then performed for the combined antecedent values and the consequent value:

$$\mu_l = \max(1-P_l, f_l) \tag{2}$$

where P is the value of the combined antecedents and f is a function describing the fuzzy membership of the consequent. Here the subscript l denotes to which hypothesis each variable belongs; it will range from 0 (the null hypothesis) to k, for k alternative hypotheses. Thus, P_2 would be the antecedents for hypothesis H_2, f_3 would be the fuzzy membership of the consequent for hypothesis H_3, etc.

Once the membership μ_0 for each x in D is determined, the values must be aggregated to determine if the values in D, taken as a whole, support H_0. This can be done in a variety of ways including arithmetic mean (each point contributes to the decision), minimum (pessimistic – if any x fail H_0, then H_0 is rejected), or maximum (optimistic – if any x pass H_0, then H_0 is accepted). For arithmetic mean, denote the overall mapping function M_k for hypothesis k:

$$M_k(D) = \frac{\sum_{i \in D} \mu_k(\bar{x}_i)}{\delta} \qquad (3)$$

where δ is the number of vectors in D that are relevant to the hypothesis under consideration.

3.4 Comparing Fuzzy Hypotheses

In the medical database, our objective is to compare between the overall mapping functions of two hypotheses:

- Hypothesis No. 1: If the age is young, then diagnosis (cause) = x
- Hypothesis No. 2: If the age is old, then diagnosis (cause) = x

If the second mapping function is significantly greater (or significantly smaller) than the first one, then we can conclude that older people have a higher (or a lower) likelihood of having that diagnosis than young people. "Significantly greater (smaller)" are fuzzy terms depending on human perception of the difference between the mapping functions. We have outlined a general approach to automated perception in [3-4]. For automating the perception of this difference, we are using here the following membership function μ_{sg}:

$$\mu_{sg} = \begin{cases} \dfrac{1}{1 + \exp(-\beta \bullet (M_2(D) - M_1(D)))}, & M_2(D) > M_1(D) \\ \dfrac{1}{1 + \exp(\beta \bullet (M_2(D) - M_1(D)))}, & otherwise \end{cases} \qquad (4)$$

where β is an adjustable coefficient representing the human confidence in the difference between frequencies, based on a given sample size. The membership function μ_{sg} increases with the value of β.

4 Analysis of the Medical Data

4.1 Hypothesis Testing

In order to create the mapping functions for each fuzzy hypothesis, the fuzzy sets corresponding to "young age" and "old age" have been determined. These fuzzy sets are shown in Fig. 2. Both sets are represented by triangular membership functions. The definition of these membership functions is completely subjective and user-dependent.

To perform an objective comparison between the fuzzy hypothesis testing and the "crisp" approach, we have used the threshold of 45 years to divide the records into "young" and "old" people. Afterwards, the proportion of each diagnosis under the age

228

of 45 has been compared statistically to the proportion of the same diagnosis for people over 45 years old. The statistical significance of the difference between proportions has been evaluated by the comparison between proportions test (see Section 2 above).

Both methods of hypothesis testing have been applied to the same random sample. The sample size has been determined by the fuzzy method of Section 3.2 above, using DoS (Degree of Satisfaction) equal to 0.90 and the constant b = 100. The number of records obtained is 20,907 (out of 33,134), including 1,908 young people and 18, 998 elderly people. For comparing fuzzy hypotheses, based on this sample size, the coefficient β = 25 has been selected.

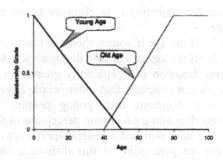

Fig. 2. Fuzzy sets "young age" and "old age"

4.2   Summary of Results

The 36 diagnoses present in the medical dataset can be divided into the following categories, by the effect of person age:
- Five diagnoses (death causes), where the difference between the young people and the elderly people is highly significant according to both the fuzzy test and the "crisp" test. These causes include: Ischaemic Heart Disease, Cerebrovascular Disease, Diseases of Pulmonary Circulation, Motor Vehicle Traffic Accidents, and Other Accidents. The likelihood increases with the age for the first three causes and decreases for the last two. From the viewpoint of the health care system, this means that older people have a higher risk of dying from the first three diseases. Consequently, this age group should be subject to frequent medical assessments as a preventive treatment. To decrease the number of traffic and other accidents in the young age group, some restrictions may be applied (and are actually applied) with respect to young drivers.
- Nineteen diagnoses, where the statistical significance of the difference is also very high (over 99.9%), but the fuzzy test has shown a relatively low significance varying from 0.50 to 0.78. For example, only 0.28% of young people have diabetes vs. 2.77% of elderly people. The significance of the fuzzy test in this case is only 0.65. However, the statistical test of comparison between proportions has provided us with a statistic z = 10.44, which has a very high significance (almost 1.00).
- Eleven diagnoses, where the significance of both tests is relatively low.
- One rare diagnosis, which has been completely omitted from the random sample.

5 Conclusions

The results of the hypothesis testing represented in Section 4 above emphasize the main drawback of statistical methods: the *statistical significance* should not be used as a synonym for *importance*. Relying solely on the results of the "crisp" testing in the above dataset would lead (actually, mislead) the analysts into concluding that almost all death causes have a strong association with the age. This could cause a wrong setting of health care priorities or even completely ignore the age for this purpose. The main contribution of Fuzzy Set Theory to this problem is the improved *differentiation* of diagnoses, starting with those completely unaffected by age, and ending with the five causes (see sub-section 4.2 above) where the age is the leading factor.

As we have shown in our work on automated perceptions [3], the potential benefit of applying fuzzy logic methods to data mining is yet to be studied. After solving one limitation of the traditional data analysis, moving from *verification* of hypotheses to their *discovery*, many data mining methods are still anchored to the statistical methods of significance testing. Consequently, a lot of unimportant (mostly, random) hypotheses are "discovered" in data. The fuzzy hypothesis testing is challenging this problem.

Acknowledgements. We thank Prof. Gabi Barabash, Director General of the Israeli Ministry of Health, and his staff, for providing data, used in this research.

References

1. Fayyad, U., Piatetsky-Shapiro, G., and Smyth, P.: From Data Mining to Knowledge Discovery: An Overview. In: Fayyad, U., Piatetsky-Shapiro, G., and Smyth, P. (eds.): Advances in Knowledge Discovery and Data Mining. AAAI/MIT Press, Menlo Park, CA (1996) 1-30
2. Kandel, A., Pacheco, R., Martins, A., and Khator, S.: The Foundations of Rule-Based Computations in Fuzzy Models. In: Pedrycz W. (ed.): Fuzzy Modelling, Paradigms and Practice. Kluwer, Boston (1996) 231-263
3. Last, M., and Kandel, A.: Automated Perceptions in Data Mining, invited paper, to appear in the proceedings of the 8th International Conference on Fuzzy Systems (FUZZ-IEEE'99), Seoul, Korea (1999).
4. Last, M., and Kandel, A.: Fuzzy Comparison of Frequency Distributions. Submitted to Publication (1999)
5. Mendenhall, W., Reinmuth, J.E., Beaver, R.J.: Statistics for Management and Economics. Duxbury Press, Belmont, CA (1993)
6. Minium, E.W., Clarke, R.B., Coladarci, T.: Elements of Statistical Reasoning. Wiley, New York (1999)
7. Schenker, A., Last, M., and Kandel, A.: Fuzzy Hypothesis Testing: Verification-Based Data Mining. Submitted to Publication (1999)
8. Wang, L.-X.: A Course in Fuzzy Systems and Control. Prentice-Hall, Upper Saddle River, NJ (1997)

Generating a Macroeconomic Fuzzy Forecasting System Using Evolutionary Search

Raouf Veliev, Alex Rubinov and Andrew Stranieri

University of Ballarat
University drive, Mount Helen, P.O. Box 663
Ballarat, Victoria, 3353, Australia
e-mail: {imcu00rv@students.ballarat.edu.au, amr@ballarat.edu.au,
a.stranieri@ballarat.edu.au}

Abstract. Fuzzy logic has not been applied to macro-economic modelling despite advantages this technique has over mathematical and statistical techniques more commonly used. The use of fuzzy logic provides a technique for modelling that makes none of the theoretical assumptions normally made in macroeconomics. However, in order to avoid making assumptions, we need to elicit fuzzy rules directly from the data. This is done using a genetic algorithm search for rules that fit the data. The technique discovered rules from artificially generated data that was consistent with the function used to generate the data. The technique was used to discover rules that predict changes to national consumption in order to explore the veracity of two economic theories that propose different causes for changes in consumption. The fuzzy rules generated illustrate a more fine-grained analysis of consumption than is predicted by either theory alone. Predictions made using the generated rules were more accurate following ten-fold cross validation than those made by a neural network and a simple linear regression model on the same data.

Introduction

Macro-economic modelling and forecasting has traditionally been performed with the exclusive use of mathematical and statistical tools. However, these tools are not always appropriate for economic modelling because of uncertainty associated with decision making by humans in an economy. The development of any economy is determined by a wide range of activities performed by humans as householders, managers, or government policy makers. Persons in each role pursue different goals and, more importantly, base their economic plans on decision-making in vague and often ambiguous terms. For example, a householder may make a decision on the proportion of income to reserve as savings according to the rule- {IF my future salary is likely to diminish, THEN I will save a greater proportion of my current salary}. Mathematical models of human decision-making impose precise forms of continuous functions and overlook the inherent fuzziness of the process.

In addition to imposing a crispness that may not be appropriate, mathematical and statistical models necessarily make assumptions that derive from economic theories. A large variety of sometimes conflicting models have emerged over the years as a consequence of this. Inferences drawn from a model hold only to the extent that the economic theoretical assumptions hold yet this is often difficult to determine.

Macroeconomic researchers solely using mathematical or statistical models are compelled to make assumptions based on their own subjective view of the world or theoretical background and beliefs. For example, hypotheses generated by researchers who accept Keynesian assumptions are quite different from hypotheses from Classical theorists. Hypotheses are not only dependent upon the subjective beliefs of their creators but can easily become obsolete. Completely different economic systems can rise in different times in different countries and be described by different models. Thus, if making assumptions and deriving hypotheses about an economy leads to subjective models, and successful theories do not last long, then the following questions arise: Is it possible to eliminate model dependence on the subjective researcher's assumptions about features and properties of the object of study?; Can there exist an approach that automatically generates a hypothetical basis for constructing a model ?; Can this approach be applied in different times to different types of economic systems ?

In this paper we introduce a modelling approach that does not rely on theoretical assumptions or subjective fine-tuning of system parameters. We apply fuzzy theory and use an evolutionary programming approach to pursue two goals:
1. To provide a user with a system, which better represents uncertainty caused by the prevalence of human decision making in an economy
2. To build a forecasting model without any initial assumptions, which aims solely to be consistent with observed economic data.

Our approach derives fuzzy rules from macro-economic data. We use an evolutionary programming approach to search for rules that best fit the data. A user can glance at the rules and visualise the main dependencies and trends between variables. Moreover, if there are exogenous variables in the model presented among input indicators, a user is able to foresee a possible impact of their simulations on other variables of interest. For example, quickly glancing at the fuzzy rules shown in Table 1, we can say that in order to obtain a large values of the output we need to increase the value of exogenous variable $x2$.

		X1	
		SMALL	LARGE
X2	SMALL	SMALL	SMALL
	LARGE	LARGE	LARGE

Table 1. Sample fuzzy rules table

We believe that fuzzy logic, though not normally used in macro-economic modelling is suitable for capturing the uncertainty inherent in the problem domain. An evolutionary approach to building the system can facilitate the design of a system free of subjective assumptions, and based only on patterns in the data.

In the following section we describe the concept of hybrid fuzzy logic and genetic algorithms. Following that we describe our method in some detail and provide an example with data that is generated from artificial rules. In Section 5 we apply the method to macro-economic data before outlining future directions.

Mining forecasting fuzzy rules with genetic algorithms

Our task is to model a macro-economic environment and capture any uncertainty in a macro-economic agent's decision-making behaviour in order to generate predictions of the economic system's development in the future. We are required to mine knowledge of this process in flexible human-like terms. The application of the fuzzy control architecture for this forecasting problem proceeds with the following modifications:
1. Macro-economic data is sourced from national repositories.
2. The fuzzy sets and defuzification methods are set as parameter features of the system. No attempt is made to automatically discover membership functions.
3. The rules governing the process are required to be discovered from the data.

Research in fuzzy control focuses on the discovery of membership functions, and independently on the fine tuning of fuzzy rules for given data. Both research strands are aimed at adjusting a fuzzy control system to the specific data. Several researchers [1], [3] used genetic algorithms to simultaneously find fuzzy rules and parameters of the membership functions. However the simultaneous search for rules and membership functions adds complexity and may not be necessary if we are dealing with economic data. With most economic indicators there is general agreement about the mapping of qualitative terms onto quantitative terms. Most economists would regard a gross domestic product (GDP) rise of 1% to be low, one of 5% to be high. There may be some disagreement surrounding a GDP value of 3% but there is expected to be little disagreement about the precise form of the function between high and low.

In order to reduce the complexity of the search problem, and, in light of the nature of economic data, we do not search for near optimal membership functions but instead determine a membership function that seems reasonable. Fuzzy rules are discovered using an evolutionary search procedure. Fuzzy rules derived by the genetic algorithm are applicable only to the pre-set membership functions but this is minor limitation.

Machine learning methods have been applied to the problem of mining fuzzy rules from data. For example, Hayashi and Imura [2] suggested a two-step procedure to extract fuzzy rules. In the first step, a neural network (NN) was trained from sample data. In the second step, an algorithm was used to automatically extract fuzzy rules from the NN. Kosko [6] interprets fuzzy rules as a mapping between fuzzy membership spaces and proposed a system, called Fuzzy Cognitive Maps, to integrate NN and fuzzy systems. Lin and Lee [7] proposed a NN-based fuzzy logic system, which consists of five layers. The first is linguistic input, the second and fourth are terms representing a membership function, the third is a set of rules and the fifth is an output. The common weaknesses of NN, however, are the lack of analytical guidance, where all relationships are hidden in the "black box" of the network connections. Furthermore, training neural networks is not deterministic and the learning process may be trapped in local solutions.

Another widely used machine learning method used is the induction of fuzzy decision trees where fuzzy entropy is used to guide the search of the most effective decision nodes [10], [11], [13]. Although, in most situations the decision tree induction works well, it has some limitations. According to Yuan and Zhuang [14] the one-step-ahead node splitting without backtracking may not be able to generate the best tree. Another limitation is that even the best tree may not be able to present the best set of rules [12]. Furthermore, this method has been found to be sub-optimal in certain types of problems such as multiplexer problems [8].

In this paper we use an evolutionary approach to find fuzzy rules from macro-economic data. Genetic algorithms have been used by Rutkowska [9] to find near-optimal fuzzy rules and learn the shapes of membership function. Karr [4] also focussed his work on looking for a high-performance membership function using genetic algorithms. Yuan and Zhuang [14] discovered fuzzy rules for classification tasks that were most correct, complete and general.

In our work, we do not seek rules that are most general, complete and correct but initially focus only finding a complete list of rules that best describe the data. The generalisation of rules is a manual process exercised if required. Often with systems as complex as dynamic economies few general rules are non-trivial and more attention is focused on specific rules. Furthermore in order to find the most general, complete and concise rules Yuan and Zhuang [14] proposed definitions of these concepts. The adoption of similar definitions with macro-economic data is one step toward re-introducing theoretical assumptions in our model and was thus avoided.

In the next section we describe the procedure used to design a genetic algorithm search for mining fuzzy rules.

Description of method

To apply the genetic algorithm search there are two main decisions to make:
1. How to code the possible solutions to the problem as a finite bit strings and
2. How to evaluate the merit of each string.

Because solutions in our case are fuzzy terms in the fuzzy rules table, we construct the solution strings as rows of the rules table.

Theoretically, it is possible to apply this coding for genetic search in any multilevel fuzzy rules space. But, the length of strings increases dramatically with an increase in number of inputs, outputs and fuzzy sets, over which these inputs and outputs are defined. We limited our examples to two inputs and one output defined over four fuzzy sets.

Although the use of binary coding is preferable by many researchers we use integer numbers. Binary coding can code variables, which can take only 2 values. To code variables, which can take more than 2 values in binary coding we have to use several genes to code each variable and deal with unused coding patterns. To avoid this complexity and to cover the possibility of the appearance of more than two values in each cell of fuzzy the rule table, we used integer coding. We assign numbers from 1 to N for each of N fuzzy sets defined for the output variable. Thus, each rule is represented with the corresponding number as a gene in the coded chromosomes.

The second task concerns the determination of the fitness function. Those chromosomes, which represent fuzzy rules that are more consistent with the data, are considered fitter then others. We calculate the sum of squared deviation between the output of the fuzzy control with a given set of rules and a real value of the output indicated in the data record. This value represents a fitness function value and is used as criteria in producing a new generation. In early trials we used a sum of modulus instead of the sum of squares of the difference between actual and predicted values to measure error, and obtained almost identical results. In order to be able to compare our system's performance to other techniques we preferred to use the sum of squared metric as the evaluation criteria.

The crossover and mutation procedures are quite common for genetic algorithms and are as follows. The current population is ranked according to the values of fitness function. The probability for each chromosome to be chosen is proportional to the place of chromosomes in the ranked list of the current population. Chromosomes are paired and either get directly copied into a new generation or produce a pair of children via a crossover operation. The newly produced children are placed in the new generation. The probability of two parents crossing over is set as a parameter of the algorithm. The crossover procedure can have one or several crossover points. We break the parent chromosomes in pre-determined places and mix the consequent parts to build a new child chromosome.

The mutation process is applied to each gene of each chromosome in all generations. The integer number in a gene randomly increases or decreases its value by one. This allows us to represent new genes in a population for a given place in chromosome, whilst avoiding huge changes in the original solution pattern so as to adjust the solution toward a mutant in the neighbourhood area.

The following section presents an implementation and tests the described method with data generated from known rules.

Example with generated data

In order to test our method we ran the system over data generated artificially. By defining the functional dependence between input and output variables we know exactly what the derived fuzzy rules should be.

Two inputs and one output were used to test the system. The same four fuzzy sets – {Negative high (NH), Negative low (NL), Positive low (PL) and Positive High (PH)} were defined over all system's variables as shown in the Figure 1. Thus, possible solutions representing the 4x4 fuzzy rule table were coded into 16 gene-length chromosomes. The most popular defuzification method, centre of gravity, was chosen. The task was to find all rules in a view {If x1 is FuzzySet(i) AND x2 is FuzzySet(j), THEN y is FuzzySet(k)}, where i,j,k=1..4 and FuzySet(i), FuzySet(j) and FuzySet(k) belong to the set - {Negative high, Negative low, Positive low and Positive High}.

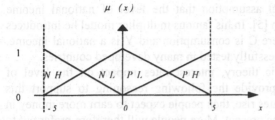

Fig. 1. Fuzzy sets for x1, x2 and y

Artificial data was constructed as follows. One hundred random values were generated from the interval (-5,5) for the variables x1 and x2. Then we put them through a pre-defined function, $y=(10*x1-x2)/11$ and stored output. The function $y=(10*x1-x2)/11$ has been chosen randomly only to demonstrate the method's performance. We ran the system with a crossover probability of 40%, mutation rate of 0.1% and a population size = 50. The genetic algorithm was run 50 times with different initial populations. In 100% of these test simulations the search converged to the solution presented in Table 3 after approximately 50 generations.

Table 3 illustrates that the search algorithm finds a "very good" solution, for data generated by the function $y=(10*x1-x2)/11$. As expected, the order of fuzzy outputs for y in the fuzzy rule table decreases with an increase in x2 and increases with an increase in x1. This fact is consistent with the positive coefficient on variable x1 and negative coefficient of variable x2. Moreover, the value of x1 is more significant in determining an output y, as we would expect given the coefficient of x1 is 10 times larger than that for x2 in the function. This fact can be observed in the first and the fourth row, where the values for y are Negative high and Positive high respectively regardless of the values of x1. The rest of the cells also confirm that positive values of x1 are more prominent in determining the value of y than negative values of x2 and visa versa.

		X2			
		NH	NL	PL	PH
X1	NH	NH	NH	NH	NH
	NL	PL	NL	NL	NH
	PL	PL	PL	PL	NL
	PH	PH	PH	PH	PH

Table 2. Fuzzy rules for the dummy data

The next section describes an example of applying the algorithm to real world economic data.

Example with economic data looking for theoretical assumptions e.g. keynesian theory

In this section, in order to test the algorithm with real economic data, we chose economic indicators with well-known interrelationships. The Keynesian General

Theory is based on a fundamental assumption that the level of national income determines the level of consumption [5]. In his famous multiplier model he introduces an increasing function $C=f(Y)$, where C is consumption and Y is a national income. This hypothesis has been quite successfully tested in many developed countries.

According to classical economic theory, interest rates impact on the level of consumption,. Classical theorists provide the following reasoning to support this hypothesis. If the level of interest rates rise, then people expect to earn more money in the future on each dollar saved in the present. More people will therefore prefer not to spend today, but wait for a future time when they will have more to spend. Provided a given level of Production Output or National Income, more savings mean less consumption.

In our study we expect to find evidence for well-known associations depicted by both Keynesian and Classical theories. Economic data, describing dynamics of these indicators in the United States was obtained from The Federal Reserve Bank of St Louise. The records were collected on a regular basis from 1960 till 1997.

We compared our fuzzy rules generation method with linear regression and feed-forward neural network on the same Federal Reserve Bank data.

Data transformation took the form transforming actual quarterly values of consumption and national income into changes in those values over a quarter. 150 records representing change from one quarter to the next was collected. This data allowed us to make our system more sensitive to changes in the modelling economic indicators.

The first input is the change over a quarter period of the level of national income. The second input is the change in the interest rate over a quarter. The output was changes in the level of real personal consumption over a quarter. The interval of real values of inputs and the output were set from minimum and maximum observed changes in the corresponding variables. The four fuzzy sets – {Negative high (NH), Negative low (NL), Positive low (PL) and Positive High (PH)} were set in a manner illustrated in Figure 1. The choice of fuzzy sets is supported by the importance in economic modelling to distinguish between an increase and a decrease in the control variable, which reflected in the negative or positive direction of the changes. Furthermore, it is valuable to distinguish between different degrees of change, therefore high and low fuzzy sets are distributed over both positive and negative sides of the variables domain.

Ten fold cross-validation was used with hold out sets of size 15 and training sets of size 135. For each cross validation set, fuzzy rules were generated as described above. The sum of square of differences between consumption predicted by the fuzzy rules and actual consumption on the test set was recorded. This was repeated with a simple linear regression model and also with a feed-forward neural network trained with back-propagation of errors (3 layer, learning rate = 0.2, no improvement in error rates after 40-55 epochs). Table 4 illustrates the median, average and standard deviation of the sum of square of the difference between predicted and actual change in consumption for the fuzzy rules, neural network and linear regression over the ten cross-validation sets.

	Fuzzy rules	Neural network	Linear regression
Mean	14.75	17.31	23.25
Std. Deviation	5.5	5.56	10.42
Median	13.59	15.4	21.98

Table 3. Comparison of fuzzy rules, neural network and linear regression

The fuzzy rules generated by the genetic algorithm method proposed here performed very well in comparison to a linear regression model. This was perhaps to be expected because the relationship between changes in national income, interest rates and consumption is expected to be more complex than a simple linear one. Neural networks can capture non-linear relationships and the networks trained performed better than the linear regression models. However, the performance of the fuzzy rules was comparable to the trained networks.

The table of rules is included in Table 4 where Y is change in national income and I is change in interest rates. The fuzzy rules predict change in consumption.

		I			
		NH	*NL*	*PL*	*PH*
Y	*NH*	**PH**	**PL**	**NH**	**NL**
	NL	**NH**	**NL**	**NL**	**NL**
	PL	**NL**	**PL**	**PL**	**NL**
	PH	**PH**	**PH**	**PH**	**NL**

Table 4. An optimal set of fuzzy rules for the data of Example 2.

The black box nature of neural networks is a distinct disadvantage for the analysis of macro-economic data. In contrast, as Table 4 illustrates, fuzzy rules generated without any theoretical assumptions can be used to explore patterns and to even assess the veracity of theories. To perform this assessment let us summarise search results in light of both theories. Firstly, taking into account that both types of economists usually assume consumption dependencies close to linear we can approximately define them in rule view as it shown in Table 5. Then, The Table 6 can be interpreted as to what degree it confirm either or both theories.

I	C		Y	C			NH	NL	PL	PH
NH	PH		NH	NH		NH	Classical	Classical	Cl. & Kn.	Cl. & Kn.
NL	PL		NL	NL		NL	Keynesian	Keynesian	Cl. & Kn.	Cl. & Kn.
PL	NL		PL	PL		PL		Cl. & Kn.	Keynesian	Classical
PH	NH		PH	PH		PH	Cl. & Kn.	Cl. & Kn.	Keynesian	Classical

Table 5. Classical and Keynesian consumption dependecies **Table 6.** Rules interpretetion

The selected part of the table is areas where two theories do not contrivers each other and both theories are confirmed by the rules. In fact, according to the table 5, when interest rates rise and national income falls, then consumption shrinks and, on the other corner, when interest rate fall and national income rise, consumption rises. The (Y-PL, I-NH) cell is the only exception to the theory predictions in these areas.

In the rest of the table we can observe that high rises in interest rates make consumption behaviour classically, while under low rises in interest rates it confirms to Keynesian theory. Regarding national income, under high decrease in national income consumption reacts in classical manner, while low decreases in national income consumption is determined in Keynesian way.

Conclusions

In this paper we demonstrated an application of fuzzy logic to macro-economic modelling. Despite benefits, fuzzy logic has not been used as widely as mathematical and statistical techniques for this purpose. Our use of fuzzy logic makes none of the theoretical assumptions normally made in macroeconomics and is more intuitive. We elicited fuzzy rules directly from macro-economic data using a genetic algorithm search for rules that best fit the data. The technique was evaluated initially with artificially generated data and then with data from the US economy. Fuzzy rules were successfully discovered that described the function used to generate the artificial data. Furthermore, fuzzy rules generated from real economic data provided a fine grained analysis of economic activity and was used to explore the relationship between two diverse economic theories. The fuzzy rules generated by this approach were compared for predictive accuracy with a linear regression model and with a neural network. The fuzzy rules out-performed both approaches.

References

[1] Hanebeck, U. D. and G.K. Schmidt. 1996. Genetic optimisation of fuzzy networks. Fuzzy Sets Systems. 79; 59-139.
[2] Hayashi Y. and A. Imura. 1990. Fuzzy neural expert system with automated extraction of fuzzy "if-then" rules from a trained neural network. 1st International Symposium on Uncertainty Modelling and Analysis. December; 489-494.
[3] Homaifar, A. and E. McCornick. 1995. Simultaneous design of membership fubnctions and rules sets for fuzzy controllers using genetic algorithms. IEEE Trans. Fuzzy Sets. 3(2); 129-139.
[4] Karr C. L. 1991. Design of an adaptive fuzzy logic controller using a genetic algorithm. Proc. 4th International Conference on Genetic Algorithms (San Diego). July; 450-457.
[5] Keynes J. 1961. M. The general theory of employment, interest and money. (MacMillan, NJ)
[6] Kosko B. 1992. Neural Networks and Fuzzy Systems (Prentice-Hall, Eaglewood Cliffs, NJ).
[7] Lin C-T. and C.S.G. Lee. 1991. Neural-network-based fuzzy logic control and decision system. IEEE trans. Computing. 12; 1320-1336.
[8] Quinlan J. R. 1986. Induction of decision trees. Machine learning. 1; 81-106.
[9] Rutkowska D. 1998. On generating fuzzy rules by an evolutionary approach. Cybernetics and Systems. 29; 391-407.
[10] Weber R. 1992. Automatic knowledge acquisition for fuzzy control applications. Proc. International Symposium on Fuzzy Systems (Japan). July; 9-12.
[11] Weber R. 1992. Fuzzy-ID3: a class of methods for automatic knowledge acquisition. Proc. 2nd International Conference on Fuzzy Logic and Neural Networks (Japan). July; 265-268.
[12] Weiss S. M. and C. A. Kulikowski 1991. Computer systems that learn: Classification and prediction methods from statistics, Neural nets, machine learning and expert systems (Morgan Kaufmann. San Mateo, CA).
[13] Yuan Y. and M. J. Shaw. 1995. Induction of fuzzy decision trees. Fuzzy sets and systems. 69; 125-139.
[14] Yuan Y. and H. Zhuang. 1996. A genetic algorithm for generating fuzzy classification rules. Fuzzy sets and systems. 84; 1-19.

Fuzzy Control of Nonlinear Systems Using Nonlinearized Parameterization

Hugang Han* and Hiroko Kawabata**

*School of Business, Hiroshima Prefectural University
Shobara-shi, Hiroshima 727-0023, Japan
**Faculty of Engineering, Kyusyu Institute of Technology
Sensui-cho, Tobata-ku, Kitakyushu 804-8550, Japan

Abstract. In the most of adaptive fuzzy control schemes presented so far still only the parameters (weights of each rule's consequent), which appear linearly in the radial basis function (RBF) expansion, were tuned. The major disadvantage is that the precision of the parameterized fuzzy approximator can not be guaranteed. Consequently, the control performance has been influenced. In this paper, we not only tune the weighting parameters but tune the variances which appears nonlinearly in the RBF to reduce the approximation error and improve control performance, using a lemma by Annaswamy *et al* (1998) which was named as concave/convex parameterization. Global boundedness of the overall adaptive system and tracking to within precision are established with the proposed adaptive controller.

1. Introduction

The application of fuzzy set theory to control problems has been the focus of numerous studies. The motivation is that the fuzzy set theory provides an alternative way into the traditional modeling and design of control systems when system knowledge and dynamic models in the traditional sense are uncertain and time varying. In spite of many successes, fuzzy control has not been viewed as rigorous approach due to the lack of formal synthesis techniques, which guarantee the basis requirements for control system such as global stability. Recently, various adaptive fuzzy control schemes have been proposed to deal with nonlinear systems with poorly understood dynamics by using the parameterized fuzzy approximator [1-3]. However, most of the schemes presented so far still only the parameters (weights of each rule's consequent), which appear linearly in the radial basis function (RBF) expansion, were tuned. The major disadvantage is that the precision of the parameterized fuzzy approximator can not be guaranteed, therefore, the control performance may be affected. In the RBF expansion, three parameter vectors are used, which are named as connection weights, variances and centers. Recently, very few results are available for adjustment of nonlinearly parameterized systems. Though the gradient approaches were used [4-6], however, the way of fusing into the adaptive fuzzy control schemes to generate a global stability is still left behind. The desirable approach will be apparently to tune the three parameter vectors simultaneously. However, it can definitely lead complicated algorithms and cost of calculation. Since the RBF expansion is just a kind of approximetor and nothing more, we can refer to neural networks, which has perfect ability of approximation as known. In the neural network, it is sufficient to tune the weights and variances in general due to improve the precision to approximation, whereas the centers are simply placed on a regular mesh covering a relevant region of system space. In this paper, using a lemma by Annaswamy *et al* which was named as concave/convex parameterization [7], we not only tune the weighting parameters, but tune the variances, which appear nonlinearly in the RBF to reduce the approximation error and improve control performance. Global boundedness of the overall adaptive system and tracking to within precision are established with the proposed adaptive controller.

2. Problem Statement

This paper focuses our attention on the design of adaptive control algorithms for a class of dynamic systems whose equation of motion can be expressed in the canonical form:

$$x^{(n)}(t) + f\left(x(t), \dot{x}(t), \cdots, x^{(n-1)}(t)\right) = b\left(x(t), \dot{x}(t), \cdots x^{(n-2)}(t)\right)u(t) \tag{1}$$

where $u(t)$ is the control input, $f(\cdot)$ and $g(\cdot)$ are unknown linear or nonlinear function and b is the control gain. It should be noted that more general classes of nonlinear control problems can be transformed into this structure [8].

The control objective is to force the state $X(t) = \left[x(t), \dot{x}(t), \cdots, x^{(n-1)}(t)\right]^T$ to follow a specified desired trajectory, $X_d(t) = \left[x_d(t), \dot{x}_d(t), \cdots, x_d^{(n-1)}(t)\right]^T$. Defining the tracking error vector, $\tilde{X}(t) = X(t) - X_d(t)$, the problem is thus to design a control low $u(t)$ which ensures that $\tilde{X}(t) \to 0$, as $t \to \infty$. For simplicity in this initial discussion, we take $b = 1$ in the subsequent development.

3. Fuzzy System

Assume that there are N rules in considered fuzzy system and each of which has the following form:

$$R_j : IF\ x_1\ is\ A_j^1\ and\ x_2\ is\ A_j^2\ and \cdots and\ x_n\ is\ A_j^n,\ THEN\ z\ is\ B_j$$

where $j = 1, 2, \cdots, N$, $x_i (i = 1, 2, \cdots, n)$ are the input variables to the fuzzy system, z is the output variable of the fuzzy system, and A_j^i and B_j are linguistic terms characterised by fuzzy membership functions $\mu_{A_j^i}(x_i)$ and $\mu_{B_j}(z)$, respectively. As in [2], we consider a subset of the fuzzy systems with *singleton fuzzifier*, *product inference*, and *Gaussian membership function*. Hence, such a fuzzy system can be written as

$$h(X) = \sum_{j=1}^{N} \omega_j \left(\prod_{i=1}^{n} \mu_{A_j^i}(x_i) \right) \tag{2}$$

where $h: U \subset R^n \rightarrow R$, $X = (x_1, x_2, \cdots, x_n) \in U$; ω_j is the point in R at which $\mu_{B_j}(\omega_j) = 1$, named as the connection weight; $\mu_{A_j^i}(x_i)$ is *the Gaussian membership function*, defined by

$$\mu_{A_j^i}(x_i) = \exp\left[-(\sigma_j^i(x_i - \xi_j^i))^2\right] \tag{3}$$

where ξ_j^i, σ_j^i are real-valued parameters. Contrary on the traditional notation, in this paper we use $1/\sigma_j^i$ to represent the variance just for the convenience of later development.

Definition 1: Define *fuzzy basis functions* (FBF's) as

$$g_j\left(\sigma_j \|X - \xi_j\|\right) = \prod_{i=1}^{n} \mu_{A_j^i}(x_i), \quad j = 1,2,\cdots,N \tag{4}$$

where $\mu_{A_j^i}(x_i)$ are the Gaussian membership functions defined in (3), $\xi_j = (\xi_j^1, \xi_j^2, \cdots, \xi_j^n)$ and $\sigma_j = (\sigma_j^1, \sigma_j^2, \cdots, \sigma_j^n)$. Then, the fuzzy system (2) is equivalent to a FBF expansion

$$h(X) = \sum_{j=1}^{N} \omega_j g_j\left(\sigma_j \|X - \xi_j\|\right) \tag{5}$$

Remark: It is obvious that $g_j(\cdot)$ is convex and $-g_j(\cdot)$ is concave with respect to σ_j. The definitions of the concave and convex can be refer to [7].

Theorem 1: For any given real continuous function f on the compact set $U \in R^n$ and arbitrary $\varepsilon_h > 0$, there exists optimal *FBF expansion* $h^*(X) \in A$ such that

$$\sup_{X \in U} |f(X) - h^*(X)| < \varepsilon_h \tag{6}$$

This theorem states that the FBF expansion (5) is universal approximator on a compact set. Since the fuzzy universal approximator is characterized by parameters ω_j, σ_j and ξ_j, the optimal $h^*(X)$ contains optimal parameters ω_j^*, σ_j^* and ξ_j^*.

Without doubt, the desirable approach is to tune the three parameter vectors simultaneously. However, it can lead to complicated algorithms and cost of calculation definitely. Since in this paper, the FBF expansion is just a kind of approximator and nothing more, we can refer to neural networks, which has perfect ability of approximation as known. In the neural network, it is sufficient to tune the weights and variances in general due to improve the precision to approximation, whereas the centers are simply places on a regular mesh covering a relevant region of system space.

To guarantee the stability of proposed adaptive fuzzy system, we lead the algorithm into a min-max problem in LP. Though solving the Min-max problem is an ordinary problem in LP and there are a lot of approaches [9-10], the most of the approaches are with a complicated procedure. In this paper, we use a lemma by Annaswamy *et al* which was named as concave/convex parameterization to develop adaptive fuzzy control system.

4. A Solution of Min-Max Problem

Let's consider a scalar function $f(\phi(t),\theta)$, which is continuous and bounded with respect to its arguments. θ is an unknown parameter vector and belongs to a known hypercube $\Theta \in R^n$, $\phi(t) \in R^n$ is a known bounded function of X, and for any $\phi(t)$, $f(\phi(t),\theta)$ is either convex or concave on Θ_s, where Θ_s is a simplex in R^n such that $\Theta_s \supset \Theta$. Suppose that vertices of Θ_s are denoted as θ_{si} ($i = 1,2,\cdots,n$). Then Θ_s may be expressed as

$$\Theta_s = \left\{\theta_s \mid \theta_s = \sum_{i=1}^{n+1}\alpha_i\theta_{si}\right\}, \quad \sum_{i=1}^{n+1}\alpha_i = 1; \quad 0 \le \alpha_i \le 1 \tag{7}$$

Theorem 2: a^* and κ^* are the solutions of the min-max optimizations as follows:

$$a^* = \min_{\omega \in R^n} \max_{\theta \in \Theta_s} \beta J(\omega,\theta) \tag{8}$$

$$\kappa^* = \arg\min_{\omega \in R^n} \max_{\theta \in \Theta_s} \beta J(\omega,\theta) \tag{9}$$

$$J(\omega,\theta) = f(\phi,\theta) - f(\phi,\hat{\theta}) + \kappa^T(\hat{\theta}-\theta) \tag{10}$$

where $\hat{\theta} \in \Theta_s$ and β is a known non-zero constant. Then

$$a^* = \begin{cases} A_1 & \text{if } \beta f \text{ is convex on } \Theta_s \\ 0 & \text{if } \beta f \text{ is concave on } \Theta_s \end{cases}, \quad \kappa^* = \begin{cases} A_2 & \text{if } \beta f \text{ is convex on } \Theta_s \\ \nabla f_\theta & \text{if } \beta f \text{ is concave on } \Theta_s \end{cases} \tag{11}$$

where $\nabla f_\theta = \partial f/\partial\theta$, $A = [A_1,A_2] = G^{-1}b$, A_1 is a scalar, $A_2 \in R^n$,

$$G = \begin{bmatrix} -1 & \beta(\hat{\theta}-\theta_{s1})^T \\ -1 & \beta(\hat{\theta}-\theta_{s2})^T \\ \vdots & \vdots \\ -1 & \beta(\hat{\theta}-\theta_{sn+1})^T \end{bmatrix}, \quad b = \begin{bmatrix} \beta(\hat{f}-f_{s1}) \\ \beta(\hat{f}-f_{s2}) \\ \vdots \\ \beta(\hat{f}-f_{sn+1}) \end{bmatrix} \tag{12}$$

and $f_{si} = f(\phi,\theta_{si})$.

Remarks:

1) The solution to such an LP problem will generally involve numerical searches over the feasible set of solutions. The above Theorem introduced the simplex, Θ_s, precisely to avoid such a search.

2) To decide the solutions in (11), the Theorem2 requires that either convex or concave of βf would be known. Moreover β is a known non-zero constant. This is a strict restriction to apply the theorem into some applications. Though the convex/concave of discussed function f is known, however, the sign of β could not be unknown generally. To deal with the problem, we introduce a concept of one-to-one mapping in the next section.

We are now ready to develop the adaptive fuzzy control system in which the parameters ω_j and σ_j will be tuning and the stability of system will be ensured.

5. Adaptive Fuzzy Control System

Firstly applying the Theorem 1, unknown function in (1) can be approximated by a fuzzy approximator $f^*(X)$,

$$f^*(X) = \sum_{j=1}^{N} \omega_j^* g_j \left(\sigma_j^* \| X - \xi_j \| \right) = \sum_{j=1}^{N} \omega_j^* g_j \left(\sigma_j^* \right) \tag{13}$$

Where $\omega_j^* \in R$ and $\sigma_j^* = (\sigma_j^{1^*}, \sigma_j^{2^*}, \cdots, \sigma_j^{n^*})$ are optimal parameters for unknown function $f(X)$ in (1). It is reasonable to suppose that it exists a known constant $\varepsilon_f > 0$, so that the approximation error ε, defined as in (18), satisfies that $|\varepsilon| \leq \varepsilon_f$.

$$\varepsilon = f(X) - f^*(X) \tag{14}$$

Normally, the unknown parameters values ω_j^* and σ_j^* are replaced by their estimates $\hat{\omega}_j$ and $\hat{\sigma}_j$, and the estimate function $\hat{f}(X) = \sum_{j=1}^{N} \hat{\omega}_j g_j \left(\hat{\sigma}_j \| X - \xi_j \| \right) = \sum_{j=1}^{N} \hat{\omega}_j g_j \left(\hat{\sigma}_j \right)$, instead of $f^*(X)$, is used to approximate the unknown function $f(X)$. The parameter in the estimate function $\hat{f}(X)$ should then be stably tuned to provide effective tracking control architecture. Define the estimation errors of the parameters as

$$\tilde{\omega}_j = \omega_j^* - \hat{\omega}_j, \quad \tilde{\sigma}_j = \sigma_j^* - \hat{\sigma}_j \tag{15}$$

As in [2], an error metric is defined as

$$s(t) = \left(\frac{t}{dt} + \lambda \right)^{n-1} \tilde{X}(t) \quad \text{with } \lambda > 0 \tag{16}$$

which can be rewritten as $s(t) = \Lambda^T \tilde{X}(t)$ with $\Lambda^T = [\lambda^{n-1}, (n-1)\lambda^{n-2}, \cdots, 1]$. The equation $s(t) = 0$ defines a time-varying hyperplane in R^n on which the tracking error vector decays exponentially to zero, so that perfect tracking can be asymptotically obtained by maintaining this condition. In this case the control objective becomes the design of controller to force $s(t) = 0$. The time derivative of the error metric can be written as

$$\dot{s}(t) = -x_d^{(n)}(t) + \Lambda_v^T \tilde{X}(t) + u - f(X) \tag{17}$$

where $\Lambda_V^T = [0, \lambda^{n-1}, (n-1)\lambda^{n-2}, \cdots, (n-1)\lambda]$.

Our adaptive control law is now described below:

$$u(t) = -k_d s(t) + x_d^{(n)}(t) - \Lambda_V^T \tilde{X}(t) + \hat{f}(X) - \left(a^* + \varepsilon_f\right)\mathrm{sgn}(s) \qquad (18)$$

$$\dot{\hat{\omega}}(t) = -\lambda_j s(t) g_j(\hat{\sigma}_j) \qquad (19)$$

$$\dot{\hat{\sigma}}_j = -\beta_j \kappa_j^* s(t) \qquad (20)$$

where λ_j and $\beta_j \in R^{Nn \times Nn}$ are rates of adaptation, $a^* \in R$ and $\kappa_j^* \in R^{Nn \times 1}$ will lead to global stability and be clear as follows.

Consider Liapunov function candidate

$$V(t) = \frac{1}{2}\left(s^2(t) + \sum_{j=1}^{N} \lambda_j^{-1} \tilde{\omega}_j^2 + \sum_{j=1}^{N} \tilde{\sigma}_j^T \beta_j^{-1} \tilde{\sigma}_j \right) \qquad (21)$$

Time derivative of V is given,

$$\dot{V}(t) = -k_d s^2(t) - \varepsilon s(t) - \varepsilon_f |s(t)| - s(t)\left[\sum_{j=1}^{N} \left(\omega_j^*\left(g_j(\sigma_j^*) - g_j(\hat{\sigma}_j)\right) + \kappa_j^*\left(\hat{\sigma}_j - \sigma_j^*\right)\right) + a^* \mathrm{sgn}(s) \right] \qquad (22)$$

We now consider three cases, (a) $s(t) = 0$, (b) $s(t) < 0$ and (c) $s(t) > 0$, and show that $\dot{V}(t) \leq 0$ in the three cases.

(a) $s(t) = 0$.

It is clear that $\dot{V}(t) = 0$.

(b) $s(t) < 0$.

It follows that $\dot{V}(t) \leq 0$ if

$$a^* \geq \sum_{j=1}^{N} \left(\omega_j^*\left(g_j(\sigma_j^*) - g_j(\hat{\sigma}_j)\right) + \kappa_j^*\left(\hat{\sigma}_j - \sigma_j^*\right)\right) \qquad (23)$$

Therefore, we choose

$$a^* = \max_{\sigma \in \Theta_s} \sum_{j=1}^{N} \left(\omega_j^*\left(g_j(\sigma_j^*) - g_j(\hat{\sigma}_j)\right) + \kappa_j^*\left(\hat{\sigma}_j - \sigma_j^*\right)\right) \qquad (24)$$

Since the form of the controller in equation (21) suggests that the quantity a^* in like a gain, we seek to find an κ_j^* so that a^* is minimized. Hence our goal is to choose κ_j^* as

$$a^* = \min_{k_j \in R^{Nn}} \max_{\sigma \in \Theta_s} \sum_{j=1}^{N} \left(\omega_j^*\left(g_j(\sigma_j^*) - g_j(\hat{\sigma}_j)\right) + \kappa_j^*\left(\hat{\sigma}_j - \sigma_j^*\right)\right) \qquad (25)$$

Performing the min-max optimization to find a^* and κ_j^* is needed to complete the controller design. As a PL problem, there are a lot of algorithms [9-10] to solve the values of a^* and κ_j^*, however it generally involves a numerical search over the feasible set of solutions with a complicated procedure. Hence, we use the theorem 2 to solve the values of a^* and κ_j^* in (25). Though at first glance, the Theorem 2 looks like providing a way to solve the values, however, in order to determine that the function $\omega_j^* g_j(\sigma_j)$ is either convex or concave, the sign of ω_j^* must be known. Namely, because $g_j(\sigma_j)$ is convex on Θ, so $\omega_j^* g_j(\sigma_j)$ will be convex when $\omega_j^* \geq 0$, and concave when $\omega_j^* < 0$. Hence it is obvious that the Theorem 2 can not apply directly, since it does not show us any clue on how to determine the sign of ω_j^*.

Since ω_j^* is the optimal weights in (6), it is reasonable to assume that the range of ω_j^* is known, i.e., $\omega_j^* \in [\omega_{min}^*, \omega_{max}^*]$. Now, we set up a new parameter $p_j \in [p_{min}, p_{max}]$, the boundaries p_{min} and p_{max} are positive constants, which can be chosen by the designer. To deal with the problem of sign of ω_j^*, we introduce the following one-to-one mapping:

$$\omega_j^* = m + n p_j; \qquad p_j > 0 \tag{26}$$

where

$$m = \omega_{min}^* - \frac{\omega_{max}^* - \omega_{min}^*}{p_{max} - p_{min}} p_{min}, \qquad n = \frac{\omega_{max}^* - \omega_{min}^*}{p_{max} - p_{min}} \quad (>0) \tag{27}$$

Substituting (26-27) to (25), we have

$$a^* = \min_{k_j \in R^{Nn}} \max_{\sigma \in \Theta_s} \sum_{j=1}^{N} \left((m + n p_j) \left(g_j(\sigma_j^*) - g_j(\hat{\sigma}_j) \right) + \kappa_j^* (\hat{\sigma}_j - \sigma_j^*) \right) = a_1^* + a_2^* \tag{28}$$

where

$$a_1^* = \min_{k_{1j} \in R^{Nn}} \max_{\sigma \in \Theta_s} \sum_{j=1}^{N} \left(\omega_{min}^* \left(g_j(\sigma_j^*) - g_j(\hat{\sigma}_j) \right) + \kappa_{1j}^* (\hat{\sigma}_j - \sigma_j^*) \right) \tag{29}$$

$$a_2^* = \min_{k_{2j} \in R^{Nn}} \max_{\sigma \in \Theta_s} \sum_{j=1}^{N} \left(n(p_j - p_{min}) \left(g_j(\sigma_j^*) - g_j(\hat{\sigma}_j) \right) + \kappa_{2j}^* (\hat{\sigma}_j - \sigma_j^*) \right) \tag{30}$$

Remark: The either convex or concave of $\omega_{min}^* g_j(\sigma_j)$ and $n(p_j - p_{min}) g_j(\sigma_j)$ could be determined in both (29) and (30) via the one-to-one mapping, since the sign of ω_{min}^* is known by the assumption and $n(p_j - p_{min}) > 0$.

Now applying the Theorem 2 straightforwardly, we can get the solutions of a_1^*, a_2^*, κ_{1j}^* and κ_{2j}^* as follows,

$$a_1^* = \begin{cases} \sum_{j=1}^{N} A_{j1} & \text{if } \omega_{min}^* \geq 0 \\ 0 & \text{if } \omega_{min}^* < 0 \end{cases}, \qquad \kappa_{1j}^* = \begin{cases} \omega_{min}^* A_{j2} & \text{if } \omega_{min}^* \geq 0 \\ \omega_{min}^* \nabla g_{j\hat{\sigma}_j} & \text{if } \omega_{min}^* < 0 \end{cases} \tag{31}$$

where $A_j = \left[A_{j1}, A_{j2} \right]^T = G_j^{-1} b_j$, A_{j1} is a scalar, $A_{j2} \in R^{Nn}$,

$$G_j = \begin{bmatrix} -1 & \omega_{min}^* (\hat{\sigma}_j - \sigma_{js1})^T \\ -1 & \omega_{min}^* (\hat{\sigma}_j - \sigma_{js2})^T \\ \vdots & \vdots \\ -1 & \omega_{min}^* (\hat{\sigma}_j - \sigma_{jsNn+1})^T \end{bmatrix}, \qquad b_j = \begin{bmatrix} \omega_{min}^* (\hat{g}_j - g_{js1}) \\ \omega_{min}^* (\hat{g}_j - g_{js2}) \\ \vdots \\ \omega_{min}^* (\hat{g}_j - g_{jsNn+1}) \end{bmatrix}, \qquad g_{jsi} = g_j(\sigma_{si}) \tag{32}$$

and

$$a_2^* = \sum_{j=1}^{N} B_{j1}, \qquad \kappa_{2j}^* = n(p_j - p_{\min})B_{j2} \qquad (33)$$

where $B_j = [B_{j1}, B_{j2}]^T = G_j^{-1}b_j$, B_{j1} is a scalar, $B_{j2} \in R^{Nn}$,

$$G_j = \begin{bmatrix} -1 & n(p_j - p_{\min})(\hat{\sigma}_j - \sigma_{js1})^T \\ -1 & n(p_j - p_{\min})(\hat{\sigma}_j - \sigma_{js2})^T \\ \vdots & \vdots \\ -1 & n(p_j - p_{\min})(\hat{\sigma}_j - \sigma_{jsNn+1})^T \end{bmatrix}, \qquad b = \begin{bmatrix} n(p_j - p_{\min})(\hat{g}_j - g_{js1}) \\ n(p_j - p_{\min})(\hat{g}_j - g_{js2}) \\ \vdots \\ n(p_j - p_{\min})(\hat{g}_j - g_{jsNn+1}) \end{bmatrix} \qquad (34)$$

(c) $s(t) > 0$.

Similar argument to (b), it follows that $\dot{V}(t) \le 0$ if

$$a^* = \min_{k_j \in R^{Nn}} \max_{\sigma \in \Theta_s} \sum_{j=1}^{N} \left[-\left(\omega_j^*(g_j(\sigma_j) - g_j(\hat{\sigma}_j)) + \kappa_j^*(\hat{\sigma}_j - \sigma_j) \right) \right] - a_1^* + a_2^* \qquad (35)$$

where

$$a_1^* = \min_{k_{1j} \in R^{Nn}} \max_{\sigma \in \Theta_s} \sum_{j=1}^{N} \left[-\left(\omega_{\min}^*(g_j(\sigma_j) - g_j(\hat{\sigma}_j)) + \kappa_{1j}^*(\hat{\sigma}_j - \sigma_j) \right) \right] \qquad (36)$$

$$a_2^* = \min_{k_{2j} \in R^{Nn}} \max_{\sigma \in \Theta_s} \sum_{j=1}^{N} \left[-\left(n(p_j - p_{\min})(g_j(\sigma_j) - g_j(\hat{\sigma}_j)) + \kappa_{2j}^*(\hat{\sigma}_j - \sigma_j) \right) \right] \qquad (37)$$

We can get the solutions of a_1^*, a_2^*, κ_{1j}^* and κ_{2j}^* as follows,

$$a_1^* = \begin{cases} 0 & \text{if } \omega_{\min}^* \ge 0 \\ \sum_{j=1}^{N} A_{j1} & \text{if } \omega_{\min}^* < 0 \end{cases}; \qquad \kappa_{1j}^* = \begin{cases} -\omega_{\min}^* \nabla g_{j\hat{\sigma}_j} & \text{if } \omega_{\min}^* \ge 0 \\ -\omega_{\min}^* A_{j2} & \text{if } \omega_{\min}^* < 0 \end{cases} \qquad (38)$$

where $A_j = [A_{j1}, A_{j2}]^T = G_j^{-1}b_j$, A_{j1} is a scalar, $A_{j2} \in R^{Nn}$,

$$G_j = \begin{bmatrix} -1 & -\omega_{\min}^*(\hat{\sigma}_j - \sigma_{js1})^T \\ -1 & -\omega_{\min}^*(\hat{\sigma}_j - \sigma_{js2})^T \\ \vdots & \vdots \\ -1 & -\omega_{\min}^*(\hat{\sigma}_j - \sigma_{jsNn+1})^T \end{bmatrix}, \qquad b_j = \begin{bmatrix} -\omega_{\min}^*(\hat{g}_j - g_{js1}) \\ -\omega_{\min}^*(\hat{g}_j - g_{js2}) \\ \vdots \\ -\omega_{\min}^*(\hat{g}_j - g_{jsNn+1}) \end{bmatrix} \qquad (39)$$

and

$$a_2^* = 0, \qquad \kappa_{2j}^* = -n(p_j - p_{\min})\nabla g_{j\hat{\sigma}_j} \qquad (40)$$

The stability of the closed-loop system described by (1), (18-20), (28), (31-35) and (38-40) is established in the following theorem.

Theorem 3: If the robust adaptive control (18-20), (28), (31-35) and (38-40) is applied to the nonlinear plant (1), then all closed-loop signals are bounded and $\tilde{X}(t) \to 0$ as $t \to \infty$.

6. Conclusion

The novel feature of results in this paper is that, thanks to Min-max's solutions of Annaswamy *et al* which could simplify the procedures of our proposed algorithm, a new adaptive fuzzy control law is presented. The adaptive fuzzy control law is capable of stably turning the parameters which appear nonlinearly in the fuzzy approximator in an effort to reduce appximation error and improve control performance. The developed controller guarantees the global stability of the resulting closed-loop system in the sense that all signals involved are uniformly bounded and tracking to within a desired precision. Hereafter, we will verify our theoretic analysis by computer simlation.

References

1. L.-X. Wang, "Stable adaptive fuzzy control of nonlinear system," *IEEE Trans. on Fuzzy Systems*, vol. 1, pp. 146-155, 1993.
2. C.-Y. Su and Y. Stepanenko, "Adaptive control of a class of nonlinear systems with fuzzy logic," *IEEE Trans. Fuzzy Systems*, Vol. 2, pp. 258-294, 1994.
3. H. Lcc and M. Tomizuka, "Robust adaptive control using a universal approximator for SISO nonlinear systems," *Technical Report*, UC Berkeley, 1998.
4. L. X. Wang and J. M. Mendel, "Back-propagation fuzzy system as nonlinear dynamic system identifiers," *Proc. of IEEE Conf. on Fuzzy Systems*, San Diego, 1992, pp. 1409-1418.
5. X. Ye and N. K. Loh, "Dynamic system identification using recurrent radial basis function network," *Proc. of the American Control Conf.*, 1993, pp. 2912-2916.
6. E. Kim, M. Park, S. Ji and M. Park, "A new approach to fuzzy modelling," *IEEE Trans. Fuzzy Systems*, Vol. 5, pp. 328-337, 1997.
7. A.M.Annaswamy, F.P.Skantze and A.-P.Loh, "Adaptive control of Continuous time systems with convex/concave parameterization," *Automatica*, Vol.34, No.1,pp.33-49, 1998.
8. S.Sastry and M.Bodson, *Adaptive Control*, Prentice Hall, 1989.
9. K.Shimizu and K.Ito, "Design of neural stabilizing controller for nonlinear systems via Lyapunov's direct method," *Trans. of the Society of Instrument and Control Engineers*, Vol.35, No.4, pp489-495, 1999.
10. K.shimizu and E.Aiyoshi, *Mathematical Programming*, Shokodo, 1984

Control of Chaotic Systems Using Fuzzy Model-Based Regulators

Keigo Watanabe[1], Lanka Udawatta[2], Kazuo Kiguchi[1], and Kiyotaka Izumi[3]

[1] Dept. of Advanced Systems Control Engineering,
Graduate School of Science and Engineering,
{watanabe,kiguchi}@me.saga-u.ac.jp
[2] Dept. of Mechanical Engineering (Production Division),
Graduate School of Science and Engineering,
98tj02@edu.cc.saga-u.ac.jp
[3] Dept. of Mechanical Engineering,
Faculty of Science and Engineering,
Saga University, 1-Honjomachi, Saga 840-8502, Japan
izumi@me.saga-u.ac.jp

Abstract. This paper presents a new approach to controlling chaotic systems using fuzzy regulators. The relaxed stability conditions and LMI (Linear Matrix Inequalities) based designs for a fuzzy regulator are used to construct a fuzzy attractive domain, in which a global solution is obtained so as to achieve the desired stability condition of the closed-loop system. In the control of chaotic systems, we use two-phases of control, first phase uses an open-loop control with inherent chaotic features of the system itself and a fuzzy model-based controller is employed under state feedback control in the second phase of control. The Henon map is employed to illustrate the above design procedure.

Keywords: Chaotic systems, Fuzzy model-based control, Evolutionary computation, Nonlinear dynamics, System stability, Lyapunov function.

1 Introduction

Recently, development of chaos theory brings up scientist into a new era in analyzing nonlinear systems. It is known that the chaos exhibits a deterministic random behavior. Yet it needs more investigation on such nonlinear systems in designing control algorithms. Edward Lorenze, the first experimenter in chaos was a meteorologist and described his model of weather prediction phenomena with a set of nonlinear differential equations in 1963. Among the various methods available to analyze such nonlinear systems, there exist fixed point analysis, linearization, Pioncare map, Lyapunov method, spectral analysis, fractal, chaos etc. [1],[2],[8]. Employing intended deterministic chaos to control nonlinear dynamical systems is an interesting in the development of control theory [1]. In this method, it is proposed that in a nonlinear dynamic system, a chaotic attraction can be formed by an appropriate usage of open-loop control until the system states converge within a specified area of attractive domain and then a state feedback control can be used to converge the system states to desired values. In this approach, the usage of state feedback control involves finding a largest level curve based on Lyapunov stability condition in advance. However this optimization problem will generally have multiple local minima. Therefore to find a global minimum, almost all such local minima have to be identified with trial and error. That implies the complexity of this approach.

Recent advances in LMI (Linear Matrix Inequalities) theory [3],[4],[5],[10] allowed to handle nonlinear control system problems, via semi-definite programming. For an example, stability condition is guaranteed by the well-known Lyapunov approach [7] for fuzzy model-regulators and the LMI tool searches the solution space subjected to various constraints. In particular, Takagi-Sugeno (T-S) fuzzy model plays an important role in designing such fuzzy regulators [4],[5],[7]. In fact, the fuzzy model-based control (FMC) can be applied very well to nonlinear dynamic systems, this attempt also implies the ability of controlling chaotic systems via FMC.

In the control of chaotic systems, there are two phases of control. First phase uses an open-loop control such that the inherent chaotic features attract the states to a desired area as studied in [1]. Once the system has entered a specified area, open-loop control is cut off and the second phase of control is adopted. In the proposed second phase, an FMC is employed under state feedback control; it has been implemented by employing a set of IF-THEN fuzzy rules and assuring the global stability of domain constructed by the total FMC. For this purpose, the state feedback gain scheduling of the control system in the second phase is achieved by solving a set of LMIs via an optimization technique based on evolutionary computation. The rest of the paper is organized as follows: In section 2, introduction of dynamic system modeling and FMC are reviewed. The proposed fuzzy-chaos hybrid control scheme is presented in section 3. Finally, the chaotic system, Henon map is taken into consideration to illustrate the design procedure and a simulation result of the fuzzy model-based regulator is presented in section 4.

2   System Modeling

Inherent chaotic characteristics can be useful in moving a system to various points in state space. In the proposed method, this feature is used to drive the system states to a pre-defined domain C. An appropriate open-loop control (OLC) input can be employed to create chaos or to use chaos in a nonlinear system itself. Once it reaches to the pre-defined fuzzy attractive domain, a fuzzy model-based controller (FMC) is employed under state feedback control to achieve desired target. This design concept is schematically given in Fig. 1 for a two dimensional case. Here it is intended to drive the system state P_1 to P_3. The feedback controller design is based on multiple linearizations around a single equilibrium point, i.e., so called off-equilibrium linearizations. It is known [7] that the method will significantly improve the transient dynamics of the control system for a general control problem. Rather, it is interesting to note that such a technique is useful for constructing a globally stable fuzzy attraction domain without trial and error.

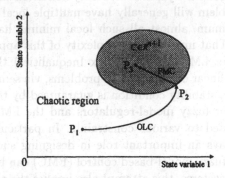

Fig. 1. Fuzzy-chaos hybrid control scheme

2.1 Off-equilibrium Linearization

Nonlinear dynamic continuous time systems (CS) can be described by nonlinear differential equations [1],[2] (or difference equations for discrete time systems (DS)) as

$$\dot{x} = F(x, u) \quad \text{for CS}$$
$$x(t+1) = F(x, u) \quad \text{for DS} \tag{1}$$

where $x \in R^n$ is the state vector and $u \in R^m$ gives the control input vector of the systems. The equilibrium points $(\bar{x}, \bar{u})$ (or fixed points) of the dynamic system satisfy

$$\varepsilon = \{(x, u) \in R^{n+m} \mid F(\bar{x}, \bar{u}) = 0 \} \quad \text{for CS}$$
$$\varepsilon = \{(x, u) \in R^{n+m} \mid \bar{x} = F(\bar{x}, \bar{u}) \} \quad \text{for DS} \tag{2}$$

In this paper, it is proposed to select a suitable set of off-equilibrium points such that all the subsystems compose a convex region $C \in R^{n+1}$ which keeps the equilibrium point $(\bar{x}, \bar{u}) \in C$. More generally, n-dimensional state space system needs at least $n + 1$ off-equilibrium points which represent the convex region to ensure the stability of a particular subsystem.

Neglecting higher order terms, we obtain a linearized model around any arbitrary point $(x_0, u_0) \in C$ as follows:

$$\dot{x} = A_0(x - x_0) + B_0(u - u_0) + F(x_0, u_0) \quad \text{for CS}$$
$$x(t+1) = A_0(x - x_0) + B_0(u - u_0) + F(x_0, u_0) \quad \text{for DS} \tag{3}$$

where

$$A_0 = \frac{\partial F}{\partial x}(x_0, u_0)$$

$$B_0 = \frac{\partial F}{\partial u}(x_0, u_0)$$

For an example, two dimensional state space model needs three off-equlibrium points such that the equilibrium point lies on the center of mass of an equilateral triangle keeping its corners on the three off-equilibrium points.

2.2 Fuzzy Models and Regulators

The dynamics of the nonlinear system are approximated near an arbitrary point $(x_0, u_0) \in C$. Then, equation (3) can be rewritten in the form:

$$\dot{x} = A_0 x + B_0 u + d_0 \quad \text{for CS}$$

$$x(t+1) = A_0 x + B_0 u + d_0 \quad \text{for DS} \tag{4}$$

where $d_0 = F(x_0, u_0) - A_0 x_0 - B_0 u_0$. Note here that an arbitrary point (x_0, u_0) need not be an equilibrium point $(\bar{x}, \bar{u})$.

Fuzzy models due to Takagi-Sugeno consist of a set of IF-THEN rules [7] for the above approximate systems. The ith plant rule of each subsystems for both continuous-time and discrete-time fuzzy systems is given by

IF $z_1(t)$ is M_{i1} and...and $z_p(t)$ is M_{ip}

THEN $\begin{cases} \dot{x}(t) = A_i x(t) + B_i u(t) + d_i & \text{for CS} \\ x(t+1) = A_i x(t) + B_i u(t) + d_i & \text{for DS} \\ y(t) = C_i x(t) & i = 1, ..., r \end{cases}$ (5)

where r is the number of fuzzy rules and M_{ij} ($i = 1, ..., r$ and $j = 1, ..., p$) are the fuzzy sets. The state vector is $x(t) \in R^n$, input vector is $u(t) \in R^m$ and the output vector is given by $y(t) \in R^q$. A_i, B_i and C_i are the system parameter matrices and d_i is the offset term of the ith fuzzy model. For a given state, $z_1(t), ..., z_p(t)$ are the premise variables (or antecedent inputs).

Subjecting to the parallel distributed compensation, we can design the following fuzzy regulators:

Regulator Rule i :

IF $z_1(t)$ is M_{i1} and...and $z_p(t)$ is M_{ip}

THEN $u(t) = -K_i[x(t) - x_r] + u_r, \quad i = 1, ..., r$

for the fuzzy models (5), where x_r is a state reference trajectory, u_r is the corresponding input trajectory, and K_i is the local feedback gain matrix. Thus, the fuzzy regulator rules have linear state-feedback laws in the consequent parts and the overall fuzzy regulator can be reduced to

$$u(t) = -\sum_{i=1}^{r} h_i(z(t)) K_i[x(t) - x_r] + u_r \tag{6}$$

where

$$z(t) = [z_1(t), ..., z_p(t)]$$

$$w_i(z(t)) = \prod_{j=1}^{p} M_{ij}(z_j(t)), \quad h_i(z(t)) = \frac{w_i(z(t))}{\sum_{l=1}^{r} w_l(z(t))}$$

for all t, in which $M_{ij}(z_j(t))$ denotes the confidence (or grade) of membership of $z_j(t)$ in M_{ij}.

3 Fuzzy-Chaos Hybrid Controller

In order to control the original nonlinear system with a chaotic input in the open-loop system or a fuzzy controller in the closed-loop system, a fuzzy-chaos hybrid control scheme is proposed here. Such a control scheme can be considered in two cases, depending on the choice of equilibrium points as the reference.

3.1 Stabilization of a Prespecified Equilibrium Point

In this case, the fuzzy-chaos hybrid control can be implemented by

$$\text{IF } \sum_{i=1}^{r} w_i(z(t)) \equiv 0 \text{ THEN}$$
$$u(t) = \hat{u}(t)$$
$$\text{ELSE} \tag{7}$$
$$u(t) = -\sum_{i=1}^{r} h_i(z(t)) K_i[x(t) - \bar{x}] + \bar{u}$$

where $\hat{u}(t)$ is an open-loop input to make the original nonlinear system chaotic and $(\bar{x}, \bar{u})$ is the prespecified equilibrium point which would be stabilized.

3.2 Stabilization of any Equilibrium Point

If the stabilized equilibrium point is arbitrary among all the equilibrium points, the above fuzzy-chaos hybrid control can be modified as follows.

$$\text{IF } \sum_{i=1}^{r} w_i(z(t)) \equiv 0 \text{ THEN}$$
$$u(t) = \hat{u}(t)$$
$$\text{ELSE} \tag{8}$$
$$i_{\max} = \max\{h_1(z(t)), ..., h_r(z(t))\}$$
$$u(t) = -K_{i_{\max}}[x - \bar{x}_{\max}] + \bar{u}_{\max}$$

where $i_{\max}$ denotes the rule number that has largest rule confidence, $K_{i_{\max}}$ is the corresponding feedback gain matrix, and $(\bar{x}_{\max}, \bar{u}_{\max})$ is an equilibrium point existing in the fuzzy attractive domain constructed by using the $i_{\max}$-th rule.

Gain scheduling of the feedback controller is determined by employing a set of LMIs [3] and eigenvalue minimization algorithm was developed to determine the positive definite and positive semi-definite matrices associated with various linear matrix inequalities, using evolutionary computation technique by making a penalty for an individual which violates the inequality condition. This optimization problem can be also efficiently solved by means of recently developed interior-point methods [5].

4 Design Example and Results

4.1 Henon Map

In this example, the chaotic system, Henon map, is presented to illustrate the proposed design procedure. The nonlinear dynamic equations of the Henon map are given by

$$x_1(t+1) = -1.4x_1^2 + x_2 + 1$$
$$x_2(t+1) = 0.3x_1 \tag{9}$$

Fixed points of the system of difference equations (9) are satisfied as the equation (2), resulting two fixed-points (0.6314, 0.1894) and (−1.1314, −0.3394). Therefore, we can design two convex regions which correspond to two fixed-points. Here we select a triple point sub-region $(A_i x(t) + B_i u(t)$ for $i = 1, 2, 3)$ such that it surrounds the fixed-point $(x_a = 0.6314, x_b = 0.1894)$ as shown in Fig. 2. The equilibrium point lies on the center of mass of an equilateral triangle having the coordinates (x_{a1}, x_{b1}), (x_a, x_{b2}) and (x_{a3}, x_{b1}) in order to determine the common P and common Q. The open-loop control input to the system (9) u is selected as in equation (10) in implementing the desired control system.

$$x_1(t+1) = -1.4x_1^2 + x_2 + 1 + u$$
$$x_2(t+1) = 0.3x_1 \tag{10}$$

Three linearized models corresponded to the first fixed-point are given below taking $(x_a - x_{a1}) = 0.2$.

$$A_1 = \begin{bmatrix} -1.4878 & 1 \\ 0.3 & 0 \end{bmatrix}, B_1 = \begin{bmatrix} 1 \\ 0 \end{bmatrix}; \quad A_2 = \begin{bmatrix} -1.7678 & 1 \\ 0.3 & 0 \end{bmatrix}, B_2 = \begin{bmatrix} 1 \\ 0 \end{bmatrix}$$

$$A_3 = \begin{bmatrix} -2.0478 & 1 \\ 0.3 & 0 \end{bmatrix}, B_3 = \begin{bmatrix} 1 \\ 0 \end{bmatrix}$$

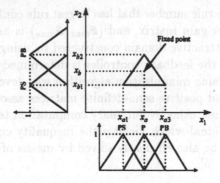

Fig. 2. Tripple point sub-system ($i = 1, 2, 3$) and its membership functions

The same procedure is repeated to select the next sub-region ($A_i x(t) + B_i u(t)$) for $i = 4, 5, 6$) around the second fixed-point and the three linearized models are given below.

$$A_4 = \begin{bmatrix} 3.4478 & 1 \\ 0.3 & 0 \end{bmatrix}, B_4 = \begin{bmatrix} 1 \\ 0 \end{bmatrix}; \quad A_5 = \begin{bmatrix} 3.1678 & 1 \\ 0.3 & 0 \end{bmatrix}, B_5 = \begin{bmatrix} 1 \\ 0 \end{bmatrix}$$

$$A_6 = \begin{bmatrix} 2.8878 & 1 \\ 0.3 & 0 \end{bmatrix}, B_6 = \begin{bmatrix} 1 \\ 0 \end{bmatrix}$$

4.2 Calculation of Feedback Gains

Gain scheduling of the above problem can be formulated as an optimization problem with the LMIs [3] and it is solved by using an optimization technique based on evolutionary computation [9]. Here we obtain the common P and common Q for the first fixed-point guaranteeing the stability. We obtained the P_1 and Q_1 as follows at $\beta = 0.6020$ ($s = 3$):

$$P_1 = \begin{bmatrix} 292.519 & 39.7689 \\ 39.76891 & 535.242 \end{bmatrix}, \quad Q_1 = \begin{bmatrix} 13.8323 & 15.8108 \\ 15.8108 & 142.855 \end{bmatrix}$$

Similarly, P_2 and Q_2 matrices associated with the second fixed-point can be obtained as follows at $\beta = 0.9581$ ($s = 3$):

$$P_2 = \begin{bmatrix} 633.277 & 29.3221 \\ 29.3221 & 656.3649 \end{bmatrix}, \quad Q_2 = \begin{bmatrix} 55.4408 & 30.5980 \\ 30.5980 & 76.9469 \end{bmatrix}$$

The gains K_1, K_2 and K_3 are obtained as below providing the global stability of the fuzzy control system for the first sub-region,

$$K_1 = [-0.3318 \quad 1.0329], \quad K_2 = [-1.2171 \quad 0.7717]$$
$$K_3 = [-2.1002 \quad 0.5644]$$

Similarly, the gains K_4, K_5 and K_6 are obtained for the second sub-region as follows:

$$K_4 = [3.4272 \quad 0.7360], \quad K_5 = [3.8298 \quad 1.5356]$$
$$K_6 = [2.6047 \quad 0.8615]$$

Based on the methodology, as proposed in section 2, the Henon map is controlled by a fuzzy-chaos hybrid controller. Since the system has been already a chaotic, it can be used for the first phase of control with no input ($\hat{u} = 0$). Once the system states reach to one of the above two sub-domains, fuzzy controller will drive the system towards the fixed point. For example, by using the above gains K_i ($i = 1, 2, 3$), the fuzzy controller is constructed from the following IF-THEN rule base:

$$R_1 : \textbf{IF } x_1 \textbf{ is PS AND } x_2 \textbf{ is PS}$$
$$\textbf{THEN } K = K_1$$
$$R_2 : \textbf{IF } x_1 \textbf{ is P AND } x_2 \textbf{ is PB}$$
$$\textbf{THEN } K = K_2 \tag{11}$$
$$R_3 : \textbf{IF } x_1 \textbf{ is PB AND } x_2 \textbf{ is PS}$$
$$\textbf{THEN } K = K_3$$

where PS, P and PB represent the words positive small, positive and positive big respectively. In order to verify the above design procedure in sections 2 and 3, the proposed fuzzy-chaos hybrid controller is applied to the chaotic system (9). Here we allow the system to drive its states towards one of the above two fixed points chaotically. The rule base (8) was employed here to construct the controller as follows:

$$\textbf{IF } \sum_{i=1}^{6} w_i(z(t)) \equiv 0$$
$$u = \hat{u} = 0$$
$$\textbf{ELSE}$$
$$u = -K_{i_{max}}[x - \bar{x}_{i_{max}}] \ (i = 1, ..., 6) \tag{12}$$

where $K_{i_{max}}$ is the gain which corresponds to $w_{i_{max}}(k)$ ($i = 1, ..., 6$). Figure 3 shows the resulting trajectory of the chaotic system controlled by the proposed controller starting from $(-0.3, 0)$.

5 Conclusions

Inherent chaotic features have been used to drive the system states to a predefined domain using an OLC. Once it reaches to the predefined fuzzy attractive domain, a fuzzy model-based controller is employed under state feedback control to achieve reference target. Eigenvalue minimization algorithm was used to determine the positive definite and positive semi-definite matrices associated with various LMIs, using evolutionary computation technique.

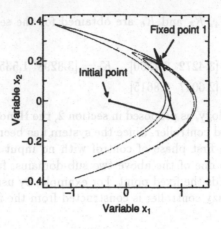

Fig. 3. Trajectory of the controlled Henon map starting from $(-0.3, 0)$.

In the example, the Henon map was driven by chaotic system itself before reaching the fuzzy attractive domain with no control input. Once it reaches to one of the two fuzzy domains, a fuzzy model-based controller drives the system towards the equilibrium point. The simulation result has shown the good tracking performance of the proposed controller in spite of the uncertainties of the chaotic system. Thus, the proposed methodology is useful for the design of nonlinear control systems which exhibit deterministic random-behaviors.

References

1. Vincent, T. L.: Control Using Chaos. IEEE Control Systems Society. **17-6** (1997) 65–76
2. Kapitaniak, T.: Chaos for Engineers—Theory, Applications and Control. Berlin: Springer-Verlag (1998)
3. Tanaka, K., Ikeda, T., Wang, H. O.: Fuzzy Regulators and Fuzzy Observers: Relaxed Stability Condition and LMI-Based Designs. IEEE Trans. on Fuzzy Systems, **6-2** (1998) 250–265
4. Limanond, S., Si, J.: Neural-Network Based Control Design: An LMI Approach. IEEE Trans. on Neural Netwirks, **9-6** (1998) 1422–1429
5. Boyd, S., Ghaoui, L. E., Feron, E., Balakrishnan, V.: Linear Matrix Inequalities in Systems and Control Theory. Philadelphia, PA: SIAM (1994)
6. Driankov, D., Palm, R.: Advances in Fuzzy Control. Berlin: Springer-Verlag (1998)
7. Palm, R., Driankov, D.: Model Based Fuzzy Control, Berlin: Springer-Verlag (1997)
8. Vincent, T. L., Grantham, W. J.: Nonlinear and Optimal Control Systems. New York, NY: Wiley (1997)
9. Nanayakkara, T., Watanabe, K., Izumi, K.: Evolving in Dynamic Environment through Adaptive Chaotic Mutation. Procs. of Fourth Int. Symposium on Artificial Life and Robotics, **2** (1997) 520–523
10. Ying, H.: Constructing Nonlinear Variable Gain Controllers via the Takagi-Sugeno Fuzzy Control. IEEE Trans. on Fuzzy Systems, **6-2** (1998) 226–234

Fuzzy Behavior-Based Control for the Obstacle Avoidance of Multi-link Manipulators

Palitha Dassanayake[1], Keigo Watanabe[2], and Kiyotaka Izumi[3]

[1] Faculty of Engineering Systems and Technology,
Graduate School of Science and Engineering,
Saga University, 1-Honjomachi, Saga 840-8502, Japan
vpalitha@hotmail.com

[2] Dept. of Advanced Systems Control Engineering,
Graduate School of Science and Engineering,
Saga University, 1-Honjomachi, Saga 840-8502, Japan
watanabe@me.saga-u.ac.jp

[3] Dept. of Mechanical Engineering,
Faculty of Science and Engineering,
Saga University, 1-Honjomachi, Saga 840-8502, Japan
izumi@me.saga-u.ac.jp

Abstract. The behavior based approach has been actively used in many applications of intelligent robots due to the advantages of dividing the control system according to the task achieving behaviors over the conventional method in which the division is based on functions. One important application that had been done is for a mobile robot to reach a target while avoiding obstacles. The objective of this paper is for a multi-link manipulator to reach a target while avoiding obstacles by using a fuzzy behavior-based control approach. The control system that had been applied to the mobile robot in the previous work, is modified to suit to the manipulator. Fuzzy behavior elements are trained by a genetic algorithm. An additional component is also introduced in order to overcome the gravitational effect. Simulation results show that the manipulator reaches the target with an acceptable solution.

Keywords: Manipulator, Fuzzy control, Behavior-based control system, Obstacle avoidance, Genetic algorithms

1 Introduction

Brooks [1] proposed a new architecture called "Subsumption Architecture" for controlling a mobile robot. Layers of control system were built to let the robot operate at increasing level of competence. Decomposition of the control system was based on task achieving behaviors. This behavior-based control has been actively applied to several intelligent robots [2–6]. Watanabe and Isumi [5] studied a fuzzy behavior-based control system for a mobile robot by applying the concept of subsumption-like architecture using soft computing techniques, in which a simple fuzzy reasoning was assigned to one elemental behavior consisting of a single input-output relation, and then two consequent results from two behavioral groups were competed or cooperated.

On the other hand, Rahnanian-shahri and Troch [7] presented a new method to on-line collision-avoidance of redundant manipulators with obstacles. Ding and Chang [8] introduced a real-time planning algorithm for avoidance of redundant robots in collision-free trajectory planning. Nearchou and Aspragathos [9] presented an algorithm for Cartesian trajectory by redundant robots in environments with obstacles. It should be noted that all the above mentioned works are based on trajectory planning while avoiding obstacles, but none of them is based on behavior-based control strategy.

A fuzzy behavior-based control method developed in references [4] and [5] is used to control a multi-link manipulator to reach a target while avoiding obstacles in this work. The basic concept used for the mobile manipulator [4,5] is applied with some modifications. Thus, a fuzzy behavior-based control system is applied to three-degree-of-freedom, three-link manipulator to reach a target from a given point while avoiding obstacles.

2 Three-Link Manipulator

It is assumed that the robot has three-degree-of-freedom and it moves in a two dimensional vertical plane. The axes are selected such a way that $O - X - Y$ is the vertical plane where the center of gravity acts opposite to the $O - Y$ axis and $O - Z$ is selected according to the right hand rule. Let $[f_x \ f_y \ f_z]^T$ and $[n_x \ n_y \ n_z]^T$ be the force vector and the moment vector at the end-effector of the robot, where the subscripts x, y and z are used to represent $O - X$ axis, $O - Y$ axis and $O - Z$ axis respectively. Since the robot is moving in $O - X - Y$ plane, f_z, n_x and n_y are zero. Link coordinate axes $O_i - X_i - Y_i$ are defined in such a way that the origin of each link coordinate system is selected at the end of the respective link and the $O_i - X_i$ axis is selected along the link. $O_i - Y_i$ axis is perpendicular to the $O_i - X_i$ axis in the counterclockwise direction and $O_i - Z_i$ axis is selected according to the right hand rule.

Length, joint angle and mass for link i are denoted by l_i, θ_i and m_i respectively. $\bar{x}_i$ and I_{ixx}, $\bar{y}_i$ and I_{iyy}, and, $\bar{z}_i$ and I_{izz} are the center of gravity and the moment of inertia for link i in $O_i - X_i$, $O_i - Y_i$ and $O_i - Z_i$ directions respectively. It is assumed that all links are homogenous and the center of gravity acts in the middle and each link is symmetrical about its center of gravity. Therefore $\bar{x}_i = -l_i/2$, $y_i = z_i = 0.0$, $I_{ixx} = 0.0$, and $I_{iyy} = I_{izz} = m_i l_i^2/12$.

To simulate this model on a computer, dynamic equations for this manipulator are derived using Newton-Euler method [11]. The dynamic equations are given by Eq. (1) in the matrix-vector form:

$$\tau(t) = D(\theta(t))\ddot{\theta}(t) + h(\theta(t), \dot{\theta}(t)) + g(\theta(t)) \tag{1}$$

where $D(\theta(t))$ is the 3×3 inertia matrix; $h(\theta(t), \dot{\theta}(t))$ is the 3×1 Coriolis and centrifugal force vector; and $g(\theta(t))$ is the 3×1 gravitational vector.

3 Behavior Model for the Manipulator

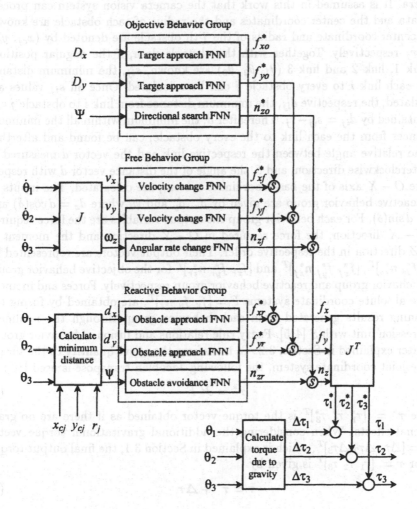

Fig. 1. Fuzzy behavior-based control system for the manipulator

Figure 1 shows the behavior-based control system consisting of three behavior groups for the manipulator with the higher level behavior group shown over the lower behavior group. Inputs for the objective behavior group are D_x, D_y and Ψ, where D_x and D_y are the distances between the end-effector point and the target in $O - X$ direction and $O - Y$ direction respectively, and Ψ is the relative angle between the moving direction of the end-effector point and the objective point. Inputs to the reactive behavior group are v_x, v_y and ω_z, where v_x and v_y are the velocities at the end-effector point in $O - X$ direction and $O - Y$ direction respectively, and ω_z is the angular velocity of the end-effector point. v_x, v_y and ω_z must be calculated by using the Jacobian matrix with the angular velocities of link 1, link 2 and link 3 ($\dot{\theta}_1$, $\dot{\theta}_2$, $\dot{\theta}_3$). All the obstacles are assumed to be circles in the $O - X - Y$ plane and they are sensed by a CCD camera. It is assumed in this work that the camera vision system can process its data and the center coordinates and the radius of each obstacle are known. The center coordinate and radius of the j-th obstacle are denoted by (x_{cj}, y_{cj}) and r_j respectively. Together with the obstacle data, if the angular positions of link 1, link 2 and link 3 (θ_1, θ_2, θ_3) are known, s_{ij}, the minimum distance from each link i to every obstacle j can be calculated. Once all s_{ij} values are calculated, the respective d_{ij}, the minimum distance from link i to obstacle j can be obtained by $d_{ij} = s_{ij} - r_j$. Thereafter d, the minimum within all the minimum distances from the each link to the every obstacle, can be found and afterthat ψ, the relative angle between the respective link and the vector d measured in counterclockwise direction, and δ, the angle of the distance vector d with respect to the $O - X$ axis of the base coordinate system, are calculated. The inputs to the reactive behavior group are given by d_x, d_y and ψ, where $d_x = d\cos(\delta)$ and $d_y = d\sin(\delta)$. For each behavior group the output variables are the force required in $O - X$ direction, the force required in $O - Y$ direction and the moment in $O - Z$ direction in the respective order. Their output vectors are represented by $[f_{xo}^* \; f_{yo}^* \; n_{zo}^*]^T$, $[f_{xf}^* \; f_{yf}^* \; n_{zf}^*]^T$ and $[f_{xr}^* \; f_{yr}^* \; n_{zr}^*]^T$ for the objective behavior group, free behavior group and reactive behavior group respectively. Forces and moment in the absolute coordinate systems $F = [f_x \; f_y \; n_z]^T$, are obtained by fusing the reasoning results generated from each behavior group through the nonlinear suppression unit with S [4,5]. Fuzzy rule relations and fusion of behavior groups are later explained in Section 3.2. To transform the above quantity to the torque in the joint coordinate system, the following Jacobian transpose is used [10]:

$$\tau^* = J^T F \tag{2}$$

where $\tau^* = [\tau_1^* \; \tau_2^* \; \tau_3^*]^T$ is the torque vector obtained as if there are no gravitational effects. When considering the additional gravitational torque vector $\Delta\tau = [\Delta\tau_1 \; \Delta\tau_2 \; \Delta\tau_3]^T$, which is explained in Section 3.1, the final output torque vector $\tau = [\tau_1 \; \tau_2 \; \tau_3]^T$ is given by

$$\tau = \tau^* + \Delta\tau \tag{3}$$

3.1 Compensation of the Gravitational Effect

Each behavior group controls the movements of the robot. For example, in the objective behavior group, the first input element D_x must be reduced in order to reach the target. This is obtained by controlling the output element f_x. Since none of the behaviors does not consider the gravitational effect, the required torque to overcome the gravity is added to the output torque from the fuzzy behavior-based control before applying to the manipulator and it is assumed in this work that the length, mass and center of gravity of the links are available. The required torque equations for this calculation are given by

$$\Delta\tau_1 = m_1 g(l_1 + \bar{x}_1)c_1 + m_2 g[l_1 c_1 + (l_2 + \bar{x}_2)c_{12}]$$
$$+ m_3 g[l_1 c_1 + l_2 c_{12} + (l_3 + \bar{x}_3)c_{123}] \tag{4}$$

$$\Delta\tau_2 = m_2g(l_2 + \bar{x}_2)c_{12} + m_3g[l_2c_{12} + (l_3 + \bar{x}_3)c_{123}] \quad (5)$$

$$\Delta\tau_3 = m_3g(l_3 + \bar{x}_3)c_{123} \quad (6)$$

where, $c_1 = \cos(\theta_1)$, $c_{12} = \cos(\theta_1 + \theta_2)$, $c_{123} = \cos(\theta_1 + \theta_2 + \theta_3)$, and g is the gravitational acceleration. However for a robot moving in a horizontal plane this additional torque is not required.

3.2 Fuzzy Rule of a Behavior Group

A simple fuzzy reasoning is applied to one behavioral element using one sensor information or processed sensor information y, and it generates one reasoning result u^*. Gaussian-type function is used as the membership function, and uses the simplified fuzzy reasoning. The resultant fuzzy reasoning consequent is obtained as

$$u^* = \sum_{i=1}^{M} p_i w_{bi} \quad (7)$$

where M denotes the total number of rules, w_{bi} the constant in the conclusion of the i-th rule, and p_i the normalized rule confidence such as

$$p_i = \frac{\mu_i}{\sum_{j=1}^{M} \mu_j} \quad (8)$$

$$\mu_i = \exp\{(\ln(0.5)(y - w_{ci})^2 w_{di}^2\} \quad (9)$$

where w_{ci} denotes the center value associated with the i-th membership function, and w_{di} denotes the reciprocal value of the deviation from the center w_{ci} to which the i-th Gaussian function of the input data on the support set has value 0.5. Consider two behavior groups i and $i+1$, where the behavior group i represents the lower behavior group and the behavior $i+1$ represents the higher behavior group. Let the two same outputs from two behavior groups be described as a and b from behavior group i and behavior group $i+1$, respectively. The fusion result is given by c,

$$c = (1 - s)a + sb \quad (10)$$

with

$$s = |\text{sat}(a) + \text{sat}(b)|/2 \quad (11)$$

Here the saturation function "sat" is given by

$$\text{sat}(x) = \begin{cases} \text{sgn}(x) & |x| > \varepsilon \\ x/\varepsilon & |x| \le \varepsilon \end{cases} \quad (12)$$

where ε denotes a positive number.

4 Learning Using Genetic Algorithm

For the tournament selection, the fittest individual is selected as the parent out of the three individuals. Two-point crossover is used with a crossover probability of 0.6. One individual has 135 training parameters, because $n_p = N_f \times N_r \times N_e \times N_b$, where n_p denotes the number of training parameters, N_f the fuzzy parameters (w_{ci}, w_{di}, w_{bi}), N_r the number of rules (5), N_e the number of elements in one group (3), and N_b the number of behavior groups (3). Since one parameter is represented by an 8 bit string, the code length is 1080 (i.e., 135×8). The number of individuals in one generation is 60, the elite number is 8, which is kept for the next generation, and the number of parents selected is 52, which generate 52 offspring. The mutation rate is equal to the 1/code length which is 1/1080.

4.1 Fitness Function

One individual run of the manipulator is over if one or more links go out of range or any of the link collides with the obstacle or the given time is over or the end-effector point moves 0.5 [m] away from the minimum distance between the end-effector point and the target during the run or the end-effector point reaches the target successfully. Fitness value, *Fitness*, is given by

$$Fitness = -d_{min} + (D_0 - d_{min})/T_D - 3.0 \times N_o - 0.1 \times N_c \qquad (13)$$

where d_{min} is the minimum distance between the end-effector of the robot and the target during the run, D_0 is the initial distance between the end-effector and the target, and T_D is the travel distance of the end-effector point during the period of that run. Here N_o is equal to the number of links went out of range and N_c is equal to the number of links collided with the obstacle. The objective of the optimization is to minimize the distance of travel to an acceptable solution while avoiding the obstacles to reach the target.

5 Simulations

5.1 Parameter Settings

The manipulator parameters are 0.5 [m], 0.4 [m] and 0.3 [m] in length, 0.48 [kg], 0.36 [kg] and 0.24 [kg] in mass and -160 to $+160$, -135 to $+135$ and -110 to $+170$ in joint range for link 1, link 2 and link 3 respectively. C programming was used to model the manipulator dynamically using the dynamic equations given by Eq. (1). Sampling time interval is 0.05 [s] and the differential equations are solved by the numerical method known as Runge-Kutta-Gill method. The maximum time allowed for one individual run is 15 [s]. Three obstacles are considered and two simulations were carried out. In the first simulation the robot is placed initially on the $O - X$ axis horizontally and in the second simulation it is placed initially on the $O - Y$ axis vertically. In each case, 1000 generations were taken into account to train the fuzzy parameters.

5.2 Results

Figure 2 shows the end-effector point path for an individual with the best fitness, whose value is 0.847344, after 1000 generations in simulation 1, and Fig. 3 shows the corresponding best and the average fitness values of each generation. Figures 4 and 5 present the similar results with the best fitness values equal to 0.968144 for simulation 2. These results show that for all cases the robot managed to reach the target with an acceptable solution. Therefore, it is confirmed from these simulations that the present approach is useful for the task control of multi-link manipulators, while avoiding obstacles.

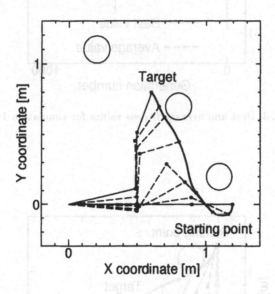

Fig. 2. End-effector point path with a best individual for simulation 1

6 Conclusions

The fuzzy behavior-based control strategy has been applied to controlling a multi-link manipulator. It was proved from simulations that such an approach is effective for complex manipulators to achieve certain tasks while avoiding obstacles. Also, this approach had the advantage of moving towards a particular point without knowing the inverse kinematics while avoiding obstacles. It means that the manipulator can be controlled to achieve the desired task with on line information and suitable fitness function without going into actual analytical details. This is very useful especially in robot manipulators because the kinematics and dynamics of the manipulators are usually complex by nature and the analysis gets more sophisticated for redundant manipulators. This approach is of course suitable not only when the relationships of the system dynamics are linear, but also when the relationships are nonlinear.

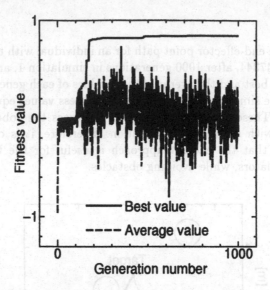

Fig. 3. Best and average fitness values for simulation 1

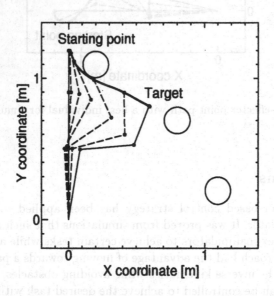

Fig. 4. End-effector point path with a best individual for simulation 2

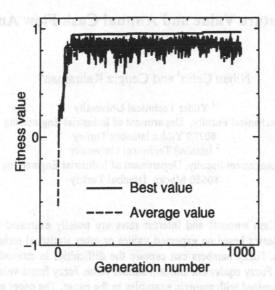

Fig. 5. Best and average fitness values for simulation 2

References

1. Brooks, R. A.: A Robust Layered Control Systems for a Mobile Robot. IEEE Journal of Robotics and Automation, **2**-1 (1986) 14–23
2. Maes, P.: Learning Behavior Networks from Experience. Procs. of the First European Conference on Artificial Life (1992) 48–57
3. Koza, J. R. Evolution of Subsumption Using Genetic Programming. Procs. of the First European Conference on Artificial Life (1992) 110–119
4. Watanabe, K., Izumi, K.: Construction of Fuzzy Behavior-Based Control Systems for a Mobile robot. Procs. of the Third Int. Symposium on Artificial Life and Robotics (AROB 3rd'98), **2** (1998) 518–523
5. Watanabe, K., Izumi, K.: A Fuzzy Neural Realization of Behavior-Based Control System for a Mobile Robot. in Soft Computing for Intelligent Robotics Systems, L. C. Jain and T. Fukuda, Eds., Heidelberg: Physica-Verlag (1998) 1–26
6. Watanabe, K., Izumi, K.: Fuzzy Behavior-Based Control for an Omnidirectional Mobile Robot. Procs. of the Third Int. Symposium on Artificial Life and Robotics (AROB 3rd'98), **2** (1998) 524–527
7. Rahmanian-shahri, N., Troch, I.: Collision-Avoidance for Redundant Robots through Control of the Self-Motion of the Manipulator. J. of Intelligent and Robotic Systems, **16**-2 (1996) 123–149
8. Ding, H., Chan, S. P.: A Real-Time Planning Algorithm for Obstacle Avoidance of Redundant Robots. J. of Intelligent and Robotic Systems, **16**-3 (1996) 229–243
9. Nearchou, A. C., Aspragathos, N. A.: Collision-Free Cartesian Trajectory Generation using Raster Scanning and Genetic Algorithms. J. of Intelligent and Robotic Systems, **23**-(2-4) (1998) 351–377
10. Craig, J. J.: Introduction to Robotics: Mechanics and Control. Reading, MA: Addison-Wesley (1989) 179–180
11. Fu, K. S., Gonazalez, R. C., Lee, C. S. G.: Robotics, Control, Sensing, Vision, and Intelligence. New York, NY: McGraw-Hill (1987) 103–118

Fuzzy Future Value and Annual Cash Flow Analyses

Nihan Çetin[1] and Cengiz Kahraman[2]

[1] Yıldız Technical University
Mechanical Faculty, Department of Industrial Engineering
80750 Yıldız İstanbul Turkey
[2] İstanbul Technical University
Management Faculty, Department of Industrial Engineering
80680 Macka İstanbul Turkey

Abstract: Cash amounts and interest rates are usually estimated by using educated guesses based on expected values or other statistical techniques to obtain them. Fuzzy numbers can capture the difficulties in estimating these parameters. Fuzzy equivalent uniform annual value, fuzzy future value are the methods examined with numeric examples in the paper. The paper also gives the ranking methods of fuzzy number

1 Introduction

To deal with vagueness of human thought, Zadeh [1] first introduced the fuzzy set theory, which was based on the rationality of uncertainty due to imprecision or vagueness. A major contribution of fuzzy set theory is its capability of representing vague knowledge. The theory also allows mathematical operators and programming to apply to the fuzzy domain.

A fuzzy number is a normal and convex fuzzy set with membership function $\mu_A(x)$ which both satisfies normality: $\mu_A(x)=1$, for at least one $x \in R$ and convexity: $\mu_A(x') \geq \mu_A(x_1) \wedge \mu_A(x_2)$, where $\mu_A(x) \in [0,1]$ and $\forall x' \in [x_1, x_2]$. '$\wedge$' stands for the minimization operator.

Quite often in finance future cash amounts and interest rates are estimated. One usually employs educated guesses, based on expected values or other statistical techniques, to obtain future cash flows and interest rates. Statements like *approximately between $ 12,000 and $ 16,000* or *approximately between 10% and 15%* must be translated into an exact amount, such as *$ 14,000* or *12.5%* respectively. Appropriate fuzzy numbers can be used to capture the vagueness of those statements.

A tilde will be placed above a symbol if the symbol represents a fuzzy set. Therefore, $\tilde{P}, \tilde{F}, \tilde{G}, \tilde{A}, \tilde{\imath}, \tilde{r}$ are all fuzzy sets. The membership functions for these fuzzy sets will be denoted by $\mu(x|\tilde{P}), \mu(x|\tilde{F}), \mu(x|\tilde{G})$, etc. A fuzzy number is a special fuzzy subset of the real numbers. The extended operations of fuzzy numbers can be found in [11, 12]. A triangular fuzzy number (TFN) is shown in Figure 1. The membership function of a TFN ($\tilde{M}$) defined by

$$\mu(x|\tilde{M}) = (m_1, f_1(y|\tilde{M}) / m_2, m_2 / f_2(y|\tilde{M}), m_3) \qquad (1)$$

where $m_1 \prec m_2 \prec m_3$, $f_1(y|\tilde{M})$ is a continuous monotone increasing function of y for $0 \le y \le 1$ with $f_1(0|\tilde{M}) = m_1$ and $f_1(1|\tilde{M}) = m_2$ and $f_2(y|\tilde{M})$ is a continuous monotone decreasing function of y for $0 \le y \le 1$ with $f_2(1|\tilde{M}) = m_2$ and $f_2(0|\tilde{M}) = m_3$. $\mu(x|\tilde{M})$ is denoted simply as $(m_1 / m_2, m_2 / m_3)$.

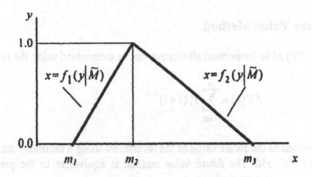

Figure 1. A Triangular Fuzzy Number, $\tilde{M}$

A flat fuzzy number (FFN) is shown in Figure 2. The membership function of a FFN, $\tilde{V}$ is defined by

$$\mu(x|\tilde{V}) = (m_1, f_1(y|\tilde{V}) / m_2, m_3 / f_2(y|\tilde{V}), m_4) \qquad (2)$$

where $m_1 \prec m_2 \prec m_3 \prec m_4$, $f_1(y|\tilde{V})$ is a continuous monotone increasing function of y for $0 \le y \le 1$ with $f_1(0|\tilde{V}) = m_1$ and $f_1(1|\tilde{V}) = m_2$ and $f_2(y|\tilde{V})$ is a continuous monotone decreasing function of y for $0 \le y \le 1$ with $f_2(1|\tilde{V}) = m_3$ and $f_2(0|\tilde{V}) = m_4$. $\mu(y|\tilde{V})$ is denoted simply as $(m_1 / m_2, m_3 / m_4)$.

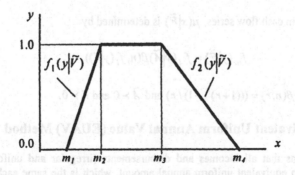

Figure 2. A FlatFuzzy Number, $\tilde{V}$

The fuzzy sets $\tilde{P},\tilde{F},\tilde{G},\tilde{A},\tilde{I},\tilde{r}$ are usually fuzzy numbers but n will be discrete positive fuzzy subset of the real numbers [2]. The membership function $\mu(x|\tilde{n})$ is defined by a collection of positive integers $n_i,\ 1 \le i \le K$, where

$$\mu(x|\tilde{n}) = \begin{cases} \mu(n_i|\tilde{n}) = \lambda_i, 0 \le \lambda_i \le 1 \\ 0, otherwise \end{cases} \tag{3}$$

2 Fuzzy Future Value Method

The future value (FV) of an investment alternative can be determined using the relationship

$$FV(r) = \sum_{t=0}^{n} P_t (1+i)^{n-t} \tag{4}$$

where $FV(r)$ is defined as the future value of the investment using a minimum attractive rate of return $(MARR)$ of $r\%$. The future value method is equivalent to the present value method and the annual value method.

Chiu and Park's [3] formulation for the fuzzy future value has the same logic of fuzzy present value formulation:

$$\{ \sum_{t=0}^{n-1} [\max(P_t^{l(y)},0) \prod_{t'=t+1}^{n} (1+r_{t'}^{l(y)}) + \min(P_t^{l(y)},0) \prod_{t'=t+1}^{n} (1+r_{t'}^{r(y)})] + P_n^{l(y)} ,$$

$$\sum_{t=0}^{n-1} [\max(P_t^{r(y)},0) \prod_{t'=t+1}^{n} (1+r_{t'}^{r(y)}) + \min(P_t^{r(y)},0) \prod_{t'=t+1}^{n} (1+r_{t'}^{l(y)})] + P_n^{r(y)} \} \tag{5}$$

Buckley's [2] membership function $\mu(x|\tilde{F})$ is determined by

$$f_i(y|\tilde{F}_n) = f_i(y|\tilde{P})(1+f_i(y|\tilde{r}))^n \tag{6}$$

For the uniform cash flow series, $\mu(x|\tilde{F})$ is determined by

$$f_{ni}(y|\tilde{F}) = f_i(y|\tilde{A})\beta(n,f_i(y|\tilde{r})) \tag{7}$$

where $i=1,2$ and $\beta(n,r) = (((1+r)^n - 1)/r)$ and $\tilde{A} \succ 0$ and $\tilde{r} \succ 0$.

3 Fuzzy Equivalent Uniform Annual Value (EUAV) Method

The $EUAV$ means that all incomes and disbursements (irregular and uniform) must be converted into an equivalent uniform annual amount, which is the same each period. The major advantage of this method over all the other methods is that it does not require making the comparison over the least common multiple of years when the alternatives have different lives [5]. The general equation for this method is

$$EUAV = A = NPV\gamma^{-1}(n,r) = NPV[\frac{(1+r)^n r}{(1+r)^n - 1}] \quad (8)$$

where NPV is the net present value. In the case of fuzziness, $\tilde{NPV}$ will be calculated and then the fuzzy $EU\tilde{A}V$ ($\tilde{A}_n$) will be found. The membership function $\mu(x|\tilde{A}_n)$ for $\tilde{A}_n$ is determined by

$$f_{nl}(y|\tilde{A}_n) = f_i(y|\tilde{NPV})\gamma^{-1}(n, f_i(y|\tilde{r})) \quad (9)$$

and $TFN(y)$ for fuzzy $EUAV$ is

$$\tilde{A}_n(y) = (\frac{NPV^{l(y)}}{\gamma(n,r^{l(y)})}, \frac{NPV^{r(y)}}{\gamma(n,r^{r(y)})}) \quad (10)$$

Example

Assume that $\tilde{NPV} = (-\$3,525.57, -\$24.47, +\$3,786.34)$ and $\tilde{r} = (3\%,5\%,7\%)$. Calculate the fuzzy $EUAV$.

$$f_{6,1}(y|\tilde{A}_6) = (3,501.1y - 3,525.57)[\frac{(1.03+0.02y)^6(0.02y+0.03)}{(1.03+0.02y)^6 - 1}]$$

$$f_{6,2}(y|\tilde{A}_6) = (-3,,810.81y + 3,786.34)[\frac{(1.07-0.02y)^6(0.07-0.02y)}{(1.07-0.02y)^6 - 1}]$$

For y=0, $f_{6,1}(y|\tilde{A}_6) = -\$650.96$

For y=1, $f_{6,1}(y|\tilde{A}_6) = f_{6,2}(y|\tilde{A}_6) = -\$4.82$

For y=0, $f_{6,2}(y|\tilde{A}_6) = +\$795.13$

It is now necessary to use a ranking method to rank the triangular fuzzy numbers such as Chiu and Park's [3], Chang's [6] method, Dubois and Prade's [7] method, Jain's [8] method, Kaufmann and Gupta's [9] method, Yager's [10] method. These methods may give different ranking results and most methods are tedious in graphic manipulation requiring complex mathematical calculation. In the following, two of the methods which does not require graphical representations are given. Chiu and Park's (3) weighted method for ranking TFNs with parameters (a, b, c) is formulated as

$$((a+b+c)/3) + wb \quad (11)$$

where w is a value determined by the nature and the magnitude of the most promising value. The preference of projects is determined by the magnitude of this sum.

Kaufmann and Gupta (9) suggest three criteria for ranking TFNs with parameters (a,b,c). The dominance sequence is determined according to priority of:

1. Comparing the ordinary number (a+2b+c)/4
2. Comparing the mode, (the corresponding most promise value), b, of each TFN.
3. Comparing the range, c-a, of each TFN.

The preference of projects is determined by the amount of their ordinary numbers. The project with the larger ordinary number is preferred. If the ordinary numbers are equal, the project with the larger corresponding most promising value is preferred. If projects have the same ordinary number and most promising value, the project with the larger range is preferred.

4 Conclusions

In this paper, capital budgeting techniques in the case of fuzziness and discrete compounding have been studied. Fuzzy set theory is a powerful tool in the area of management when sufficient objective data has not been obtained. Appropriate fuzzy numbers can capture the vagueness of knowledge. The other financial subjects such as replacement analysis, income tax considerations, continuous compounding in the case of fuzziness can be also applied [11], [12]. Comparing projects with unequal lives has not been considered in this paper. This will also be a new area for a further study.

References

[1] Zadeh, L. A., Fuzzy Sets, Information and Control, Vol. 8, pp. 338-353, (1965).
[2] Buckley, J. U., The Fuzzy Mathematics of Finance, Fuzzy Sets and Systems, Vol. 21, pp. 257-273, (1987).
[3] Chiu, C.Y., Park, C.S., Fuzzy Cash Flow Analysis Using Present Worth Criterion, The Engineering Economist, Vol. 39, No. 2, pp. 113-138, (1994).
[4] Ward, T.L., Discounted Fuzzy Cash Flow Analysis, in 1985 Fall Industrial Engineering Conference Proceedings, pp. 476-481, December (1985).
[5] Blank , L. T., Tarquin, J. A., Engineering Economy, Third Edition, McGraw-Hill, Inc., (1978).
[6] Chang, W., Ranking of Fuzzy Utilities with Triangular Membership Functions, Proc. Int. Conf. of Policy Anal. and Inf. Systems, pp. 263-272, (1981).
[7] Dubois, D., Prade, H., Ranking Fuzzy Numbers in the Setting of Possibility Theory, Information Sciences, Vol. 30, pp. 183-224, (1983).
[8] Jain, R., Decision Making in the Presence of Fuzzy Variables, IEEE Trans. on Systems Man Cybernet, Vol. 6, pp. 698-703, (1976).
[9] Kaufmann, A., Gupta, M. M., Fuzzy Mathematical Models in Engineering and Management Science, Elsevier Science Publishers B. V., (1988).
[10] Yager, R. R., On Choosing Between Fuzzy Subsets, Kybernetes, Vol. 9, pp. 151-154, (1980).
[11] Kahraman, C., Ulukan, Z., Continuous Compounding in Capital Budgeting Using Fuzzy Concept, in the Proceedings of 6th IEEE International Conference on Fuzzy Systems (FUZZ-IEEE'97), Bellaterra-Spain, July 1-5, (1997).
[12] Kahraman,C., Ulukan, Z., Fuzzy Cash Flows Under Inflation, in the Proceedings of Seventh International Fuzzy Systems Association World Congress (IFSA'97), University of Economics, Prague, Czech Republic, Vol. IV, pp. 104-108, June 25-29, (1997).
[13] Zimmermann, H. -J., Fuzzy Set Theory and Its Applications, Kluwer Academic Publishers, (1994).

Semi Linear Equation with Fuzzy Parameters

Said Melliani

Departement of Applied Mathematics and Informatic
Faculty of Sciences and Technics, B.P 523 Béni Mellal Morocco
E-mail melliani@fstbm.ac.ma

Abstract. In this paper we introduce the concepts of fuzzy solution for a semi linear equation with fuzzy parameters. The extension principle described by L. A. Zadeh [5] provides a natural way for obtaining the notion of fuzzy solution. The fuzzy extension of the solution operator is shown to provide the unique solution in the formar case.

1 Introduction

Fuzzy sets theory is a powerful tool for modelling uncertainty and for processing vague or subjective information in mathematical models. While its main directions of development have been information theory, data analysis, artificial intelligence, decision theory, control, and image processing (see e.g. [1], [6], [7]), fuzzy set theory is increasingly used as a means for modelling and evaluating the influence of imprecisely known parameters in mathematical, technical, physical models. The purpose of this paper is to work out this approch when the models are constitued by partial differential equations.

Based on the fuzzy description of parameters and mathematical objects, we shall be concerned here with partial differential equation in the scalar case of the form

$$u_t + \lambda u_x = au$$
$$u(x, 0) = u_0(x)$$

Here the parameters a and λ will be fuzzy numbers. The solution $u(x, t)$ at any fixed point (x, t) will be a fuzzy number as well.

2 Partial differential equations

Consider a semi-linear equation :

$$u_t + \lambda u_x = au \tag{1}$$

for a function $u = u(x, t)$, where $\lambda =$ const. > 0. Along a line of the family

$$x - \lambda t = \xi = \text{const.}$$

("characteristic line "in the xt-plane) we have for a solution u of (1)

$$\frac{du}{dt} = \frac{d}{dt}u(\lambda t + \xi, t) = \lambda u_x + u_t = au \qquad (2)$$

Hence u is constant along a line, and depends only on the parameter ξ which distinguishes different lines. The general solution of (1) has the form

$$u(x, t) = u_0(\xi)\exp(at) = u_0(x - \lambda t)\exp(at) \qquad (3)$$

Formula (3) represents the general solution u uniquely in terms of its initial values

$$u(x, 0) = u_0(x)$$

Conversely every u of the form (3) is a solution of (1)with initial values u_0 provided u_0 is of class $C^1(\mathbb{R})$. We notice that the value of u at any point (x, t) depends only on the initial value u_0 at the single argument $\xi = x - \lambda t$, the abscissa of the point of intersection of the characteristic line through (x, t) with the initial line, the x-axis. We consider $\Omega = \mathbb{R} \times \mathbb{R}_+$
introducing the equation operator

$$E : C^1(\Omega) \to C(\Omega) : v \to [(x, t) \to v_t + \lambda v_x - av],$$

the restriction operator $\forall (x, t) \in \Omega$

$$R_{x,t} : C^1(\Omega) \to \mathbb{R} : v \to v(x, t) = u_0(x - \lambda t)e^{at},$$

and the solution operator

$$L : \mathbb{R}^2 \to C^1(\Omega) : (\lambda, a) \to L(\lambda, a).$$

3 Fuzzy sets and fuzzy numbers

Geven a set X, a fuzzy set A over X is a map

$$m_A : X \longrightarrow [0, 1]$$

called the membership of A (it is convenient to distinguish between A and its membership functions m_A to be able to employ the usual language of sets theory). Thus given $x \in A$, $m_A(x)$ is considered the degree to which, respectively the possibility that, x belong to A (In calssical sets theory, m_A would correspond to the characteristic function of A). This concept allows to model uncertainty in situations where more information than just upper and lower bounds is available (in contrast to interval analysis), but no probability distribution are available. This situation often arises e.g. in engineering practice, when parameters are estimated partially in subjective way.

We denote the family of fuzzy sets over X by $\mathbb{F}(X)$. The α-level sets are the classical sets

$$A^{\alpha \geq} = \{x \in X : m_A(x) \geq \alpha\}$$

A fuzzy real number is an element $A \in \mathbb{F}(\mathbb{R})$ such that all level sets $A^{\alpha \geq}$ are compact intervals $(0 < \alpha \leq 1)$ and $A^{1 \geq}$ is not empty. The graph of m_A has a monotonically increasing left branch, a central point or plateau of membership degree one, and a monotonically decreasing right branch. Similary, one can define fuzzy vectors, fuzzy functions etc. The *extension principle* introduced by [5] allows the evaluation of functions on e.g. fuzzy numbers according to the following definition : Let

$$f : \mathbb{R}^n \to \mathbb{R}$$

be a function, define the extension [1], [3]

$$f : (\mathbb{F}(\mathbb{R}))^n \to \mathbb{F}(\mathbb{R})$$

by

$$m_{f(a_1,\ldots,a_n)}(y) = \sup_{y=f(x_1,\ldots,x_n)} \inf(m_{a_1}(x_1),\ldots,m_{a_n}(x_n))$$

It can be shown that in case f is continuous, $f(a_1,\ldots,a_n)$ is a fuzzy number as well, and

$$f(a_1,\ldots,a_n)^{\alpha \geq} = f\left(a_1^{\alpha \geq},\ldots,a_n^{\alpha \geq}\right)$$

the set theoretic image of the level sets. Thus the upper and lower endpoints of the interval $f(a_1,\ldots,a_n)^{\alpha \geq}$ can be obtained by minimizing / maximizing f over $a_1^{\alpha \geq} \times \ldots \times a_n^{\alpha \geq}$.

we denote by 0 the crisp zero function in $\mathbb{F}(\mathbb{R}^n)$, that is,

$$m_0(f) = \begin{cases} 1 \text{ if } f = 0 \\ 0 \text{ otherwise} \end{cases}$$

Definition 1. Let A a fuzzy set

- A is normalized if there exists an element x in A such that $m_A(x) = 1$.
- The α-level sets $A^{\alpha \geq}$ for $(0 < \alpha < 1)$ of a fuzzy set A are the classical (crisp) sets.
- A is convex if and only if its α-level are convexs.
- A fuzzy number is a convex, normalized fuzzy subset of the domain A

The concept of fuzzy number is an extension of the notion of real number : its encodes approximate but non probabilistic quantitative knowledge [2].

4 Fuzzy semi-linear equation

Let us consider a fuzzy semi-linear equation

$$\begin{cases} \tilde{u}_t + \tilde{\lambda} u_x = \tilde{a} \tilde{u} \\ \tilde{u}(x,0) = u_0(x) \end{cases} \tag{4}$$

with $\tilde{\lambda}$ and $\tilde{a}$ are two fuzzy numbers, the initial condition u_0 is a classic function in $C(\mathbb{R})$.

by the extension principle

$$E \; : \; \mathbb{F}\left(C^1(\Omega)\right) \to \mathbb{F}\left(C(\Omega)\right)$$
$$R_{x,t} : \mathbb{F}\left(C^1(\Omega)\right) \to \mathbb{F}(\mathbb{R})$$
$$L \; : \; \mathbb{F}\left(\mathbb{R}^2\right) \to \mathbb{F}\left(C^1(\Omega)\right)$$

$\forall\, (x,t) \in \mathbb{R} \times \mathbb{R}_+$

$$L_{x,t} : \mathbb{F}\left(\mathbb{R}^2\right) \to \mathbb{F}(\mathbb{R})$$
$$\left(\tilde{\lambda}, \tilde{a}\right) \; \mapsto \; \tilde{u}(x,t) \text{ solution of (4)}$$

The fuzzy value $L_{x,t}\left(\tilde{\lambda}, \tilde{a}\right)$ may be computed by the extension principle in this way

$$m_{L_{x,t}(\tilde{\lambda},\tilde{a})}(y) = \sup\left\{\inf\left(m_{\tilde{\lambda}}(\lambda), m_{\tilde{a}}(a)\right) : y = L_{x,t}(\lambda, a)\right\}$$

Lemma 2. *We have*

$$R_{x,t} \circ L = L_{x,t} \quad in \; \mathbb{F}\left(\mathbb{R}^2\right)$$

Proof (of lemma).

$$m_{L_{x,t}(\tilde{\lambda},\tilde{a})}(y) = \sup_{\substack{(\lambda,a)\,\in\,\mathbb{R}^2 \\ y = L_{x,t}(\lambda,a)}} \min\left(m_{\tilde{\lambda}}(\lambda), m_{\tilde{a}}(a)\right)$$

$$m_{L(\tilde{\lambda},\tilde{a})}(f) = \sup_{\substack{(\lambda,a)\,\in\,\mathbb{R}^2 \\ f(x,t) = L_{x,t}(\lambda,a)}} \min\left(m_{\tilde{\lambda}}(\lambda), m_{\tilde{a}}(a)\right)$$

$$m_{R_{x,t}\circ L(\tilde{\lambda},\tilde{a})}(y) = \sup_{\substack{(\lambda,a)\,\in\,\mathbb{R}^2 \\ y = R_{x,t}\circ L(\lambda,a)}} \min\left(m_{\tilde{\lambda}}(\lambda), m_{\tilde{a}}(a)\right)$$

$$= \sup_{\substack{(\lambda,a)\,\in\,\mathbb{R}^2 \\ y = u_0(x - \lambda t)e^{at}}} \min\left(m_{\tilde{\lambda}}(\lambda), m_{\tilde{a}}(a)\right)$$

$$= \sup_{\substack{(\lambda,a)\,\in\,\mathbb{R}^2 \\ y = L_{x,t}(\lambda,a)}} \min\left(m_{\tilde{\lambda}}(\lambda), m_{\tilde{a}}(a)\right)$$

$$= m_{L(\tilde{\lambda},\tilde{a})}(y)$$

□

we have

$$m_L(f) = \sup\left\{\inf\left(m_{\tilde{\lambda}}(\lambda), m_{\tilde{a}}(a)\right) : f = L(\lambda, a)\right\}$$

and

$$m_{L_{x,t}(\tilde{\lambda},\tilde{a})}(y) = \sup\left\{m_L(f) : f \in C^1(\Omega) \text{ with } y = f(x,t)\right\}$$
$$= \sup_{\substack{(\lambda,a)\,\in\,\mathbb{R}^2 \\ y = L_{x,t}(\lambda,a)}} \min(m_{\tilde{\lambda}}(\lambda), m_{\tilde{a}}(a))$$

Definition 3. An element $\tilde{u} \in \mathbb{F}\left(C^1(\Omega)\right)$ is called a fuzzy solution with the initial data $u_0 \in C(\Omega)$, if $E(\tilde{u}) = 0$ in $\mathbb{F}(C(\Omega))$, $R_{x,t}(\tilde{u}) = u_0 \left(x - \tilde{\lambda}t\right) e^{\tilde{a}t}$ in $\mathbb{F}(\mathbb{R})$

Proposition 4. *Given $\tilde{\lambda}$, $\tilde{a}$ in $\mathbb{F}(\mathbb{R})$, $\tilde{u} = L\left(\tilde{\lambda}, \tilde{a}\right)$ is a fuzzy solution of (4)*

Proof (of proposition). To show that $\tilde{u} = L\left(\tilde{\lambda}, \tilde{a}\right)$ solves the fuzzy partial differential equation, we compute :

$$m_{E(\tilde{u})=L(\tilde{\lambda},\tilde{a})}(w) = \sup\left\{m_{\tilde{u}=L(\tilde{\lambda},\tilde{a})}(v) : w = E(v)\right\}$$
$$= \sup\left\{\sup\left\{\inf\left(m_{\tilde{\lambda}}(\lambda), m_{\tilde{a}}(a)\right) : v = L(\lambda, a)\right\} : w = E(v)\right\}$$

if $w \neq 0$ and $w = E(v)$; then $\left\{(\lambda, a) \in \mathbb{R}^2 : v = L(\lambda, a)\right\} = \emptyset$, so the inner supremum is zero and $m_{EL(\tilde{\lambda},\tilde{a})}(w) = 0$.

if $w = 0$, we may take $(\lambda, a) \in \mathbb{R}^2$ with $m_{\tilde{\lambda}}(\lambda) = m_{\tilde{a}}(a) = 1$ and $v = L(\lambda, a)$. Then $E(w) = 0$, and so the supremum equals 1.

Let $S = \left\{u \in C^1(\Omega) : E(u) = 0\right\}$. We can view $\mathbb{F}(S)$ as a subset of $\mathbb{F}\left(C^1(\Omega)\right)$, setting the membership degree of any $u \in C^1(\Omega) \backslash S$ to some $\tilde{u} \in F(S)$ equal to zero. □

Lemma 5. *If $\tilde{u} \in \mathbb{F}\left(C^1(\Omega)\right)$ is a solution to (4) , then $\tilde{u}$ belongs to $\mathbb{F}(S)$*

Proof (of lemma). we have that

$$m_{E(\tilde{u})}(v) = \sup\left\{m_{\tilde{u}}(w) : v = E(w)\right\}$$

suppose there exist $v \notin S$, such that $m_{(\tilde{u})}(v) > 0$. Putting $w = E(v)$ we have $m_{E(\tilde{u})}(w) \geq m_{\tilde{u}}(v) > 0$, contradicting the hypothesis that $E(\tilde{u}) = 0$. □

Proposition 6. *The fuzzy solution $\tilde{u} \in \mathbb{F}\left(C^1(\Omega)\right)$ to (4) is unique.*

Proof (of proposition). Since $L : \mathbb{R}^2 \to S$ is bijective, the same is true of the extension $L : \mathbb{F}\left(\mathbb{R}^2\right) \to \mathbb{F}(S)$. If $\tilde{u} \in \mathbb{F}\left(C^1(\Omega)\right)$ is a solution, it belongs to $\mathbb{F}(S)$ by the lemma and hence is uniquely determined by the initial data. □

References

1. Dubois D. and Prade H. : Fuzzy sets and systems, Academic Press New York (1980).
2. Kaufmann A. and Gupta M.: Introduction to fuzzy arithmetic : theory and applications Van Nostran d Rehinold New York NY (1985).
3. Nguyen H. T.: A note on the extension principle for fuzzy sets J. Math. Anal. Appl. **64** (1978) 369-380.
4. Sanchez E.: Solution of fuzzy equations with extended operations Fuzzy Sets and Systems North-Holland **12** (1984) 237-248.
5. Zadeh L. A.: Fuzzy sets Information and Control **8** (1965) 338-353.
6. Zimmermann, H. J. : Fuzzy sets theory and its applications Kluwer Dardrecht 1993.
7. Zimmermann, H. J. : Fuzzy sets, decesion marketing and expert systems Kluwer Dardrecht 1993.

First Order Rough Logic-Revisited

T.Y. Lin[1,2] and Qing Liu[3]

[1] Department of Mathematics and Computer Science
San Jose State University, San Jose, California 95192
[2] Department of Electrical Engineering and Computer Science
University of California, Berkeley, California 94720
tylin@cs.sjsu.edu, tylin@cs.berkely.edu
[3] Department of Computer Science
NanChang University NanChang, Jiangxi P.R.China

Abstract. Based on axiomatic rough set theory, first order rough logic was developed earlier. In this paper, a new model theory for that logic is introduced. With this new semantic, first order rough logic is shown to be equivalent to first order $S5$, and hence consistent and complete.

1 Introduction

One of the most important studies of rough set theory is the study of the lower and upper approximations of equivalence relations. Many interesting properties have been reported [4]. In 1993, we presented an axiomatic approach to such a study, namely, we showed two abstractor operators which are characterized by certain axioms are *the* lower and upper approximations of an equivalence relation [1]. By translating these axioms into logic terms, we constructed first order rough logic. The syntax is similar to the modal logic $S5$, its semantics, however, is different [2]; we showed the model is consistent, but we had not proved its completeness. In this paper, we revisit the model and propose a new semantics. With this new semantics, we show that first order rough logic is equivalent to first order $S5$, and hence it is consistent and complete.

2 Possible Worlds Semantics - An Informal Overview

In this section we shall describe the relationships between rough logic and rough set theory. Rough set theory is based on a known equivalence relation (indiscernibility relation). However, in applications, the equivalence relations are often unknown, so the proposed rough logic is based solely on the notions of "lower" and "upper" approximations without using an *explicit* equivalence relation. However, in order to see clearly the relationships between rough logic and rough set theories, we explain the idea using *explicit* equivalence relations. Subsequent expositions, the use of equivalence relations is avoided.

2.1 Observable Worlds

Let E be the universe of discourse and P an equivalence relation. Note that we should view P as the one induced by the axioms of two abstract operators. By abuse of notation, we will use P to denote the corresponding partition. The collection of all equivalence classes is called *quotient space*, denoted by Q. Based on the available information, all elements in the same equivalence class are indistinguishable. To each observer, an equivalence class is a multi-set (bag) that consists of multiple copies of one element. We should, however, also note that different observers may see the multiple copies of different element(but in the same equivalence class). This led us to define *an observable world as the collection of one representative from each equivalence class.* Different observers have different collections of representatives. We should also point out that not all mathematical combinations of representatives will be observed by some one. In other words, a collection of all observable worlds is a subset of all possible combinations.

The intent of rough logic is

1. to describe E as much as we can, using
2. only the available imperfect observations (observable worlds).

Example. The universe $E = \{1, 2, 3, 4, 5, 6, 7, 8, 9\}$ has a partition:

$$H_1 = \{3, 6, 9\}, \qquad H_2 = \{2, 5, 8\}, \qquad H_3 = \{1, 4, 7\}.$$

Then, the quotient space is $Q = \{H_1, H_2, H_3\}$

Observers may see E as, for example,

$$W^1 = \{1, 1, 1, 2, 2, 2, 3, 3, 3\}, \qquad W^2 = \{1, 1, 1, 2, 2, 2, 6, 6, 6\}$$

If we use set notations, they are

$$W^1 = \{1, 2, 3\}, \qquad W^2 = \{1, 2, 6\},$$

Other possible observable worlds are

$$W^3 = \{1, 2, 9\}, \qquad W^4 = \{1, 5, 3\},$$
$$\cdots,$$
$$W^{26} = \{7, 8, 6\}, \qquad W^{27} = \{7, 8, 9\}$$

Each W^h is a set of representatives and equivalent to the quotient space Q. The relational structure on each W^h is induced from E by restriction; see next subsection.

2.2   Induced Structures on Observable Worlds

In this section, we will explain how observable worlds get their relational structures. The material in this section is slightly different from the corresponding structures given in [2]. The structure on relations is the same. For the structure of functions is slightly different; in the cited paper, the function values that are out of "range" was treated as has missing values; in this paper, we replace the missing values by their equivalent ones. Each W^h is a subset of E, hence all relations and functions on E can be interpreted to W^h by restricting their domain to W^h. Intuitively, each W^h represents one particular imperfect observation. These induced functions may be distorted and relations may have missing values.

Functions

Let $E = \{1, 2, 3, 4, 5, 6, 7, 8, 9\}$ be the universe of discourse. Let $f(-)$ be a function defined by

$$f(x) = 9 - [x/2],$$

where $[z]$ represents the integral part of z. The function $f(-)$ induces a new function on each W^h, for example,

(1) In Observable World $W^1 = \{1, 2, 3\}$, the function values are:

$$f(1) = 9,\ f(2) = 8,\ f(3) = 8,$$

since these values lie outside of W^1, we replace them by their equivalent values, namely,

$$f(1) = 3,\ f(2) = 2,\ f(3) = 2,$$

Such a new function is the induced view of f on W^1.

(2) In Observable World $W^2 = \{1, 2, 6\}$, the same function will be replaced by their equivalent values in W^2, i.e.,

$$f(1) = 6,\ f(2) = 2,\ f(6) = 6.$$

It is the induced view of f in W^2.

So the same function f is distorted into a different, yet equivalent, function on each observable world. These distorted functions are equivalent in the sense that each function induces the same function in the quotient space Q. Intuitively, each distorted function represents an imperfect observation of the "true" function. The goal of rough logic is to recapture some essential features of the function f via these distorted versions.

Relations

Let R denote the collection of all relations in E. Let r be an n-ary relation, and r^h its restriction to W^h; some values in r may not appear in r^h. The collection of these restricted relations will be denoted by R^h. We do *require R^h is an non-empty set*; see next subsection.

2.3 Impossible World

First we should note that *not* all mathematical W^h's would be an observable world. In order for it to be qualified as an observable world, at least, one of these n-ary relations r^h's is an non-empty relation. Informally, an impossible world is a world in which all predicates are evaluated to false in all instances. So we do require

no observable world is an impossible world.

Let $P(x_1, x_2, \ldots)$ be a n-ary predicate and the variable x_i are assigned to $e_i \in E, i = 1, 2, \ldots$. Then the predicate $P(e_1, e_2, \ldots)$ is evaluated to truth at W^h if the relation r^h that interprets P contains the tuple, $(e_1, e_2, \ldots)$. A world in which R^h is an empty set is an impossible world.

3 Axiomatic Rough Set Theory

In last few subsections, we explain the possible world semantics using an explicit equivalence relations. In applications, such explicit equivalence relations may not be available. We recall here the axiomatic rough set theory, in which only upper and lower approximate operators are available.

Pawlak introduced rough sets via equivalence relations. He derived many interesting properties of upper and lower approximation. In [1], we showed that Pawlak's lower and upper approximation can be characterized axiomatically by the following "Six" Axioms: Let E be the universe of discourse, $X \subseteq E$, and $C(X) = E \sim X$. Let L and H be the lower rough and upper (higher) rough operators.

$$(1) H(\emptyset) = \emptyset; \qquad (2) L(X) \subseteq X;$$
$$(3) L(X) \subseteq L(L(X)); \qquad (4) H(X) \bigcup H(Y) = H(X \bigcup Y);$$
$$(5) L(C(X)) = C(H(X)); \qquad (6) L(X) = H(L(X));$$
$$(6a) H(X) = L(H(X)); \qquad (6a) H(X) = L(H(X));$$

These seven axioms are not minimal. Since this is not a mathematical paper, we will not digress on it. The axioms consist of the Kuratowski's axioms of topological spaces and one additional axiom that declare open sets are close sets and vise versa. Essentially, the seven axioms characterize clopen spaces. It is easy to see that a clopen space induces a partition, and hence an equivalence relation. The two abstract operators H and L turns out to be *the* upper and lower approximations of the induced equivalence relation.

4 Rough and S_5 Models

The language and axioms of rough and $S5$ logic are not specifically referenced in this paper, so we refer readers to [2] for details.

In this section, we will show that the models of rough and $S5$ logic are equivlent. Following [3], a frame for $S5$, or for short $S5$-frame, is:

$$SM = (D, N, W, B, \gamma)$$

where

1. D is a non-empty set, called domain of SM.
2. N is a subset of D, called constants.
3. W is a non-empty set of possible worlds.
4. B is a binary relation of "accessibility" on W; for $S5$ logic, B can be the trivial one, namely, every worlds are accessible to each other.
5. γ is a function which assigns to each pair consisting of an n-ary predicate symbol $n \geq 0$, differ from "$=$", and an element w of W, an n-ary relation on D, to the symbol "$=$", if present in the language, the identity relation on D, and to each n-ary functions symbol, and to each n-ary function symbol a function from D^n to D.

γ is called an interpretation. Intuitively D is "equivalent" to the quotient set Q.

4.1 Rough Model

A Rough Model is a 7-tuple

$$RM = (E, N, R, F, RO, W, \gamma)$$

where:

1. E is a set of entities $\{e, e_1, \ldots\}$;
2. N is a set of distinguished entities $\{n, n_1, \ldots\}$, called the domain of constants; moreover $H(n_i) = n_i, i = 1, 2, \ldots$.
3. R is a set of non-empty relations $\{r, r_1, \ldots\}$, each of which is defined on E;
4. F is a set of functions $\{f, f_1, \ldots\}$, each of which is defined on E;
5. RO is a set of rough operators satisfying six axioms, i.e., $RO = \{H\}$;
6. γ is a function which assigns to each n-ary predicate symbol $n \geq 0$, differ from "$=$", an n-ary relation on E, to the symbol "$=$", if present in the language, the identity relation on E, and to each n-ary functions symbol a function from E^n to E.
7. W is a collection of observable worlds which are constructed from E and RO as explained in Section 2,

$$W^h = (W^h, N^h, R^h, F^h),$$

where we require that R^h is *non-empty*; note that this condition *excludes the impossible worlds*.

Roughly, the new model is the same as the one in [2], except we exclude the "impossible worlds."

Two rough models

$$RM = (E, N, R, F, RO, W, \gamma)$$
$$RM = (E', N', R',' F, RO', W', \gamma')$$

are said to be equivalent, if there is a map F between two models such that F induces isomorphism between the two families of observable worlds.

or more formally, F induces an isomorphism between

$$W^h = (W^h, N^h, R^h, F^h), \text{ and}$$
$$W'^h = (W'^h, N'^h, R'^h, F'^h)$$

in the sense there is four family of one-to-one maps

$$\forall h, \quad W^h \to W'^h,$$
$$\forall h, \quad N^h \to N'^h,$$
$$\forall h, \quad R^h \to R'^h,$$
$$\forall h, \quad F^h \to F'^h$$

Remark: It should be noted we have not required the two model RM and RM' to be isomorphic.

4.2 The Equivalence of Two Models

Next we will have the most important result of this paper

Proposition. A rough model induces an $S5$-frame and vice versa.

Proof: We will prove this proposition in three steps. First we will show that a rough model induces an $S5$-frame, then in the second step we construct a rough model from a given $S5$-frame. In third and fourth steps, we show that the compositions these two steps are identity.

(1) Step One: Assume we are given a rough model RM, namely,

$$RM = (E, N, R, F, RO, W, \gamma)$$

Note that the rough model has the following family W of observable worlds,

$$W = \{ W^h \mid h \text{ an index set } \}, \text{ where } W^h = (W^h, N^h, R^h, F^h),$$

We will show that the family W determines an $S5$-frame:

Fist, note that all W^h's are "isomorphic" to each other, that is $W^h \equiv W^k$; see Section 2.1; so are N^h's and F^h's respectively. We will identify them via respective isomorphisms.

$$D_{S5} \equiv W^h, \ \forall h, \qquad N_{S5} \equiv N^h, \ \forall h, \qquad F_{S5} \equiv F^h, \ \forall h.$$

Let us write,

$$W_{S5} = \{NAME(W^h) \mid \forall h\}$$

We will show $D_{S5}, N_{S5}, F_{S5}, W_{S5}$ together with an equivalence relation B_{S5} and an interpretation γ_{S5}, both to be defined below, do form an $S5$-frame.

A) Construction of B_{S5}: First, we note that the isomorphism among W^h's $\forall$ h defines an equivalence relation on W_{S5}; it is a trivial one, i.e., there is only one equivalence class; We will denote this equivalence relation by B_{S5}.

B) Define the induced interpretation γ_{S5}: Let R'^h (in domain D_{S5}) be the isomorphic copy of R^h(in domain W^h). The union of all those R'^h, $\forall h$ is denoted by R.

Then the induced interpretation can be defined as follows: In rough model the interpretation γ assigns a predicate symbol to a relation $r \in R$. In trun, r induces a relation r^h on each W^h $\forall h$; see Section 2. So the predicate is interpreted to W^h $\forall h$ via such a route. The interpretations of function symbols is the same; See Section 2.2.

Thus we have an $S5$-frame,

$$SM = (D, N, W, B, \gamma).$$

(2) *Step Two*: Conversely, given an $S5$-frame SM, we will construct a rough model A) Construction of E, N and RO: First, we need to set up the notations for symbols in SM. We write

$$D = \{d^k \mid k \text{ is an index }\} \qquad W = \{w^h \mid h \text{ is an index }\}.$$

Next, we consider $E' = D \times W$, and write

$$E'^k = \{(d^k, w) \mid d^k \text{ is a fixed element in } D \text{ and } w \text{ varies through } W\}$$

The collection of E'^k, $\forall k$ forms a partition of E', we call it vertical partition and each E'^k a vertical equivalence class. Now, we will consider a quotient set E as follows: First note that N is a subset of D. If $d^k = n^k$ is an element of N, then we collapse the vertical equivalence class $E'k$ to an element, denoted by $(n^k, \emptyset)$ or simply n^k. This new set is denoted by E; the collapsing map is denoted by $Q : E' \longrightarrow E$. E inherits a partition from E': if $d^k \in N$, the equivalence class is singleton, if $d^k \in (D \setminus N)$ then the equivalence class is the vertical equivalence class. By abuse of terminology, we will refer to such "collapsed" vertical partition as vertical partition; similarly, each equivalence class still be called vertical equivalence class. Finally, we observe that the vertical partition of E gives rise to the upper approximation operator H, which satisfies the six axioms; see Section 3. This constitutes the component RO.

B) Construction of R. Let p be an n-ary predicate symbol, and its interpretation in ($S5$-frame SM) be

$$\gamma(p, w^h) = r^h,$$

where r^h is an n-ary relation in D, that is,

$$r^h = \{(d_r^1, d_r^2 ..., d_r^n) \mid d_r^k \in D\} \subseteq D^n$$

We should stress that those d_r^k are associated with w^h. We shall "embed" r^h into a relation on E through the following consideration. First, we "embed" it to E'

$(*)$ $r^h \times w^h = \{((d_r^1, w^h), (d_r^2 w^h), \ldots, (d_r^n, w^h)) \mid (d_r^1, d_r^2 ..., d_r^n) \in r^h$ for a fixed h $\}$

Then we apply the collapsing map Q. $Q(r^h \times w^h)$ is the induced relation on E; it will be denoted by r, i.e., $r = Q(r^h \times w^h)$ Next, we consider the union (varying h),

$$r = \bigcup_{\forall h} (r^h \times w^h)$$

r is a subset of E^n, hence is a relation on E. This r is an interpretation of p in E. Let the collection of all such r's be denoted by R; a required component of RM.

C)Construction of F: F consists of all functions that are the Cartesian product of a function $d : D^n \longrightarrow D$ and identity map on W.

D)Construction of W: The observable worlds will be induced from E, and its vertical partition, as explained in Section 2.1

Combining A), B), C) and D), we have constructed RM.

(3) Step Three: Now, we need to complete "the cycle." By *Step Two*, we have constructed RM_2 from SM_1; please note the index. By *Step One* RM_2 is transformed back to SM_3; finally, we need to show two SM's are equivalent. Let X be an SM and is transformed to Y by *Step Two*. In the model Y, E has a vertical partition. To get an observable world, we select a representative from each vertical equivalence class. The selection can be expressed by the composition of a map $d' \longrightarrow (d, w)$ and Q. We write the composition, $d \longrightarrow (d', w)) \longrightarrow w$, by f, understanding that when $d = n \in N$, $w = d$. We will say f is a *constant* map, if $f(d) = w_o$ is a constant on $D \setminus N$, D minus N. The observable world selected will be denoted by W^f From *Step One*, we know that each W^f is an observable world, if R^f is non-empty. However, from the equation $(*)$ in *Step Two*, one observes that if f is not a *constant* map, then R^f is an empty set, hence W^f are not included in the family of observable worlds. So the observable worlds are precisely the same as those given in SM. This completes the proof of cycle one.

(4) Step Four: Let Y be an RM_1. By *Step One*, it is transformed to X, an SM_2. We need to show, by *Step two* X will be transformed back to RM_3; we need to show that two RM are equivalent(not necessarily equal); see Section 4.1 Remark 4.1. In *(3) Step 3*, we have shown that X is transformed by *Step Two*, then *Step One* back to X. Now we take Y and transform it to X by

Step One. By it Step Two, we get Y'. Now note that the observable worlds of Y' is precisley X, so Y and Y' are equivalent. QED.
Since $S5$-frame is complete, we have

Theorem. Rough logic with this new interpretation is sound and complete.

5 Conclusion

Rough set theory models uncertainty by equivalence relations (indiscernibility relations). In real world applications, such a precise knowledge of equivalence relations is often unavailable. However, one could often observe the approximations, in other words, the knowledge of approximate operators are reasonably available. Based on such belief, we developed the axiomatic rough set theory and the first order rough logic [2] without explicit equivalence relation.

The model, namely RM, proposed is too rich in semantics for the syntax. The language can not completely determine the model. So in this paper, we introduce the equivalence relation among those models;see Section 4.1. Then the equivalence class contains the right amount of information to be characterized by the syntax. In other words, RM is the "ideal world" that the syntax intend to address. However, due to insufficient information, the syntax can only determine an equivalence class of the "ideal" worlds; there are uncertainty. From this aspect, rough logic is richer than $S5$.

References

1. T. Y. Lin and Q. Liu (1994),"Rough Approximate Operators-Axiomatic Rough Set Theory," in: Rough Sets, Fuzzy Sets and Knowledge Discovery, ed by W. Ziarko, Springer-Verlag, 1994, 256-260.
2. T. Y. Lin and Q. Liu (1996), First Order Rough Logic I: Approximate Reasoning via Rough Sets, Fundamenta Informaticae. Volume 27, Number 2,3, 137-153
3. W. Lukaszewicz (1990), Non-Monotonic Reasoning-Formalization of Commonsense Reasoning, Institute of Information, Ellis Horwood limited, 1990
4. Z. Pawlak, *Rough Sets (Theoretical Aspects of Reasoning about Data).* Kluwer Academic, Dordrecht, 1991.

A Generalized Decision Logic in Interval-Set-Valued Information Tables*

Y.Y. Yao[1] and Qing Liu[2]

[1] Department of Computer Science, University of Regina
Regina, Saskatchewan, Canada S4S 0A2
E-mail: yyao@cs.uregina.ca
[2] Department of Computer Science, NanChang University
NanChang, JiangXi 330029, P.R. China
E-mail: qliu@263.net

Abstract. A generalized decision logic in interval-set-valued information tables is introduced, which is an extension of decision logic studied by Pawlak. Each object in an interval-set-valued information table takes an interval set of values. Consequently, two types of satisfiabilities of a formula are introduced. Truth values of formulas are defined to be interval-valued, instead of single-valued. A semantics model of the proposed logic language is studied.

1 Introduction

The theory of rough sets is commonly developed and interpreted through the use of information tables, in which a finite set of objects are described by a finite number of attributes [10, 11]. A decision logic, called DL-language by Pawlak [11], has been studied by many authors for reasoning about knowledge represented by information tables [8, 11]. It is essentially formulated based on the classical two-valued logic. The semantics of the DL-language is defined in Tarski's style through the notions of a model and satisfiability in the context of information tables. A strong assumption is made about information tables, i.e., each object takes exactly one value with respect to an attribute. In some situations, this assumption may be too restrictive to be applicable in practice. Several proposals have been suggested using much weaker assumptions. More specifically, the notion of set-based information tables (also known as incomplete or nondeterministic information tables) has been introduced and studied, in which an object can take a subset of values for each attribute [3, 14, 16, 20]. Based on the results from those studies, the main objective of this paper is to introduce the notion of interval-set-valued information tables by incorporating results from studies of interval-set algebra [17, 19]. A generalized decision logic GDL is proposed, which is similar to modal logic, but has a different semantics interpretation.

* This study is partially supported by the National Natural Science Foundation of China and Natural Science Foundation of JiangXi Province, China.

This paper reports some of our preliminary results. In Section 2, we first briefly review Pawlak's decision logic DL, and then introduce the notions of α-degree truth and α-level truth. In Section 3, the notion of interval-set-valued information tables is introduced. A generalized decision logic DGL is proposed and interpreted based on two types of satisfiabilities. The concepts of interval-degree truth and interval-level truth are proposed and studied. Inference rules are discussed. In Section 4, two related studies are commented.

2 A Decision Logic in Information Tables

The notion of an information table, studied by many authors [3, 10, 11, 16, 21], is formally defined by a quadruple:

$$S = (U, At, \{V_a \mid a \in At\}, \{I_a \mid a \in At\}),$$

where

U is a finite nonempty set of objects,

At is a finite nonempty set of attributes,

V_a is a nonempty set of values for $a \in At$,

$I_a : U \longrightarrow V_a$ is an information function.

Each information function I_a is a total function that maps an object of U to exactly one value in V_a. Similar representation schemes can be found in many fields, such as decision theory, pattern recognition, machine learning, data analysis, data mining, and cluster analysis [11].

With an information table, a decision logic language (DL-language) can be introduced [11]. In the DL-language, an atomic formula is given by (a, v), where $a \in At$ and $v \in V_a$. If ϕ and ψ are formulas in the DL-language, then so are $\neg\phi$, $\phi \wedge \psi$, $\phi \vee \psi$, $\phi \rightarrow \psi$, and $\phi \equiv \psi$. The semantics of the DL-language can be defined in Tarski's style through the notions of a model and satisfiability. The model is an information table S, which provides interpretation for symbols and formulas of the DL-language. The satisfiability of a formula ϕ by an object x, written $x \models_S \phi$ or in short $x \models \phi$ if S is understood, is given by the following conditions:

(a1). $x \models (a, v)$ iff $I_a(x) = v$,

(a2). $x \models \neg\phi$ iff not $x \models \phi$,

(a3). $x \models \phi \wedge \psi$ iff $x \models \phi$ and $x \models \psi$,

(a4). $x \models \phi \vee \psi$ iff $x \models \phi$ or $x \models \psi$,

(a5). $x \models \phi \rightarrow \psi$ iff $x \models \neg\phi \vee \psi$,

(a6). $x \models \phi \equiv \psi$ iff $x \models \phi \rightarrow \psi$ and $x \models \psi \rightarrow \phi$.

For a formula ϕ, the set $m_S(\phi)$ defined by:

$$m_S(\phi) = \{x \in U \mid x \models \phi\}, \tag{1}$$

is called the meaning of the formula ϕ in S. If S is understood, we simply write $m(\phi)$. Obviously, the following properties hold [8, 11]:

$$(b1). \quad m(a, v) = \{x \in U \mid I_a(x) = v\},$$
$$(b2). \quad m(\neg \phi) = -m(\phi),$$
$$(b3). \quad m(\phi \wedge \psi) = m(\phi) \cap m(\psi),$$
$$(b4). \quad m(\phi \vee \psi) = m(\phi) \cup m(\psi),$$
$$(b5). \quad m(\phi \rightarrow \psi) = -m(\phi) \cup m(\psi),$$
$$(b6). \quad m(\phi \equiv \psi) = (m(\phi) \cap m(\psi)) \cup (-m(\phi) \cap -m(\psi)).$$

The meaning of a formula ϕ is therefore the set of all objects having the property expressed by the formula ϕ. In other words, ϕ can be viewed as the description of the set of objects $m(\phi)$. Thus, a connection between formulas of the DL-language and subsets of U is established.

A formula ϕ is said to be true in an information table S, written $\models_S \phi$ or $\models \phi$ for short when S is clear from the context, if and only if $m(\phi) = U$. That is, ϕ is satisfied by all objects in the universe. Two formulas ϕ and ψ are equivalent in S if and only if $m(\phi) = m(\psi)$. By definition, the following properties hold [11]:

$$(c1). \quad \models \phi \text{ iff } m(\phi) = U,$$
$$(c2). \quad \models \neg\phi \text{ iff } m(\phi) = \emptyset,$$
$$(c3). \quad \models \phi \rightarrow \psi \text{ iff } m(\phi) \subseteq m(\psi),$$
$$(c4). \quad \models \phi \equiv \psi \text{ iff } m(\phi) = m(\psi).$$

Thus, we can study the relationships between concepts described by formulas of the DL-language based on the relationships between their corresponding sets of objects.

The previous interpretation of DL-language is essentially based on classical two-valued logic. One may generalize it to a many-valued logic by introducing the notion of degrees of truth [4, 5]. For a formula ϕ, its truth value is defined by [4, 5]:

$$v(\phi) = \frac{|m(\phi)|}{|U|}, \tag{2}$$

where $|\cdot|$ denotes the cardinality of a set. This definition of truth value is probabilistic in natural. Thus, the generalized logic is in fact a probabilistic logic [7]. When $v(\phi) = \alpha \in [0, 1]$, we say that the formula ϕ is α-degree true. By definition, we immediately have the properties:

$$(d1). \quad \models \phi \text{ iff } v(\phi) = 1,$$
$$(d2). \quad \models \neg\phi \text{ iff } v(\phi) = 0,$$
$$(d3). \quad v(\neg\phi) = 1 - v(\phi),$$
$$(d4). \quad v(\phi \wedge \psi) \leq \min(v(\phi), v(\psi)),$$
$$(d5). \quad v(\phi \vee \psi) \geq \max(v(\phi), v(\psi)),$$
$$(d6). \quad v(\phi \vee \psi) = v(\phi) + v(\psi) - v(\phi \wedge \psi).$$

Properties (d3)-(d6) follow from the probabilistic interpretation of truth value. Similar to the definitions of α-cuts in the theory of fuzzy sets [2], we define α-level truth. For $\alpha \in [0, 1]$, a formula ϕ is said to be α-level true, written $\models_\alpha \phi$, if $v(\phi) \geq \alpha$, and ϕ is strong α-level true, written $\models_{\alpha+}$, if $v(\phi) > \alpha$. From (d1)-(d6), for $0 \leq \alpha \leq \beta \leq 1$ and $\gamma \in [0, 1]$ we have:

(e1). $\models_0 \phi$,

(e2).   If $\models_\beta \phi$, then $\models_\alpha \phi$,

(e3). $\models_\alpha \neg\phi$ iff not $\models_{(1-\alpha)+} \phi$,

(e4).   If $\models_\alpha \phi \wedge \psi$, then $\models_\alpha \phi$ and $\models_\alpha \psi$,

(e5).   If $\models_\alpha \phi$, then $\models_\alpha \phi \vee \psi$,

(e6).   If $\models_\alpha \phi$ and $\models_\gamma \psi$, then $\models_{\max(\alpha,\gamma)} \phi \vee \psi$.

Property (e5) is implied by properties (e2) and (e6).

With the concept of α-level truth, we have the probabilistic *modus ponens* rule [15]:

$$\frac{\models_\alpha \phi \to \psi}{\models_\beta \phi} \qquad \frac{v(\phi \to \psi) \geq \alpha}{v(\phi) \geq \beta}$$
$$\overline{\models_{\max(0,\alpha+\beta-1)} \psi} \qquad \overline{v(\psi) \geq \max(0, \alpha + \beta - 1)}.$$

Given conditions $v(\phi \to \psi) \geq \alpha$ and $v(\phi) \geq \beta$, from properties (d3) and (d6), we have:

$$v(\phi \to \psi) \geq \alpha$$
$$\implies v(\neg\phi \vee \psi) \geq \alpha$$
$$\implies v(\neg\phi) + v(\psi) - v(\neg\phi \wedge \psi) \geq \alpha$$
$$\implies (1 - v(\phi)) + v(\psi) \geq \alpha$$
$$\implies v(\psi) \geq \alpha + v(\phi) - 1$$
$$\implies v(\psi) \geq \alpha + \beta - 1.$$

Since the value $v(\psi)$ must be non-negative, we can conclude that the proposed *modus ponens* rule is correct. Similar properties and rules can be expressed in terms of strong α-level truth.

3   A Generalized Decision Logic

Let $\mathcal{X}$ be a finite set and $2^{\mathcal{X}}$ be its power set. A subset of $2^{\mathcal{X}}$ of the form:

$$\mathcal{A} = [A_1, A_2] = \{X \in 2^{\mathcal{X}} \mid A_1 \subseteq X \subseteq A_2\} \tag{3}$$

is called a closed interval set, where it is assumed $A_1 \subseteq A_2$. The set of all closed interval sets is denoted by $I(\mathcal{X})$. Degenerate interval sets of the form $[A, A]$ are equivalent to ordinary sets. Thus, interval sets may be considered as an extension of standard sets. In fact, interval-set algebra may be considered as a set-theoretic counterpart of interval-number algebra [6]. A detailed study of interval-set algebra can be found in papers by Yao [17, 19].

An interval-set-valued information table generalizes a standard information table by allowing each object to take interval sets as its values. Formally, this can be described by information functions:

$$I_a : U \longrightarrow I(V_a). \tag{4}$$

For an object $x \in U$, its value on an attribute $a \in At$ is an interval set $I_a(x) = [I_{a*}(x), I_a^*(x)]$. The object x *definitely* has properties in $I_{a*}(x)$, and *possibly* has properties in $I_a^*(x)$. With the introduction of interval-set-valued information tables, a generalized decision logic language, called *GDL*-language, can be established. The symbols and formulas of the *GDL*-language is the same as that of the *DL*-language. The semantics of the *GDL*-language can be defined similarly in Tarski's style using the notions of a model and two types of satisfiabilities, one for necessity and the other for possibility. If an object x *necessarily* satisfies formula ϕ, we write $x \models_* \phi$, and if x *possibly* satisfies ϕ, we write $x \models^* \phi$. The semantics of $\models_*$ and $\models^*$ are defined as follows:

(f1). $x \models_* (a, v)$ iff $v \in I_{a*}(x)$,

   $x \models^* (a, v)$ iff $v \in I_a^*(x)$,

(f2). $x \models_* \neg\phi$ iff not $x \models^* \phi$,

   $x \models^* \neg\phi$ iff not $x \models_* \phi$,

(f3). $x \models_* \phi \wedge \psi$ iff $x \models_* \phi$ and $x \models_* \psi$,

   $x \models^* \phi \wedge \psi$ iff $x \models^* \phi$ and $x \models^* \psi$,

(f4). $x \models_* \phi \vee \psi$ iff $x \models_* \phi$ or $x \models_* \psi$,

   $x \models^* \phi \vee \psi$ iff $x \models^* \phi$ or $x \models^* \psi$,

(f5). $x \models_* \phi \rightarrow \psi$ iff $x \models_* \neg\phi \vee \psi$,

   $x \models^* \phi \rightarrow \psi$ iff $x \models^* \neg\phi \vee \psi$,

(f6). $x \models_* \phi \equiv \psi$ iff $x \models_* \phi \rightarrow \psi$ and $x \models_* \psi \rightarrow \phi$,

   $x \models^* \phi \equiv \psi$ iff $x \models^* \phi \rightarrow \psi$ and $x \models^* \psi \rightarrow \phi$,

The following property follows immediately from definition:

(g1).   If $x \models_* \phi$, then $x \models^* \phi$.

Although the introduced notions of necessity and possibility are similar in nature to the notions in modal logic [1], our semantics interpretation is different. There is a close connection between the above formulation and three-valued logic [19].

In *GDL*, with respect to an interval-set-valued information system S, the meaning of a formula ϕ is the interval set $m(\phi)$ defined by:

$$m(\phi) = [\{x \in U \mid x \models_* \phi\}, \{x \in U \mid x \models^* \phi\}] = [m_*(\phi), m^*(\phi)]. \tag{5}$$

It can be verified that the following properties hold:

(h1). $m(a, v) = [\{x \in U \mid x \models_* \phi\}, \{x \in U \mid x \models^* \phi\}]$,

(h2). $m(\neg\phi) = \backslash m(\phi)$,

(h3). $m(\phi \wedge \psi) = m(\phi) \sqcap m(\psi)$,

(h4). $m(\phi \vee \psi) = m(\phi) \sqcup m(\psi)$,

(h5). $m(\phi \rightarrow \psi) = \backslash m(\phi) \sqcup m(\psi)$,

(h6). $m(\phi \equiv \psi) = (\backslash m(\phi) \sqcup m(\psi)) \sqcap (m(\phi) \sqcup \backslash m(\psi))$,

where $\backslash$, $\sqcap$, and $\sqcup$ are the interval-set complement, intersection, and union given by [17]: for two interval sets $\mathcal{A} = [A_1, A_2]$ and $\mathcal{B} = [B_1, B_2]$,

$$\backslash \mathcal{A} = \{-X \mid X \in \mathcal{A}\} = [-A_2, -A_1],$$
$$\mathcal{A} \sqcap \mathcal{B} = \{X \cap Y \mid X \in \mathcal{A}, Y \in \mathcal{B}\} = [A_1 \cap B_1, A_2 \cap B_2],$$
$$\mathcal{A} \sqcup \mathcal{B} = \{X \cup Y \mid X \in \mathcal{A}, Y \in \mathcal{B}\} = [A_1 \cup B_1, A_2 \cup B_2]. \qquad (6)$$

The meaning of a formula ϕ is therefore the interval set of objects, representing those that definitely have the properties expressed by the formula ϕ, and those that possibly have the properties.

Given the meaning of formulas in terms of interval sets, we define the interval-valued truth for a formula ϕ by extending equation (2):

$$v(\phi) = \left[\frac{|m_*(\phi)|}{|U|}, \frac{|m^*(\phi)|}{|U|} \right] = [v_*(\phi), v^*(\phi)]. \qquad (7)$$

Both lower and upper bounds of $[v_*(\phi), v^*(\phi)]$ have probabilistic interpretation, hence we have a probability related interval-valued logic [18]. Properties corresponding to (d3)-(d6) are given by:

(i1). $v_*(\neg\phi) = 1 - v^*(\phi)$,
$\quad v^*(\neg\phi) = 1 - v_*(\phi)$,

(i2). $v_*(\phi \wedge \psi) \leq \min(v_*(\phi), v_*(\psi))$,
$\quad v^*(\phi \wedge \psi) \leq \min(v^*(\phi), v^*(\psi))$,

(i3). $v_*(\phi \vee \psi) \geq \max(v_*(\phi), v_*(\psi))$,
$\quad v^*(\phi \vee \psi) \geq \max(v^*(\phi), v^*(\psi))$,

(i4). $v_*(\phi \vee \psi) = v_*(\phi) + v_*(\psi) - v_*(\phi \wedge \psi)$,
$\quad v^*(\phi \vee \psi) = v^*(\phi) + v^*(\psi) - v^*(\phi \wedge \psi)$.

The formula ϕ is said to be $[v_*(\phi), v^*(\phi)]$-degree true. For a sub-interval $[\alpha_*, \alpha^*]$ of the unit interval $[0, 1]$, a formula ϕ is $[\alpha_*, \alpha^*]$-level true, written $\models_{[\alpha_*, \alpha^*]} \phi$, if $\alpha_* \leq v_*(\phi) \leq v^*(\phi) \leq \alpha^*$, and ϕ is strong $[\alpha_*, \alpha^*]$-level true, written $\models_{[\alpha_*, \alpha^*]+} \phi$, if $\alpha_* < v_*(\phi) \leq v^*(\phi) < \alpha^*$. For sub-intervals $[\alpha_*, \alpha^*] \subseteq [\beta_*, \beta^*] \subseteq [0,1]$ and $[\gamma_*, \gamma^*] \subseteq [0,1]$, the following properties hold:

(j1). $\models_{[0,1]} \phi$,

(j2). If $\models_{[\alpha_*, \alpha^*]} \phi$, then $\models_{[\beta_*, \beta^*]} \phi$,

(j3). $\models_{[\alpha_*, \alpha^*]} \neg\phi$ iff not $\models_{[1-\alpha^*, 1-\alpha_*]+} \phi$,

(j4). If $\models_{[\alpha_*, \alpha^*]} \phi \wedge \psi$, then $\models_{[\alpha_*, 1]} \phi$ and $\models_{[\alpha_*, 1]} \psi$,

(j5). If $\models_{[\alpha_*, \alpha^*]} \phi$, then $\models_{[\alpha_*, 1]} \phi \vee \psi$,

(j6). If $\models_{[\alpha_*, \alpha^*]} \phi$ and $\models_{[\gamma_*, \gamma^*]} \psi$, then $\models_{[\max(\alpha_*, \gamma_*), 1]} \phi \vee \psi$,

(j7). If $\models_{[\alpha_*, \alpha^*]} \phi \vee \psi$, then $\models_{[0, \alpha^*]} \phi$ and $\models_{[0, \alpha^*]} \psi$,

(j8). If $\models_{[\alpha_*, \alpha^*]} \phi$, then $\models_{[0, \alpha^*]} \phi \wedge \psi$,

(j9). If $\models_{[\alpha_*, \alpha^*]} \phi$ and $\models_{[\gamma_*, \gamma^*]} \psi$, then $\models_{[0, \min(\alpha^*, \gamma^*)]} \phi \wedge \psi$.

They follow from (i2) and (i3). In fact, properties (j4)-(j6) are the properties (e4)-(e6) of the DL-language. Properties (j4)-(j6) show the characteristics of the lower bound, while (j7)-(j9) state the characteristics of the upper bound.

The generalized interval-based *modus ponens* rule is given by:

$$\frac{\models_{[\alpha_*, \alpha^*]} \phi \to \psi}{\models_{[\beta_*, \beta^*]} \phi} \qquad \frac{\alpha_* \leq v_*(\phi \to \psi) \leq v^*(\phi \to \psi) \leq \alpha^*}{\beta_* \leq v_*(\phi) \leq v^*(\phi) \leq \beta^*}$$
$$\frac{}{\models_{[\max(0,\alpha_*+\beta_*-1),\alpha^*]} \psi} \qquad \frac{}{\max(0,\alpha_*+\beta_*-1) \leq v_*(\psi) \leq v^*(\psi) \leq \alpha^*} .$$

The part concerning the lower bound is in fact the probabilistic modus ponens rule introduced in Section 2. The upper bound can be seen as follows. From $v^*(\phi \to \psi) \leq \alpha^*$ and (i3), we can conclude that:

$$v^*(\psi) \leq v^*(\neg\phi \vee \psi) = v^*(\phi \to \psi) \leq \alpha^*.$$

Thus, the interval-based *modus ponens* rule is correct. Finally, it should be pointed out that the logic of Section 2 is a special case of interval-valued logic. More specifically, α-level truth can be translated into the $[\alpha, 1]$-level truth.

4 Comments on Related Studies

An interval-valued logic can also be introduced in the standard information tables through the use of lower and upper approximations of the rough set theory [5, 9]. For each subset of the attributes, one can define an equivalence relation on the set of objects in an information table. An arbitrary set is approximated by equivalence classes as follows: the lower approximation is the union of those equivalence classes that are included in the set, while the upper approximation is the union of those equivalence classes that have an nonempty intersection with the set. Thus, for a formula ϕ with interpretation $m(\phi)$, we have a pair of lower and upper approximations $\underline{apr}(m(\phi))$ and $\overline{apr}(m(\phi))$. An interval-valued truth can be defined as:

$$v(\phi) = \left[\frac{|\underline{apr}(m(\phi))|}{|U|}, \frac{|\overline{apr}(m(\phi))|}{|U|} \right] = [v_*(\phi), v^*(\phi)]. \tag{8}$$

Based on this interpretation of interval-valued truth, Parsons *et al.* [9] introduced a logic system RL for rough reasoning. Their inference rules are related to, but different from, the inference rules introduced in this paper. A problem with RL is that the interpretation of the rough measure is not entirely clear. The measure is not fully consistent with the definition of truth value given by equation (8). It may be interesting to have an in-depth investigation of the interval-valued logic based on equation (8). An important feature of such a logic is its non-truth-functional logic connectives. This makes it different from the interval set algebra related systems GDL and RL.

In a recent paper, Pawlak [12] introduced the notion of *rough modus ponens* in information tables. The logical formula $\phi \to \psi$ is interpreted as a decision rule. A certainty factor is associated with $\phi \to \psi$ as follows:

$$\mu_S(\phi, \psi) = \frac{|m(\phi) \cap m(\psi)|}{|m(\phi)|}. \tag{9}$$

It can in fact be interpreted as a conditional probability. The rough modus ponens rule is given by:

$$\frac{\phi \to \psi : \mu_S(\phi, \psi)}{\psi : v(\neg\phi \wedge \psi) + v(\phi)\mu_S(\phi, \psi)}.$$

This rule is closely related to Bayes' theorem [13]. One may easily generalize the rough modus ponens if α-level truth values are used. The main difference between two modus ponens rules stems from the distinct interpretations of the logical formula $\phi \to \psi$.

5 Conclusion

Two generalizations of Pawlak's information table based decision logic DL are introduced and examined. One generalization is based on the notion of degree of truth, which extend DL from two-valued logic to many-valued logic. The other generalization relies on interval-set-based information tables. In this case, two types of satisfiabilities are used, in a similar spirit of modal logic. They lead to interval-set interpretation of formulas. Consequently, interval-degree truth and interval-level truth are introduced as a generalization of single-valued degree of truth. The truth values of formulas are associated with probabilistic interpretations. The derived logic systems are essentially related to probabilistic reasoning. In particularly, probabilistic *modus ponens* rules are studied.

In this paper, we only presented the basic formulation and interpretation of the generalized decision logic. As pointed out by an anonymous referee of the paper, a formal proving system is needed and applications need to be explored. It may also be intersting to analyze other non-probabilistic interpretations of truth values.

References

1. Chellas, B.F., *Modal Logic: An Introduction*, Cambridge University Press, Cambridge, 1980.
2. Klir, G.J. and Yuan, B., *Fuzzy Sets and Fuzzy Logic, Theory and Applications*, Prentice Hall, New Jersey, 1995.
3. Lipski, W. Jr. On databases with incomplete information, *Journal of the ACM*, **28**, 41-70, 1981.
4. Liu, Q. The OI-resolutions of operator rough logic, LNAI 1424, LNAI 1424, Springer-Verlag, Berlin, 432-436, 1998.
5. Liu, Q. and Wang, Q. Rough number based on rough sets and logic values of λ operator (in Chinese), *Journal of Software*, **7**, 455-461, 1996.
6. Moore, R.E. *Interval Analysis*, Prentice-Hall, New Jersey, 1966.
7. Nilsson, N.J. Probabilistic logic, *Artificial Intelligence*, **28**, 71-87, 1986.
8. Orlowska, E. Reasoning about vague concepts, *Bulletin of Polish Academy of Science, Mathematics*, **35**, 643-652, 1987.
9. Parsons, S., Kubat, M. and Dohnal, M. A rough set approach to reasoning under uncertainty, *Journal of Experimental and Theoretical Artificial Intelligence*, **7**, 175-193, 1995.
10. Pawlak, Z. Information systems – theoretical foundations, *Information Systems*, **6**, 205-218, 1981.
11. Pawlak, Z. *Rough Sets, Theoretical Aspects of Reasoning about Data*, Klumer Academic Publishers, Boston, 1991.
12. Pawlak, Z. Rough modus ponens, *Proceedings the 7th International Conference on Information Processing and Management of Uncertainty in Knowledge-Based Systems*, 1162-1166, 1998.
13. Pawlak, Z. Data mining – a rough set perspective, LNAI 1574, Springer-Verlag, Berlin, 3-12, 1999.
14. Pomykala, J.A. On definability in the nondeterministic information system, *Bulletin of the Polish Academy of Sciences, Mathematics*, **36**, 193-210, 1988.
15. Prade, H. A computational approach to approximate and plausible reasoning with applications to expert systems, *IEEE Transactions on Pattern Analysis and Machine Intelligence*, **PAMI-7**, 260-283, 1985.
16. Vakarelov, D. A modal logic for similarity relations in Pawlak knowledge representation systems, *Fundamenta Informaticae*, **XV**, 61-79, 1991.
17. Yao, Y.Y. Interval-set algebra for qualitative knowledge representation, *Proceedings of the Fifth International Conference on Computing and Information*, 370-374, 1993.
18. Yao, Y.Y. A comparison of two interval-valued probabilistic reasoning methods, *Proceedings of the 6th International Conference on Computing and Information*, special issue of *Journal of Computing and Information*, 1, 1090-1105 (paper number D6), 1995.
19. Yao, Y.Y. and Li, X. Comparison of rough-set and interval-set models for uncertain reasoning, *Fundamenta Informaticae*, **27**, 1996.
20. Yao, Y.Y. and Noroozi, N. A unified framework for set-based computations, *Soft Computing: the Proceedings of the 3rd International Workshop on Rough Sets and Soft Computing*, 252-255, 1995.
21. Yao, Y.Y. and Zhong, N. *Granular Computing Using Information Tables*, manuscript, 1999.

Many-Valued Dynamic Logic for Qualitative Decision Theory

Churn-Jung Liau

Institute of Information Science
Academia Sinica, Taipei, Taiwan
liaucj@iis.sinica.edu.tw

Abstract. This paper presents an integration of the dynamic logic semantics and rational decision theory. Logics for reasoning about the expected utilities of actions are proposed. The well-formed formulas of the logics are viewed as the possible goals to be achieved by the decision maker and the truth values of the formulas are considered as the utilities of the goals. Thus the logics are many-valued dynamic logics. Based on different interpretations of acts in the logics, we can model different decision theory paradigms, such as possibilistic decision theory and case-based decision theory.

1 Introduction

Rational decision theory is a very important research topic in many academic fields such as economics, politics, and philosophy. Recently, it has also received more and more attention of the AI community due to the development of intelligent agent systems. The basic execution loop of an intelligent agent consists of three phases: perception, deliberation, and action. In the perception phase, the agent senses the status of the environment and receives information from other agents. Then, in the deliberation phase, the agent reasons with the observed and received information and plans its actions for achieving its goals. Finally, in the action phase, the plan is really executed. The capabilities of both reasoning about actions and decision making are crucial to the success of the deliberation phase since it has to know the possible effects of actions and select the appropriate actions for achieving its goals.

A variety of formalisms for reasoning about actions have been developed in AI, theoretical computer science, and philosophical logic. Among them, dynamic logic is originally proposed for reasoning about program behavior[10], and subsequently adopted for reasoning about actions by the AI community. Though the advantages of using dynamic logic for reasoning about actions have been emphasized in [6], the traditional dynamic logic has only limited capability in handling uncertainty.

In dynamic logic, a formula $[\alpha]\varphi$ denotes that φ holds after the execution of (possibly compound) action α, so in principle, if the agent's goal is φ and $[\alpha]\varphi$ can be derived from the description of the initial situation, then α is a feasible plan for achieving the goal. This gives rise to a decision theoretic reading of dynamic logic semantics. Since nondeterministic actions are allowed in dynamic logic, it may capture the agent's ignorance on the possible effects of actions to some extent. However, uncertainty pervades the whole deliberation phase, so further extensions of dynamic logic for handling different forms of uncertainty are needed. In general, there are three forms of uncertainty in the deliberation process.

- The perception of the agent may be imperfect and its received information may be incomplete and faulty, so its knowledge in the status of the environment is uncertain. Sometimes, the probabilistic instead of the exact knowledge is available. However, in other times, even probabilistic knowledge is not available, so more general consideration is also needed.
- In multi-agent systems, our agent may not be the only one which can cause the change of the world, so it has only partial knowledge about the possible effects of the actions. The classical dynamic logic may handle the case when the knowledge is *imprecise* (i.e. the effect of an action is represented as a set of states). However, the knowledge may be also probabilistic or possibilistic (i.e. the effect of an action are represented as a probability or possibility distribution on the set of states). We recall that a possibility distribution on a set X is a mapping $\pi : X \to [0,1]$ such that $\pi(x)$ measures the extent to which x is likely to be the actual consequence[14]. The dynamic logic should be extended to cover such cases.
- Since dynamic logic is two-valued, the goal for an agent to achieve must be crisp and non-flexible. A goal is either satisfied or non-satisfied. However, sometimes, we may want to describe more flexible goals. A goal may be satisfied to some degree. In decision theory, this is in general described by a real-valued utility function. Recently, more general notions of ordinal preference are considered[1, 13]. To represent the flexible goals in dynamic logic, we will generalize its semantics to a many-valued one.

On the other hand, in the decision theoretic contexts, far richer notions of uncertainty have been explored. Besides classical decision theory, in which the notions of probability and expected utility are of central importance, some alternatives, such as possibilistic decision theory[1], case-based decision theory[7], and belief function-based decision theory[11] have been proposed and axiomatically justified in different settings recently. The main concern of decision theory is to choose an action which will maximize the expected utility of performing the action given some knowledge on the effects of the action and the desirability of these effects. In the extreme case that the utility function is two-valued and the available knowledge is imprecise, this is just a rephrasing of the decision theoretic interpretation of dynamic logic. The only difference is that in general the set of available acts in decision theory is not algebraically structured as in dynamic logic. Thus, due to the usefulness of dynamic logic in reasoning about actions and the rich notions of uncertainty in decision theories, the combination of decision theory and dynamic logic semantics will have the advantages of cross-fertilization.

In this paper, we suggest a kind of integration between decision theoretic notions and dynamic logic semantics. The dynamic logic semantics is enhanced to a many-valued one, so the truth value of a formula in a state plays a two-fold role. One is the degree of satisfaction of the formula and the other is the utility of the state. Since each formula corresponds to a goal and we can consider more than one formulas in the same time, this means that we can easily describe the multiple objective decision-making in the logic. In classical dynamic logic, each action is interpreted as a binary transition relation on the set of states. Here, depending on the different uncertainty handling formalisms, we can generalize it to a fuzzy relation, a set of probability distributions, or a set of possibility distributions generated from a similarity relation. Thus, we will try to develop several many-valued dynamic logics for reasoning with different uncertainty formalisms.

In what follows, we will first review the basic notions from classical decision theory and some recent proposals of qualitative alternatives. Then the dynamic logics for possibilistic decision theory and case-based decision theory are considered respectively. Finally, conclusion is given and some possible research directions are suggested.

2 Review of Some Decision Theories

Classical quantitative decision theory considers expected utility maximization (EUM) as the criteria of rational choice. In the theory, a decision framework is a 4-tuple (D, X, μ, u), where D is a set of available decision acts, X is a set of possible outcomes, $\mu : D \times X \to [0, 1]$ assigns to each decision act $d \in D$ a probability distribution $\mu(d, \cdot)$ on X, and $u : X \to \Re$ is the utility function. Then the expected utility of a decision d is defined as

$$U(d) = \Sigma_{x \in X} \mu(d, x) \cdot u(x),$$

and the decision maker will choose d_0 such that $U(d_0) = \max_{d \in D} U(d)$.

While the computation of $E(d)$ relies on the arithmetic operations (mainly $+$ and $\cdot$) on real numbers, qualitative decision theory concentrates more on the decision maker's ordinal preference and uncertainty about the possible outcomes. Recently, a qualitative decision theory(PODT) based on possibilistic logic is proposed[1]. In the theory, a possibilistic decision framework is a 4-tuple (D, X, π, u), where D and X are defined as above, $\pi : D \times X \to T_1$ assigns to each decision act $d \in D$ a possibility distribution $\pi(d, \cdot) : X \to T_1$, and $u : X \to T_2$ is the utility assignment function. Here, T_1 and T_2 are linearly ordered scales, and under the commensurability assumption, we can assume $T_1 = T_2 = T$ without loss of generality. Typical examples of T are [0,1] or a subset of [0,1]. Let $n : T \to T$ be an order-reversing map on T, then two qualitative expected utilities for a decision d can be defined. For the risk-averse decision maker, the pessimistic expected utility is

$$U_*(d) = \min_{x \in X} \max(n(\pi(d, x)), u(x)),$$

and for a risk-prone decision maker, the optimistic expected utility is

$$U^*(d) = \max_{x \in X} \min(\pi(d,x), u(x)).$$

The PODT is particularly suitable for complicate situations in which complete probabilistic information is rarely available.

Another recent proposed alternative decision theory is the case-based one. According to [7], the purpose of case-based decision theory(CBDT) is to model decision making under uncertainty by formalizing reasoning by analogies. It suggests that decision makers tend to choose actions which performed well in similar past cases. Each case is viewed as a triple of a situation (i.e. a decision problem), the action chosen in it, and the consequence of performing the action. Thus, a case-based decision framework is a 6-tuple (P, D, X, C, s, u), where D, X, and u are defined as above, P is a set of situations, $C \subseteq P \times D \times X$ is a finite set of cases, called the memory of the decision maker, and $s : P \times P \to [0,1]$ is a similarity function which measures the similarity between situations. Then, for a given situation p, the expected utility for a decision d is defined as

$$U_c(d) = \sum_{(q,d,x) \in C} s(p,q)u(x).$$

However, it is also pointed out that U_c is cumulative in nature, so the number of times a certain act was chosen in the past will affect perceived desirability. Thus, an act that was chosen repeatedly producing bad results may be considered superior to an act that was chosen only once but producing good result. To overcome the difficulty, the average utility is considered in [7], namely, in the above equation, the similarity function s is replaced by s' which is defined as

$$s'(p,q) = \begin{cases} s(p,q) / \sum_{(q',d,x) \in C} s(p,q') & \text{if well--defined} \\ 0 & \text{otherwise} \end{cases}$$

In [2], a more qualitative version of expected utility is considered which can also eliminate the cumulation assume the utility values are normalized to the range $[0,1]$. The definition is analogous to the pessimistic and optimistic expected utility in the PODT.

$$U_{c*}(d) = \min_{(q,d,x) \in C} \max(n(s(p,q)), u(x)),$$

$$U_c^*(d) = \max_{(q,d,x) \in C} \min(s(p,q), u(x)).$$

While in PODT, it is observed that the criterion of maximizing $U^*(d)$ is sometimes over-optimistic[5], it seems that for CBDT, the pessimistic utility has some counterintuitive results. For example, if an act a was only adopted in the past for the cases that are completely different with the present situation, i.e., for all $(q, a, x) \in C$, $s(p.q) = 0$, then we will have $U_{c*}(a) = 1$ which is the maximum value. This phenomenon is due to the fact that the case memory is only a partial description of the world, so we may encounter a novel situation in which no past experience can be followed. In CBDT, we tend to find the most similar past case and apply its solution to the new situation, so the optimistic criterion seems more suitable.

3 Possibilistic Dynamic Logic

The possibilistic dynamic logic (*PoDL*) provides the integrated treatment of possibilistic decision theory and dynamic logic. The syntax and semantics of *PoDL* is an extension of that for dynamic logic[10] and fuzzy modal logic *SLMV* in [8], which in turn bases on rational Pavelka logic proposed in [9].

3.1 Syntax

The alphabet of *PoDL* consists of

1. A set of propositional letters, $PV = \{p, q, \ldots\}$,
2. a set of atomic actions, $A = \{a, b, c, \ldots\}$,
3. the set of truth constants r for each rational $r \in [0, 1]$, and
4. the logical symbols $\sim$, $\rightarrow$, $\wedge$, $\vee$, [,], ;, *, $\cup$, and ?.

The set of well-formed formulas(Φ) and the set of action expressions(Ξ) are defined inductively in the following way.

1. Φ is the smallest set such that
   - $PV \subseteq \Phi$ and $r \in \Phi$ for all rational $r \in [0, 1]$, and
   - if $\varphi, \psi \in \Phi$ and $\alpha \in \Xi$, then $\sim \varphi, \varphi \rightarrow \psi, \varphi \wedge \psi, \varphi \vee \psi, [\alpha]\varphi \in \Phi$.
2. Ξ is the smallest set such that
   - $A \subseteq \Xi$, and
   - if $\alpha, \beta \in \Xi$ and $\varphi \in \Phi$, then $\alpha; \beta, \alpha \cup \beta, \alpha^*, \varphi? \in \Pi$.

Some abbreviations of *PoDL* include $\neg\varphi = \varphi \rightarrow 0$, $\varphi \otimes \psi = \neg(\varphi \rightarrow \neg\psi)$, $\varphi \oplus \psi = \neg\varphi \rightarrow \psi$, and $\langle\alpha\rangle\varphi = \neg[\alpha]\neg\varphi$.

3.2 Semantics

The semantics of *PoDL* is defined relative to a given Kripke structure $M = (W, \tau, |\cdot|, w_0)$, where W is a set of possible worlds, $\tau : W \times PV \rightarrow [0, 1]$ is the truth valuation function, $|\cdot| : A \rightarrow (W \times W \rightarrow [0, 1])$ is the action denotation function, and $w_0 \in W$ is a designated world. The mappings τ and $|\cdot|$ is extended to Φ and Ξ as follows.

1. $\tau(w, r) = r$,
2. $\tau(w, \sim \varphi) = \begin{cases} 0 \text{ if } \tau(w, \varphi) = 1 \\ 1 \text{ otherwise} \end{cases}$
3. $\tau(w, \varphi \rightarrow \psi) = I(\tau(w, \varphi), \tau(w, \psi))$,
4. $\tau(w, \varphi \wedge \psi) = \min(\tau(w, \varphi), \tau(w, \psi))$,

5. $\tau(w, \varphi \vee \psi) = \max(\tau(w, \varphi), \tau(w, \psi))$,
6. $\tau(w, [\alpha]\varphi) = \inf_{x \in W} \max(1 - \|\alpha\|(w, x), \tau(x, \varphi))$,
7. $\|\alpha; \beta\| = \|\alpha\| \circ \|\beta\|$, i.e,, composition of two fuzzy relations $\|\alpha\|$ and $\|\beta\|$,
8. $\|\alpha \cup \beta\| = \|\alpha\| \cup \|\beta\|$,
9. $\|\alpha^*\| = \|\alpha\|^*$, i.e., the reflexive and transitive closure of $\|\alpha\|$,
10. $\|\varphi?\|(w, x) = \begin{cases} \tau(w, \varphi) & \text{if } w = x \\ 0 & \text{otherwise} \end{cases}$

where $I : [0, 1] \times [0, 1] \to [0, 1]$ is an implication function. Typical implication functions include material implication $I(x, y) = \max(1 - x, y)$, Łukasiewicz's implication $I(x, y) = \min(1, 1 - x + y)$, and Gödel implication $I(x, y) = 1$ if $x \leq y$ and $= y$ if $x > y$. Here we will let I denote the Łukasiewicz's implication.

Let $w \models_M \varphi$ denote $\tau(w, \varphi) = 1$, then φ is true in M, written as $M \models \varphi$ if $w_0 \models_M \varphi$ and for a set Σ of wffs, we write $M \models \Sigma$ if $M \models \varphi$ for all $\varphi \in \Sigma$. Furthermore, φ is said to be an (external) logical consequence of Σ, denoted by $\Sigma \models \varphi$, if for any model M, $M \models \Sigma$ implies $M \models \varphi$. When Σ is the empty set, this is abbreviated as $\models \varphi$ and φ is said to be valid. Moreover, φ is satisfiable if $\sim \varphi$ is not valid and weakly satisfiable if $\neg\varphi$ is not valid.

3.3 Discussion

In the semantics above, if we consider Ξ as the set of available acts, and Φ the set of goals, then a Kripke structure is a generalization of possibilistic decision framework. It can be seen that $\tau(w, [\alpha]\varphi)$ and $\tau(w, \langle\alpha\rangle\varphi)$ are respectively the pessimistic and optimistic utility of doing α under state w with respect to the decision objective φ.

To see how the $PoDL$ model generalize a possibilistic decision framework, let us consider the formal correspondence between them. Let $M = (W, \tau, \|\cdot\|, w_0)$ be a $PoDL$ model and φ be a fixed wff, then $\mathcal{D}(M, \varphi) = (D, X, \pi, u)$ is a possibilistic decision framework, where

- $D = \Xi$,
- $X = W$,
- $\pi : D \times X \to [0, 1]$, $\pi(\alpha, w) = \|\alpha\|(w_0, w)$ for all $\alpha \in D$ and $w \in X$,
- $u : X \to [0, 1]$, $u(w) = \tau(w, \varphi)$ for all $w \in X$.

The difference is made explicit from the formal correspondence. First, in a possibilistic decision framework, the set of available decision acts is taken as primitive, so it is only needed to specify the possible effects of each act from the initial situation w_0, which is implicitly assumed. On the other hand, in $PoDL$, the set of actions is composed from some atomic ones, so to know the effects of an action under the initial situation, we have to know also effects of its constitutive actions under different situations. In other words, the decision maker choose plans instead of a single action in $PoDL$ model. Second, in a possibilistic decision framework, a utility function is given, which is implicitly assumed to correspond to a goal of the decision maker, whereas in $PoDL$, we have a bundle of utility functions, each corresponding to a wff of the language. Thus, we can know the utilities of not only some primitive goals p and q, but also the conjunctive goal $p \wedge q$, the aggregated goal $p \otimes q$, the negated goal $\neg p$, and so on. This is in particular suitable for multiple objectives decision making. Moreover, because of the use of rational Pavelka logic, the prioritized or thresholded goals mentioned in [3] can also be expressed in *PoDL*. For example, we can write goals like $\varphi \vee (1 - r)$ or $r \rightarrow \varphi$.

When we can completely specify a possibilistic decision framework or a *PoDL* model M, then the decision making process amount to the model checking problem in M. For example, if $M \models r \rightarrow [\alpha]\varphi$, then we know that for goal φ, the plan α has pessimistic expected utility at least r. If $M \models \sim ([\alpha]\varphi \rightarrow [\beta]\varphi)$, then α is a better plan than β for satisfying φ according to the criterion of maximizing U_*. Sometimes, it is not easy to have a complete specification of the decision framework. Instead, we may have only some partial description of the status of the environment and the preconditions and effects of the primitive actions. Some typical sentences for the description are non-modal wffs of *PoDL*, or formulas of the form $\varphi \rightarrow (r \rightarrow [a]\psi)$ where a is atomic and φ and ψ are non-modal. In this case, assume $\Sigma \subseteq \Phi$ is the set of descriptions, then the decision making process amount to deduction problem in the logic. We must try to derive the formulas like $r \rightarrow [\alpha]\varphi$ or $\sim ([\alpha]\varphi \rightarrow [\beta]\varphi)$ from Σ by proof methods of the logic. Though the development of proof methods for the logic is beyond the purely semantic concern of the paper, it is indeed a very interesting direction for further research.

4 Case-based Dynamic Logic

Analogous to *PoDL*, in this section, we develop a case-based dynamic logic(*CbDL*) for reasoning about actions and decisions according to CBDT. Though the syntax of *CbDL* is similar to that of *PoDL*, we will add a similarity-based modal operator which is of independent interest to fuzzy reasoning[4, 8]. Furthermore, a more classical dynamic operator $\{\cdot\}$ is used for describing primitive actions in case bases.

Thus, the alphabet of *CbDL* consists of those for *PoDL* and three additional logical symbols $\{,\}, \nabla$ and the following formation rules are added to those for *PoDL*,

- if $\varphi \in \Phi$ and $a \in A$, then $\{a\}\varphi$ and $\nabla \varphi \in \Phi$.

To define the semantics for *CbDL*, we first recall the definition of fuzzy similarity relation. A fuzzy relation $S : X \times X \rightarrow [0, 1]$ is a similarity one if it satisfies the following three properties, for all $x, y \in X$,

1. reflexivity: $S(x, x) = 1$,
2. symmetry: $S(x, y) = S(y, x)$, and
3. transitivity: $S(x, y) \geq \sup_{z \in X} \min(S(x, z), S(z, y))$.

Then a *CbDL* model is a 5-tuple $M = (W, \tau, \| \cdot \|_0, S, w_0)$, where W, τ, and w_0 are as above, $\| \cdot \|_0 : A \rightarrow 2^{W \times W}$, and $S : W \times W \rightarrow [0, 1]$ is a similarity relation.

The mapping $\| \cdot \| : A \to (W \times W \to [0,1])$ is defined by

$$\|a\|(x,y) = \sup_{z \in W, (z,y) \in \|a\|_0} S(x,z).$$

That is, $\|a\| = S \circ \|a\|_0$. Then the mappings τ and $\| \cdot \|$ is extended to Φ and Ξ as in $PoDL$ with the extra rules for wffs of the forms $\nabla\varphi$ and $\{a\}\varphi$,

10. $\tau(w, \nabla\varphi) = \inf_{x \in W} \max(1 - S(w,x), \tau(x,\varphi))$,
11. $\tau(w, \{a\}\varphi) = \inf_{x \in W, (w,x) \in \|a\|_0} \tau(x,\varphi)$.

The definitions of logical consequence, validity, and satisfiability, etc. are analogous to those of $PoDL$. Sometimes, to distinguish the logical consequence relations between $PoDL$ and $CbDL$, we will add the subscripts to them.

The mapping $\| \cdot \|_0$ is to model the case memory, The definition is such that a case (x, a, y) is in the memory iff $(x,y) \in \|a\|_0$. This restricts that the actions appearing in the memory must be atomic. This restriction is not so restrictive as it seems at the first glance. Imagine that the agent has a detail trace of the execution of the actions in the past cases. Then for a compound action like $\alpha;\beta$, if we know the intermediate state after the execution of α, then we can decompose a case $(x, \alpha;\beta, y)$ into two cases (x, α, z) and (z, β, y), and store the latter two on the memory. According to the original restriction of case memory in [7], there do not exist two cases (p, a, x) and (p, a', x') in the memory such that $a \neq a'$ or $x \neq x'$, so we can also require that $\|a\|_0$ is a partial function in the $CbDL$ model. However, for generality, we do not impose the restriction on the models.

The definition of $\| \cdot \|$ from $\| \cdot \|_0$ and S make the following a valid axiom schema in $CbDL$. That is,

$$\models [a]\varphi \equiv \nabla\{a\}\varphi,$$

for $a \in A$ and $\varphi \in \Phi$.

Apparently, the $CbDL$ and $PoDL$ models have some correspondence. In the semantics for $CbDL$, we have constructed $\| \cdot \|$ from $\| \cdot \|_0$ and S, if we then ignore the latter two components, a $PoDL$ model is obtained. Since the language of $PoDL$ is a sublanguage of $CbDL$, this means that $CbDL$ is an extension of $PoDL$. Namely, if Σ is a set of wffs of $PoDL$ and φ is a wff of $PoDL$, then $\Sigma \models_{PoDL} \varphi$ implies $\Sigma \models_{CbDL} \varphi$.

However, unlike $PoDL$, we can not find a direct correspondence between $CbDL$ models and CBDT framework. This is due to the fact that in CBDT, the set of situation P and the set of consequence X are not necessarily the same, whereas in $CbDL$ models, we model the past cases by a set of binary relations on W, so W plays both the roles of P and X. To transform a CBDT framework (P, D, X, C, s, u) into a $CbDL$ model, we can let $W = P \cup X$, and then extend the similarity function s to $W \times W$ and the utility function u to W. However, is is likely that s is not well-defined outside $P \times P$ and u is not definable in P. In this case, a simple approach is just let $u(p) = 0$ for all $p \in P$ and $s(x,y) = 0$ for $x \in X$ or $y \in X$. Then the extended s is just the similarity fuzzy relation S and for the extended u, $u(w)$ is the truth value $\tau(w, \varphi)$ for some fixed goal

φ. The mapping $\| \cdot \|_0$ is derived from the case memory C, i.e., $(p, x) \in \|a\|_0$ if $(p, a, x) \in C$. If the process of extending s and u does not distort the original CBDT framework, then we have a complete specification of $CbDL$ model. This reduces the process of reasoning about actions and decision in the original CBDT framework to the model checking problem in $CbDL$ as in the case of $PoDL$.

On the other hand, from a more practical viewpoint, we may have only a partial description of the whole CBDT framework from the beginning. In this case, we assume there is a subset of crisp propositional symbols $PV_0 \subseteq PV$ for describing the case memory. Let Φ_0 is the set of sentences resulting from Boolean combinations of symbols in PV_0. Then in general, we have four sets of proper axioms for the description of the framework. The first one Σ_s is to describe the similarity function, so each sentence in it is of the form $\varphi \rightarrow (r \rightarrow \nabla \psi)$ for some $\varphi, \psi \in \Phi_0$, the second is Σ_c for the case memory, so its sentences are in the form of $\varphi \rightarrow \{a\} \psi$ for some $\varphi, \psi \in \Phi_0$ and $a \in A$, the third is Σ_u which specifies the utility functions, so each sentence is of the form $\varphi \rightarrow (r \rightarrow \psi)$ for some $\varphi \in \Phi_0$ and $\psi \in \Phi$, and the last is $\Sigma_0 = \{p \lor \neg p \mid p \in PV_0\}$ to enforce that each $p \in PV_0$ is two-valued. These four sets are proper axioms instead of premises because we require that they are true not only in w_0 but also in all possible worlds of a model. Let $\Omega = \Sigma_s \cup \Sigma_c \cup \Sigma_u \cup \Sigma_0$ and suppose the agent faces a new problem described by a set of (possibly just propositional) wffs Σ, then our problem is a theorem-proving one. For example, if we have $\Sigma \models_{\Omega} \sim (\langle \alpha \rangle \varphi \rightarrow \langle \beta \rangle \varphi)$, then α will be a better plan for the goal φ with respect to the criterion of maximizing U_c^*, where $\models_{\Omega}$ means the logical consequence relation in a $CbDL$ system with Ω as the set of proper axioms.

5 Future Works and Conclusion

We have outlined two logical languages for reasoning about actions and decisions based on the dynamic logic semantics and decision theory framework. One is based on possibilistic decision theory and the other on case-based decision theory. The language for $PoDL$ is a subset of $CbDL$ and it is shown that $CbDL$ is a conservative extension of $PoDL$. Both logics can be used in two ways. When we have a complete specification of the decision frameworks, the logic can be used in a model checking way and if we have only a partial description of the problem, then the logic should be used in a theorem proving way.

As mentioned above, since theorem proving is in general more difficult than model checking, the development of theorem proving methods for both logics is the first demanding problem for further research.

Second, while the logics developed here are mainly based on qualitative decision theories, we would also like to develop similar logics for quantitative decision theories. In particular, probabilistic dynamic logics [12] should be a good starting point. However, since these logics are aimed at reasoning about the behavior of probabilistic algorithm, they are still two-valued, so the generalization to many-valued ones are needed. If this is successful, we can model the classical decision framework as well as the original quantitative criterion of maximizing U_c in CBDT.

In conclusion, the results reported in this paper is just at the early stage of a long-term goal to integrate logical reasoning and decision theories. We expect the cross-fertilization of both fields can result from the research.

References

1. D. Dubois and H. Prade. "Possibility theory as a basis for qualitative decision theory". In *Proc. of the 14th International Joint Conference on Artificial Intelligence*, pages 1924–1930. Morgan Kaufmann Publishers, 1995.
2. D. Dubois and H. Prade. "Constraint satisfaction and decision under uncertainty based on qualitative possibility theory". In *Proc. of the 6th IEEE International Conference on Fuzzy Systems*, pages 23–30, 1997.
3. D. Dubois and H. Prade. "Possibilistic logic in decision". In *Proc. of the ECAI'98 Workshop: Decision Theory Meets Artificial Intelligence–Qualitative and Quantitative Approaches*, pages 11–21, 1998.
4. F. Esteva, P. Garcia, L. Godo, and R. Rodriguez. "A modal account of similarity-based reasoning". *International Journal of Approximate Reasoning*, pages 235–260, 1997.
5. H. Fargier, J. Lang, and R. Sabbadin. "Towards qualitative approaches to multi-stage decision making". In *Proc. of the 6th International Conference on Information Processing and Management of Uncertainty in Knowledge-Based Systems*, pages 31–36, 1996.
6. G. De Giacomo and M. Lenzerini. "PDL-based framework for reasoning about actions". In M. Gori and G. Soda, editors, *THe 4th Congress of the Italian Association for Artificial Intelligence*, LNAI 992, pages 103–114. Springer-Verlag, 1995.
7. I. Gilboa and D. Schmeidler. "Case-based decision: An extended abstract". In H. Prade, editor, *Proc. of the 13th European Conference on Artificial Intelligence*, pages 706–710. John Wiley & Sons Ltd., 1998.
8. L. Godo and R. Rodriguez. "A similarity-based fuzzy modal logic". Preprint, 1998.
9. P. Hájek. *Metamathematics of Fuzzy Logic*. Kluwer Academic Publisher, 1998.
10. D. Harel. "Dynamic logic". In D.M. Gabbay and F. Guenthner, editors, *Handbook of Philosophical Logic, Vol II: Extensions of Classical Logic*, pages 497–604. D. Reidel Publishing Company, 1984.
11. J.Y. Jaffray and P. Wakker. "Decision making with belief functions: compatibility and incompatibility with the sure-thing principle". *Journal of Risk and Uncertainty*, 8:255–271, 1994.
12. D. Kozen. "A probabilistic PDL". In *Proc. of the 15th ACM Symposium on Theory of Computing*, pages 291–297, 1983.
13. C.-J. Liau. "A logic for reasoning about action, preference, and commitment". In *Proc. of the 13th European Conference on Artificial Intelligence*, pages 552–556. John Wiley & Sons Ltd., 1998.
14. L.A. Zadeh. "Fuzzy sets as a basis for a theory of possibility". *Fuzzy Sets and Systems*, 1(1):3–28, 1978.

Incorporating Fuzzy Set Theory and Matrix Logic in Multi-layer Logic
– A Preliminary Consideration

Hiroyuki YAMAUCHI[1] and Setsuo OHSUGA[2]

[1] Research Center for Advanced Science and Technology, University of Tokyo
4-6-1 Komaba, Meguro-ku, Tokyo 153-8904, JAPAN
yama@ai.rcast.u-tokyo.ac.jp
[2] Dept. of Information and Computer Science, Waseda University
3-4-1 Okubo, Shinjuku-ku, Tokyo 169-8555, JAPAN
ohsuga@ohsuga.info.waseda.ac.jp

Abstract. This paper is concerned with a preliminary consideration to provide the formal specification of language of knowledge processing system SKAUS (Super Knowledge Acquisition and Utilization System) which incorporates uncertain knowledge processing and non-symbolic information processing units in the system. SKAUS is planned as a super set of KAUS developed by the authors. KAUS implement multi-layer logic (MLL for short) based on classical set theory. SKAUS is intended to have additional capabilities of KAUS, such as representing uncertain knowledge in the forms of language used in fuzzy set theory and probability theory. In addition to this extension, we try to incorporate matrix logic into our extension so as to process non-symbolic information in corporation with neural networks.

1 Introduction

For the practical AI systems, the ability of reasoning with uncertainty is very important [1–3]. For example, in the application of AI technology to problem domains of diagnosis, control and prediction, the systems are required to have the facility of reasoning with uncertainty because these domains are usually ill-defined and so the problems and the solving method could not be well described. Another example is seen in intelligent information retrieval systems. The users sometimes pose ambiguous queries to the retrieval systems. In this case, the systems have to resolve ambiguities involved in the queries so that the systems can retrieve the users' surely desired data.

The aim of this paper is to give a preliminary consideration to extend MLL [5, 6] so that we can handle reasoning with uncertainty by incorporating fuzzy sets originated by Zadeh [4] into the extended MLL. In addition to this extension, we try to incorporate matrix logic into our system so as to process non-symbolic information in corporation with neural networks. We have a plan to extend KAUS

[10] as SKAUS (super knowledge acquisition and utilization system) which enables us soft computing with these extensions. We introduce here MLL and KAUS shortly. More details of MLL are discussed in the literatures [5, 6].

In 1985 we have formalized multi-layer logic (MLL for short) which is an extended version of first order logic. MLL was formulated as a formal system for constructing general purpose knowledge processing systems. Though ordinary first order logic does not assume any structural constraints to variable and constant terms, these in MLL may be structured in the set hierarchy. The sets treated in MLL are crisp sets based on axiomatic set theory, specifically, based on admissible sets [7].

Adopting MLL as the theoretical basis, we have developed KAUS (knowledge acquisition and utilization system) and it is used as a tool for building knowledge-based systems. Until now we have applied KAUS to various model building and evaluation by computer [8, 9].

Rules described in KAUS language are not restricted Horn clauses but arbitrary AND–OR clauses. Variables appearing in KAUS clauses may be universally quantified or existentially quantified with type restrictions. For example, (1) *Every boy likes a girl*, (2) *The age of John is 24 or 25*, (3) *If a person X does not have his own car, then X is not a car driver or X is a paper-driver*, and (4) *If each member Y of the students X who is a group interested in computer science learns a programming language* are respectively expressed in KAUS as follows.

```
(1).[AX/boy][EY/girl](like X Y).
(2).(| (age john 24) (age john 25)).
(3).[AX/person][ACar/car]
        (| (| ~(carDirver X) (paper-driver X) ) (have X Car)).
(4).[AX/*student][AY/X][EL/programmingLanguage]
        (| (learn Y L) ~(interestedGoroup X computerScience)).
```

As seen above, we represent a clause $A \rightarrow B$ by $\neg A \vee B$.

We have implemented inference rules given in [5, 6] with the unification algorithm based on resolution principle, in which if the two variables to be unified are typed variables (as seen in the above example), type unification is also performed [10]. Relating to uncertain reasoning, disjunctive logic programming [11], though in the limited way, and building models of possible worlds using the world constructor of KAUS are possible.

2 Incorporating Fuzzy Set Theory into MLL

The most primitive concept in a set theory and so in MLL is the membership relation that an element x belongs to a set A. In the classical set theory we write this relation as $x \in A$. The truth value of $x \in A$ is often described using its characteristic function $\phi(x)$ such that $\phi(x) = 1$ if and only if $x \in A$, and $\phi(x) = 0$ if and only if $x \notin A$. For example, *John is a student* will be described as $John \in student$ and $\phi_{student}(John) = 1$. How about *John is a tall student*? We might write it as $John \in tall\_student$ and $\phi_{tall\_student}(John) = 1$. However, if we want to classify Jim as $Jim \in tall\_student$ from the fact *Jim is a student and his height is 180cm*, we have to clarify the concept *tall*. Since *'tall'* is a vague concept, we cannot define exactly what is meant by *tall*. In the approach of classical logic and so MLL, we can only heuristically or subjectively define tallness in a logical form such as

$$(\forall x \in student)[height(x, h) \land h \geq 175 \leftrightarrow tall(x)]$$

From this we can say that the set of all tall students (denoted by *tall_student*) is the subset of the set *student* that satisfies the above relation. This is a definition from the intensionallity of a set. Another definition is possible from the extensionality of a set. In the extensional definition of a set, we explicitly enumerate all elements of the set. For example, *tall_student* = {*John, Jim, ...*}. The enumerated elements are thought of having the common properties and attributes. There is no ambiguity in the definitions of intensionality and extensionality of sets in the classical set theory. Furthermore it seems that classical set theory is enough for describing all things including ambiguous and vague concepts in such a way. All descriptions by the classical theory can be evaluated exactly true or false. So, one could say that MLL is enough and there is no problem in MLL. However if we describe ambiguous and vague concepts in MLL cooperating with fuzzy set theory, such MLL will become more practical theory because fuzzy set theory is very practical and intuitive theory for real applications. In the following we describe extended MLL from the point of views describe above.

2.1 Extending Set Relationships in MLL

Our criterion of incorporating fuzzy sets into MLL is that set relationships defined in MLL are special cases in that the fuzzy set membership functions are restricted to the extreme points {0,1} of [0,1]. Because of this we adopt α-*level* sets to define set relationships. The α-*level* set A_α of a fuzzy set A is defined as follows [12, 13].

Definition 1. *(α-cuts) Given $\alpha \in [0,1]$ and the membership function μ_A of a fuzzy set A, we define α-level sets A_α of A from the following α-cuts (a) or strong α-cuts (b).*

$$\begin{aligned} (a). \ A_\alpha &= \{x \in U | \mu_A(x) \geq \alpha\} \\ (b). \ A_\alpha &= \{x \in U | \mu_A(x) > \alpha\} \end{aligned} \tag{1}$$

We can reconstruct μ_A from the family of α-*level* sets A_α of A :

$$\mu_A(x) = sup\{\alpha | x \in A_\alpha\} \tag{2}$$

An α-*level* set is a crisp set and the $\in$ -*relation* used in (2) is the ordinary membership relation. Since as described earlier in this chapter the membership relation is the most fundamental relation in a set theory, we define the similar $\in$ -*relation* for fuzzy sets.

Definition 2. *($\in_\alpha$ -relation) We write $x \in_\alpha A$ iff x belongs to an α-level set A_α of a fuzzy set A, and $x \notin_\alpha A$ iff x does not belong to an α-level set A_α of a fuzzy set A.*

$$\begin{cases} x \in_\alpha A \text{ iff } x \in A_\alpha \\ x \notin_\alpha A \text{ iff } x \notin A_\alpha \end{cases} \tag{3}$$

The inclusion relation between fuzzy sets is defined using α-cuts of fuzzy sets.

Definition 3. *(inclusion relation)*

$$A_\alpha \subseteq B_\beta : \quad (\forall x)[x \in A_\alpha \rightarrow x \in B_\beta] \leftrightarrow \beta \leq \alpha \tag{4}$$

Union, intersection, and complementation operators are defined as follow.

Definition 4. *(union and intersection operators) Given X whose elements are fuzzy sets and $\alpha, \beta \in [0, 1]$, we define union and intersection operators as follows.*

$$union \cup X : \quad [x \in_\alpha \cup X \leftrightarrow (\exists Z)(\exists \beta)[\beta \geq \alpha \wedge x \in_\beta Z \wedge Z \in X]] \tag{5}$$

$$intersection \cap X : \quad [x \in_\alpha \cap X \leftrightarrow (\forall Z)(\exists \beta)[\beta \leq \alpha \wedge x \in_\beta Z \wedge Z \in X]] \tag{6}$$

For example, if $X = \{A, B\}$, $\cup X = A \cup B$. The membership function of $A \cup B$ is defined from the α-cuts given in (1) and (2). Typically, the membership function of $A \cup B$ is a s-norm (t-conorm) such that $\mu_{A \cup B}(x) = \oplus(\mu_A(x), \mu_B(x))$ where each membership function of A and B is defined from some β_A-cuts and β_B-cuts such that $\beta_A \leq \alpha$ and $\beta_B \leq \alpha$. The membership function of $A \cap B$ is given by the t-norm such that $\mu_{A \cap B}(x) = \otimes(\mu_A(x), \mu_B(x))$.

Definition 5. *(complementation) Given A as a fuzzy set and $\alpha \in [0, 1]$, we define the complement set of A as follows.*

$$complementation \ \overline{A} : \quad [x \in_\alpha \overline{A} \leftrightarrow x \in_{1-\alpha} A] \tag{7}$$

Definition 6. *(powerset) Given a fuzzy set X, we define the powerset of X as follows.*

$$powerset \ *X : \quad [Y \in *X \leftrightarrow Y \subseteq X] \tag{8}$$

These definitions (1) – (8) are used for defining the inference rules of the extended MLL.

2.2 Extending Inference Rules in MLL

In a fuzzy system, a typical pattern of the fuzzy inference rules is expressed like

$$\begin{array}{ccc} \text{if } x \text{ is } A, & then & x \text{ is } B \\ \underline{a \text{ is } A^*} & & \\ & & a \text{ is } B^* \end{array} \tag{9}$$

where x is a variable and a is a constant. A, A^*, B and B^* represent fuzzy predicates (fuzzy sets) [14]. For example, from (9), if *'if x is tall, then x is heavy'* and 'John is very tall' are given as premises, we can conclude that 'John is very heavy'. Zadeh defined such a generalized modus pones rule [14]. However, it should be noted here that there exists the difference between 'if x is A, x is B' and 'if the more x is A, the more x is B' and it should be clarified in the inference system [14]. We agree to this and our inference system will reflect this agreement. As for fuzzy unification, unification rules between fuzzy predicates have been discussed and formulated [15–18]. Until now some real tools for developing fuzzy systems have been also developed [19, 20].

In this section we attempt to extend MLL so that we can handle such a fuzzy inference as (9) in the extended MLL. In MLL predicates are assumed 2-valued predicates. The variables may be quantified like $(\forall x / X)$ indicating $(\forall x \in X)$ and $(\exists x / X)$ indicating $(\exists x \in X)$. The values of variables may be sets but they are assumed crisp sets. Constant terms may also be sets but they are assumed crisp sets. The inference rules of MLL are formalized in [5, 6]. We show two of these here.

$$\{(\forall x / X) P[x], a \in X\} \vdash P[a]. \tag{10}$$

$$\{(\forall x / X) P[x], (\forall y / Y)[P[y] \rightarrow Q[y]], Y \supseteq X\} \vdash (\forall x / X) Q[x]. \tag{11}$$

(11) is the modus pones rule in MLL. To incorporate fuzzy inference rule into MLL, we need additional inference rules. Relating to (10) and (11), we need

$$\{(\forall x / X_\alpha) P[x], a \in_\alpha X\} \vdash P[a]. \tag{12}$$

$$\{(\forall x / X_\alpha) P[x], (\forall y / Y_\beta)[P[y] \rightarrow Q[y]], Y_\beta \supseteq X_\alpha\} \vdash (\forall x / X_\gamma) Q[x]. \tag{13}$$

where γ in the conclusion of (13) is conditioned to be $\gamma = \otimes\{\mu_X(x), \mu_Y(x)\}$.

The unification rule of fuzzy constants is also required to the extended MLL. For example, unification between $height(x, tall)$ and $height(x, very.tall)$ should be possible. Unification between a literal value and a numerical value of a fuzzy constant is also considered.

In the following, we illustrate a simple example of inference involving fuzzy predicates. We transform following (14) step by step into the form in the extended MLL.

$$
\begin{array}{c}
(\forall x)[person(x) \wedge tall(x) \rightarrow heavy(x)] \\
(\forall x)[boy(x) \rightarrow person(x)] \\
\underline{boy(john), very.tall(john)} \\
very.heavy(john)
\end{array} \quad (very) \tag{14}
$$

In (14), 'very' is attached to the horizontal line to indicate that the inference is performed with fuzzy unification between 'tall' and 'very.tall'. It shows that the grade of tallness should be transmitted to the grade of heaviness in the conclusion.

We transform (14) to (15) using the axiom of intensionality of a set and α-cuts of a fuzzy predicate (fuzzy set). we first rewrite $person(x)$ by $x \in PERSON$, $boy(x)$ by $x \in BOY$, and $boy(john)$ by $john \in BOY$. Next we rewrite each fuzzy predicate. For example, $tall(x)$ by $tall_\alpha(x)$ indicating that 'tall' is a fuzzy predicate and its fuzziness is determined by the fuzzy membership function induced from the α-cuts $TALL_\alpha$ of the fuzzy set $TALL$. Finally we rewrite $very.tall(john)$ by $tall_{very}(john)$. The fuzziness γ of the conclusion is calculated by the composition rule attached to the horizontal line:

$$
\frac{
\begin{array}{c}
(\forall x)[x \in PERSON \wedge tall_\alpha(x) \rightarrow heavy_\beta(x)] \\
(\forall x)[x \in BOY \rightarrow x \in PERSON] \\
john \in BOY, tall_{very}(john)
\end{array}
}{
heavy_\gamma(john)
} \quad (\gamma = \oplus\{\otimes\{\alpha, very\}, \beta\})
\tag{15}
$$

We next transform (15) into (16) using the notations given in (17).

$$
\frac{
\begin{array}{c}
(\forall x/PERSON)[tall_\alpha(x) \rightarrow heavy_\beta(x)] \\
BOY \subset PERSON \\
john \in BOY, tall_{very}(john)
\end{array}
}{
heavy_\gamma(john)
} \quad (\gamma = \oplus\{\otimes\{\alpha, very\}, \beta\})
\tag{16}
$$

where

$$
\begin{aligned}
(\forall x/X)p(x) &\equiv (\forall x)[x \in X \rightarrow p(x)] \\
(\exists x/X)p(x) &\equiv (\exists x)[x \in X \wedge p(x)]
\end{aligned}
\tag{17}
$$

In (16), the fuzzy predicates are left as these are, but the second premise in (15) is replaced by the set inclusion relation. Similar to (15), the calculation of fuzziness γ of the conclusion is performed using the composition rule attached to the horizontal line.

Finally we transform (16) into the complete extended MLL form. Until now we have represented a fuzzy predicate such as $P_\alpha(x)$. Here we introduce a new notation $p : \alpha$ in order to declare that a predicate p is a fuzzy predicate or a fuzzy term having α as its fuzzy parameter. If we write $p : \alpha(x)$, we assume that $p : \alpha$ is a fuzzy predicate identifier. Otherwise, namely, if we write simply $p : \alpha$, we assume it is a fuzzy term. This results in the extension of the well formed formulas in MLL. α may be a variable or a constant given either literally or numerically. We note here that the truth value of $p : \alpha(x)$ is determined as follows.

$$
p : \alpha(x) = \begin{cases} true & \text{iff } x \in P_\alpha \cap X \text{ where } P_\alpha \text{ is an } \alpha\text{-cut of } P \\ false & \text{otherwise} \end{cases}
\tag{18}
$$

where X denotes the domain of x. Then for the first premise of (16), $tall_\alpha(x)$ is written as $tall : \alpha(x)$, $heavy_\beta(x)$ as $heavy : \beta(x)$. Using this notation (16) can be rewritten as follows.

$$
\frac{
\begin{array}{c}
(\forall x/PERSON)[tall : \alpha(x) \rightarrow heavy : \beta(x)] \\
BOY \subset PERSON \\
john \in BOY, tall : very(john)
\end{array}
}{
heavy : \gamma(john)
} \quad (\gamma = \oplus\{\otimes\{\alpha, very\}, \beta\})
\tag{19}
$$

As a result, if $\alpha = \beta$ in (19) and $\otimes = min$ and $\oplus = max$, we can easily conclude in $\gamma = very$ in $heavy : \gamma(john)$, that means 'John is very heavy'. The extended MLL intends to use such a notation as (19).

2.3 Probabilistic Reasoning

We can apply probability theory to reasoning with uncertainty. However the mixed use of probability theory and fuzzy set theory is dangerous. There exist critical differences between them. Fuzzy set theory deals with vague and imprecise notions and defines partial degrees of truth. On the other hand, probability theory deals with crisp notions and does not define partial degrees of truth but defines the degree of belief on truth [21]. Consider the following assertion.

$$\text{If } x \text{ is } tall \text{ (A) and } x \text{ is } heavy \text{ (B), then } x \text{ is } strong \text{ (C).} \qquad (20)$$

In fuzzy set theory, if A and B are partially satisfied with some evidence A^* and B^*, then the graded truth value of the conclusion C^* can be assigned with the composition rules between the membership functions of A^* and B^*. In probability theory, (20) can be rewritten using probabilities such as

$$\text{If } A \text{ is } p\text{-probable and } B \text{ is } q\text{-probable, then } C \text{ is } r\text{-probable.} \qquad (21)$$

where p, q and r are probabilities of A, B and C respectively. These probabilities define the degrees of beliefs of A, B and C. If some evidences $p^*(A)$ and $q^*(B)$ are given, then we can conclude $r^*(C)$ by some probabilistic composition rules. For example, plausible and possibility reasoning are formulated by using belief functions in Dempster-Shafer's evidence theory [1, 22] and probabilistic measures described in [23]. The Bayes approach [22] using conditional probabilities is strictly probabilistic approach. Baldwin et al. have relaxed probability measures and they have formulated support pairs of necessity measures and possibility measures [19] as probabilistic measures for formulas. Dubois et al. also have relaxed probability measures by using mass functions as probabilities [18]. In any way, we can apply probabilistic approach for reasoning with uncertainty under certain restrictions. The main problems would be what norms of uncertainty and composition rules are available to reasoning with uncertainty in real applications. The real implementation of SKAUS which can reason with uncertainty should incorporate a selection mechanism of appropriate uncertainty measures and composition rules.

3 Incorporating Matrix Logic into MLL

In August Stern's matrix logic [25], logical truth values and connectives are represented by logic vectors and matrix operators. In this formulation, not only the ordinary 2-valued logic but also many valued logic including fuzzy logic, modal logic and probabilistic logic are uniformly treated in the same framework. Matrix logic is closely related to neural network computing because of its algebraic treatment of objects. By incorporating matrix logic into SKAUS as a meta-predicate, we could expect that the fusion of symbolic processing and non-symbolic processing is realized.

3.1 Matrix Logic

In this section, we introduce matrix logic by August Stern shortly. First we give some notations used in matrix logic.

$$\text{row logic vector}: \quad < p| = (\bar{p}, p) \quad \text{(called bra vector)}$$
$$\text{column logic vector}: \quad |q> = \begin{pmatrix} \bar{q} \\ q \end{pmatrix} \quad \text{(called ket vector)} \tag{22}$$

where p and q in the left side of (22) are atomic formulas, whereas p and q in the right side represent truth values. In 2-valued logic, if p is true, then $< p| = (0,1)$. The inner product and outer product of logic vectors are written as follows.

$$\text{inner product}: < x|y > = (\bar{x}, x) \begin{pmatrix} \bar{y} \\ y \end{pmatrix} = \bar{x}\bar{y} + xy$$

$$\text{outer product}: |x> < y| = \begin{pmatrix} \bar{x}\bar{y} & \bar{x}y \\ x\bar{y} & xy \end{pmatrix} \quad (= \Omega), \| \Omega \| = \bar{x}\bar{y} + \bar{x}y + x\bar{y} + xy = 1 \tag{23}$$

Ω is called universal operator. A matrix operator L is written as follows.

$$< x|L|y > = < x| \begin{pmatrix} L_{11} & L_{12} \\ L_{21} & L_{22} \end{pmatrix} |y> = x_1 L_{11} y_1 + x_1 L_{12} y_2 + x_2 L_{21} y_1 + x_2 L_{22} y_2 \tag{24}$$

Note that $\overline{< x|L|y >} = < x|\bar{L}|y >$, $\bar{x} = 1 - x$, $|\bar{x}> = \mathbf{1} - |x> = \neg|x>$, and $\bar{L} = \hat{1} - L \neq \neg L$, where $\mathbf{1}$ and $\hat{1}$ indicate a special vector and an operator respectively, and these components are all one. Another point is

$$< x| \wedge |y > = < x|1 > < 1|y > = < 1|x > < y|1 > = < 1|L(|x> < y|)|1 > \tag{25}$$

where $|1 >$ is a column vector of $< 1| = (0,1)$, and $L(|x> < y|)$ is a variable logic operator, taking different shapes, depending on the values obtained by the vectors $|x >$ and $< y|$.

Some examples of matrix operators are shown below.

$$\wedge = \begin{pmatrix} 0 & 0 \\ 0 & 1 \end{pmatrix}, \vee = \begin{pmatrix} 0 & 1 \\ 1 & 1 \end{pmatrix}, \rightarrow = \begin{pmatrix} 1 & 1 \\ 0 & 1 \end{pmatrix}, \neg = \begin{pmatrix} 0 & 1 \\ 1 & 0 \end{pmatrix}, \downarrow = \begin{pmatrix} 1 & 1 \\ 1 & 0 \end{pmatrix}, \uparrow = \begin{pmatrix} 1 & 0 \\ 0 & 0 \end{pmatrix} \tag{26}$$

$\downarrow$ and $\uparrow$ are *nand* and *nor* operators respectively. We see $\rightarrow = \neg\vee$ from (26). In matrix logic, the *modus ponens* rule is represented like

$$(x \wedge (x \rightarrow y)) \rightarrow y = \overline{< x|1 > < x|} \rightarrow |y> < 0|y > = 1 \tag{27}$$

Note here that (27) is obtained using (25) and $< 0|y > = \neg|y >$.

3.2 Neural Networks

In this section we consider the problem of adjusting membership functions of fuzzy sets using neural network techniques. As well known, by combining fuzzy systems with neural networks, we can add the learning ability to fuzzy systems.

312

On the other hand, difficulties of the plain explanation of neural computations are overcome [25].

We consider here a fuzzy inference scheme given in (9). We encode (9) in the neural network using matrix logic as follows.

$$
\begin{array}{ccc}
\text{input from neuron X} & \text{hidden layer} & \text{output to neuron Y} \\
& (r): < A(x)| \rightarrow |B(x) > & \\
\underline{(a): < A^*(a)|} & & \\
& (b): |B^*(a) > & (\beta^*)
\end{array}
\tag{28}
$$

The hidden layer receives input (a) and computes (b) by applying (r). In each computing cycle of (b), the output function of the hidden neuron adjust membership functions used in (r) by applying a fuzzy version of (27), such that the following equation is satisfied.

$$
(1-\alpha^*,\alpha^*)\begin{pmatrix}0\\1\end{pmatrix}(1-\alpha,\alpha)\begin{pmatrix}1&1\\0&1\end{pmatrix}\begin{pmatrix}1-\beta\\\beta\end{pmatrix}=\beta^*
\tag{29}
$$

We note here that we used the algebraic product as t-norm, and the algebraic sum as s-norm for calculating β^* in (29). We also assumed fuzzification of input at (a) and defuzzification of output at (b) in (28) are performed as the preprocess and postprocess respectively. Furthermore, we note that in a fuzzy version of (27) under (29), $< \beta^*| \wedge |\overline{\beta^*} >=< \beta^*| \wedge \neg|\beta^* >= \beta^*(1-\beta^*) \neq 0$ in general.

4 Conclusion

We have applied an idea of α-cuts of fuzzy sets to formulate the extended MLL which can perform reasoning with uncertainty. We have also considered matrix logic as a tool for the fusion of symbolic computation and non-symbolic (numerical) computation with relation to neural networks. SKAUS which is based on the extended MLL and having a meta-predicate of matrix logic computation will be expected to enlarge applications under the real environments.

References

1. F. Voorbraak, Reasoning with Uncertainty in AI, LNCS 1093, 52-89, 1996.
2. Henry E. Kyburg, Jr., Uncertain Inferences and Uncertain Conclusions, Logic programming and Soft Computing, T.P. Martin and F. Arcelli Fontana (eds.), 365-372, Research Studies Press Ltd., 1998.
3. P. Krause and D. Clark, Representing Uncertain Knowledge, Kluwer Academic, 1993.
4. L.A. Zadeh: Fuzzy Sets, information and control 8, 338-353, 1965.
5. S. Ohsuga and H. Yamauchi: Multi-layer Logic - A predicate logic including data structure as knowledge representation language, New Generation Computing 3, 1985.

313

6. H. Yamauchi and S.Ohsuga: Modelling Objects by Extensions and Intensions – A Theoretical Background of KAUS -, Information Modelling and Knowledge Bases III, S. Ohsuga et al (eds.), IOS Press, 1992.
7. J. Barwise: Admissible Sets and Structures, Springer, 1975.
8. H. Yamauchi and S. Ohsuga: KAUS as a tool for model building and evaluation, Proc. 5th Internat. Workshop Expert Systems and their Applications, 1985.
9. S.Ohsuga: Toward truly intelligent information systems – from expert systems to automatic programming, Knowledge-Based Systems 10, 363-396, Elsevier, 1998.
10. H. Yamauchi: KAUS User's Manual Version 6.502, 1999.
11. J. Lobo, J. Minker, and A. Rajasekar, Foundations of Disjunctive Logic Programming, The MIT Press, 1992.
12. D. Dubois and H. Prade: Fuzzy Sets and Systems: Theory and Applications, Academic Press, 1980.
13. Y.Y. Yao, A comparative study of fuzzy sets and rough sets, Information Science 109, 227-242, 1998.
14. P. Magrez and P. Smets, Fuzzy Modus Ponens: A New Model Suitable for Applications in Knowledge-Based Systems, Readings in Fuzzy Sets for Intelligent Systems, D. Dubois, H. Prade, and R.R. Yager (eds.), 565-574, Morgan Kaufmann, 1993.
15. R.C.T. Lee, Fuzzy Logic and the Resolution Principle, Readings in Fuzzy Sets for Intelligenct Systems, D. Dubois, H. Prade, and R.R. Yager (eds.), 442-452,Morgan Kaufmann, 1993.
16. F.A. Fontana, F. Formato and G. Gerla: Fuzzy Unification as a Foundation of Fuzzy Logic Programming, Logic programming and Soft Computing, T.P. Martin and F. Arcelli Fontana (eds.), Research Studies Press Ltd., 1998.
17. Xu-Han, Sheng-Li Shi, Fuzzy Reasoning with Fuzzy Quantifiers, Proceedings of IEEE International Symposium on Multiple Valued Logic, 36-40, Washington DC, USA, 1989.
18. D. Dubois, H. Prade and S.A. Sandri: Possibilistic Logic with Fuzzy Constants and Fuzzy Restricted Quantifiers, Logic programming and Soft Computing, T.P. Martin and F. Arcelli Fontana (eds.), Research Studies Press Ltd., 1998.
19. J.F. Baldwin and T.P. Martin: The Management of Uncertainty in Logic Programs using Fril, Logic programming and Soft Computing, T.P. Martin and F. Arcelli Fontana (eds.), Research Studies Press Ltd., 1998.
20. M. Mukaidono, Z. Shen, and L. Ding, Fundamentals of Fuzzy Prolog, Readings in Fuzzy Sets for Intelligenct Systems, D. Dubois, H. Prade, and R.R. Yager (eds.), 454-460, Morgan Kaufmann, 1993.
21. P. Hajek, L. Godo and F. Esteva, Fuzzy Logic and Probability, Logic programming and Soft Computing, T.P. Martin and F. Arcelli Fontana (eds.), 237-244, Research Studies Press Ltd., 1998.
22. Judea Pearl, Probabilistic Reasoning in Intelligenct Systems, Morgan Kaufmann, 1988.
23. NG. Raymond and V.S. Subrahmanian, Probabilistic Logic Programming, Information and Computation 101, 150-201, 1992.
24. August Stern, Matrix Logic and Mind, North-Holland/Elsevier, 1992.
25. Constantin von Altrock, Fuzzy Logic & Neurofuzzy Applications Explained, Prentice Hall, 1995.

Fuzzy Logic as Interfacing Media for Constraint Propagation Based on Theories of Chu Space and Information Flow

Ken Sato[1], Tadashi Horiuchi[2], Toshihiro Hiraoka[1],
Hiroshi Kawakami[1] and Osamu Katai[1]

[1] Dept. of Systems Science, Graduate School of Informatics, Kyoto University
Yoshida Honmachi, Sakyo-ku, Kyoto 606-8501, JAPAN
[2] Institute of Scientific and Industrial Research, Osaka University
8-1 Mihogaoka, Ibaraki, Osaka 567-0047, JAPAN

Abstract. In this paper, we will introduce a novel perspective on Fuzzy Logic by referring to the theories of Chu Space and Information Flow, i.e., Channel Theory, which results in a deep insight on the interaction and coordination of agents with environments. First, a constraint-oriented interpretation of fuzzy set is introduced yielding the notion of Constraint-Interval Fuzzy Set (CoIFS). Then the above theories are introduced, which elucidate the basic structures of fuzzy inference as constraint propagation yielding the spaces of Coordination and Interaction. Also, the structure of Information Transmission Channel of constraint propagation is clarified together with its relevance with the results by the theory of Chu Space. All the results can be used to elucidate the basic structure of "interfacing (interfacing media)" between agents and environments.

1 Introduction

Various artificial systems are concerned complicatedly with human beings, societal systems, and environments. It becomes more and more important to manage these complexities and to improve the quality of interactions among them. We will focus ourselves to the boundaries, media, and mechanisms of interactive systems concerned with the structural coupling between subjects and environments. Here we will introduce constraint-oriented perspectives on these interactions and the symbolization of continuously valued quantities. For this symbolization, we will introduce the notion of Constraint-Interval Fuzzy Set (CoIFS) which we have introduced to tie up Fuzzy Logic and the traditional two-valued logic which underlies the traditional AI. Namely, we encode action space of the subject and sensation signal space from the environment into constraint-interval fuzzy sets and also introduce spaces for representing the background structures of situation−action relation, where the correspondence of "constraint-intervals" with "constraint-levels" between CoIFSs is treated.

Moreover, by introducing the theories of Chu Space and Information Flow to Fuzzy Logic, we will provide two novel perspectives on the theory of CoIFS's.

Introducing Chu Space provides us with not only formal treatment of constraint propagation among fuzzy sets but also a new relation among fuzzy sets. Also, introducing Information Flow gives new perspective on the constraint propagation as flow of information and on systems of fuzzy sets as distributed and decentralized systems.

2 Fuzzy Logic-Based Coding of Constraints

Based on the constraint-oriented perspectives on problem solving, a *constraint-interval fuzzy set* (CoIFS) is given as an ordered collection of intervals (constraint-intervals). We have also proposed fuzzy logic-based operations and the defuzzification for the CoIFS and also elucidated the specificities of CoIFS, i.e.,

- Symbolic (hard) inference can be related to fuzzy (soft) inference on the same ground [1].
- CoIFS involves chaotic characteristics in the interaction of constraint propagation via symbolic reasoning [2].
- CoIFS plays a role of distributed, concurrent and self-organizing module for "symbiotic" problem solving [3].

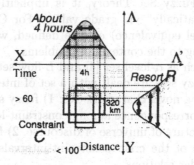

Fig. 1. An example of constraint-interval fuzzy set

As shown in the $X - \Lambda$ space of Fig. 1, a CoIFS is given as an ordered collection of "crisp" intervals on the universe of discourse (i.e., the space X) each of which represents a constraint called *interval constraint*. The grade axis (in the traditional Fuzzy Set Theory) is now regarded to be an "ordinal" scale axis. By introducing this notion of fuzzy set, a continuous variable can be coded into symbols via fuzzy sets (fuzzy labels), which divides the domain of a variable (universe of discourse) into intervals. Fuzziness in each variable is derived by decomposing a "joint constraint" on several variables into componential (marginal) constraints through "projection" of the joint constraint onto componential variables.

Relations between constraint intervals of two fuzzy sets are represented by "rectangles" in the $X - Y$ space as shown in Fig. 1 [1]. Namely, if we have a crisp constraint relation $\mathcal{C}$ on a pair of variables (say x and y), we can approximate constraint $\mathcal{C}$ by introducing an appropriate constraint interval fuzzy set on x and that on y, respectively.

Suppose that we are now planning to go to a tourist resort R for the next holiday. Now we have to decide a sightseeing spot in the outskirts of R to visit. The constraint now is the speed limit on the freeway ($60 \leq s \leq 100$). We have about 4 hours for one way, but it depends on the time when all of us get up. In this case, constraint $\mathcal{C}$ is represented as the meshed area shown in Fig. 1, and the "time for one way" and the "distance from our city to the sightseeing spot" are represented by fuzzy sets. A portion of the constraint region is approximated by "rectangles" which represent the relations between constraint-intervals of two fuzzy sets. The vertically oblong rectangles show that if all of us assemble at the meeting time conscientiously, there may be many alternatives for visiting spots. Otherwise, the alternatives are limited.

CoIFS is a kind of "topological" representation of fuzzy sets. Departing from the direct representation of fuzzy sets by numeric "membership grade value" results in the following characteristics:

- The "shape" of a fuzzy set is of no use, and only their "types" such as symmetricity are meaningful.
- In the traditional Fuzzy Set Theory, it is implicitly assumed that fuzzy sets are related "statically" via grade values. For CoIFSs, the same relation (constraint-level equivalence) can be defined, which is "dynamically" changeable according to the contexts of problems.
- Several concepts such as α-level set [4] and L-flow set [5] have been already proposed, where fuzzy sets are dealt with a set of intervals. However, CoIFS involves the following novel characteristics: 1) fuzzy sets are related to each other through the correspondence among constraint-levels in order to reflect the topological structure of universe of discourse. 2) fuzzy sets are regarded as the organization of the crisp constraint-intervals, which is changeable according to the order relations.

The $\Lambda - \Lambda'$ space of Fig. 1 shows the constraint-level equivalence relations between two fuzzy sets shown in the $X - \Lambda$ and $Y - \Lambda'$ spaces in the figure.

3 Introducing Chu Space to Fuzzy Logic

In this section, we will introduce the notion of Chu Space [6,7] to Fuzzy Logic for the formal treatment of constraint propagation among two fuzzy sets.

3.1 A brief review of Chu Space

A Chu space $\mathcal{A} = (I, A, R)$ is a binary relations between two sets I and A, where $R : I \times A \to \Sigma$ gives a binary relation, and Σ is the set $\{0, 1\}$. A Chu space can be represented as a binary matrix of dimension $|A| \times |I|$ (Chu map).

Given a Chu space $\mathcal{A} = (I, A, R)$, functions $\hat{R} : I \to \Sigma^A$ and $\check{R} : A \to \Sigma^I$ satisfying $\hat{R}(i)(a) = R(i, a) = \check{R}(a)(i)$ are the representations of I and A respectively [1]. $\hat{R}(i) : A \to \Sigma$ and $\check{R}(a) : I \to \Sigma$ are called column and row of $\mathcal{A}$, respectively. $\hat{R}(i)$ represents i as a function from A to Σ, and $\check{R}(a)$ represents a as a function from I to Σ.

Given two Chu spaces $\mathcal{A} = (I, A, R)$ and $\mathcal{B} = (J, B, S)$, a pair of functions $f : I \to J$ and $g : B \to A$ is called "Chu transform" from $\mathcal{A}$ to $\mathcal{B}$, provided that the following *adjointness condition* holds:

$$(\forall i \in I, \forall b \in B) \ S(f(i), b) = R(i, g(b)). \tag{1}$$

The *dual* of $\mathcal{A} = (I, A, R)$ is defined as $\mathcal{A}^\perp = (A, I, R^\cup)$, where $R^\cup : A \times I \to \Sigma$. The dual of Chu transform $\langle f, g \rangle$ from $\mathcal{A}$ to $\mathcal{B}$ is a Chu transform $\langle g, f \rangle$ from $\mathcal{B}^\perp$ to $\mathcal{A}^\perp$.

The composition of $\hat{S} : J \to \Sigma^B$ and $f : I \to J$, denoted as $\hat{S}f$, represents f, since $\hat{S}(f(i))$ represents the image of f as a function from B to Σ. Later on, we will write $\hat{S}f$ as $\hat{\phi} : I \to \Sigma^B$ such that $\phi(i, b) = \hat{S}(f(i))(b) = S(f(i), b)$. The Chu space $\mathcal{F} = (I, B, \phi)$ represents f. On the other hand, the composition of $\check{R} : A \to \Sigma^I$ and $g : B \to A$, denoted as $\hat{\phi}^\cup : B \to \Sigma^I$, satisfies the relation: $\phi^\cup(b, i) = \check{R}(g(b))(i) = R(g(b), i)$. The Chu space $\mathcal{F}^\perp = (B, I, \phi^\cup)$ represents g.

From the fact that the relation: $\phi(i, b) = \phi^\cup(b, i)$ implies $S(f(i), b) = R(i, g(b))$, we can interpret the adjointness conditions as follows:

> Given two Chu spaces $\mathcal{A} = (I, A, R)$ and $\mathcal{B} = (J, B, S)$, $\langle f, g \rangle$ is a Chu transform from $\mathcal{A}$ to $\mathcal{B}$, where $\mathcal{F} = (I, B, \phi)$ represents the function $f : I \to J$ from $\mathcal{A}$ to $\mathcal{B}$, and its dual $\mathcal{F}^\perp = (B, I, \phi^\cup)$ represents the function $g : B \to A$ from $\mathcal{B}^\perp$ to $\mathcal{A}^\perp$.

3.2 Coding fuzzy sets and their constraints by Chu Space

We introduce an interpretation of fuzzy set as a Chu space which consists of the set X of universe of discourse, a constraint-level set Λ and their relation $R : X \times \Lambda \to \Sigma$.

As shown in Fig. 1, a portion of the constraint C on a pair of variables X and Y is approximated by a set of rectangles, which transmit intervals from one fuzzy set to another. In the case where two fuzzy sets are represented by Chu spaces $\mathcal{A} = (X, \Lambda, R)$ and $\mathcal{B} = (Y, \Lambda', S)$, the constraint propagation (transmission) is represented as $\mathcal{A} \to \mathcal{B}^\perp$, which can be seen as a Chu transform $\langle f, g \rangle$ from $\mathcal{A} = (X, \Lambda, R)$ to $\mathcal{B}^\perp = (\Lambda', Y, S^\cup)$, where two functions $f : X \to \Lambda'$ and $g : Y \to \Lambda$ satisfy the following condition:

$$(\forall x \in X, \forall y \in Y) \ S^\cup(f(x), y) = R(x, g(y)). \tag{2}$$

The Chu transform $\langle f, g \rangle : \mathcal{A} \to \mathcal{B}^\perp$ itself also constitutes a Chu space $\mathcal{F} = (X, Y, \phi)$ which is represented as a Chu map with dimension $|Y| \times |X|$. Namely,

---

[1] Σ^A is the collection of all maps from A to Σ

318

Chu space $\mathcal{F}$ means a portion of the constraint region of space $X - Y$ which is approximated by rectangles.

Since $\mathcal{F}$ represents g and $\mathcal{F}^{\perp}$ represents f, we can interpret the adjointness condition as follows:

$$(\forall x \in X, \forall y \in Y) \quad \phi(x,y) = S^{\cup}(f(x),y), \tag{3}$$
$$(\forall x \in X, \forall y \in Y) \quad \phi^{\cup}(y,x) = R(x,g(y)). \tag{4}$$

The map between row y of $\mathcal{F}$ and the corresponding row $g(y)$ of $\mathcal{A}$ is given by g, and that of column x of $\mathcal{F}$ and the corresponding column $f(x)$ of $\mathcal{B}^{\perp}$ is given by f. In other words, f and g provide projection of $\mathcal{F}$ onto $\mathcal{A}$ and $\mathcal{B}^{\perp}$ as shown in the left part of Fig. 2.

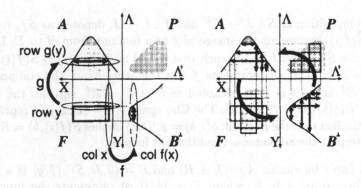

Fig. 2. Chu transform between two fuzzy sets with adjointness relation

The function f associates each point $x \in X$ with an interval represented by $\lambda' \in \Lambda'$ which equivalently represents constraint propagation. Conversely, function g transforms a point $y \in Y$ to $\lambda \in \Lambda$ showing a constraint propagation to an interval, where constraint propagation is given as yielding an interval derived by the intersection of associated intervals for points in the original interval.

Based on Chu transform $\langle f,g \rangle : \mathcal{A} \to \mathcal{B}^{\perp}$, we can define a Chu transform $\mathcal{B}^{\perp} \to \mathcal{A}$ which is a pair of function $p : \Lambda' \to X$ and $q : \Lambda \to Y$, denoted by $\langle p,q \rangle$, such that the following condition holds:

$$(\forall a \in \Lambda, \forall b \in \Lambda') \quad R(p(b),a) = S^{\cup}(b,q(a)). \tag{5}$$

The Chu transform $\mathcal{B}^{\perp} \to \mathcal{A}$ is a Chu space denoted as $\mathcal{P} = (\Lambda',\Lambda,\psi)$ as shown in the upper-right part of Fig. 2.

While space $\mathcal{F}$ represents the propagation of horizontal intervals in fuzzy sets, the space $\mathcal{P}$ represents those of vertical intervals, i.e., the corresponding constraint-levels. The matrix representation of $\mathcal{F}$ does not correspond directly to the rectangular representation of constraint combination of fuzzy sets. The rectangles are generated from the matrix representation of $\mathcal{F}$ by referring to $\mathcal{P}$.

A pair of fuzzy sets that are connected by Chu transform $\langle f, g \rangle = \mathcal{A} \to \mathcal{B}^\perp$ and $\langle p, q \rangle = \mathcal{B}^\perp \to \mathcal{A}$ can be interpreted as a Chu transform from $\mathcal{F}$ to $\mathcal{P}$. In other words, it is a transformation from the space of constraint-intervals to that of constraint-levels, and vice versa.

We will call $\mathcal{A}$ and $\mathcal{B}^\perp$ *Fuzzy Spaces*, respectively, and $\mathcal{F}$ and $\mathcal{P}$ will be called *Interaction Space* and *Coordination Space*, respectively. These four spaces constitute a unified construction of the interaction and coordination involved in the setting of problems.

4 Introducing Channel Theory to Fuzzy Logic

In this section, we will introduce the notion of Channel Theory[8],[9] to Fuzzy Logic for organizing distributed systems which consist of CoIFSs.

4.1 A brief review of Channel Theory on Information Flow

Barwise and Seligman have proposed *Channel Theory* which gives a mathematical framework to Dretske's *qualitative theory of information*. Channel Theory is a qualitative theory of information which treats the content of information rather than its amount. Based on the notion of *classification* and *infomorphism*, Channel Theory involves the concepts of *information channel* and *local logic*.

A "classification" $\mathcal{A} = \langle A, \Sigma_A, \models_A \rangle$ consists of a set A of objects to be classified, called "tokens" of A, a set Σ_A of objects used to classify the tokens, called "types" of $\mathcal{A}$, and a binary relations $\models_A$ between A and Σ_A indicating the types to which the tokens to be classified into.

For any classification $\mathcal{A}$, a pair $\langle \Gamma, \Delta \rangle$ of sets of types Σ_A is called a "sequent" of $\mathcal{A}$. A token a of A *satisfies* $\langle \Gamma, \Delta \rangle$ provided that if a is of type α for every $\alpha \in \Gamma$ then a is of type α for some $\alpha \in \Delta$. If every token a of $\mathcal{A}$ satisfies $\langle \Gamma, \Delta \rangle$, it will be written as $\Gamma \vdash_A \Delta$.

Given $\mathcal{A} = \langle A, \Sigma_A, \models_A \rangle$ and $\mathcal{C} = \langle C, \Sigma_C, \models_C \rangle$, a pair $\langle f^\wedge, f^\vee \rangle$ of functions is called an "infomorphism" from $\mathcal{A}$ to $\mathcal{C}$, provided that the following holds:

$$
\begin{aligned}
&f^\wedge : \Sigma_A \to \Sigma_C, \\
&f^\vee : C \to A, \\
&\forall c \in C, \forall \alpha \in \Sigma_A, \\
&\quad f^\vee(c) \models_A \alpha \Leftrightarrow c \models_C f^\wedge(\alpha).
\end{aligned}
$$

An *information channel* consists of an indexed family $\mathcal{C} = \{f_i : \mathcal{A}_i \to \mathcal{C}\}_{i \in I}$ of infomorphisms with a common codomain $\mathcal{C}$ as shown in Fig.3. The classification $\mathcal{C}$ is called the "core" of the channel and its token is called "connection" among tokens $a_i = f_i^\vee, i \in I$.

A *local logic* $\mathcal{L} = \langle \mathcal{A}, \vdash_{\mathcal{L}}, N_{\mathcal{L}} \rangle$ consists of the following three components:

1. a classification $\mathcal{A} = \langle A, \Sigma_A, \models_A \rangle$.
2. a set $\vdash_{\mathcal{L}}$ of sequents of $\mathcal{A}$.

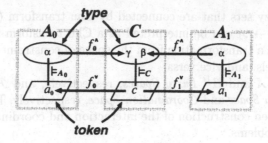

Fig. 3. An example of an information channel

3. a subset $N_{\mathcal{L}} \subseteq A$, called the normal tokens of $\mathcal{L}$, each element (token) of which satisfy all the sequents of $\vdash_{\mathcal{L}}$.

According to the definition of normal tokens, there exit several local logics in a classification in general. In such a sense, local logic has "locality" in the core C of the channel.

4.2 Coding fuzzy sets and their constraints by Channel Theory

By regarding a collection I_X of intervals over the universe of discourse X and a constraint-level set Λ as tokens and types, respectively, we interpret a fuzzy set as classification $\mathcal{A} = \langle I_X, \Lambda, \models_{\mathcal{A}} \rangle$. A token which is associated with a type represents a "constraint interval" and the collection of these tokens constitute the fuzzy set.

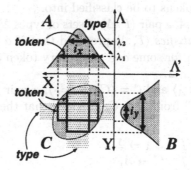

Fig. 4. A channel theoretic model of constraint propagation

Given two fuzzy sets $\mathcal{A} = \langle I_X, \Lambda, \models_{\mathcal{A}} \rangle$ and $\mathcal{B} = \langle I_Y, \Lambda', \models_{\mathcal{B}} \rangle$ as being classifications, the "constraint region" generated by these fuzzy sets is represented as the following classification $\mathcal{C}$:

1. The tokens of $\mathcal{C}$ is the Cartesian product of I_X and I_Y. More precisely, the tokens of $\mathcal{C}$ are pairs $\langle i_x, i_y \rangle$ of tokens ($i_x \in I_X, i_y \in I_Y$) and represent

"rectangular region" which is the product of interval i_x over X and interval i_y over Y.

2. The types of C is the disjoint union of Λ and Λ'. For simplicity, the types of C are represented as pairs $\langle i, \alpha \rangle$, where $i = 0$ and $\alpha \in \Lambda$ or $i = 1$ and $\alpha \in \Lambda'$.

3. The classification relation $\models_C$ is defined by

$$\langle i_x, i_y \rangle \models_C \langle 0, \lambda \rangle \quad \text{iff} \quad i_x \models_A \lambda,$$
$$\langle i_x, i_y \rangle \models_C \langle 1, \lambda' \rangle \quad \text{iff} \quad i_y \models_B \lambda'.$$

The infomorphism $\langle \sigma_A\hat{}, \sigma_A\check{} \rangle : A \to C$ and $\langle \sigma_B\hat{}, \sigma_B\check{} \rangle : B \to C$ is defined as follows:

1. $\sigma_A\hat{}(\lambda) = \langle 0, \lambda \rangle$ for $\forall \lambda \in \Lambda$,
2. $\sigma_B\hat{}(\lambda') = \langle 1, \lambda' \rangle$ for $\forall \lambda' \in \Lambda'$, and
3. for each pair $\langle i_x, i_y \rangle$,
   $$\sigma_A\check{}(\langle i_x, i_y \rangle) = i_x \text{ and } \sigma_B\check{}(\langle i_x, i_y \rangle) = i_y.$$

We can describe "regularity" and "order formation" among conclusion relations over several spaces through infomorphism. More precisely, when certain relation is given as a conclusion of other relations, each infomorphism maps the types representing these relations onto the core and the relations among mapped images (types) is formulated as the statement that a token satisfies a "sequent", that is a kind of proposition on C. These relation is represented by the sequent $\langle \Gamma, \Delta \rangle$ which consists of a pair of $\Gamma = \{\sigma_A(\lambda)\}$ and $\Delta = \{\sigma_B(\lambda')\}$, where $\lambda \in \Lambda$ and $\lambda' \in \Lambda'$. The sequents $\langle \Gamma, \Delta \rangle$ to which every token in C should be subject can be considered as rectangular regions generated by constraint intervals of two fuzzy sets. Furthermore, the whole order formation given by these propositions is formalized as a "local logic".

5 New Perspective on Fuzzy Logic by Introducing Theories of Chu Space and Information Flow

We have introduced Chu Space and Channel Theory to Fuzzy Logic in the previous two sections. In this section, we will discuss the place where the structural coupling between subjects and environments takes place, what we call *interfacing media*. This structural coupling in the interfacing media will be, we think, elucidated from two points of view, i.e., from Chu Space and Channel Theory.

We encode action space of the subject and the sensation space from the environment into constraint-interval fuzzy sets, and also introduce spaces for representing the background structures of situation−action relation, where the correspondence of constraint-intervals with constraint-levels between two fuzzy sets is treated.

By introducing Chu Space Theory to Fuzzy Logic, we can clarify that CoIFS can naturally represents the *Coordination Space* which stands for the coordination relation between constraint-intervals in two fuzzy sets. A coupling of sensation and action is represented as the adjointness relation among constraint intervals in the *Interaction Space* and the adjointness relation among constraint levels in the *Coordination Space*. These spaces prescribe each other, which in turn leads to the "stabilization of coupling" as shown in Fig.5.

While each of these spaces represents adjointness relation as a Chu transform, they have adjointness relation of sensation and action in *meta level* when these spaces have the correspondence with each other. Given certain fluctuation on the side of the subject or that of the environment, the adjointness relation in meta level is newly formed. Such a stabilization of coupling makes the interaction with the environment smooth and to form certain order.

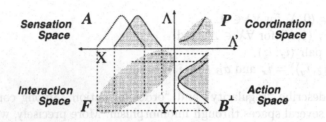

Fig. 5. Stabilization of the structural coupling

On the other hand, Channel Theory treats information flow qualitatively, i.e., it puts emphasis on the "content" of information rather than its amount. Constraint propagation among fuzzy sets propagates the constraint information (intervals) with preserving the relational structure among constraint levels. In this sense, constraint propagation can be seen as information flow through channel as shown in Fig.6. CoIFSs with constraint relations among them can be interpreted as a distributed and decentralized system on the medium of the channel.

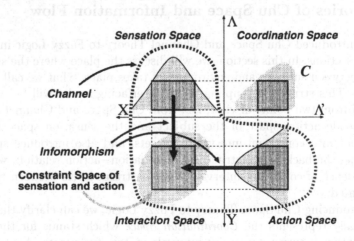

Fig. 6. Formation of the structural coupling

Roughly speaking, an infomorphism which constitutes a channel is equivalent to a Chu transform which connects Chu spaces. Chu Space and Channel Theory respectively can be regarded as a local view and a global view on the distributed and decentralized system.

6 Conclusion

Introducing the theories of Chu Space and Information Flow has shown to provide us with a new perspective on Fuzzy Logic. Particularly, the notion of CoIFS prescribes the coupling structure as constraint region between two variables such as sensation and action, and Chu Space can be used to clarify the new relational structure in the coupling, which we have called *Coordination Space*. Channel Theory gives the form of constraint propagation as flow of information and also provides us with a novel view that elucidates CoIFSs connected by constraint relations as a distributed and decentralized system.

There will be the following things as future works. Namely, we have to examine theoretically whether our treatment of *sequents* is suitable for Fuzzy Logic or not. Since the fundamental relationship of Channel Theory with Fuzzy Logic is still not so clear, we have to examine more detailed relationship with Shannon's Information Theory.

References

1. O. Katai, M. Ida, T.Sawaragi and S. Iwai, "Treatment of Fuzzy Concept by Order Relations and Constraint-Oriented Fuzzy Inference", *Proc. of NAFIPS'90, Quarter Century of Fuzziness*, pp.300-303, 1990.
2. O. Katai, T. Horiuchi, et al., "Chaotic Behavior of Fuzzy Symbolic Dynamics and Its Relation to Constraint-Oriented Problem Solving with Hierarchical Structures Reflecting Natural Life", *Proc. of FUZZ-IEEE/IFES'95*, Vol.3, pp.1955-1962, 1995.
3. O. Katai, T. Horiuchi and T. Sawaragi, "The Cause and Meaning of Chaos Phenomena in Symbiotic Problem Solving Uniting Natural System Redundancy and Artificial System Logicality", *Proc. of International Symposium on System Life*, pp.253-262, 1997.
4. L.A. Zadeh, "The Concept of a Linguistic Variable and Its Application to Approximate Reasoning", Part 1, *Information Sciences*, Vol.8, pp.199-249, 1975.
5. C.V. Negoita and D.A. Ralescu, "Representation Theorems for Fuzzy Concepts", *Kybernetes*, Vol.4, pp.169-174, 1975.
6. V. Pratt, "Chu Spaces from the Representational Viewpoint", *Parikh Festschrift*, 1997.
7. V. Gupta, "Chu Spaces: A Model of Concurrency", *PhD Thesis, Stanford University*, 1994.
8. J. Barwise, J. Seligman, *"Information Flow : The Logic of Distributed Systems"*, Cambridge University Press, 1997.
9. A. Shimojima, "What we can do with Channel Theory", *Journal of Japan Society for Fuzzy Theory and Systems*, Vol.10, No.5, pp.775-784, 1998 (in Japanese).

Pattern Reasoning:
A New Solution for Knowledge Acquisition Problem

Hiroshi Tsukimoto

Research & Development Center, Toshiba Corporation
70, Yanagi-cho, Saiwai-ku, Kawasaki 210 Japan
tukimoto@ssel.toshiba.co.jp

Abstract. This paper presents pattern reasoning, that is, the logical reasoning of patterns. Pattern reasoning is a new solution for the knowledge acquisition problem. Knowledge acquisition tried to acquire linguistic rules from patterns. In contrast, we try to modify logics to reason patterns. Patterns are represented as functions, which are approximated by neural networks. Therefore, pattern reasoning is realized by logical reasoning of neural networks. A few non-classical logics can reason neural networks, because neural networks can be basically regarded as multilinear functions and the logics are complete for multilinear function space, therefore, the logics can reason neural networks. This paper explains intermediate logic LC as an example of the logics and demonstrates how neural networks can be reasoned by LC.

1 Introduction

This paper presents pattern reasoning, that is, the logical reasoning of patterns. An example of pattern reasoning is presented. For example, expert doctors diagnose using a lot of images like brain images, electrocardiograms and so on, which can be formalized as follows:

Rule 1: If a brain image is a pattern, then an electrocardiogram is a pattern.

Rule 2: If an electrocardiogram is a pattern, then an electromyogram is a pattern.

Using the above two rules, we can reason as follows:

If a brain image is a pattern, then an electromyogram is a pattern.

This is a pattern reasoning. Symbols can be regarded as special cases of patterns. For example, let a rule be

If a brain image is a pattern, then a subject has a disease.

The right side of the rule is a symbol. The rule can be regarded as a special case of pattern reasoning.

The pattern reasoning is a new solution for knowledge acquisition problem. The explanation is as follows. Since it is important to simulate human experts by computer software, expert systems have been studied to simulate human experts by computer software. Many expert systems are based on classical logic or something like classical logic (Hereinafter " classical logic" is used for simplification).

Knowledge acquisition is necessary, because the obscure knowledge of human experts cannot be reasoned by classical logic, while linguistic rules can be reasoned by classical logic. Knowledge acquisition means the conversion from obscure knowledge of human experts to linguistic rules. Knowledge acquisition has been studied by many researchers, but the results have not been successful, that is, the results show that knowledge acquisition is very difficult.

Generally speaking, a processing consists of a method and an object. For example, logical reasoning consists of the reasoning as the method and the symbols as the object. The methods of the processings by human experts are a kind of reasoning, which are different from the reasoning by classical logic. The objects of the processings by human experts are patterns (and symbols). That is, a lot of the processings by human experts can be regarded as the pattern reasonings.

Therefore, the pattern reasoning is a solution for the knowledge acquisition problem based on conversion of the knowledge acquisition to a completely different problem. However, readers may think that it is impossible to reason patterns. This paper shows that patterns can be reasoned by non-classical logics.

There are several possible definitions for patterns. Patterns such as images can be basically represented as functions. For example, two-dimensional images can be represented as the functions of two variables. Patterns are functions. Since it is desirable to be able to deal with any function, 3-layer feedforward neural networks, which can basically approximate any function[5], are studied.

Therefore, pattern reasonings are realized as logical reasonings of neural networks. However, classical logic cannot reason neural networks, while a few non-classical logics can reason neural networks. For example, intermediate logic LC[1],[4], product logic[4], and Lukasiewicz logic[4] can reason neural networks. The reason why the above three logics can reason neural networks is as follows: Neural networks are multilinear functions in the discrete domain and are well approximated to multilinear functions in the continuous domain, and the three logics are complete for multilinear function space, therefore, the three logics can reason neural networks.

The key is the multilinear function space. In the domain $\{0, 1\}$, the multilinear function space is an extension of Boolean algebra of Boolean functions. The space is the linear space expanded by the atoms of Boolean algebra of Boolean functions and can be made into a Euclidean space. Logical operations are represented as vector operations, which are numerical computations. In the domain [0,1], continuous Boolean functions can be obtained. Roughly speaking, continuous Boolean functions consist of conjunction, disjunction, direct proportion and inverse proportion. The multilinear function space of the domain [0,1] is the linear space of the atoms of Boolean algebra of continuous Boolean functions and can be made into a Euclidean space.

As explained above, multilinear function space is a model of three logics, but due to space limitations, intermediate logic LC(Hereinafter, LC for short) is explained in this paper. Intermediate logics are logics which are stronger than intuitionistic logic and weaker than classical logic. The multilinear function space is an algebraic model of intuitionistic logic, but intuitionistic logic is not complete for the space. For intuitionistic logic, refer to [1]. LC, which is stronger than intuitionistic logic and weaker than classical logic, is complete for the space. Therefore, multilinear functions can be regarded as propositions of LC. Neural networks which can be basically regarded as multilinear functions, can also be regarded as propositions of LC. Therefore, neural networks can be logically reasoned.

Section 2 explains multilinear function space which is the theoretical foundation for the logical reasoning of neural networks. Section 3 explains LC. Section 4 describes how neural networks can be reasoned by LC. Section 5 states several remarks on pattern reasoning.

2 Multilinear function space

First, the multiliner functions are explained. The domains are divided into discrete domains and continuous domains. The discrete domain is reduced to $\{0,1\}$ and the continuous domain is normalized to $[0,1]$. Therefore, $\{0,1\}$ domain and $[0,1]$ domain are discussed. Second, it is shown that multilinear function space of the domain $\{0,1\}$ is a Euclidean space spanned by the atoms of Boolean algebra of Boolean functions. Third, it is explained that the space of the domain $[0,1]$ can be made into a Euclidean space. Fourth, the vector representations are explained. Finally, the relationship between neural networks and multilinear functions is explained.

2.1 Multilinear functions

Definition 1 *Multilinear functions of n variables are as follows:*
$$\sum_{i=1}^{2^n} a_i x_1^{e_{i1}} \cdots x_n^{e_{in}},$$
where a_i is real, x_i is a variable, and e_i is 0 or 1.

In this paper, n stands for the number of variables.
Example Multilinear functions of 2 variables are as follows:
$axy + bx + cy + d$.
Multilinear functions do not contain any terms such as
$x_1^{k_1} x_2^{k_2} \cdots x_n^{k_n}$,
where $k_i \geq 2$. A function $f : \{0,1\}^n \to \mathbf{R}$ is a multilinear function, because $x_i^{k_i} = x_i$ holds in $\{0,1\}$ and so there is no term like $x_1^{k_1} x_2^{k_2} \cdots x_n^{k_n}$ $(k_i \geq 2)$ in the functions. In other words, multilinear functions are functions which are linear when only one variable is considered and the other variables are regarded as parameters.

2.2 Multilinear function space of the domain $\{0,1\}$

Definition 2 *The atoms of Boolean algebra of Boolean functions of n variables are as follows:*

$\phi_i = \prod_{j=1}^{n} e(x_j) \ (i = 1, ..., 2^n),$
where $e(x_j) = \overline{x_j}$ or x_j.

Example The atoms of Boolean algebra of Boolean functions of 2 variables are
as follows: $x \wedge y, \quad x \wedge \overline{y}, \quad \overline{x} \wedge y, \quad \overline{x} \wedge \overline{y}.$

Theorem 1 *The space of multilinear functions $(\{0,1\}^n \to \mathbf{R})$ is the linear space
spanned by the atoms of Boolean algebra of Boolean functions. The proof can be
found in [11],[12].*

Definition 3 *The inner product is defined as follows:*
$< f, g > = \sum_{\{0,1\}^n} fg.$
The sum in the above formula is done over the whole domain.

Definition 4 *Norm is defined as follows:*
$|f| = \sqrt{< f, f >} = \sqrt{\sum_{\{0,1\}^n} f^2}.$

Theorem 2 *The multilinear function space is a Euclidean space with the above
norm. The proof can be found in [11].*

2.3 Multilinear functions of the domain [0,1]

Multilinear functions of the domain [0,1] are briefly described in this subsection.
For details, see [8].

Definition 5 *Definition of* τ
    Let $f(x)$ be a real polynomial function. Consider the following formula:
    $f(x) = p(x)(x - x^2) + q(x),$
*where $q(x) = ax + b$, where a and b are real, that is, $q(x)$ is the remainder. τ_x is
defined as follows:*
    $\tau_x : f(x) \to q(x).$
The above definition implies the following property:
    $\tau_x(x^n) = x.$
In the case of n variables, τ is defined as follows:
    $\tau = \prod_{i=1}^{n} \tau_{x_i}.$

Example $\tau(x^2y^3 + y + 1) = xy + y + 1.$

Theorem 3 *The space of multilinear functions $([0,1]^n \to R)$ is a Euclidean
space with the following inner product:*
    $< f, g > = 2^n \int_0^1 \tau(fg)dx.$
The proof can be found in [8].

Definition 6 *Logical operations are defined as follows:*
AND: $\tau(fg)$, *OR:* $\tau(f + g - fg)$, *NOT:* $\tau(1 - f)$.

Theorem 4 *The functions obtained from Boolean functions by extending the domain from $\{0,1\}$ to $[0,1]$ can satisfy all axioms of classical logic with the logical operations defined above. The proof can be found in [9].*

Therefore, in this paper, the functions obtained from Boolean functions by extending the domain from $\{0,1\}$ to $[0,1]$ are called continuous Boolean functions. **Example** x, $1 - x(= \bar{x})$, and xy are continuous Boolean functions, where $x, y \in [0,1]$. x means a direct proportion and $\bar{x}$ means an inverse proportion.

2.4 Vector representations of logical operations

Multilinear functions are divided into Boolean functions and the others. The others can also be regarded as logical functions, which will be explained later. The vector representations of logical functions are called logical vectors. $\mathbf{f}((f_i)), \mathbf{g}((g_i)), ..$ stand for logical vectors. Note that f stands for a function, while f_i stands for an component of a logical vector $\mathbf{f}$.

Vector representations of logical operations are as follows:

$\mathbf{f} \wedge \mathbf{g} = (Min(f_i, g_i))$, $\mathbf{f} \vee \mathbf{g} = (Max(f_i, g_i))$, $\bar{\mathbf{f}} = (1 - f_i)$.

When multilinear functions are Boolean functions, the above vector representations of logical operations are the same as the representations below.

$\mathbf{f} \wedge \mathbf{g} = (f_i g_i)$, $\mathbf{f} \vee \mathbf{g} = (f_i + g_i - f_i g_i)$, $\bar{\mathbf{f}} = (1 - f_i)$.

2.5 The relationship between neural networks and multilinear functions

Theorem 5 *When the domain is $\{0,1\}$, neural networks are multilinear functions.*

Proof As described in 2.1, a function whose domain is $\{0,1\}$ is a multilinear function. Therefore, when the domain is $\{0,1\}$, neural networks are multilinear functions.

When the domain is $[0,1]$, neural networks are approximately multilinear functions with the following:

$x^k = x(k \leq a), = 0(k > a)$,

where a is a natural number. When $a = 1$, the above approximation is the linear approximation.

3 Intermediate logic LC and multilinear function space

The section briefly explains an intermediate logic LC and multilinear function space. Intermediate logics are weaker than classical logic and stronger than intuitionistic logic. An explanation of intermediate logics can be found in [1]. LC is an intermediate logic, which was presented by Dummett[2]. The logic is defined as follows[1].

LC=intuitionistic logic $+ (x \rightarrow y) \vee (y \rightarrow x)$,

329

where x and y are logical formulas. LC stands for Logic of Chain, which comes from the fact that the model of the logic is a chain, that is, a linearly ordered set.

First, it is explained that LC is complete for the interval[0,1]. The proof for the completeness of LC for the model cannot be described due to space limitations, and so an intuitive explanation is given. Second, it is explained that the multilinear function space is an algebraic model of LC. Finally, an example of logical reasoning of multilinear functions by LC are given.

3.1 An intuitive explanation for LC

Interval $[0, 1]$ is a Heyting algebra, which is the algebraic model of intuitionistic logic, with the following definitions:
$x \wedge y = Min(x, y)$, $x \vee y = Max(x, y)$,
$$x \supset y = \begin{cases} 1(x \leq y) \\ y(x > y), \end{cases}$$
$\top = 1$, $\quad \bot = 0$, where x and y stand for points.
The above fact can be easily verified. Let x and y stand for two points, then
$(x \leq y) \vee (y \leq x)$
holds. Roughly speaking, by replacing $\leq$ in the above formula by $\rightarrow$, the following formula is obtained.
$(x \rightarrow y) \vee (y \rightarrow x)$,
where x and y are propositions. The above formula does not hold in intuitionistic logic. In other words, intuitionistic logic is not complete for an interval.

If the above formula is added to intuitionistic logic, a logic which is complete for an interval is obtained. The logic is LC.
$(x \rightarrow y) \vee (y \rightarrow x)$,
holds in LC, therefore, LC is complete for an interval.

The completeness of LC for an interval can be proved using the fact that LC is complete for linearly ordered Kripke models[1] and the correspondence between Kripke models and algebraic models [3].

3.2 The multilinear function space is an algebraic model of LC

It is explained that the multilinear function space is an algebraic model of LC as follows[10].

1. If an interval is a model of a logic, the direct sum of the intervals is also a model of the logic[6]. The logical operations are done componentwise. Therefore, since an interval [0,1] is an algebraic model of LC, a direct sum of intervals $[0, 1]^m$ (m is dimension) is also an algebraic model of LC.
2. The multilinear function space is a linear space, therefore, a subset of the space $[0, 1]^m$ is a direct sum of the interval $[0,1]$.
3. From item 1 and 2, the subset $[0, 1]^m$ of the multilinear function space is an algebraic model of LC.

Theorem 6 *LC is complete for the hypercube* $[0,1]^m$ *of the space. The definitions are as follows:*

$$\mathbf{f} \leq \mathbf{g} = \forall i(f_i \leq g_i), \ \mathbf{f} \wedge \mathbf{g} = (Min(f_i, g_i)), \ \mathbf{f} \vee \mathbf{g} = (Max(f_i, g_i)),$$
$$\mathbf{f} \supset \mathbf{g} = (f_i \supset g_i)$$
$$f_i \supset g_i = \begin{cases} 1(f_i \leq g_i) \\ g_i(f_i > g_i), \end{cases}$$
$$\overline{\mathbf{f}} = (f_i \supset 0)$$

where **f** *and* **g** *stand for logical vectors. This theorem is understood from the above discussions. The proof is omitted.*

Example
$$f = 0.6xy + 0.1x + 0.1y + 0.1$$
is transformed to
$$0.9xy + 0.2x\overline{y} + 0.2\overline{x}y + 0.1\overline{xy},$$
therefore, $\mathbf{f} = (0.9, 0.2, 0.2, 0.1)$.
In the same way,
$$\overline{\mathbf{f}} = (f_i \supset 0) = (0.9 \supset 0, 0.2 \supset 0, 0.2 \supset 0, 0.1 \supset 0) = (0, 0, 0, 0).$$
Therefore, from Theorem 6,
$$\mathbf{f} \vee \overline{\mathbf{f}} = (0.9, 0.2, 0.2, 0.1),$$
which means $\mathbf{f} \vee \overline{\mathbf{f}} \neq 1$
This example shows that the law of excluded middle $f \vee \overline{f} = 1$,
which holds in classical logic, does not holds in LC. If f is limited to Boolean functions, the law of excluded middle holds. For example, let f be xy, that is,
$$f = 1.0xy + 0.0x + 0.0y + 0.0 = 1.0xy + 0.0x\overline{y} + 0.0\overline{x}y + 0.0\overline{xy},$$
then $\mathbf{f} = (1, 0, 0, 0)$,
$$\overline{\mathbf{f}} = (f_i \supset 0) = (1 \supset 0, 0 \supset 0, 0 \supset 0, 0 \supset 0) = (0, 1, 1, 1).$$
Therefore, $\mathbf{f} \vee \overline{\mathbf{f}} = (1, 1, 1, 1)$, that is, $f \vee \overline{f} = 1$.

4 Logical reasoning of neural networks by LC

Neural networks are multilinear functions and the multilinear function space is an algebraic model of LC. Therefore, neural networks can be reasoned by LC. The domain is $\{0, 1\}^n$, where n is the number of variables.

Let N_1 and N_2 be two trained neural networks, which have 3 layers, two inputs x and y, two hidden units, and one output. The output function of each unit is a sigmoid function. The following tables show the training results of weight parameters and biases of N_1 and the training results of weight parameters and bias of N_2.

unit	w1(w3, w5)	w2(w4,w6)	bias	unit	w1(w3, w5)	w2(w4,w6)	bias
hidden 1	-4.87	-4.86	-6.70	hidden 1	4.80	4.72	-2.31
hidden 2	-2.86	-2.88	3.50	hidden 2	-3.49	-3.56	1.67
output	7.61	-3.83	4.50	output	5.81	-4.62	-0.42

N_1 is as follows:
$$S(7.61S(-4.87x - 4.86y - 6.70) - 3.83S(-2.86x - 2.88y + 3.50) + 4.50),$$

From the above formula, the logical vector is calculated as follows:
$(0.98, 0.01, 0.01, 0.00)$.
The logical vector of N_2 is calculated in the same way as follows:
$(0.02, 0.98, 0.98, 0.99)$.
The logical conjunction of the two logical vector is as follows:
$(0.02, 0.01, 0.01, 0.00)$,
which is nearly equal to 0.
The multilinear function is as follows:
$0.02xy + 0.01x(1 - y) + 0.01(1 - x)y + 0.00 = 0.01x + 0.01y$.

The function is nearly equal to 0. The above result shows that the logical conjunction of two trained neural networks is almost false, which cannot seen from the training results of neural networks. N_1 has been trained using $x \wedge y$ and N_2 has been trained using the negation of $x \wedge y$. Therefore, the logical conjunction of N_1 and N_2 is as follows:
$N_1 \wedge N_2 \simeq (x \wedge y) \wedge \overline{(x \wedge y)} = 0$.
As seen in the above example, the logical reasoning of neural networks shows the logical relations among neural networks. If the components of logical vectors are 0 or 1, the calculation can be done by Boolean algebra, that is, classical logic. However, even if the training targets are Boolean functions, the training results of neural networks are not 0 or 1, but are values like 0.01 or 0.98. These numbers cannot be calculated by Boolean algebra, but can be calculated by LC. In the above example, the training targets are Boolean functions for simplification. However, any function can be the training target of neural networks and any trained neural network can be reasoned by LC. The logical implication between a neural network and another neural network can be calculated in a similar way as in the above example.

The computational complexity of the logical reasoning is exponential. Therefore, efficient algorithms are needed, which have been developed[12]. However, due to space limitations, the efficient algorithms cannot be explained in this paper. They will be explained in another paper.

5   Remarks on pattern reasoning

Patterns can be regarded as functions and the functions can be approximated by neural networks. Neural networks can be reasoned by a few logics such as LC, Łukasiewicz logic and product logic. Therefore, pattern reasoning can be realized by logical reasoning of neural networks. However, there are a lot of open problems for pattern reasoning to be applied to real data.

1. Computational complexity
   A basic algorithm is exponential in computational complexity, therefore, a polynomial algorithm is needed. A polynomial algorithm for a unit in a neural network has been presented. For networks, an algorithm which uses only big weight parameters has been presented. The reduction of computational complexity is included in future work.

2. Appropriate logics for pattern reasoning

Probability calculus is similar to a reasoning for patterns, although it does not have the formal system. Probability calculus does not satisfy the contraction rule[7]:

$$\text{contraction } \frac{x, x \to y}{x \to y}.$$

Therefore, appropriate logics for pattern reasoning should not satisfy the contraction rule. From this viewpoint, Łukasiewicz logic and product logic, which do not satisfy the contraction rule, are more appropriate than LC, which satisfies the contraction rule. It is desired that probability calculus be formalized logically, but this is very difficult. We are investigating appropriate logics for pattern reasoning.

3. Typical patterns

There are countless patterns, and some patterns are appropriate for pattern reasoning, while other patterns are inappropriate. Therefore a dictionary of patterns is necessary. The patterns included in the dictionary are typical patterns, which cannot be described linguistically. The typical patterns can be gathered by various methods, but we do not have to be seriously concerned with gathering typical patterns, because pattern reasoning is flexible as explained in the next item. However, gathering typical patterns are important for efficient pattern reasoning.

4. A difference between pattern reasoning and symbol reasoning in the reasoning mechanism

In symbol reasoning, when the left side of a rule is not matched, the rule does not work, while, in pattern reasoning, even when the left side of a rule is not matched, the rule works. For example, let a rule be $a \to b$ and the left side of the rule be a'. If a is very similar to a', the truth value of the rule is almost 1. On the other hand, if a is very different from a', the truth value of the rule is almost 0. Pattern reasoning works like this, because the pattern reasoning makes use of continuously valued logics. There are several other methods which deal with matching degrees of the left sides of rules. However, the methods are basically arbitrary, whereas the pattern reasoning presented in this paper includes the matching degrees in the system.

5. Formal system

In pattern reasoning, for example, a question like "Is this pattern logically deduced from the set of rules of patterns?" should be answered. Therefore, formal systems are needed for pattern reasoning.

6. Incompleteness

In mathematical logic, completeness is important. In reality, humans cannot reason or prove true things, that is, humans are incomplete. Therefore, pattern reasoning should deal with incompleteness.

7. The relationship with probability theory

Probability calculus deals with continuous values, but probability events are not continuous, that is, the objects of probability theory are not continuous, while the objects of pattern reasoning are continuous. Therefore, pattern reasoning can be regarded as an extension of probability calculus.

8. Experimental study
   The most typical patterns are images, therefore the final target is the reasoning of images. We have to begin experiments with simple examples. We have tried to realize pattern reasoning for one-dimensional data, for example, time series data, by logical reasoning of neural networks using LC, Łukasiewicz logic or product logic. The results show that the logical reasoning of neural networks works well, which will be reported in another paper.

6 Conclusions

This paper has presented pattern reasoning, which is a new solution for the knowledge acquisition problem. Knowledge acquisition tried to acquire linguistic rules from patterns. In contrast, we have tried to modify logics to reason patterns. Patterns are represented as functions, which are approximated by neural networks. Therefore, the logical reasonings of neural networks have been studied. A few logics can reason neural networks. This paper has explained intermediate logic LC. There are a lot of open problems, therefore the author strongly encourages the readers to join the research field.

References

1. D.V. Dalen: Intuitionistic Logic, *Handbook of Philosophical Logic III*, D. Gabbay and F.Guenthner eds., pp.225-339, D.Reidel, 1984.
2. M. Dummett: A Propositional Calculus with Denumerable Matrix, *The Journal of Symbolic Logic*, Vol.24, No.2, pp.97-106, 1959.
3. M. C. Fitting: *Intuitionistic Logic-Model Theory and Forcing*, North Holland, 1969.
4. P. Hájek: *Metamathematics of Fuzzy Logic*, Kluwer, 1998.
5. K.Hornik: Multilayer Feedforward Networks are Universal Approximators, *Neural Networks*, Vol.2 pp.359-266, 1989.
6. T. Hosoi and H. Ono:Intermediate Propositional Logics(A Survey),*Journal of Tsuda College*, Vol.5, pp.67-82, 1973.
7. H. Ono and Y. Komori: Logics without the contraction rule, *J. Symbolic Logic* 50, pp.169-201, 1985.
8. H. Tsukimoto and C. Morita: The discovery of propositions in noisy data, *Machine Intelligence 13*, pp.143-167, Oxford University Press, 1994 .
9. H. Tsukimoto: Continuously Valued Logical Function Satisfying All Axioms of Classical Logic, *Systems and Computers in Japan*, Vol.25, No.12, pp.33-41, SCRIPTA TECHNICA, INC., 1995.
10. H. Tsukimoto: The space of multi-linear functions as models of logics and its applications, *Proceedings of the 2nd Workshop on Non-Standard Logic and Logical Aspects of Computer Science*, 1995.
11. H. Tsukimoto and Chie Morita: Efficient algorithms for inductive learning-An application of multi-linear functions to inductive learning, *Machine Intelligence 14*, pp.427-449, Oxford University Press, 1995.
12. H.Tsukimoto:Symbol pattern integration using multilinear functions, in *Deep Fusion of Computational and Symbolic Processing*, (eds.) Furuhashi,T., Tano,.S., and Jacobsen,H.A. Springer Verlag, 1999. (To appear)

Probabilistic Inference and Bayesian Theorem Based on Logical Implication

Yukari Yamauchi[1] and Masao Mukaidono[1]

Dept. of Computer Science, Meiji University,
Kanagawa, Japan
{yukari, masao}@cs.meiji.ac.jp

Abstract. Probabilistic reasoning is an essential approach of approximated reasoning to treat uncertain knowledge. Bayes' theorem based on the interpretation of a If-Then rule as the conditional probability is widespread in applications of probabilistic reasoning. A new type of Bayes theorem based on the interpretation of a If-Then rule as the logical implication is introduced in this paper, where addition and subtraction are employed in the probabilistic operations instead of multiplication and division employed for the conditional probability of the traditional Bayes' theorem. Inference based on both interpretations of the If-Then rules, conditional probability and logical implication, are discussed.

1 Introduction

In propositional logic, the truth values of propositions are given either 1(true) or 0(false). Inference based on propositional (binary) logic is done using inference rule : Modus Ponens, shown in Fig. 1. This rule implies that if an If-Then rule "$A \to B$" and proposition A are given true(1) as premises, then we come to a conclusion that proposition B is true(1).

$$A \to B$$
$$\frac{A}{B}$$

Fig. 1. Modus Ponens

The inference rule based on propositional logic is extended to probabilistic inference based on probability theory in order to treat uncertain knowledge. The truth values of propositions are given as the probabilities of events that take any value in the range of $[0, 1]$. Here, U is the sample space (universal set), $A, B \subseteq U$ are events, and the probability of "an event A happens", $\mathbf{P}(A)$ is defined as $\mathbf{P}(A) = |A|/|U|$ ($|U| = 1$, $|A| = a \in [0, 1]$) under the interpretation of randomness. Thus the probabilistic inference rule can be written as Fig. 2 adapting the style of Modus Ponens.

$$\mathbf{P}(A \to B) = i$$
$$\mathbf{P}(A) = a$$

$$\overline{\mathbf{P}(B) = b} \quad i, a, b \in [0, 1]$$

Fig. 2. Probabilistic Inference

If the probability of $A \to B$ and A are given 1 ($i = a = 1$), then b is 1, since the probabilistic inference should be inclusive of modus ponens as a special case. Our focus is to determine the probability of B from the probabilities of $A \to B$ and A that take any value in $[0, 1]$. $A \to B$ is interpreted as "if A is true, then B is true" in meta-language. Traditional Bayes' theorem applied in many probability system adopts conditional probability as the interpretation of If-Then rule. However, the precise interpretation of the symbol "$\to$" is not unique and still under discussion among many researchers.

E. Trillas and S. Cubillo [1] remarked the inequality $x \cdot (x \to y) \leq y$ (where $x \to y = x^\sim \vee y$) valid in an arbitrary Boolean algebra ($\mathbf{B}, \vee, \cdot, ^\sim, 0, 1$), and determined Boolean variants of modus ponens by replacing conjunction ($\cdot$) and implication ($\to$) by other truth functions. Nilsson [2] presented a semantical generalization of ordinary first-order logic in which the truth values of sentences can range between 0 and 1. He established the foundation of *probabilistic logic* through a possible-world analysis and probabilistic entailment. However, in most cases, we are not given the probabilities for the different sets of possible worlds, but must induce them from what we are given.

Our goal is to deduce a conclusion and its associated probability from given rules and facts and their associated probabilities through simple geometric analysis. The probability of the sentence "if A then B" is interpreted in two ways: conditional probability and the probability of logical implication. In this paper, we define the probabilistic inferences based on the two interpretations of "If-Then" rule, conditional probability and logical implication, and introduce a new variant of Bayes' theorem based on the logical implication.

2 Inference Based on Probability Theory

2.1 Conditional Probability

Conditional probability, "how often B happens when A is already (or necessary) happens", only deals with the event space that A certainly happens. Thus the sample space changes from U to A.

$$\mathbf{P}(A \to B) = \mathbf{P}(B|A) = |A \cap B| / |A|, \tag{1}$$
$$i_c = \mathbf{P}(A \cap B)/a. \quad (a \neq 0) \tag{2}$$

Since $\mathbf{P}(A \cap B) = i_c \times a$ from Equation (2), the possible size of B is restricted from $|A \cap B| = i_c \times a$ to $|A^c \cup B| = 1 - (a - a \times i_c)$. Thus the probabilistic inference based on the interpretation of if-then rule as the conditional probability determines $\mathbf{P}(B)$ from given $\mathbf{P}(A \to B)$ and $\mathbf{P}(A)$ by the following inference style in Fig. 3.

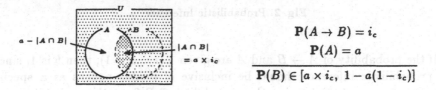

$$\mathbf{P}(A \to B) = i_c$$
$$\mathbf{P}(A) = a$$
$$\mathbf{P}(B) \in [a \times i_c, \ 1 - a(1 - i_c)]$$

Fig. 3. Conditional Probability

Note, $\mathbf{P}(B)$ can not be determined uniquely from $\mathbf{P}(A \to B)$ and $\mathbf{P}(A)$ thus expressed as the interval probability [3]. When the condition, $a \times i_c = 1 - a(1 - i_c)$ (thus $a = 1$), holds, $\mathbf{P}(B)$ is unique and equal to $\mathbf{P}(A \to B)$.

$$\mathbf{P}(A \to B) = i_c$$
$$\mathbf{P}(A) = 1$$
$$\mathbf{P}(B) \in [1 \times i_c, \ 1 - 1 + 1 \times i_c] = [i_c, \ i_c] = i_c$$

Fig. 4. Conditional Probability $a = 1$

2.2 Logical Implication

The interpretations of $\to$ (implication) in logics: propositional (binary or Boolean)logic, multi-valued logic, fuzzy logic, etc., are not unique in each logic. However, the most common interpretation of $A \to B$ is $\sim A \vee B$.

$$\mathbf{P}(A \to B) = \mathbf{P}(A^c \cup B) = |A^c \cup B|/|U|, \tag{3}$$
$$i_l = \mathbf{P}(A \cap B) + (1 - a). \quad (a + i_l \geq 1) \tag{4}$$

In order to avoid contradiction in premises, the relationship between a and i_l must hold the condition: $a + i_l \geq 1$.

Since $\mathbf{P}(A \cap B) = a - (1 - i_l)$ from Equation (4), the possible size of B is restricted from $|A \cap B| = a - (1 - i_l)$ to $|A^c \cup B| = i_l$. The probabilistic inference based on the interpretation of if-then rule as the logical implication determines $\mathbf{P}(B)$ as the interval probability from given $\mathbf{P}(A \to B)$ and $\mathbf{P}(A)$ as shown in Fig. 5.

Similar to the conditional probability case shown in Fig. 4, $\mathbf{P}(B)$ is unique and equal to $\mathbf{P}(A \to B)$ when the condition $i_l + a - 1 = i_l$ (thus $a = 1$) holds.

337

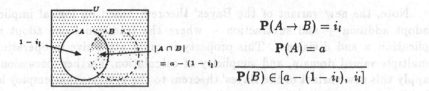

Fig. 5. Logical Implication

$$P(A \to B) = i_l$$
$$P(A) = a$$
$$P(B) \in [a - (1 - i_l),\ i_l]$$

3 Bayes' Theorem

Bayes' theorem is widespread in application since it is a powerful method to trace a cause from effects. The relationship between a priori probability $P(A \to B)$ and a posteriori probability $P(B \to A)$ is expressed in the following equation by eliminating $P(A \cap B)$ from the definitions.

$P(A \to B) = P(B|A) = P(A \cap B)/P(A),$
$P(B \to A) = P(A|B) = P(A \cap B)/P(B),$

$$P(B \to A) = P(A) \times P(A \to B)/P(B) \tag{5}$$

Theorem 3.01 *The interpretation of $\to$ as the logical implication satisfies the following equation.*

$$P(B \to A) = P(A) + P(A \to B) - P(B) \tag{6}$$

Proof. Given $P(A \to B) = i_l$, $P(A) = a$, and $P(B) = b$,

$$\begin{aligned}
P(B \to A) &= P(B^c \cup A) \\
&= 1 - P(B \cap A^c) \\
&= 1 - (b - (a - 1 + i_l)) \\
&= 1 - (b - a + 1 - i_l) \\
&= a + i_l - b \\
&= P(A) + P(A \to B) - P(B).
\end{aligned}$$

□

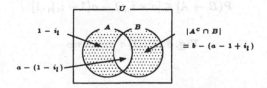

Fig. 6. Bayes' Theorem with Logical Implication

Note, the new variant of the Bayes' theorem based on logical implication adopt addition + and subtraction − where the traditional one adopt multiplication × and division /. This property is quite attractive in operations on multiple-valued domain, and simplicity of calculation. Farther discussion is to apply this new variant of the Bayes' theorem to the systems that employ logical implication.

4 Inference Applying Bayes' Theorem

4.1 Bayes' Inference Based on Conditional Probability

Now, we apply Bayes' theorem as the inference rule, and define $\mathbf{P}(B \to A)$ from $\mathbf{P}(A \to B)$, $\mathbf{P}(A)$, and $\mathbf{P}(B)$. The inference based on the traditional Bayes' theorem (conditional probability) is shown in Fig. 7.

$$\mathbf{P}(A \to B) = i_c$$
$$\mathbf{P}(A) = a$$
$$\mathbf{P}(B) = b$$
$$\overline{\mathbf{P}(B \to A) = a \times i_c/b}$$

Fig. 7. Bayes' Inference - Conditional Probability

The condition, $\max((a+b-1), 0)/a \leq i_c \leq b/a$, must be satisfied between the probabilities a, b, and i_c. Since $i_c = \mathbf{P}(A \cap B)/a$ thus $\max(a + b - 1, 0) \leq \mathbf{P}(A \cap B) \leq \min(a, b)$.

From $\mathbf{P}(A \to B)$ and $\mathbf{P}(A)$, $\mathbf{P}(B)$ is determined as the interval probability by the inference rule, Fig. 3 in the previous discussion. Thus $\mathbf{P}(B \to A)$ can be determined as follows when $\mathbf{P}(B)$ is unknown.

$$\mathbf{P}(A \to B) = i_c$$
$$\mathbf{P}(A) = a$$
$$\mathbf{P}(B) \in [a \times i_c, \ 1 - a(1 - i_c)]$$
$$\overline{\mathbf{P}(B \to A) \in [a \times i_c/1 - a(1 - i_c), \ 1]}$$

Fig. 8. $\mathbf{P}(B)$: unknown

$\mathbf{P}(B \to A)$ is unique ($\mathbf{P}(B \to A) = 1$) when $a \times i_c = 1 - a(1 - i_c)$, that is $a = 1$. Note, $\mathbf{P}(B \to A)$ does not depends on $i_c = \mathbf{P}(A \to B)$.

$$\mathbf{P}(A \rightarrow B) = i_c$$
$$\mathbf{P}(A) = 1$$
$$\overline{\mathbf{P}(B \rightarrow A) = 1}$$

Fig. 9. $\mathbf{P}(A) = 1$

4.2 Bayes' Inference Based on Logical Implication

Similarly, applying the new variant of Bayes' theorem based on logical implication, we get the following inference rule in Fig. 10.

$$\mathbf{P}(A \rightarrow B) = i_l$$
$$\mathbf{P}(A) = a$$
$$\mathbf{P}(B) = b$$
$$\overline{\mathbf{P}(B \rightarrow A) = a + i_l - b}$$

Fig. 10. Bayes' Inference - Logical Implication

The condition, $\max(b, 1-a) \leq i_l \leq 1 - a + b$, must be satisfied between the probabilities a, b, and i_l. Since $i_l = \mathbf{P}(A \cap B)/a$ thus $\max(a + b - 1, 0) \leq \mathbf{P}(A \cap B) \leq \min(a, b)$.

$\mathbf{P}(B)$ is determined from $\mathbf{P}(A \rightarrow B)$ and $\mathbf{P}(A)$ by the inference rule, Fig. 5. Thus $\mathbf{P}(B \rightarrow A)$ can be determined as follows when $\mathbf{P}(B)$ is unknown. Note, the result of inference does not depend on the probability of $\mathbf{P}(A \rightarrow B)$. Clearly, $\mathbf{P}(B \rightarrow A)$ is unique ($\mathbf{P}(B \rightarrow A) = 1$) when $a = 1$.

$$\mathbf{P}(A \rightarrow B) = i_l$$
$$\mathbf{P}(A) = a$$
$$\mathbf{P}(B) \in [a + i_l - 1, i_l]$$
$$\overline{\mathbf{P}(B \rightarrow A) \in [a, 1]}$$

Fig. 11. $\mathbf{P}(B)$ unknown

5 Generalization on Interval Probability

Since the results of inferences are given as interval probability, we shall discuss the inference methods when the probability of sentences are given as the interval probabilities. Given the set of all interval value $\mathcal{I}$,

$$\mathcal{I} = \{[a, b] \mid 0 \le a \le b \le 1\}$$

the interval probability of "A happens" is $\mathbf{P}(A) \in [a_1, a_2]$.

5.1 Interval Probabilistic Inference

In the previous section, the probabilistic inference based on the conditional probability determines $\mathbf{P}(B) \in [a \times i_c, 1 - a(1 - i_c)]$ from given $\mathbf{P}(A \to B) = i_c$ and $\mathbf{P}(A) = a$. Thus, given $\mathbf{P}(A \to B)$ and $\mathbf{P}(A)$ as interval probabilities $[ic_1, ic_2]$ and $[a_1, a_2]$, the possible probability of $\mathbf{P}(B)$ is minimum $a_1 \times ic_1$ and maximum $1 - a_1(1 - ic_2)$.

$$\mathbf{P}(A \to B) \in [ic_1, ic_2]$$
$$\mathbf{P}(A) \in [a_1, a_2]$$
$$\overline{\mathbf{P}(B) \in [a_1 \times ic_1, \ 1 - a_1(1 - ic_2)]}$$

Fig. 12. Probabilistic Inference Based on Conditional Probability

Similarly, the probabilistic inference based on the interpretation of if-then rule as the logical implication determines $\mathbf{P}(B)$ from given $\mathbf{P}(A \to B) \in [il_1, il_2]$ and $\mathbf{P}(A) \in [a_1, a_2]$ as shown in Fig. 13. $\mathbf{P}(B)$ is minimum if both $\mathbf{P}(A \to B)$ and $\mathbf{P}(A)$ takes minimum value il_1 and a_1. However, in order to avoid contradiction, the condition $a + il \ge 1$ must be satisfied between any combination of the probabilities a and il. Thus the minimum value of $\mathbf{P}(B)$ is restricted to $\max(a_1 + il_1, 1) - 1 = \max(a_1 + il_1 - 1, 0)$.

$$\mathbf{P}(A \to B) \in [il_1, il_2]$$
$$\mathbf{P}(A) \in [a_1, a_2]$$
$$\overline{\mathbf{P}(B) \in [a_1 + il_1 - 1, \ il_2]}$$

Fig. 13. Probabilistic Inference Based on Logical Implication

Note, in both cases, the results of inference $\mathbf{P}(B)$ does not depends on a_2 (the maximum value of $\mathbf{P}(A)$).

5.2 Inference on Interval Probability based on Bayes' Theorem

Now, we apply Bayes' theorem as the inference rule on interval probabilities. Given $\mathbf{P}(A \to B)$ as the conditional interval probability, $\mathbf{P}(B \to A)$ is determined from $\mathbf{P}(A \to B) \in [ic_1, ic_2]$, $\mathbf{P}(A) \in [a_1, a_2]$ and $\mathbf{P}(B) \in [b_1, b_2]$ by the Bayes' theorem as shown in Fig. 14.

$$P(A \to B) \in [ic_1, ic_2]$$
$$P(A) \in [a_1, a_2]$$
$$P(B) \in [b_1, b_2]$$

$$\overline{P(B \to A) \in [a_1 \times ic_1/b_2, a_2 \times ic_2/b_1]}$$

Fig. 14. Bayes' Inference - Conditional

Note, the same condition in previous section 4.1, $\max((a + b - 1), 0)/a \leq i_c \leq b/a$, must be satisfied between any combinations of the probabilities a, b, and ic.

$P(B)$ is determined as the interval probability from $P(A \to B)$ and $P(A)$ by the inference rule as shown in Fig. 12 in the previous discussion. Thus $P(B \to A)$ can be determined as follows when $P(B)$ is unknown.

$$P(A \to B) \in [ic_1, ic_2]$$
$$P(A) \in [a_1, a_2]$$
$$P(B) \in [a_1 \times ic_1, \, 1 - a_1(1 - ic_2)]$$

$$\overline{P(B \to A) \in [a_1 \times ic_1/1 - a_1(1 - ic_1), \, 1]}$$

Fig. 15. Bayes' Inference - Conditional: $P(B)$ unknown

$P(B \to A) = 1$ when $a_1 \times ic_1 = 1 - a_1(1 - ic_1)$, that is $a_1 = 1$. $P(B \to A)$ does not depends on $P(A \to B)$.

Similarly, applying the new variant of Bayes' theorem based on logical implication on interval probabilities, $P(A \to B)$, $P(A)$ and $P(B)$, we get the following inference rule in Fig. 16.

$$P(A \to B) \in [il_1, il_2]$$
$$P(A) \in [a_1, a_2]$$
$$P(B) \in [b_1, b_2]$$

$$\overline{P(B \to A) \in [a_1 + il_1 - b_2, a_2 + il_2 - b_1]}$$

Fig. 16. Bayes' Inference - Logical Implication

The same condition in previous section 4.2, $\max(b, 1 - a) \leq i_l \leq 1 - a + b$ must be satisfied between the probabilities a, b, and i_l.

$P(B)$ is determined as the interval probability from $P(A \to B)$ and $P(A)$ by the inference rule as shown in Fig. 13 in the previous discussion. Thus $P(B \to A)$ can be determined as follows when $P(B)$ is unknown.

$$P(A \to B) \in [il_1, il_2]$$
$$P(A) \in [a_1, a_2]$$
$$P(B) \in [a_1 + il_1 - 1, il_2]$$

$$P(B \to A) \in [a_1, 1]$$

Fig. 17. Bayes' Inference - Logical Implication: $P(B)$ unknown

Note, the result of inference does not depend on the probability of $P(A \to B)$. $P(B \to A) = 1$ when $a_1 = 1$.

6 Conclusion

Inference based on probability theory is discussed as a method of approximated reasoning that treat uncertain knowledge. A new type of Bayesian theorem based on the interpretation of a If-Then rule as the logical implication is introduced. The new variant of the Bayes' theorem based on logical implication adopt addition $+$ and subtraction $-$ where the traditional one adopt multiplication $\times$ and division $/$. This property is quite attractive in consideration of operations on multiple-valued domain, and simplicity of calculation. Interesting topic for farther discussion should be to apply this new variant of the Bayes' theorem to the systems that employ logical implication.

References

1. E.Trillas, S.Cubillo, "Modus ponens on Boolean algebra revisited," *Mathware & Soft Computing 3* (1996), 105-112.
2. N.J.Nilsson, "Probabilistic Logic", *Artificial Intelligence*, Vol.28, No.1, 71-78, (1986)
3. Y.Yamauchi, M.Mukaidono, "Interval and Paired Probabilities for Treating Uncertain Events," IEICE Transactions of Information and Systems, Vol.E82-D, No.5, 955-961, (1999)

Reasoning with Neural Logic Networks

Ramin Yasdi

GMD FIT, German National Research Center for Information Technology,
Institute for Human-Computer Interaction, 53754 Sankt Augustin, Germany,
ramin.yasdi@gmd.de

Abstract. This article presents a neural network approach for human reasoning. It is based on a three-valued Boolean logic. We will first laying down the foundations for study of a neural logic and represent it by a neural logic network. We than realize the process of reasoning by the structure of a neuro model. The nodes represents the function of reasoning and the connection weights the parameter of reasoning. The model is close to realization of the particular application. The goal of this research is to develop a reasoning system capable of human reasoning based on neural logic network.

1 Introduction

Rule-based systems have been successfully applied in many domains. However, the current rule-based technology is generally lacking in learning capability and parallism. It is also weak in dynamic reasoning, control, and uncertainty processing. The reasoning proceeds through a pre defined tree. Neural logic networks, on the other hand, lend themselves very well to learning and parallelism due to their self-organization features. In additions, it is also capable of incorporating temporal reasonings, and certainty factors with ease. In spite of these capabilities, however, neural logic networks basically are confined to relatively simpler problem. They generally are deficient in the in-depth reasoning afforded by the rule based-based systems. In view of the above, one of the goals of current neural logic networks research is to bridge the gap between symbolic and sub-symbolic approaches with the fusion of hybrid systems. We propose a computional model that combines the best of the both worlds by integrating an ordinary rule-based system into a neural logic network architecture. The model contains a neural logic based inference engine that *dynamically chains active rules* together into a neural logic network. Parallel execution of rules becomes possible when active rules are being chained along different path. The system further exhibits learning capability by allowing weights to be adjusted during training sessions.

Logic has traditionally been one of the foundations for symbolic paradigm. A neural logic network model that represents propositional truth values as neural activations and logical operations as connection weights, has been proposed by [4, 2] to represent logic and perform logical inference by the structure and dynamics of the network respectively. The underlying neural logic network demonstrates a multitude of logical operations, besides the standard operators such as AND, OR, NOT, NOR and NAND. Particularly, the user is free to define any operations to meet any specific needs. The strengths of neural logic thus provide a much greater expressive power for the systems's rules syntax than a ordinary rule-based systems. We present a rule inferencing system whereby the internal representation and the inferencing mechanism of propositional rules are driven by a neural logic network. We base our discussion on a framework for deductive systems in A.I., namely the logic level, the calculus level, the representation level, and the control level. We will show that neural logic does enrich the meaning logic with DONT KNOW truth value and human-like logical operators which are more appropriate for knowledge processing and decision making.

The paper is organized as follows. We start next section with introducing the basic of neural logic network and give the definitions of standard and non standard logical operators. We describe than the human logical reasoning. We conclude with a discussion of the presented method and look ahead to extension and future directions of this work.

2 Neural Logic Network

Neural logic network is a class of artificial neural network which is used to model human intelligence by computing systems. It can model classical two-valued Boolean logic effectively. This logic is in fact a good model to study human logic which is multivalued, fuzzy and biased. The neural logic network considered in this work is inspired from [4].

A Neural Logic Network (NLN for short) is a finite directed graph. It contains a set of input nodes and output nodes. Every node can take one of the three order pair activation values: (1,0) for *true* (0,1) for *false* and (0,0) for *unknown*. Every edge in net is also associated with an ordered pair weight (t, f) where t and f are real numbers of positive, negative or zero value.

2.1 Mathematical definition of NLN

An abstract neural logic network is a mathematical system with following features:

- It is a finite directed graph consisting of a set of nodes N and a set of links E;

- A non-empty subset I of N is chosen as input nodes. Another non-empty subset O of N is chosen as output nodes. Other nodes are called hidden nodes;

- An algebraic system $< R, +, \times >$ which is satisfying the axioms of a ring. An association of a set of links to a set of R is defined by a mapping φ_1
  $$\varphi_1 : \qquad E \to R$$
  That is to say, every links of the directed graph is given a value from the chosen ring $< R, +, \times >$;

- A subset of A is chosen from the ring R together with a specially chosen mapping φ_2 from the set of all non-input nodes (i.e. $N \sim I$) to the set A
  $$\varphi_2 : \qquad (N \sim I) \to A$$
  That is to say, non-input node of N is given a value in A. The elements in A are to be called *truth-values*;

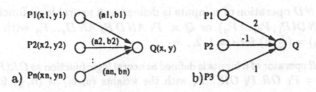

Figure 1: a) 3-valued NLN, b) Boolean NLN

- A mapping from R to A i.e.
  $$f: \qquad R \to A$$
  called the threshold function.

An abstract neural logic can now be denoted by

$$Net =< N, E, I, O, R, A, \varphi_1, \varphi_2, f >$$

By changing ring R and the threshold function different sub-classes of neural logic network can be obtained. Three-valued NLN, Boolean NLN and Fuzzy NLN are the three main sub-class of NLN. For instance the 3-valued NLN can be obtained by representing the value of input $P_i (i = 1, ... n)$ by ordered pairs (x_i, y_i) with the weights (a_i, b_i) and the value of output Q by an ordered pair (x, y). Letting $A = A^{(T)} = \{(1,0), (0,1), (0,0)\}$, (where $A^{(T)}$ means the truth value set in 3-valued NLN) be the truth value set, where $(1,0), (0,1), (0,0)$ represent *true, false* and *don't know* respectively, a_i, b_i be any real numbers and the threshold function can be defined as

$$(x,y) = \begin{cases} (1,0) & if & \sum_{i=1}^{n} a_i x_i - \sum_{i=1}^{n} b_i y_i \geq 1 \\ (0,1) & if & \sum_{i=1}^{h} b_i y_i - \sum_{i=1}^{n} a_i x_i \geq 1 \\ (0,0) & otherwise \end{cases}$$

Fig. 1 a) shows the general structure of a single 3-valued NLN.

Boolean NLN is the simplest type of NLN. Its theory plays a special role because of its link to other well known neural networks such as *multi-layer perceptron* [3], *Kohonen nets* [1], etc. In a boolean NLN, single real numbers a_i are used for weights of the links and boolean numbers 1 or 0 are used for value of input P_i ($i = 1, ... n$) and output Q. The truth value set is then denoted by $A^{(B)}$ (where B for Boolean NLN) and $A^{(B)} - \{0, 1\}$. The threshold function is defined as

$$Q = \begin{cases} 1 & if & \sum_{i=1}^{n} a_i P_i \geq 1 \\ 0 & if & \sum_{i=1}^{h} a_i P_i < 1 \end{cases}$$

Suppose we are given a boolean neural logic network and suppose Q is one of its nodes with incoming links such as Fig. 1 b)

To find the value of Q we need to find the current values of nodes at P_1, P_2, P_3 , say $\alpha_1, \alpha_2, \alpha_3$ respectively. Then we find sum $x = 2\alpha_1 - \alpha_2 + \alpha_3$ and put this number x into the threshold function $f(x)$ to decide whether it should be 1 or 0.

The choice of weights associated with NLN offers a great variety of different logic operations. In theory, for a network with two inputs, total of 3^9 distinct meaningful binary logical operations are possible. The definition of AND, OR and NOT are as follows:

- An AND operation of n inputs is defined as a neural logic function, written as $AND(P_1, P_2, ...P_n)$ or $Q = P_1\,AND\,P_2\,AND,...P_n$ with the weights $(a_i, b_i) = (\frac{1}{n}, n)$ for $i = 1, 2, ...n$

- An OR operator of n inputs is defined as neural logic function as $OR(P_1, P_2, ...P_n)$ or $Q = P_1\,OR\,P_2\,OR,...P_n$ with the weights $(a_i, b_i) = (n, \frac{1}{n})$ for $i = 1, 2, ...n$

- A NOT operation of 1 input is defined as a neural logic function, written as $NOT(P)$ or $Q = NOT(P)$ with the weight $(a, b) = (-1, -1)$.

Fig. 2 shows several useful operations in 3-valued NLN.

3 Logical Reasoning

In section 2, we have introduced the Neural Logic model and its capability to incorporate the local inference of Boolean logic. The interconnection of this model called Neural Logic Network (for short NLN). In this section we introduce a rule inferencing system based on neural logic model for propositional knowledge base.

A proposition is represented as a neural logic neuron labelled Q. The truth value of Q, denoted as $t(Q)$, is given by neuron's activation. The truth values: *true, false,* and *don't-know* are denoted by ordered pair (1, 0), (0, 1) and (0, 0) respectively. The connection weight from a neuron denoting proposition Q is also extended to ordered pair (x, y) where x and y are real numbers that can be viewed as the truth and false value or as the strength of the support and opposition respective of proposition P for proposition Q.

Definition Given proposition $P_1, P_2, ...P_n$ with truth values $(x_1, y_1), (x_2, y_2), ...(x_n, y_n)$, which are connected to proposition Q with the weights $(a_1, b_1), (a_2, b_2), ...(a_n, b_n)$ respectively. The $Netinput(Q)$ is defined as, $Netinput(Q) = \sum (a_i x_i - b_i y_i)$. The activation of neuron is defined as follows:

$$Act(Q) = \begin{cases} (1, 0) & if & Netinput(Q) \geq \lambda \\ (0, 1) & if & Netinput(Q) \leq -\lambda \\ (0, 0) & otherwise \end{cases}$$

where λ is threshold, usually set to 1. The P_i's and Q in Fig. 1 are referred as inputs nodes and output node respectively.

3.1 Neural Logic Element

A Neural Logic Element (for short Netelm) can be seen as one layer or maximum two layers neural network with n input nodes and one output node, and an optional layer of hidden nodes. With reference to Fig. 1 the definition of a neural logic element can be given as follows.

Definition A neural logic element of n inputs with the proposition $P_1, P_2, ...P_n$ connected to proposition Q is defined as
$Netelm: \quad \{(1,0), (0,1), (0,0)\}^n \to \{(1,0), (0,1), (0,0)\}$

$$t(Q) = Netelm(P_1, P_2, ...P_n) = \begin{cases} (1,0) & if & Netinput(Q) \geq 1 \\ (0,1) & if & Netinput(Q) \leq -1 \\ (0,0) & otherwise \end{cases}$$

347

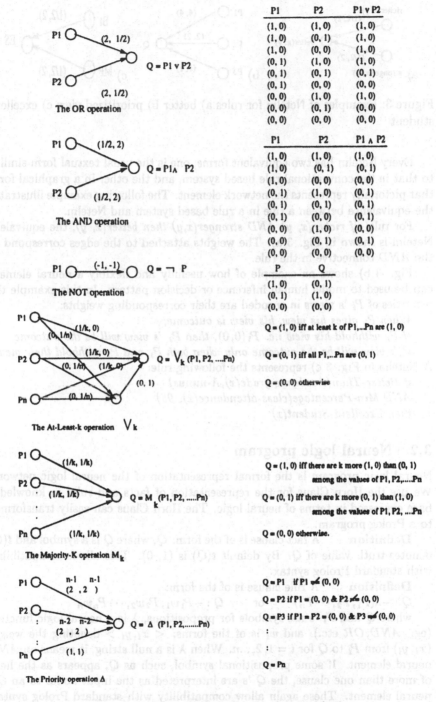

P1	P2	P1 v P2
(1, 0)	(1, 0)	(1, 0)
(1, 0)	(0, 1)	(1, 0)
(1, 0)	(0, 0)	(1, 0)
(0, 1)	(1, 0)	(1, 0)
(0, 1)	(0, 1)	(0, 1)
(0, 1)	(0, 0)	(0, 0)
(0, 0)	(1, 0)	(1, 0)
(0, 0)	(0, 1)	(0, 0)
(0, 0)	(0, 0)	(0, 0)

P1	P2	P1 ∧ P2
(1, 0)	(1, 0)	(1, 0)
(1, 0)	(0, 1)	(0, 1)
(1, 0)	(0, 0)	(0, 0)
(0, 1)	(1, 0)	(0, 1)
(0, 1)	(0, 1)	(0, 1)
(0, 1)	(0, 0)	(0, 1)
(0, 0)	(1, 0)	(0, 0)
(0, 0)	(0, 1)	(0, 1)
(0, 0)	(0, 0)	(0, 0)

P	Q
(1, 0)	(0, 1)
(0, 1)	(1, 0)
(0, 0)	(0, 0)

Figure 2: Operations in 3-valued neural logic

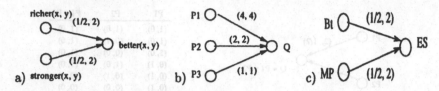

Figure 3: Examples of Netelm for rules a) better b) prioritized view c) excellent student

Every Netelm has two equivalent forms; one is the usual textual form similar to that in the conventional rule based system, and the other in a graphical form that pictorially represents the network element. The following example illustrates the equivalence between a rule in a rule based system and Netelm.

For rule: *if richer(x, y) AND stronger(x,y) then better(x, y)*, the equivalent Netelm is shown in Fig. 3 a). The weights attached to the edges correspond to the *AND* connective in the rule.

Fig. 3 b) shows an example of how usefully and flexibly a neural element can be used to model human inference or decision pattern. In this example the priorities of P_i's views in encoded are their corresponding weights:

When P_1 gives his view, his view is outcome;

If P_1 withhold his view i.e. $P_1(0,0)$, then P_2's view will be the outcome.

P_3's view will be the outcome only when both P_1 and P_2 withhold their views.

A Netelm in Fig. 3 c) represents the following rule:

if Better-Than(Academic-grade(x),A-minus)

AND Min-Percentage(class-attendance(x), 95)

then Excellent-Student(x)

3.2 Neural logic program

Neural logic program is the formal representation of the neural logic network. We use the Horn Claus for the representation of facts and rules in knowledge base expressed in terms of neural logic. The Horn Claus can easily transformed to a Prolog program.

Definition A fact clause is of the form: Q, where Q is a symbol and $t(Q)$ denotes truth value of Q. By default $t(Q)$ is $(1, 0)$. This allows compatibility with standard Prolog syntax.

Definition A rule clause is of the form:

$$Q : -\lambda(P_1, P_2, \cdots P_n) \qquad \text{or} \qquad Q : -P_1w_1, P_2w_2, \cdots P_nw_n$$

where Q and P_i's are symbols for propositions, λ is a neural logic function (eg. AND, OR etc.) and w_i is of the forms: $< x_i, y_i >$ denoting the weight (x_i, y_i) from P_i to Q for $i = 1, 2, ...n$. When λ is a null string, it means an AND neural element. If some propositional symbol, such as Q, appears as the head of more than one clause, the Q's are interpreted as the input nodes to an OR neural element. These again allow compatibility with standard Prolog syntax. w_i in the formula is for specifying arbitrary weight to define any neural logic function. The neural logic network representation for a rule clause will be the same as one in fig.1.

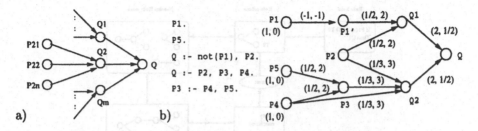

Figure 4: Neural logic network program for a) a fragment clauses, b) an example

Definition A program clause is either a fact clause or a rule clause. A neural logic program is a collection of program clauses. It will be represented as a forest of neural logic net trees, each consists of a number of neural elements joined as follows: the output node of a neural element becomes an input node of another neural element if they both denote the same proposition. The truth values of the proposition as specified by the fact clauses are the activation attached beside the relevant neurons. Thus, every neural logic program has a unique neural logic network representation and the number of distinct symbols (propositions) in the program equals the number of nodes in the neural logic minus the number of hidden nodes in the neural elements.

To further illustrate the implicit AND and OR operators in standard Prolog program and its corresponding neural logic network representation, consider the following program fragment:

$$Q^1 : -P_{11}, P_{12}, \cdots P_{1n}$$
$$Q^2 : -P_{21}, P_{22}, \cdots P_{2n}$$
$$\vdots$$
$$Q^m : -P_{m1}, P_{m2}, \cdots P_{mn}$$

Where $Q^1, Q^2, \cdots Q^m$ refer to the same syntactical symbol Q. m hidden nodes, $Q^1, Q^2, \cdots Q^m$ will be created, each of which is an output from AND neural element of n inputs $(P_{i1}, P_{i2}, \cdots P_{in})$ as well as an input node to an OR neural element of m input as shown in Fig. 4 a)

Fig. 4 b) shows a simple neural logic program in standard Prolog's syntax and its corresponding neural logic network representation.

3.3 The proposed system

A schematic diagram of the system is shown in Fig. 5). The system allows conversion from *if..then...* rules into a neural logic program. Every rule is represented in the knowledge base by netelm. Standard logic operations in conventional rules are readily transformed into netelms with fixed weights assigned for the corresponding logic operators.

The *Rule Editor* and *Query Manager* combine two forms of user interaction with the system. Besides providing a friendly environment for the user to convert, create and maintain netelm knowledge base, they also derive the conclusion.

The netelm rule base is the depository area of knowledge to be used in the inference process. It is made of the rules from three main sources: conventional

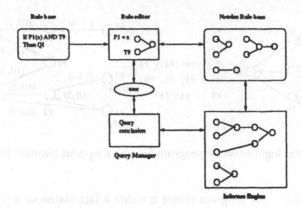

Figure 5: Architecture of the reasoning system

rules being transformed into netelms by the rule editor; rules added/modified by the uses; and, meta rules learned by the inference engine through consultations.

The *Inference Engine* dynamically links up relevant rules from the netelm rule base in the process of deriving at a conclusion for the user. The inference engine may run in consultation or training mode. In either case it chains active rules together into a neural logic network like a tree structure. With the netelm schema discussed earlier, tree structure is essentially made of fragments of neural logic network to be fitted together dynamically during the inferencing process, similar to the chaining of rules in the working memory of a conventional rule-based system.

3.4 Learning

Since all logical operations in the system are presented by arc weights, this provides a mean of modifying the rules by adjustment of weights. This is especially useful for rules that involve non standard logic operations such as that of human logic. In such cases, weight may initialy be created on the basis of some intuition but later tuned by training with examples.

The system may be trained in two ways. First, learning mechanism allows alternation of the logical operation of a rule by training a corresponding body of connected netelms with known exemplary deductions. Second, learning mechanism allows fine tuning of weights without altering the basic logical operations defined in the netelms. In the dynamic linking of netelms, either forward or backward, paths that fail to achieve the desired conclusions will have their weights decreased. Other paths that confirm the desired conclusions may have their weights increased. In training mode, when a training example is presented to the system, it will chain up an inference tree from the input nodes to the desired goal node. Relevant netelm rules are selected from the knowledge base depending on the input variables in the training example. After the activation and propagation, the activation state of the network output node is compared to the desired value in the training example. If the network is not able to derive the desired conclusion, the error is back propagated and the netelm in the inference tree will have their connection weights adjusted. However, the system assumes that the netelm rules articulated by the human experts are sufficiently close to the global minima of the neural network representing the domain knowledge. The constituent netelm therefore only require small weight adjustment - perhaps output nodes of certain netelms in the inference network fall slightly short for the threshold of the neural logic network activation. This means that the iterative process for error back propagation will only need to occur a small number of times.

4   Conclusion

We have described a rule inferencing system based on neural logic model. The model provides a richer set of logical operations which are close to human reasoning and decision making that is not easy if not impossible to be modeled by classical logic. We define a neural logic program for representation of the specific application. We also suggested the search strategy with heuristic search and an adaptive strategy for standard operators. Comparing to the rule based systems the knowledge is usually constructed in a hierarchy. However, the predefined generalization hierarchy limits the system flexibility. It is difficult to update those assumptions that are no longer significant. In this approach, when a query is assigned to the system, it will be mapped dynamically to a neural network. In doing so, the topology of the network is reduced to the size of query.

Furthermore, the conventional rule based systems often lack learning ability. Finally, the system is made more resilient to the brittleness problem of conventional rule based system which could fail abruptly in the face of fuzzy data

From the above discussion and examples, it is not difficult to envisage the power of neural logic model consisting of chain of neural logic elements to represent interesting and realistic human logic which is not possible in classical logic. We are at the beginning of the project, therefore we can not report about a real world application. This is a subject to future work. We will explore the limitations of this approach on a number of domains and we hope to show that this idea is extendible to many other AI problems.

References

1. T. Kohonen. *Self-Organization and Associative Memory*. Springer-Verlag, 1988.

2. T. Quah, K. Raman, C Tan, and H Teh. A shell environment for developing connectionist decision support systems. *Expert Systems*, 11(4), 1994.

3. E. Rumelhart, G.E. Hinton, and R.J. Williams. Learning internal representation by error propogation. In D.E. Rumelhart and J.L. Mc Clelland, editors, *Parallel distibuted processing*. Cambridge MIT press, 1986.

4. H. Teh. *Neural Logic Networks*. World Scientific, 1995.

The Resolution for Rough Propositional Logic with Lower (L) and Upper (H) Approximate Operators *

Qing Liu

(Department of Computer Science,NanChang University,NanChang,JiangXi
330029,P.R.China,email:qliu@263.net)

Abstract

Based on First-Order Rough Logic Studied by Lin and Liu, this paper establishes rough propositional logical system with rough lower (L) and upper (H) approximate operators. It discusses the resolution principle in the system. The soundness of resolution deduction, soundness and completeness of the refutation are also studied in the paper.

Keywords:Rought Propositional Logic, Resolution Principle, Resolution Deduction, Soundness Theorem, Refutaion.

1 Introduction

Lin and Liu studied a first-order rought logic based on six topological properties, in particularly, using the axioms of Kuratowski's closure (H) and interior (L) operators. Thus, the first-order rough logic system with operators L and H is developed[1]. The revision studied further by Lin and Liu considers a map f, it is defined as the map of one to one between boundary-line region in the logic and undefinable region in classical logic. Hence, the logic is proved to be sound and complete with the new intertation in the revision[2].

We bear in mind the idea of studying resolution reasoning. This paper will first establish a rough propositional logical system (RPLS) with operators (L) and (H) which are defined rough lower and upper approximate operators[1]; Next, the paper describes the resolution principle and the soundness of resolution deduction in the logic. Hence, this paper is different from other systems, it focuses in the resolution reasoning, but not the logical systems based on rough concept.

2 Rough Propositional Logical System

Let w be a rough propositional formula, We will call $m(w) \subseteq U$ the meaning of w. Meaning sets of the formulas Lw and Hw with rough lower and upper approximate operators are defined as follows:

(1). x satisfies w iff $x \epsilon m(w)$;

0 *The study is supported by national natural science fund and JiangXi Province natural science fund in China

(2). x satisfies Lw iff $\forall y(y \epsilon H(x) \rightarrow y \epsilon m(w))$, i.e., $H(x) \subseteq m(w)$;

(3). x satisfies Hw iff $\exists y(y \epsilon H(x) \wedge y \epsilon m(w))$, i.e., $H(x) \bigcap m(w) \neq \emptyset$.

Where $H(x)$ is a equivalent calss containing $x^{[1,2]}$.

Well-formed formulas(Wffs):

(1). All atomic formulas are Wff;

(2). If w and w_1 are Wff, then so are $\sim w$, $w \rightarrow w_1$ and Lw;

(3). The only Wff are those obtainable by finite applications of $(1) - (2)$ in the above.

Other logical connectives $\vee, \wedge, \leftrightarrow$ are defined by $\sim$ and $\rightarrow$ and operator H is defined by $\sim$ and $L^{[1,3]}$.

Axiom schemas

A_1. $\vdash w \rightarrow (w_1 \rightarrow w)$;

A_2. $\vdash (w \rightarrow (w_1 \rightarrow w_2)) \rightarrow (w \rightarrow w_1) \rightarrow (w \rightarrow w_2)$;

A_3. $\vdash (\sim w_1 \rightarrow \sim w_2) \rightarrow (w_2 \rightarrow w_1)$;

A_4. $\vdash L(w_1 \rightarrow w_2) \rightarrow (Lw_1 \rightarrow Lw_2)$;

A_5. $\vdash Lw \rightarrow w$;

A_6. $\vdash HLw \rightarrow Lw$.

Rules of inference

R_1. Modus Ponens (MP):From $\vdash w_1 \rightarrow w_2$ and $\vdash w_1$, we have $\vdash w_2$;

R_2. L insertion (LI):From $\vdash w$, we have $\vdash Lw$.

Where R_2 means that Lw is valid if w is valid for all obserable world[1]

The semantics of the logic

The semantic model of formulas in $RPLS$ is defined as a triple:

$$M = < W, R, m >$$

where W is a non-empty set of observable worlds[1], if each observable world are viewed as a state, then W is a state set[3]. R is a binary relation on W, such that $\forall s \epsilon W$, $\exists s' \epsilon W$, $(s, s') \epsilon R$; m is a meaning function that assigns to each propositional variable p a subset $m(p)$ of W.

Given a model M we say that formula w is satisfied by a state s in model M, written by $\models_s w$ iff the following conditions are satisfied:

(1). $M \models_s p$ iff $s \epsilon m(p)$, where p is a propositional variable;

(2). $M \models_s \sim w$ iff $\sim M \models_s w$;

(3). $M \models_s w_1 \vee w_2$ iff $M \models_s w_1 \vee M \models_s w_2$;

(4). $M \models_s w_1 \wedge w_2$ iff $M \models_s w_1 \wedge M \models_2 w_2$;

(5). $M \models_s w_1 \rightarrow w_2$ iff $M \models_s \sim w_1 \vee w_2$;

(6). $M \models_s w_1 \leftrightarrow w_2$ iff $M \models_s (w_1 \rightarrow w_2) \wedge (w_2 \rightarrow w_1)$;

(7). $M \models_s Lw$ iff $\forall s' \epsilon W$, if $(s, s') \epsilon R$ then $M \models_{s'} w$;

(8). $M \models_s Hw$ iff $\exists s' \epsilon W$, $(s', s) \epsilon R \wedge M \models_{s'} w$.

Given a model M, for each formula w in $RPLS$, which is assigned a set of states in model M, detoned by $m(w) = \{s \epsilon W : M \models_s w\}$.

We introduce truth and validity of formulas. A formula w is true in a model M iff $m(w) = W$; A formula w is valid in $RPLS$ iff w is true in every model in $RPLS$; a formula w is satisfiable iff for some model M and state s, $M \models_s w$; If w includes operators L and H, the description is also validable by (7) and (8).

3 Conjunctive Normal Form (CNF)

Let w be a formula in $RPLS$, then there is a CNF corresponding to $w^{[3,4]}$

$$C_1 \wedge C_2 \wedge \ldots \wedge C_n$$

where $n \geq 1$ and each clause C_i is a disjunction of the general form:

$$C_i = p_1 \vee \ldots \vee p_{n_1} \vee Lq_1 \vee \ldots \vee Lq_{n_2} \vee Hr_1 \vee \ldots \vee Hr_{n_3},$$

where each p_i is a literal; each q_j is a disjunction, it possesses the general form of the clauses; r_t is a conjunction, where each conjunct possesses the general form of the clauses.
For examples, The following formulas are conjuctive normal forms:
(1). $L(p \vee q \vee H(r \wedge t))$;
(2). $H((p \vee q) \wedge \sim p)$;
(3). $\sim p \vee p \vee L(r \vee s) \vee H((p \vee Lr) \wedge e)$;
(4). $\sim p \vee L(Lp \vee (H(q \wedge Lr)) \vee H(L(H((Lq \vee Ht) \wedge r) \vee Lt) \wedge p)$.
Any formula w in $RPLS$ is transformed eqivalently into the conjunctive normal form. For example, $w = L(p \wedge H(q \vee L(r \wedge t)) \wedge (p \rightarrow L(q \wedge Ht)))$
The followings are the procedure that is trasformed into CNF:
(1). $Lp \wedge LH(q \vee (Lr \wedge Lt)) \wedge (p \rightarrow L(q \wedge Ht))$;
(2). $Lp \wedge LH((q \vee Lr) \wedge (q \vee Lt)) \wedge (p \rightarrow (L(q \wedge Ht)))$;
(3). $Lp \wedge LH((q \vee Lr) \wedge (q \vee Lt)) \wedge (\sim p \vee L(q \wedge Ht))$;
(4). $Lp \wedge LH((q \vee Lr) \wedge (q \vee Lt)) \wedge (\sim p \vee (Lq \wedge LHt))$;
(5) $Lp \wedge LH((q \vee Lr) \wedge (q \vee Lt)) \wedge (\sim p \vee Lq) \wedge (\sim p \vee LHt)$.
Where each conjunct is the general form of the clause.

4 The Resolutions in the $RPLS$

For any two clauses C_1 and C_2, if there is literal p_1 in C_1, that is complementary to a literal p_2 in C_2, then delete p_1 and p_2 from C_1 and C_2 respectively, and construct the disjunction of remaining clauses. Therefore, we have resolution rule:

$$\frac{C_1 \text{ with } p_1 \epsilon C_1 \quad C_2 \text{ with } p_2 \epsilon C_2}{(C_1 - \{p_1\}) \cup (C_2 - \{p_2\})}$$

(I)

It is possible there are literals with operators L and H in the clauses of RPL, and Lp is complementary pair of literals to $H \sim p$. Hence following resolution is valid:

$$C_1 \quad \text{with } Lp\epsilon C_1$$
$$C_2 \quad \text{with } H \sim p\epsilon C_2$$

$$(C_1 - \{Lp\}) \cup (C_2 - \{H \sim p\})$$

(II)

The forms of lower of level line of (I) and (II) are called resolvent obtained from C_1 and C_2. Hence the resolution principles in $RPLS$ consist of (I) and (II). As an example, consider the following deductive resolution:

(1) $L(p \vee q) \vee C_1$ premise
(2) $H \sim p \vee C_2$ premise
(3) $L \sim q \vee C_3$ premise

(4) $Hq \vee C_1 \vee C_2$ using (1) and (2)
(5) $C_1 \vee C_2 \vee C_3$ using (3) and (4)

Where fourth step (4) has a Hq, since premise (1): $\sim (L(p \vee q) \leftrightarrow (Lp \vee Lq))$, but $L(p \vee q) \wedge H \sim p \rightarrow H(p \vee q \wedge \sim p)$. Hence, the resolution of using $L(p \vee q)$ and $H \sim p$ gets the resolvent Hq.

5 Transformable strategies of the Resolutions in $RPLS$

Let C_1 and C_2 be two clausee in $RPLS$, we can transform for them, so that we find out the complementary pair of literals in C_1 and C_2. Therefore, we give the following transformable strategies:
(1). $T(p, \sim p) = R(p, \sim p)$;
(2). $T((C_1 \vee C_2), C_3) = R(C_1, C_3) \vee C_2$;
(3). $T(C_1 \wedge C_2 \wedge C_3 \wedge C_4) = R(C_1, C_3) \wedge C_2 \wedge C_4$;
(4). $T(LC_1, LC_2) = LR(C_1, C_2)$;
(5). $T(LC_1, HC_2) = HR(C_1, C_2)$;
(6). $T(LC_1, C_2) = R(C_1, C_2)$;
(7). $T(C_1 \vee C, C_2 \vee C') = R(C_1, C_2) \vee C \vee C'$;
(8). Substitution: $\emptyset$ for every occurrence of $(\emptyset \wedge C)$; C for every occurrence of $(\emptyset \vee C)$; $\emptyset$ for every occurrence of $L\emptyset$ or $H\emptyset$.
Where $R(X, Y)$ denotes that X and Y is resolvable.

6 Soundness of Resolution in $RPLS$

Theorem 1 (soundness theorem) If there is a deduction of resolution of a clause C from a set of clauses, $\triangle$, then $\triangle$ logically implies C.
Proof The proof is achieved by simple induction on the longer of resolution

deduction. For the induction, we need to show only that any given resolution step is sound. Suppose, C_1 and C_2 are arbitrary clauses, resolution of them produces a new clause $C : (C_1 - \{p_1\}) \cup (C_2 - \{p_2\})$ by (I) or $C' : (C_1 - \{Lp\}) \cup (C_2 - \{H \sim p\})$ by (II). By induction assumption, $\models_s C_1$ and $\models_s C_2$, that is, C_1 and C_2 are true, we prove $\models_s C$ and $\models_s C'$, where $s \epsilon W$, namely C and C' are also true.

If $\models_s Lp$, then $\sim\models_s H \sim p$, because Lp and $H \sim p$ is a complementary pair of literals in RPL, and so $\models_s (C_2 - \{H \sim p\})$. If $\models_s H \sim p$, then $\sim\models_s Lp$, and so $\models_s (C_1 - \{Lp\})$. But then $\models_s C'$, that is, $\models_s (C_1 - \{Lp\}) \cup (C_2 - \{H \sim p\})$. Similarly, we obtain $\models_s C$, that is, $\models_s (C_1 - \{p_1\}) \cup (C_2 - \{p_2\})$.

Given a set of clauses, it can derive empty using resolution deduction, we call the resolution deduction a refutation. Such as, given the three clauses Lp, $H(\sim p \vee q)$ and $L \sim q$, the deductive steps of them are as following:

(1) Lp premise
(2) $H(\sim p \vee q)$ premise
(3) $L \sim q$ premise

----

(4) Hq using (1) and (2)
(5) $\emptyset$ using (3) and (4).

The theorem of soundness and completeness for Resolution refutation is vaild, that is, a set of clauses, $\triangle$, is unsatisfiable iff $\triangle$ is refutable.

7 Conclusion

We study the resolution of RPL, the aim is in order to establish a rough reasoning system using resolution method. The operators L and H in the paper come from rough lower and upper approximate operator defined by Lin and Liu in the references[1,2], they are different to necessary ($\Box$) and possible ($\Diamond$) operators in Modal Logic in the interpretation of semantis.

References

[1] T.Y.Lin and Q.Liu, First-Order Rough Logic I:Approximate Reasoning via Rough Sets, Fundamenta Infomaticae,vol. 27, Nos. 2,3 (1996),137-154.
[2] T.Y.Lin and Q.Liu, First-Order Rough Logic Revised. Manuscript, 5 (1999).
[3] Luis Farinas-del-Cerro, Resolution Modal Logic, Proceedings of the Eighth International Coference on Atomata Deduction, (1985).
[4] Q.Liu, The OI-Resolution of Operator Rough Logic , LNAI 1424, 6 (1998), 432-435.
[5] Q.Liu, Operator Rough Logic and Its Resolution Principles, Journal of Coputer, (Chinese), 5 (1998).

Information Granules in Distributed Environment

Andrzej Skowron[1] and Jaroslaw Stepaniuk[2]

[1] Institute of Mathematics, Warsaw University,
Banacha 2, 02-097 Warsaw, Poland,
E-mail: skowron@mimuw.edu.pl
[2] Institute of Computer Science, Bialystok University of Technology,
Wiejska 45A, 15-351 Bialystok, Poland,
E-mail: jstepan@ii.pb.bialystok.pl

Abstract. We propose to use complex information granules to extract patterns from data in distributed environment. These patterns can be treated as a generalization of association rules.

1 Introduction

Notions of granule [15], [9] and granule similarity (inclusion or closeness) are very natural in knowledge discovery. The exact interpretation between granule languages of different information sources (agents) often does not exist. Hence closeness (rough inclusion) of granules is considered instead of their equality.

For example, the left and right hand sides of association rules [1] describe granules and the *support* and *confidence* coefficients specify the inclusion degree of granule represented by the formula on the left hand side into the granule represented by the formula on the right hand side of the association rule.

Reasoning in distributed environment requires a construction of interfaces between agents for learning of concepts definable by different agents. In this paper we suggest one solution based on exchanging views of agents on objects with respect to a given concept. An agent delivering concept is giving positive and negative examples (objects) with respect to a given concept. The agent receiving this information can describe objects using its own attributes. In this way a data table (called a *decision table*) is created and the approximate description of concept can be extracted by the receiving agent.

An analogous method can be used in case of the *customer-agent* (agent specifying tasks) searching for a top-level cooperating agent (*root-agent*). The customer-agent is presenting examples and counter examples of objects with respect to her/his concept. The concept specified by customer-agent is approximated by agents and an agent returning the best approximation of the customer-agent concept is chosen to be the root agent. The goal of cooperating agents is to produce a concept sufficiently close (or included) to the concept specified by the customer-agent. This concept has to be constructed from some elementary concepts available for agents called *inventory* or *leaf-agents* [8]. This is realized by searching for an agent scheme [8]. The schemes are represented in the paper by expressions called *terms*.

We emphasize the fact of approximate (vague) understanding of concepts received by any agent from other agents. Our solution is based on rough set approach. We point out that our approach can be treated as an approach for extracting generalized association rules in distributed environment.

2   Rough Sets and Approximation Spaces

We recall general definition of approximation space [11], [13].

Definition 1. *A parameterized approximation space is a system* $AS_{\#,\$} = (U, I_\#, \nu_\$)$, *where*

- *U is a non-empty set of objects,*
- $I_\# : U \to P(U)$ *is an uncertainty function and* $P(U)$ *denotes the powerset of U,*
- $\nu_\$: P(U) \times P(U) \to [0,1]$ *is a rough inclusion function.*

The uncertainty function defines for every object x a set of similarly described objects. A constructive definition of uncertainty function can be based on the assumption that some metrics (distances) are given on attribute values. For example, if for some attribute $a \in A$ there is a metric $\delta_a : V_a \times V_a \longrightarrow [0, \infty)$, where V_a is the set of all values of attribute a then one can define the following uncertainty function

$$y \in I_a^{f_a}(x) \text{ if and only if } \delta_a(a(x), a(y)) \le f_a(a(x), a(y)),$$

where $f_a : V_a \times V_a \to [0, \infty)$ is a given threshold function.

A set $X \subseteq U$ is *definable* in $AS_{\#,\$}$ if it is a union of some values of the uncertainty function.

The rough inclusion function defines the value of inclusion between two subsets of U [11], [9].

Now we can define the lower and the upper approximations of subsets of U.

Definition 2. *For a parameterized approximation space* $AS_{\#,\$} = (U, I_\#, \nu_\$)$ *and any subset* $X \subseteq U$ *the lower and the upper approximations are defined by*
   $LOW(AS_{\#,\$}, X) = \{x \in U : \nu_\$(I_\#(x), X) = 1\}$,
   $UPP(AS_{\#,\$}, X) = \{x \in U : \nu_\$(I_\#(x), X) > 0\}$.

Approximations of concepts (sets) are constructed on the basis of background knowledge. Obviously, concepts are also related to unseen so far objects. Hence it is very useful to define parameterized approximations with parameters tuned in the searching process for approximations of concepts. This idea is crucial for construction of concept approximations using rough set methods. In our notation #, \$ are denoting vectors of parameters which can be tuned in the process of concept approximation.

The presented above definition of approximation space can be treated as a semantic part of the approximation space definition. Usually there is also specified a set of formulas Φ expressing properties of objects. Hence we assume that together with the approximation space $AS_{\#,\$}$ there are given

- a set of formulas Φ over some language,
- semantics $\|\bullet\|$ of formulas from Φ, i.e., a function from Φ into the power set $P(U)$.

Let us consider an example [7]. We define a language L_{IS} used for elementary granule description, where $IS = (U, A)$ is an information system. The syntax of L_{IS} is defined recursively by

1. $(a \in V) \in L_{IS}$, for any $a \in A$ and $V \subseteq V_a$.
2. If $\alpha, \beta \in L_{IS}$, then $\alpha \wedge \beta \in L_{IS}$.
3. If $\alpha, \beta \in L_{IS}$, then $\alpha \vee \beta \in L_{IS}$.

The semantics of formulas from L_{IS} with respect to an information system IS is defined recursively by

1. $\|a \in V\|_{IS} = \{x \in U : a(x) \in V\}$.
2. $\|\alpha \wedge \beta\|_{IS} = \|\alpha\|_{IS} \cap \|\beta\|_{IS}$.
3. $\|\alpha \vee \beta\|_{IS} = \|\alpha\|_{IS} \cup \|\beta\|_{IS}$.

A typical method used by classical rough set approach [7] for constructive definition of the uncertainty function is the following: for any object $x \in U$ there is given information $Inf_A(x)$ (information vector, attribute value vector of x) which can be interpreted as conjunction of selectors $a = a(x)$ for $a \in A$ and the set $I_\#(x)$ is equal to $\|\bigwedge_{a \in A} a = a(x)\|_{IS}$. One can consider a more general case taking as possible values of $I_\#(x)$ any set $\|\alpha\|_{IS}$ containing x. Next from the family of such sets the resulting neighborhood $I_\#(x)$ can be selected. One can also use another approach by considering more general approximation spaces in which $I_\#(x)$ is a family of subsets of U [2], [6].

3 Mutual Understanding of Concepts by Agents

One of the important task for Knowledge Discovery and Data Mining (KDD) [1], [4] in distributed environment is to develop tools for modeling mutual understanding of concepts definable by different agents. Mutual understanding through communication is one of the key issues to enable collaboration among agents [5]. We assume agents specify their knowledge using data tables.

3.1 Understanding of Concept Definable by Single Agent

Let us consider two agents. There are two data tables $IS_1 = (U, A_1)$ and $IS_2 = (U, \{a\})$ corresponding to agents. We assume that $a : U \to \{0, 1\}$ is a characteristic function of a concept $X = \{x \in U : a(x) = 1\}$.

In this, typical for rough set approach, situation the first agent is specifying the characteristic function of its concept on examples of objects. The second agent is trying to describe the concept using values of its own attributes from A_1 on objects considered by the first agent. In this way it is constructed a decision table with condition attributes from A_1 and the decision a. Next it is computed the lower and the upper approximation of the decision class X. The size of the boundary region of X with respect to A_1 can be used a measure of uncertainty in understanding X by the agent with attributes A_1.

Closeness of X to its approximations in the language used by the first agent can be represented by *accuracy of approximation*, i.e., by the coefficient

$$\alpha\left(AS_{A_1}, X\right) = \frac{card\left(LOW\left(AS_{A_1}, X\right)\right)}{card\left(UPP\left(AS_{A_1}, X\right)\right)}.$$

The presented above approach can be used for learning by one agent of concepts definable by another agent.

Let us consider again two agents. There are two data tables $IS_1 = (U, A_1)$ and $IS_2 = (U, A_2)$ corresponding to agents. We assume that in both data tables there is the same set of objects U and $A_1 = \left\{a_1^1, \ldots, a_l^1\right\}$, and $A_2 = \left\{a_1^2, \ldots, a_k^2\right\}$ are two sets of attributes, where $l > 0$ and $k > 0$ are given natural numbers. Let us consider concepts definable by attributes from the set A_2. For example suppose that we consider concept defined by formula $\left(a_1^2 = 1 \wedge a_2^2 = 1\right) \vee a_3^2 = 1$. This is a concept definable by the second agent. Hence this agent can compute values of the characteristic function of the concept on objects from U and the first agent can find approximations of the concept following the procedure described above.

In this way we define approximations by the first agent of concepts definable by the second one.

Let us mention that the approximation operations are in general not distributive with respect to disjunction or conjunction. Hence one can not expect to construct concept approximations of the good quality from approximation of atomic concepts (e.g. descriptors).

3.2 Understanding of Concept Definable by Team of Agents

Assume that a set of agents $Ag = \{ag_1, \ldots, ag_p\}$ where $p > 0$ is a given natural number. Let us consider a data table $IS_{ag} = (U, A_{ag})$ for any agent $ag \in Ag$. We assume any agent from Ag is defining a concept X using the above procedure. One can construct a decision table DT with condition attributes being the characteristic functions of the lower and upper approximations of X defined by all agents from Ag and the decision being the characteristic function of X on given examples of objects. The lower and upper approximation of X with respect to condition attributes of DT describe the vagueness in understanding of X by agents from Ag. One can also use other features summarizing the result of voting by different agents. Examples of such features are the majority voting feature, accepting object as belonging to concept if the number of voting agents is greater than a given threshold or the characteristic function of the intersection of the upper approximations $\bigcap_{ag \in Ag} UPP\left(AS_{A_{ag}}, X\right)$ or the intersection of the lower approximations $\bigcap_{ag \in Ag} LOW\left(AS_{A_{ag}}, X\right)$. One can observe that in some cases the above intersections can be undefinable by single agent.

The described problem is analogous to resolving conflict between decision rules voting for decision when they are classifying new objects.

One can extend our approach to the case when e.g. one agent is trying to understand concepts definable by the second agent on the basis of understanding these concepts by the third agent. Common knowledge of a given team of agents about concepts definable by members of this team [3], [14], [10] can also be considered in this framework.

One can also consider the discussed above new features as the characteristic functions of concepts definable in some new approximation spaces constructed from approximation spaces of agents from Ag.

4 Rough Sets in Distributed Systems

In this section we consider operations on approximation spaces which seem to be important for approximate reasoning in distributed systems. We consider a set of agents Ag. Each agent is equipped with some approximation spaces. Agents are cooperating to solve a problem specified by a special agent called *customer-agent*. The result of cooperation is a scheme of agents. In the simplest case the scheme can be represented by a tree labeled by agents. In this tree leaves are delivering some concepts and any non-leaf agent $ag \in Ag$ is performing an operation $o\,(ag)$ on approximations of concepts delivered by its children. The root agent returns a concept being the result of computation by the scheme on concepts delivered by leaf agents. It is important to note that different agents use different languages. Hence concepts delivered by one agent can be only perceived in an approximate sense by another agent.

We assume any non leaf agent is equipped with an operation

$o\,(ag) : U_{ag}^{(1)} \times \ldots \times U_{ag}^{(k)} \rightarrow U_{ag}^{(0)}$. Any agent ag together with an operation $o\,(ag)$ has different approximation spaces $AS_{ag}^{(1)}, \ldots, AS_{ag}^{(k)}, AS_{ag}^{(0)}$ with universes $U_{ag}^{(1)}, \ldots, U_{ag}^{(k)}, U_{ag}^{(0)}$, respectively. We assume that the agent ag is perceiving objects by measuring values of some available attributes. Hence some objects can become indiscernible [7]. This influences the specification of any operation $o\,(ag)$. We consider a case when arguments and values of operations are represented by attribute value vectors. Hence instead of the operation $o\,(ag)$ we have its inexact specification $o^*\,(ag)$ taking as arguments $I_{ag}^{(1)}\,(x_1), \ldots, I_{ag}^{(k)}\,(x_k)$ for some $x_1 \in U_{ag}^{(1)}, \ldots, x_k \in U_{ag}^{(k)}$ and returning the value $I_{ag}^{(0)}\,(o(ag)(x_1, \ldots, x_k))$ if $o(ag)(x_1, ..., x_k)$ is defined, otherwise the empty set. This operation can be extended to the operation $o^*(ag)$ with domain equal to the Cartesian product of families of definable sets (in approximation spaces attached to arguments) and with values in the family of definable set (in the approximation space attached to the result)

$$o^* (ag) (X_1, \ldots, X_k) = \bigcup_{Y_1 \subseteq X_1, \ldots, Y_k \subseteq X_k} o^* (ag) (Y_1, \ldots, Y_k),$$

where $Y_1, \ldots, Y_k$ are neighborhoods of some objects in definable sets $X_1, \ldots, X_k$, respectively. In the sequel, for simplicity of notation, we write $o(ag)$ instead of $o^*(ag)$.

This idea can be formalized as follows. First we define terms representing schemes of agents.

Let $X_{ag}, Y_{ag}, \ldots$ be agent variables for any leaf-agent $ag \in Ag$. Let $o(ag)$ denote a function of arity k. We have mentioned that it is an operation from Cartesian product of $Def\_Sets(AS_{ag}^{(1)}), \ldots, Def\_Sets(AS_{ag}^{(k)})$ into $Def\_Sets(AS_{ag}^{(0)})$, where $Def\_Sets(AS_{ag}^{(i)})$ denotes the family of sets definable in AS_{ag}^i. Using the above variables and functors we define in a standard way terms, for example $t = o(ag)(o(ag_1)(X_{ag_1}, Y_{ag_1}), o(ag_2)(X_{ag_2}, Y_{ag_2}))$. Such terms can be treated as description of complex information granules. By a valuation we mean any function val defined on the agent variables with values being definable sets satisfying $val(X_{ag}) \subseteq U_{ag}$ for any leaf-agent $ag \in Ag$. Now we can define the lower and the upper values of any term t under the valuation val with respect to a given approximation space of an agent ag.

1. If t is of the form $X_{ag'}$ then $val\left(LOW, AS_{ag}^{(0)}\right)(t) = LOW\left(AS_{ag}^{(0)}, val(t)\right)$ and $val\left(UPP, AS_{ag}^{(0)}\right)(t) = UPP\left(AS_{ag}^{(0)}, val(t)\right)$ if $val(t) \subseteq U_{ag}$, otherwise the lower and the upper values are undefined.

2. If $t = o(ag)(t_1, \ldots, t_k)$, where $t_1, \ldots, t_k$ are terms and $o(ag)$ is an operation of arity k then

$$val\left(LOW, AS_{ag}^{(0)}\right)(t) = LOW\left(AS_{ag}^{(0)}, o(ag)\left(val\left(LOW, AS_{ag}^{(1)}\right)(t_1),\right.\right.$$
$$\left.\left. \ldots, val\left(LOW, AS_{ag}^{(k)}\right)(t_k)\right)\right),$$
$$val\left(UPP, AS_{ag}^{(0)}\right)(t) = UPP\left(AS_{ag}^{(0)}, o(ag)\left(val\left(UPP, AS_{ag}^{(1)}\right)(t_1),\right.\right.$$
$$\left.\left. \ldots, val\left(UPP, AS_{ag}^{(k)}\right)(t_k)\right)\right).$$

if $val\left(LOW, AS_{ag}^{(i)}\right)(t_i), val\left(UPP, AS_{ag}^{(i)}\right)(t_i) \subseteq U_{ag}^{(i)}$ for $i = 1, \ldots, k$, otherwise $val\left(LOW, AS_{ag}^{(0)}\right)(t)$ and $val\left(UPP, AS_{ag}^{(0)}\right)(t)$ are undefined.

Example 1. We assume $Ag = \{ag, ag_1, ag_2\}$ and $o(ag)$ is a binary operation of ag. Two information systems IS_{ag_1}, IS_{ag_2} presented in Tables 1(a),(b) describe input information granules. We also consider operation $o(ag)$ described in Table 3. Two data tables $DT_1 = (U_1, A_1 \cup \{d_1\})$ and $DT_2 = (U_2, A_2 \cup \{d_2\})$ described in Tables 2(a) and 2(b) characterize interfaces between agents ag_1, ag_2 and ag.

	d_1
y_1	1
y_2	0
y_3	1
y_4	0

	d_2
z_1	1
z_2	1
z_3	1
z_4	0

Table 1. (a) Information System IS_{ag_1} (b) Information System IS_{ag_2}

U_1	a_1^1	a_2^1	a_3^1	d_1
y_1	yes	yes	no	1
y_2	no	yes	no	0
y_3	no	yes	no	1
y_4	no	no	yes	0

U_2	a_1^2	a_2^2	a_3^2	d_2
z_1	yes	yes	yes	1
z_2	yes	yes	yes	1
z_3	no	no	yes	1
z_4	no	no	yes	0

Table 2. (a) Data Table DT_1 (b) Data Table DT_2

The first four columns of Table 2(a) (2(b)) define information system $IS_{ag}^{(1)}$ ($IS_{ag}^{(2)}$) corresponding to the approximation space $AS_{ag}^{(1)}$ ($AS_{ag}^{(2)}$).

Let $t = o\,(ag)\,(X_{ag_1}, X_{ag_2})$ and $val\,(X_{ag_1}) = \{y_1, y_3\}$. Hence

$$val\left(LOW, AS_{ag}^{(1)}\right)(X_{ag_1}) = LOW\left(AS_{ag}^{(1)}, \{y_1, y_3\}\right) = \{y_1\},$$

$$val\left(UPP, AS_{ag}^{(1)}\right)(X_{ag_1}) = UPP\left(AS_{ag}^{(1)}, \{y_1, y_3\}\right) = \{y_1, y_2, y_3\}.$$

Let $val\,(X_{ag_2}) = \{z_1, z_2, z_3\}$. Hence

$$val\left(LOW, AS_{ag}^{(2)}\right)(X_{ag_2}) = LOW\left(AS_{ag}^{(2)}, \{z_1, z_2, z_3\}\right) = \{z_1, z_2\},$$

$$val\left(UPP, AS_{ag}^{(2)}\right)(X_{ag_2}) = UPP\left(AS_{ag}^{(2)}, \{z_1, z_2, z_3\}\right) = \{z_1, z_2, z_3, z_4\}.$$

We obtain $o\,(ag)\,(\{y_1\}, \{z_1, z_2\}) =$

$$o\,(ag)\left(\left\|a_1^1 = yes \wedge a_2^1 = yes \wedge a_3^1 = no\right\|_{IS_{ag}^{(1)}}, \left\|a_1^2 = yes\right\|_{IS_{ag}^{(2)}}\right) \subseteq \left\|d = +\right\|_{IS_{ag}^{(0)}}.$$

$o\,(ag)$	a_1^1	a_2^1	a_3^1	a_1^2	a_2^2	a_3^2	d
(y_1, z_1, w_1)	yes	yes	no	yes	yes	yes	+
(y_1, z_3, w_2)	yes	yes	no	no	no	yes	+
(y_2, z_2, w_3)	no	yes	no	yes	yes	yes	+
(y_3, z_4, w_4)	no	yes	no	no	no	yes	-
(y_4, z_1, w_5)	no	no	yes	yes	yes	yes	-
(y_4, z_4, w_6)	no	no	yes	no	no	yes	-

Table 3. Operation $o(ag)$

The support of the rule **if** t **then** $d = +$ under the valuation val with respect to the lower approximations is equal to $card\left(val\left(LOW, AS_{ag}^{(0)}\right)(t) \cap \|d = +\|_{IS_{ag}^{(0)}}\right)$ $= 1$ and the confidence is also equal to 1.

We also obtain $o\left(ag\right)\left(\{y_1, y_2, y_3\}, \{z_1, z_2, z_3, z_4\}\right) =$

$$o\left(ag\right)\left(\left\|a_2^1 = yes\right\|_{IS_{ag}^{(1)}}, \left\|a_1^2 = yes \vee a_2^2 = no\right\|_{IS_{ag}^{(2)}}\right) = \{w_1, w_2, w_3, w_4\}.$$

The support of the rule **if** t **then** $d = +$ under the valuation val with respect to the upper approximations is equal to $card\left(val\left(UPP, AS_{ag}^{(0)}\right)(t) \cap \|d = +\|_{IS_{ag}^{(0)}}\right)$ $= 3$ and the confidence is equal to 0.75.

Let us observe that the set $val(UPP, AS_{ag}^{(0)})(t) - val(LOW, AS_{ag}^{(0)})(t)$ can be treated as the boundary region of t under val. Moreover, in the process of term construction we have additional parameters to be tuned for obtaining sufficiently high support and confidence, namely the approximation operations.

A concept X specified by the customer-agent is *sufficiently close to t under a given set Val of valuations* if X is included in the upper approximation of t under any $val \in Val$ and X includes the lower approximation of t under any $val \in Val$ as well as the size of the boundary region of t under Val, i.e.,

$$card\left(\bigcap_{val \in Val} val\left(UPP, AS_{ag}^{(0)}\right)(t) - \bigcup_{val \in Val} val\left(LOW, AS_{ag}^{(0)}\right)(t)\right),$$

is sufficiently small relatively to $\bigcap_{val \in Val} val\left(UPP, AS_{ag}^{(0)}\right)(t)$.

We conclude by formulating some examples of basic algorithmic problems.

- *Synthesis of generalized association rules.* Searching for a scheme (term t) over a given set Ag of agents and for a valuation val such that the rule **if** t **then** α, where α is a concept description specified by customer-agent, has the support at least s and the confidence at least c under the valuation val.
- *Synthesis of complex concepts close to the concept specified by the customer-agent.* Searching for a scheme (term t) over a given set Ag of agents and a set Val of valuations such that the concept specified by the customer-agent is sufficiently close to t under Val and the total size of the term t and the set Val is minimal.

Conclusions

Our approach can be treated as a step towards understanding of complex information granules in distributed environment. The approximate understanding of concepts definable by agents in the language of other agents is an important aspect of our approach for calculus on information granules. In our next paper we will present bounds on the complexity of the above formulated problems as well as heuristics for solving them.

Acknowledgments. This research was supported by the Research Grant of the European Union - ESPRIT-CRIT 2 No. 20288 and the grant No. 8 T11C 023 15 from the State Committee for Scientific Research.

365

References

1. Agrawal R., Mannila H., Srikant R., Toivonen H., Verkano A.: Fast Discovery of Association Rules, Fayyad U.M., Piatetsky-Shapiro G., Smyth P., Uthurusamy R. (Eds.): Advances in Knowledge Discovery and Data Mining, The AAAI Press/The MIT Press 1996, pp. 307-328.
2. Cattaneo G.: Abstract Approximation Spaces for Rough Theories, L. Polkowski, A. Skowron (Eds.), Rough Sets in Knowledge Discovery 1. Methodology and Applications, Physica-Verlag, Heidelberg, 1998, pp. 59-98.
3. Fagin R., Halpern J.Y., Moses Y., Vardi M.: Reasoning about Knowledge, MIT Press, 1996.
4. Fayyad U.M., Piatetsky-Shapiro G., Smyth P., Uthurusamy R. (Eds.): Advances in Knowledge Discovery and Data Mining, The AAAI Press/The MIT Press 1996.
5. Huhns M.N., Singh M.P.(Eds.): Readings in Agents, Morgan Kaufmann, San Mateo, 1998.
6. Lin T.Y.: Granular Computing on Binary Relations I Data Mining and Neighborhood Systems, L. Polkowski, A. Skowron (Eds.), Rough Sets in Knowledge Discovery 1. Methodology and Applications, Physica-Verlag, Heidelberg, 1998, pp. 107-121.
7. Pawlak Z.: Rough Sets. Theoretical Aspects of Reasoning about Data, Kluwer Academic Publishers, Dordrecht, 1991.
8. Polkowski, L., Skowron, A.: Rough Mereology: A New Paradigm for Approximate Reasoning, International Journal of Approximate Reasoning, Vol. 15, No 4, 1996, pp. 333-365.
9. Polkowski L., Skowron A.: Towards Adaptive Calculus of Granules, Proceedings of the FUZZ-IEEE'98 International Conference, Anchorage, Alaska, USA, May 5-9, 1998.
10. Rauszer C.: Knowledge Representation Systems for Groups of Agents, J. Wolenski (Ed.), Philosophical Logic in Poland, Kluwer Academic Publishers, Dordrecht, 1994, pp. 217-238.
11. Skowron A., Stepaniuk J.: Tolerance Approximation Spaces, Fundamenta Informaticae, Vol. 27, 1996, pp. 245-253.
12. Skowron A., Stepaniuk J.: Information Granules and Approximation Spaces, Proceedings of the 7th International Conference on Information Processing and Management of Uncertainty in Knowledge-based Systems, IPMU'98, Paris, France, July 6-10, 1998, pp. 1354-1361.
13. Stepaniuk J.: Approximation Spaces, Reducts and Representatives, L. Polkowski, A. Skowron (Eds.), Rough Sets in Knowledge Discovery 2. Applications, Case Studies and Software Systems, Physica-Verlag, Heidelberg, 1998, pp. 109-126.
14. Wong S.K.M.: A Rough-Set Model for Reasoning about Knowledge, L. Polkowski, A. Skowron (Eds.), Rough Sets in Knowledge Discovery 1. Methodology and Applications, Physica-Verlag, Heidelberg, 1998, pp. 276-285.
15. Zadeh L.A.: Toward a Theory of Fuzzy Information Granulation and Its Certainty in Human Reasoning and Fuzzy Logic, Fuzzy Sets and Systems Vol. 90, 1997, pp. 111-127.

Evolving Granules for Classification for Discovering Difference in the Usage of Words

Tetsuya Yoshida[1] Teruyuki Kondo[1] Shogo Nishida[1]

Dept. of Systems and Human Science, Grad. School of Eng. Science, Osaka Univ.,
1-3 Machikaneyama-cho, Toyonaka, Osaka 560-8531, Japan
yoshida@sys.es.osaka-u.ac.jp

Abstract. This paper proposes an evolutionary approach for discovering difference in the usage of words to facilitate collaboration among people. When people try to communicate their concepts with words, the difference in the meaning and usage of words can lead to misunderstanding in communication, which can hinder their collaboration. In our approach each granule of knowledge in classification from users is structured into a decision tree so that difference in the usage of word can be discovered as the difference in the structure of tree. By treating each granule of classification knowledge as an individual in Genetic Algorithm (GA), evolution is carried out with respect to the classification efficiency of each individual and diversity as a population so that difference in the usage of words will emerge as the difference in the structure of decision tree. Experiments were carried out on motor diagnosis cases with artificially encoded difference in the usage of words and the result shows the effectiveness of our evolutionary approach.

Keywords: usage of words, classification, decision tree, evolutionary approach

1 Introduction

In accordance with the need for dealing with large-scale and complex problems, it is required to support collaborative works among people through the interaction between people and machine since people often form a team to tackle such a problem. Generally different people seem to have different ways of conception and thus can have different concepts even on the same thing. When people try to communicate their concepts with abstract or vague words, such a difference in the meaning or usage of words can lead to misunderstanding in communication, which can hinder their collaboration. Difference in the usage of words can also be reflected on the description of data in large scale databases, which are often be constructed with the participation of many people.

This paper proposes an evolutionary approach for discovering difference in the usage of words to facilitate collaboration among people. Although words can be utilized to represent meaning at various level, i.e., abstract or concrete meaning in general, this paper focuses on dealing with the difference in the usage of symbol for concept. When users specify their classification knowledge as diagnosis cases, which are represented with the symbols for attributes, values and classes, they are structured into decision trees so that the trees reflect their conceptual structure of diagnostic knowledge. By treating each granule of classification knowledge as an individual in Genetic Algorithm (GA), this paper proposes to carry out evolution with respect to the classification efficiency of each individual and diversity as a population so that the difference in the usage of words will emerge as the difference in the structure of decision tree.

2 Framework of Discovering Difference in the Usage of Words

2.1 Difference in the Usage of Words

This paper focuses on dealing with the difference in the usage of words which represent conceptual meaning at the symbol level. Hereafter we call such a difference as "conceptual difference". Usually different symbols are used to denote different concepts, however, the same symbol can be used to denote different concepts depending on the viewpoint in which the symbol is used. In contrast, different symbols can be used to represent the same concept. The types of conceptual difference dealt with in this paper are defined as follows:

- **Type 1: different symbols are used to denote the same concept.**
- **Type 2: the same symbol is used to denote different concepts.**

Suppose there are an expert in electric engineering called Adam and an expert in mechanical engineering called Bob. When they carry out the diagnosis of motor failure, Adam might point out "anomaly in voltage frequency" and Bob might point out "anomaly in revolutions" for the same symptom. The above two symbols or terms represent different concepts in general, however, they can be considered as denoting the same concept when they are used in the context of the diagnosis of motor failure.

2.2 Discovering Conceptual Difference

The kind of problem which can be dealt with in our approach is classification such as diagnosis and the *class* of cases is determined based on the *attributes* and *values*, which characterizes the cases [3]. The system tries to construct the decision tree for cases which is most effective for their classification based on the information theory. A node in decision trees holds the attribute to characterize cases. Each link below a node holds the value for the attribute at the node and cases are divided gradually by following links. The class of cases is determined at the leaf which is reached as the result of link following. We utilize ID3 algorithm [5] to construct decision trees since it is fast and thus is suitable for interactive systems.

The system architecture which incorporates the descovering method in this paper is shown in Fig 1. Currently the system requires that two users represent their knowledge as cases with their respective symbols for them. By accepting the cases as input the system constructs decision trees for them and tries to detect conceptual differences in attributes, values and classes based on the structural differences in trees. Since there are 2 types of conceptual differences for 3 entities, the system tries to discover 6 kinds of conceptual differences and shows the candidates for them in the descending order of the possibility to users. The system also displays the decision trees for cases. Visualizing their knowledge as concrete decision trees is expected to help them modify their knowledge and to stimulate further conception. Based on the result from the system users discuss each other to change their concepts toward reducing conceptual differences and modify input data to the system. The above processes are repeated interactively to remove conceptual difference gradually. In future we plan to extend the system so that it be applicable to more than two users.

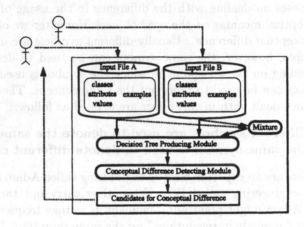

Fig. 1. System Architecture

2.3 Discovery Algorithms for Conceptual Difference

The algorithms are designed based on the role of the symbols for the classification of cases, which is discovered from the structural characteristics of decision trees This section briefly explain the key idea for discovering conceptual difference in our approach. The details of the algorithm are described in [4, 2, 7]. Hereafter the set of cases from each user and that of synthesized cases are called as A, B, A+B, respectively.

Conceptual difference for class sysmbol is defined as:

C1 : different symbols are used to denote the same class
C2 : the same symbol is used to denote the different classes

These are discovered based on the "inconsistency in the classification knowledge" for cases in A+B. **C1** is discovered when different sysmbols are used as the class symbol for the cases with the same value for each attribute. **C2** is discovered when the same symbol is used as the class symbol for the cases with the different value and/or attribute.

The algorithms for

A1 : different symbols are used to denote the same attribute
A2 : the same symbol is used to denote the different attributes
V1 : different symbols are used to denote the same value
V2 : the same symbol is used to denote the different values

are defined similarly based on the structure of decision tree [7].

3 Discovery Method for Conceptual Differences based on Diverse Structures

3.1 Problems in Utilizing ID3

In general there can be several decision trees with the equivalent classification for a set of cases, however, ID3 algorithm constructs the most simple tree in the sense that classification is carried out as fast as possible at the upper nodes in a decision tree by reducing the redundant information held in the cases. Since the time complexity of ID3 algorithm is not heavy, it is suitable for interactive systems. However, reducing redundant information held in a set of cases cannot always be advantageous for the aim of our approach, namely, discovering conceptual differences among people. With ID3 algorithm, attributes which do not contribute to classifying cases are not represented in the decision trees, even if they are utilized in the representation of the cases to denote the knowledge held by users. Since such an attribute is not represented in the decision tree, it is impossible to discover conceptual difference for the attribute or the values for it by comparing the decision trees. This implies that sometimes it would be impossible to discover the difference in concpet as the difference in structure when the structure is constructed by ID3 algorithm.

3.2 Constructing Decision Trees with Diverse Structures

Genetic Algorithm (GA) is utilized in our approach to construct decision trees with diverse structure so that effective decision trees can be sought in search by preserving the accuracy in the classification of cases. Hereafter, in decision trees the node in which the class of cases is determined is called a "leaf node", and the node in which the attribute of case is tested is called a "condition node".

Coding, Crossover and Mutation We employ GA with tree structure, which is often used in Genetic Programming, for constructing decision trees [6, 1] A node in a decision tree is represented as a gene in the coding of genetic information. The gene for a condition node contains the information for the position in a decision tree and the attribute to judge the branch in the decision tree. The gene for a branch contains the value for the attribute above the branch, which indicates the branch to follow in the decision tree.

Crossover is carried out by exchanging partial trees, as shown in Figure 2. A partial tree is defined as the tree which has a condition node as its root. Mutaion

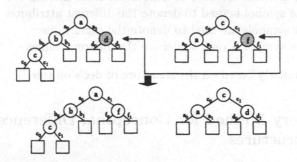

Fig. 2. Crossover by exchanging partial trees

is carried out by randomly selecting several childen which are constructed by crossover and coping them. Arbitrary nodes in these copied trees are specified as crossover points and the partiall trees below the nodes are replaced with either other partial tree or a leaf node.

Survival Strategy It is desirable to leave the set of decision trees for the next generation, each of which has high accuracy for the classification of cases and their structures are as different as possible. The following two criteria are utilized as the survival strategy in GA to select decision trees for the next generation.

(1) Error Rate
"Error rate" is defined as the index which

– decreases when cases are classified into one class at shallower depth
– increases when cases are not classified into one class even at deeper depth

The decision tree with small error rate has better capability for classification.

Error rate is calculated as shown in Fig 3. First, cases are classfied by a decision tree and stored into the leaf nodes in the tree at which the class of case is determined. Error rate for a leaf node is determined depending on the capability of the classification of cases along the path from the root to the leaf node and assigned as:

– "0" when a leaf node has only the cases with the same class.
– "n_c - 1" when a leaf node has cases with multiple classes (#class is n_c).
– "N_c" (the number of all classes) when a leaf node has no case.

The above calculation of error rate treats the leaf node with no case as the worst one and penalizes a leaf node more severely when it has multiple classes.

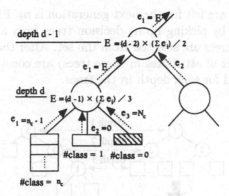

Fig. 3. Calculation of error rate

As for a condition node, all the error rates assigned to its childlen are summed and then divided by the number of branches, which is number of values for the attribute at the condition node. Taking the average of error rates below a condition node contributes to alleviating the difference in error rates, which will arise due to the difference in the number of values for each attribute. The average is multiplied by "$d-1$" (d is the depth of the condition node in the tree) and assigned as the error rate for the condition node. The depth of root in a tree is treaed as 1. Utilizing the depth of node as multiplying the average by "$d-1$" contributes to reflecting the efficiency of classification.

The equations for calculating error rate are summarized as follows:

– **Error rate at leaf nodes**
  number of class 1 $\cdots$ 0
  number of class $n_c (\geq 2)$ $\cdots$ $n_c - 1$
  number of class 0 $\cdots$ $N_c (\geq n_c)$ (the number of all classes)

– **Equations for error rate at condition nodes**

$$E = (d-1) \times (\textstyle\sum_{v=0}^{n_v} e_v)/n_v \ (d \neq 1)$$
$$E = \qquad (\textstyle\sum_{v=0}^{n_v} e_v)/n_v \qquad (d = 1)$$

d: depth of node n_v: number of values e_v: error rate for each child node

(2) Mutual Distance between Decision Trees
"Mutual distance between decision trees" is utilized to measure the degree of divergence in structure for a set of decision trees. It is calculated by summing up the distance for each pair of decision trees in the set. A set of decision trees with larger mutual distance has more divergence in structure and thus has the possibility of including the attribute or value which cannot be represented in a single decision tree.

An example of the calculation of distance between decision trees is shown in Figure 4. Suppose the number of all the decision trees is N_t and the number of decision trees which are left for the next generation is n_t. First, a set of decision trees is constructed by picking up n_t decision trees from all the decision trees. Then, two decision trees are selected from the set. After that, vectors of size n_a (which is the number of attributes in these trees) are constructed for each tree. Vectors are prepared for each depth in the trees.

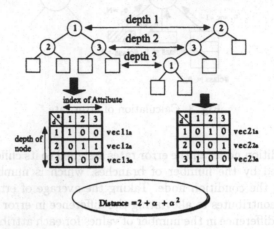

Fig. 4. Calculation of the distance between decision trees.

The value of element in a vector is set to "1" when the attribute for the element exists at the condition node with the depth; otherwise it is set to "0". Then, the vectors with the same depth are compared and the number of elements with different value is counted. Since the number of condition nodes grows in accordance with the depth in general, the attribute for the condition node with small depth is considered as significant. Thus, the result of counting is multiplied by the weight which is in inverse proportion to the depth of vector to reflect the degree of significance of each attribute. The above operations are carried out for all the combination $_{n_t}C_2$ of each pair of decision trees and the result is treated as the mutual distance for a set of decision trees.

Equation for the mutual distance of decision trees

$$Dist. = \sum_{d=1}^{n_t C_2} \sum_{d=1}^{D} \{\alpha^{d-1} \sum_{a=1}^{n_a} |vec1_{d,a} - vec2_{d,a}|\}$$

n_t: #decision trees in one generation n_a: #attributes in decision trees
$\alpha(<1)$: weight for the depth of node $D(=n_a)$: maximum depth of condition node

4   Experiments and Evaluations

A prototype system has been implemented on the UNIX workstation with C language. The experiments on motor diagnosis cases were carried out to evaluate the approach in this paper. In experiments two persons specified their knowledge in the form of 100 cases (as shown in Fig ??), which were composed of six attributes, two or three values and five classes, respectively. Conceptual differences are artificially encoded into the cases in B by modifying the original cases.

Experiments were carried out in the condition that two kinds of conceptual difference occured at the same time in the test cases to see the interaction and/or interference between the algorithms. As the quantitative ealuation, the number of discovery and its probability of discovery up to the third candidate were collected in the experiments both for the system with ID3 and that with GA. As described in Section 2.2, conceptual difference is resolved by repeating the interaction between the suggestion by the system and the modification of cases by the user in our approach, however, the result for the first cycle of suggestion by the system was focused on in the experiments. Summary of the result of discovery is shown in Table 4 and Table 4. The result shows that the system with ID3 can

Table 1. Result with ID3.   **Table 2.** Result with GA.

	number of trials	1st	2nd	3rd	probability of discovery		number of trials	1st	2nd	3rd	probability of discovery
C1	20	20	0	0	100%	C1	20	20	0	0	100%
C2		18	0	0	90%	C2		18	0	0	90%
C1	30	17	1	0	60%	C1	30	30	0	0	100%
A2		5	3	7	50%	A2		19	6	2	90%
C1	30	30	0	0	100%	C1	30	30	0	0	100%
V2		52	0	0	87%	V2		52	8	0	100%
C2	30	22	0	1	77%	C2	30	23	1	0	80%
V1		24	4	0	93%	V1		30	0	0	100%
A1	30	12	13	3	93%	A1	30	14	10	4	93%
A2		6	3	7	53%	A2		20	8	2	100%
A1	30	12	11	7	100%	A1	30	20	9	1	100%
V2		45	2	6	88%	V2		38	22	0	100%

accurately discover conceptual difference for **C1** and **A1**. However, it cannot discover other kinds of conceptual difference with high accuracy, for instance, the probability of discovery remains at 50 % for **A2**. It is noticed that conceptual difference is suggested as the first to third candidate. On the other hand, the system with GA can discover conceptual difference more accurately in general, and conceptual difference is suggested as the higher rank in candidates. These results show that the structures which are suitable for our discovery algorithms are not necessarily represented in the decision trees with ID3. Thus, diverse structure with GA can be said to contribute to improving the peformance of discovery of conceptual difference. Suggesting conceptual difference as the first candidate will also contribute to reducing the possibility of suggesting conceptual difference erroneously. Moreover, utilizing the average of discovery over multiple decision trees might make the system with GA more robust for noise due to the statistical effect of averaging.

The experiments show that utilizing diverse structures with GA is superior to that with ID3 for the construction of decision trees with respect to the precision of discovery for conceptual difference. On the other hand, with respect to the computation complexity, ID3 takes much less times than GA and thus is suitable for the interactive system.

5 Conclusion

This paper has proposed an evolutionary approach for discovering difference in the usage of words to facilitate collaboration among people. In our approach knowledge of users is structured into decision trees and candidates for conceptual difference are suggested based on the structural characteristics of decision trees. By pointing out the problem in utilizing deterministic approach for the construction of decision trees, this paper has proposed to carry out evolution with respect to the classification efficiency of each decision tree and diversity as a population. Experiments were carried out on motor diagnosis cases with artificially encoded conceptual difference. The result shows that our approach is effective to some extent as the first step for dealing with the issue of conceptual difference toward facilitating collaboration among people.

Acknowledgments

This research is partially supported by Grant in Aid for Scientific Research from the Japanese Ministry of Education, Science and Culture (10875080, 09450159).

References

1. David E. Goldberg. *Genetic algorithms in search, optimization, and machine learning.* Addision-Wesley, 1989.
2. T. Kondo, T. Yoshida, and S. Nishida. Design of the Interfaces to Detect Conceptual Difference among Different People. *Information Processing Society of Japan,* 39(5):1195–1202, May 1998. in Japanese.
3. D. Michie, D. J. Spiegelhalter, and C. C. Taylor, editors. *Machine Learning, Neural and Statistical Classification.* Ellis Horwood, Chichester, England, 1994.
4. S Nishida, T. Yoshida, and T. Kondo. Interactive Interfaces to Detect Conceptual Difference for Group Knowledge Acquisition. In *Aritificial Intelligence in Real-Time Control (AIRTC)'97,* pages 210–214, 9 1997.
5. J. R. Quinlan. Induction of decision trees. *Machine Learning,* 1:81–106, 1986.
6. John R.Koza. *Genetic programming II: automatic discovery of reusable programs.* MIT Press, 2nd edition, 1994.
7. T. Yoshida, T. Kondo, and S. Nishida. Discovering Conceptual Differences among People from Cases. In *The First International Conference on Discovery Science,* pages 162 – 173, 1998.

Interval Evaluation by AHP
with Rough Set Concept

Kazutomi Sugihara   Yutaka Maeda   Hideo Tanaka

Department of Industrial Engineering,
Osaka Prefecture University,
Gakuenchou 1-1, Sakai, Osaka 599-8531, JAPAN
{ksugi,maeda,tanaka}@ie.osakafu-u.ac .jp
http://www.ie.osakafu-u.ac.jp/

Abstract. In AHP, there exists the problem of pair-wise consistency where evaluations by pair-wise comparison are presented with crisp value. We propose the interval AHP model with interval data reflecting Rough Set concept. The proposed models are formulated for analyzing interval data with two concepts (necessity and possibility). According to necessity and possibility concepts, we obtain upper and lower evaluation models, respectively. Furthermore, even if crisp data in AHP are given, it is illustrated that crisp data should be transformed into interval data by using the transitive law. Numerical examples are shown to illustrate the interval AHP models reflecting the uncertainty of evaluations in nature.
Key-word: AHP, Evaluation, Rough sets concept, Upper and lower models, Intervals

1   Introduction

AHP(Analytic Hierarchy Process) proposed by T.L.Satty[1] has been used to evaluate alternatives in multiple criteria decision problems under a hierarchical structure and has frequently been applied to actual decision problems. Satty's AHP method is based on comparing n objects in pairs according to their relative weights. Let us denote the objects by $X_1, \ldots, X_n$ and their weights by $w_1, \ldots, w_n$. The pair-wise comparisons can be denoted as the following matrix:

$$A = \begin{pmatrix} & X_1 & X_2 & \ldots & X_n \\ \hline X_1 & \frac{w_1}{w_1} & \frac{w_1}{w_2} & \cdots & \frac{w_1}{w_n} \\ X_2 & \frac{w_2}{w_1} & \frac{w_2}{w_2} & \cdots & \frac{w_2}{w_n} \\ \vdots & \vdots & \vdots & \ddots & \vdots \\ X_n & \frac{w_n}{w_1} & \frac{w_n}{w_2} & \cdots & \frac{w_n}{w_n} \end{pmatrix}$$

which satisfies the reciprocal property $a_{ji} = \frac{1}{a_{ij}}$. If the matrix A satisfies the cardinal consistency property $a_{ij}a_{jk} = a_{ik}$, A is called consistent. Generally, A is called a reciprocal matrix.

According to Satty's method, we have

$$Aw = \lambda w$$

where a weight vector w can be obtained by solving the above eigenvalue problem.

Now suppose that the pair-wise comparison ratios are given by intervals, although they are real numbers in the conventional AHP. Intervals scales are estimated by an individual as approximations. In AHP, the ratio scale for pair-wise comparisons ranges from 1 to 9 to represent judgment entries where 1 is equally important and 9 is absolutely more important. It should be noted that the reciprocal values $a_{ji} = \frac{1}{a_{ij}}$ are always satisfied. As an example of interval ratios, we can give an interval $A_{ij} = [3,5]$ and then, A_{ji} must be $[\frac{1}{5}, \frac{1}{3}]$. AHP with fuzzy scales has been studied by C.H Cheng and D.L. Mon[2] where fuzzy scales are transformed into ordinal scales. Considering fuzziness of scales, sensitivity analysis for AHP has been done in [3].

In this paper, we propose an interval AHP model, given interval scales as pair-wise comparison ratios. Dealing with interval data, we can obtain the upper and lower models for AHP which are similar to Rough Sets Analysis[4]. Even when crisp data are given, interval data can be obtained from crisp data by using the transitive law. Thus, our proposed method can be described as reflecting the uncertainly of evaluations in nature. Our method resorts to linear programming so that interval scales can easily be handled. This approach to uncertain phenomena has been used in regression analysis and also identification of possibility distributions[5]. Numerical examples are shown to illustrate the interval AHP models.

2 Interval AHP

Let us begin with interval scales in a reciprocal matrix denoted by $A_{ij} = \left[\underline{a_{ij}}, \overline{a_{ij}}\right]$ where $\underline{a_{ij}}$ and $\overline{a_{ij}}$ are the lower and upper bounds of the interval A_{ij}. The reciprocal property is represented as

$$\underline{a_{ij}} = \frac{1}{\overline{a_{ji}}}, \overline{a_{ij}} = \frac{1}{\underline{a_{ji}}} \tag{1}$$

Reflecting interval scales, let us suppose that weights are found as interval weights W_{ij} by

$$W_{ij} = \left[\frac{\underline{w_i}}{\overline{w_j}}, \frac{\overline{w_i}}{\underline{w_j}}\right] \tag{2}$$

where $\underline{w_i}$ and $\overline{w_i}$ are the lower and upper bounds of the interval weight $W_i = \left[\underline{w_i}, \overline{w_i}\right]$. Given an interval matrix $[A]$ denoted as

$$[A] = \begin{pmatrix} A_{11} & A_{12} & \ldots & A_{1n} \\ A_{21} & A_{22} & \ldots & A_{2n} \\ \vdots & \vdots & \ddots & \vdots \\ A_{n1} & A_{n2} & \ldots & A_{nn} \end{pmatrix} \tag{3}$$

the problem under consideration is to find out interval weights $W_i = [\underline{w_i}, \overline{w_i}]$ which can be an approximation to the given interval matrix $[A]$ of (3) in some sense. Since we deal with interval data, we can consider two approximations shown in Fig.1 as follows. The lower and upper approximations should satisfy

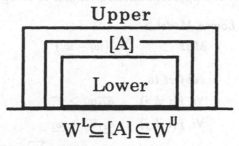

$$W^L \subseteq [A] \subseteq W^U$$

Fig. 1. Upper and lower approximations

the following constrain conditions

$$W_{ij}^L \subseteq [A_{ij}] \ (Lower \, Approximation) \tag{4}$$

$$W_{ij}^U \supseteq [A_{ij}] \ (Upper \, Approximation) \tag{5}$$

where W_{ij}^L and W_{ij}^U are the estimations of lower and upper intervals. (4) and (5) can be rewritten as

$$W_{ij}^L \subseteq [A_{ij}] \longleftrightarrow \underline{a_{ij}} \leq \frac{\underline{w_i}}{\overline{w_j}} \leq \frac{\overline{w_i}}{\underline{w_j}} \leq \overline{a_{ij}}$$

$$\longleftrightarrow \overline{a_{ij}} \, \underline{w_j} \geq \overline{w_i}, \ \underline{a_{ij}} \, \overline{w_j} \leq \underline{w_i} \tag{6}$$

$$W_{ij}^U \supseteq [A_{ij}] \longleftrightarrow \frac{\underline{w_i}}{\overline{w_j}} \leq \underline{a_{ij}} \leq \overline{a_{ij}} \leq \frac{\overline{w_i}}{\underline{w_j}}$$

$$\longleftrightarrow \underline{a_{ij}} \, \overline{w_j} \leq \underline{w_i}, \ \overline{w_j} \, \overline{a_{ij}} \geq \overline{w_i} \tag{7}$$

Now let us consider how to normalize an interval vector $(W_1, \ldots, W_n)$, although in the conventional AHP a weight vector is normalized so that its components sum to unity. The conventional normalization can be extended into the interval normalization[6] defined as follows:

An interval weight vector $(W_1, \ldots, W_n)$ is said to be normalized if and only if

$$\sum_i \overline{w_i} - \max_j \left(\overline{w_j} - \underline{w_j} \right) \geq 1 \tag{8}$$

$$\sum_i \underline{w_i} + \max_j \left(\overline{w_j} - \underline{w_j} \right) \leq 1 \tag{9}$$

(8) and (9) can be described as

$$\forall j \ \underline{w_j} \geq 1 - \sum_{i \in \Omega - \{j\}} \overline{w_i} \tag{10}$$

$$\forall j \quad \overline{w_j} \le 1 - \sum_{i \in \Omega - \{j\}} \underline{w_i} \tag{11}$$

where $\Omega = [1, \ldots, n]$. Using the concepts of "Greatest Lower Bound" and "Least Upper Bound", we can formulate the lower model and the upper model, respectively. The concept of two approximations is similar to Rough Set concept.

< Lower Model >

$$Max \quad \sum_i \left(\overline{w_i} - \underline{w_i} \right) \tag{12}$$

subject to

$$\forall i, j \, (i \ne j) \quad \overline{a_{ij}} \underline{w_j} \ge \overline{w_i}$$

$$\forall i, j \, (i \ne j) \quad \underline{a_{ij}} \overline{w_j} \ge \underline{w_i}$$

$$\forall j \quad \underline{w_j} \ge 1 - \sum_{i \in \Omega - \{j\}} \overline{w_i}$$

$$\forall j \quad \overline{w_j} \le 1 - \sum_{i \in \Omega - \{j\}} \underline{w_i}$$

$$\forall i \quad \underline{w_i} \le \overline{w_i}$$

$$\forall i \quad \underline{w_i}, \overline{w_i} \ge 0 \quad \forall i$$

< Upper Model >

$$Min \quad \sum_i \left(\overline{w_i} - \underline{w_i} \right) \tag{13}$$

subject to

$$\forall \, i, j \, (i \ne j) \quad \overline{a_{ij}} \underline{w_j} \le \overline{w_i}$$

$$\forall \, i, j \, (i \ne j) \quad \overline{w_j} \underline{a_{ij}} \ge \underline{w_i}$$

$$\forall j \quad \underline{w_j} \ge 1 - \sum_{i \in \Omega - \{j\}} \overline{w_i}$$

$$\forall j \quad \overline{w_j} \le 1 - \sum_{i \in \Omega - \{j\}} \underline{w_i}$$

$$\forall i \quad \underline{w_i} \le \overline{w_i}$$

$$\forall i \quad \underline{w_i}, \overline{w_i} \ge 0$$

Example1:
The pair-wise comparisons matrix is given as:

$$[A] = \begin{pmatrix} 1 & [1,3] & [3,5] & [5,7] & [5,9] \\ [\frac{1}{3},1] & 1 & [\frac{1}{2},2] & [1,5] & [1,4] \\ [\frac{1}{5},\frac{1}{3}] & [\frac{1}{2},2] & 1 & [\frac{1}{3},3] & [2,4] \\ [\frac{1}{7},\frac{1}{5}] & [\frac{1}{5},1] & [\frac{1}{3},3] & 1 & [1,3] \\ [\frac{1}{9},\frac{1}{5}] & [\frac{1}{4},1] & [\frac{1}{4},\frac{1}{2}] & [\frac{1}{3},1] & 1 \end{pmatrix} \tag{14}$$

Using the lower and the upper models, we obtained the interval weights shown in Table 1.

Table 1 Interval weights obtained by two models (Example1)

Alternatives	Lower model	Upper model
W_1	$[0.4225, 0.5343]$	$[0.3333, 0.3750]$
W_2	$[0.1781, 0.2817]$	$[0.1250, 0.3333]$
W_3	$[0.1408, 0.1408]$	$[0.0417, 0.2500]$
W_4	$[0.0763, 0.0845]$	$[0.0536, 0.1250]$
W_5	$[0.0704, 0.0704]$	$[0.0417, 0.1250]$

It can be found from Table 1 that the interval weights obtained by the lower model satisfy (6) and the interval weights obtained by the upper model satisfy (7) . The obtained interval weights can be said to be normalized, because (8) and (9) hold.

3 Interval scales by transitivity

If a consistent matrix A is given, the following consistency property holds:

$$\forall i, j \quad a_{ij} = a_{ik} a_{kj} \tag{15}$$

However, this property does not hold in general. Therefore, interval scales can be obtained by transitivity from crisp scales. Denote an interval scales A_{ij} as $A_{ij} = [\, \underline{a_{ij}}, \overline{a_{ij}} \,]$ and an interval matrix $[A]$ as $[A] = [A_{ij}]$. Given a crisp matrix A, the interval matrix $[A]$ can be obtained as follows:

$$\underline{a_{ij}} = \min_V (a_{ik} \cdots a_{lj}) \tag{16}$$

$$\overline{a_{ij}} = \max_V (a_{ik} \cdots a_{lj}) \tag{17}$$

where V is a set of all possible chains from i to j without any loops.

Example2:
Let us start with a crisp scale matrix as follows.

$$A = \begin{pmatrix} 1 & 3 & 3 & 5 & 7 \\ \frac{1}{3} & 1 & 1 & 2 & 3 \\ \frac{1}{3} & 1 & 1 & 1 & 3 \\ \frac{1}{5} & \frac{1}{2} & 1 & 1 & 3 \\ \frac{1}{7} & \frac{1}{3} & \frac{1}{3} & \frac{1}{3} & 1 \end{pmatrix} \tag{18}$$

Using (16) and (17), we obtained the interval matrix $[A]$.

$$[A] = \begin{pmatrix} 1 & [\frac{7}{6}, 5] & [\frac{7}{6}, 6] & [\frac{7}{3}, 6] & [\frac{9}{2}, 18] \\ [\frac{1}{5}, \frac{6}{7}] & 1 & [\frac{3}{5}, \frac{18}{7}] & [\frac{7}{9}, \frac{15}{7}] & [\frac{7}{5}, 6] \\ [\frac{1}{6}, \frac{6}{7}] & [\frac{7}{18}, \frac{5}{3}] & 1 & [\frac{7}{9}, \frac{18}{7}] & [\frac{7}{6}, 6] \\ [\frac{1}{6}, \frac{3}{7}] & [\frac{7}{15}, \frac{9}{7}] & [\frac{7}{18}, \frac{9}{7}] & 1 & [\frac{7}{6}, 3] \\ [\frac{1}{18}, \frac{9}{2}] & [\frac{1}{6}, \frac{5}{7}] & [\frac{1}{6}, \frac{6}{7}] & [\frac{1}{3}, \frac{6}{7}] & 1 \end{pmatrix} \tag{19}$$

We applied the interval matrix (19) to the lower and the upper models and obtained the interval weights shown in Table 2.

Table 2 Interval weights obtained by two models (Example2)

Alternatives	Lower model	Upper model
W_1	$[0.2972, 0.5868]$	$[0.2967, 0.4723]$
W_2	$[0.1528, 0.2547]$	$[0.0945, 0.2543]$
W_3	$[0.0991, 0.2547]$	$[0.0787, 0.2543]$
W_4	$[0.1189, 0.1274]$	$[0.0787, 0.1272]$
W_5	$[0.0425, 0.0660]$	$[0.0262, 0.0675]$

Example3:

Let us consider the binary problem shown in Fig.2 where ① → ① means that

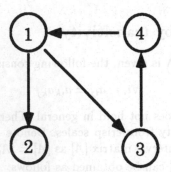

Fig. 2. Binary Problem

i won against j. It is assumed that the value of 2 is assigned to wins and also the value of $\frac{1}{2}$ is assigned to defeats. Then we obtain the matrix A with unknown scales denoted as $*$.

$$A = \begin{pmatrix} 1 & 2 & 2 & \frac{1}{2} \\ \frac{1}{2} & 1 & * & * \\ \frac{1}{2} & * & 1 & 2 \\ 2 & * & \frac{1}{2} & 1 \end{pmatrix} \tag{20}$$

Using transitivity (16) and (17), we have

$$[A] = \begin{pmatrix} 1 & 2 & [\frac{1}{4}, 2] & [\frac{1}{2}, 4] \\ \frac{1}{2} & 1 & [\frac{1}{8}, 1] & [\frac{1}{4}, 2] \\ [\frac{1}{2}, 4] & [1, 8] & 1 & [\frac{1}{4}, 2] \\ [\frac{1}{4}, 2] & [\frac{1}{2}, 4] & [\frac{1}{2}, 4] & 1 \end{pmatrix} \tag{21}$$

We applied the interval matrix (21) to the lower and upper models and obtained the interval weights shown in Table 3.

381

Table 3 Interval weights obtained by two models (Example3)

Alternatives	Lower model	Upper model
W_1	$[0.2500, 0.2500]$	$[0.0909, 0.3636]$
W_2	$[0.1250, 0.1250]$	$[0.0455, 0.1818]$
W_3	$[0.1250, 0.3723]$	$[0.0909, 0.3636]$
W_4	$[0.2527, 0.5000]$	$[0.0909, 0.3636]$

4 Concluding Remarks

In the conventional AHP, pair-wise comparisons range from 1 to 9 as ration scale. Therefore the scales range from $\frac{1}{9}$ to 9. If we use transitivity (16) and (17), the upper and lower bounds of interval scales obtained by (16) and (17) may be not within the maximal interval $[\frac{1}{9}, 9]$. Thus, instead of (16) and (17), we can use

$$\underline{a_{ij}} = \min_V f\left(a_{ik} \cdots a_{lj}\right) \tag{22}$$

$$\overline{a_{ij}} = \max_V f\left(a_{ik} \cdots a_{lj}\right) \tag{23}$$

where the function $f(x)$ is defined by

$$f(x) = \begin{cases} \frac{1}{9}; \text{ for x which is less than } \frac{1}{9} \\ x; \text{ for x which is within}\left[\frac{1}{9}, 9\right] \\ 9; \text{ for x which is larger than 9} \end{cases} \tag{24}$$

Instead of the function f, the geometric mean can be used to obtain interval scales.

References

1. T.L. Satty; The Analytic Hierarchy Process, McGraw-Hill, 1980
2. C.H.Cheng and D.L.Mon; Evaluating weapon system by Analytical Hierarchy Process basedon fuzzy scales, Fuzzy Sets and System, Vol.63, p.p.1-10, 1994
3. S.Onishi, H.Imai, M.Kawaguchi; Evaluation of a stability on weights of fuzzy analytic hierarchy process using a sensitivity analysis, J. of Japan Society for Fuzzy Theory and Systems, Vol.9, No.1, p.p.140-147, 1997 (in Japanese)
4. Z.Pawlak; Rough Sets -Theoretical Aspects of Reasoning about Data- , Kluwer Academic Publichers, Dordrecht, 1991
5. H.Tanaka and P.Guo; Possibilistic Data Analysis for Operations Research, Physica-Verlag(A Springer-verlag Company), Heidelberg, 1999
6. Y.Maeda and H.Tanaka; Non-additive probability measures with interval density functions, J. of Japan Society for Fuzzy Theory and Systems, to appear (in Japanese)

Interval Density Functions in Conflict Analysis

Yutaka Maeda, Kazuya Senoo, and Hideo Tanaka

Department of Industrial Engineering,
Osaka Prefecture University,
Gakuenchou 1-1, Sakai, Osaka 599-8531, JAPAN
{maeda, kazuya, tanaka}@ie.osakafu-u.ac.jp
http://www.ie.osakafu-u.ac.jp/

Abstract. Interval density functions are non-additive probability measures representing sets of probability density functions. Pawlak proposed a novel approach called conflict analysis based on rough set theory. In this paper, we propose a new approach of presenting expert's knowledge with interval importances and apply it to conflict analysis. It is assumed that the importance degrees are given for representing expert's knowledge. Using conditions of interval density functions, we represent many experts' knowledge as interval importance degrees. A simple example of the new introduced concepts is presented.

Keyword: Interval density functions; Decision analysis; Rough sets; Conflict analysis

1 Introduction

Interval density functions (IDF)[1] are non-additive probability measures representing sets of probability density functions. An interval density function consists of two density functions by extending values of conventional density function to interval values, which do not satisfy additivity.

Conflict analysis plays an important role in many real fields such as business, labor-management negotiations, military operations, etc. The mathematical models of conflict situations have been proposed [2][3] and investigated. Conflicts are one of the most characteristics attributes of human nature and a study of conflicts is important theoretically and practically. It seems that fuzzy sets and rough sets [4] are suitable candidates for modeling conflict situations under the presence of uncertainty.

In this paper, we propose a new approach of presenting expert's knowledge with interval importances and apply it to conflict analysis. It is assumed that an expert's knowledge is given as a relative importance for each attribute. When there are plural experts, their knowledge is formulated as an interval importance using interval density functions. Then, a conflict degree between two agents has an interval value.

2 Interval Density Functions

In this section, we introduce the concept of interval density functions [1]. Probability distributions have one to one correspondence with their density functions. A probability density function $d : X \to \Re$ on the disjoint finite universe X is defined as:

$$\forall x \in X, \ d(x) \geq 0, \ \sum_{x \in X} d(x) = 1.$$

Then the probability of the event A is gives as:

$$\forall A \subseteq X, \ P(A) = \sum_{x \in A} d(x).$$

For all $A, B \subseteq X$, the additivity holds as follows:

$$A \cap B = \emptyset \ \Rightarrow \ P(A \cup B) = P(A) + P(B).$$

Interval density functions being non-additive probability measures are defined as follows:

Definition 1 (A interval density function on the disjoint finite universe): A pair of functions (h_*, h^*) satisfying the following conditions is called an interval density function (IDF):

$h_*, h^* : X \to \Re; \ \forall x' \in X, \ h^*(x') \geq h_*(x') \geq 0,$

(I) $\sum_{x \in X} h_*(x) + (h^*(x') - h_*(x')) \leq 1,$

(II) $\sum_{x \in X} h^*(x) - (h^*(x') - h_*(x')) \geq 1.$

The conditions (I) and (II) can be transformed as:

(I') $\sum_{x \in X} h_*(x) + \max_x [h^*(x) - h_*(x)] \leq 1,$

(II') $\sum_{x \in X} h^*(x) - \max_x [h^*(x) - h_*(x)] \geq 1.$

Then, we have the following theorem.

Theorem 1 For any IDF, there exists a probability density function $h'(\cdot)$ satisfying that

$$h_*(x) \leq h'(x) \leq h^*(x), \ \sum_{x \in X} h'(x) = 1.$$

To illustrate an interval density function let us consider the case shown in Fig.1 where the number 6 is most likely occurred comparatively with the number 1 to 5. Interval density functions for the number 1 to 5 are $(h_*, h^*) = (1/10, 1/6)$, and interval density function for the number 6 is $(h_*, h^*) = (1/6, 1/2)$.

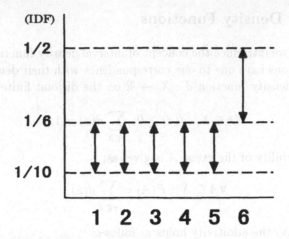

Fig. 1. Example for interval density functions

It is clear that these interval density functions satisfy Definition 1. Taking the number 6 for x',

$$\sum_{x \in X} h_*(x) + (h^*(x') - h_*(x')) = \frac{4}{6} + \frac{2}{6} \leq 1$$

$$\sum_{x \in X} h^*(x) - (h^*(x') - h_*(x')) = \frac{8}{6} - \frac{2}{6} \geq 1$$

Using this functions (h_*, h^*), we can define two distribution functions as follows:

(Lower boundary function (LB) and upper boundary function (UB) of IDF): For h' satisfying $h_*(x) \leq h'(x) \leq h^*(x)$, $\forall A \subseteq X$

$$LB(A) = \min_{h'} \left(\sum_{x \in A} h'(x) \right),$$

$$UB(A) = \max_{h'} \left(\sum_{x \in A} h'(x) \right).$$

Then, lower and upper boundary functions have the following properties. $\forall A \subseteq X$

$$LB(A) = \sum_{x \in A} h_*(x) \vee \left(1 - \sum_{x \in \bar{A}} h^*(x) \right)$$

$$UB(A) = \sum_{x \in A} h^*(x) \wedge \left(1 - \sum_{x \in \bar{A}} h_*(x) \right)$$

And, the duality of LB and UB holds.

$$1 - UB(\bar{A}) = 1 - \sum_{x \in \bar{A}} h^*(x) \wedge \left(1 - \sum_{x \in A} h_*(x)\right)$$

$$= \sum_{x \in A} h_*(x) \vee \left(1 - \sum_{x \in \bar{A}} h^*(x)\right) = LB(A)$$

Importance degrees from experts will be formulated as interval importance degrees using interval density functions in Section 4.

3 Conflict analysis

In this section, we will outline about conflict analysis from Pawlak [3]. In a conflict, at least two parties, called agents, are in dispute over some issues. The relationship of each agent to a specific issue can be clearly represented in the form of a table, as shown in Table 1. This table is taken from [3].

U	a	b	c	d	e
1	-1	1	1	1	1
2	1	0	-1	-1	-1
3	1	-1	-1	-1	0
4	0	-1	-1	0	-1
5	1	-1	-1	-1	-1
6	0	1	-1	0	1

Table 1. Example of infomation system

Table 1 is called an information system in rough sets theory [4]. The table rows of information systems are labelled by objects, the table columns are labelled by attributes and the entries of table are values of attributes, which are uniquely assigned to each object and each attribute. Then, the information system, S, is given as (U, Q, V) where U is the set of objects, Q is the set of attributes and V is the set of attribute values. In conflict analysis, a conflict situation is represented as a form of restricted information system. Then, objects correspond to agents and attributes correspond to issues. So, in Table 1, $U = \{1, \ldots, 6\}$ is a set of agents and $Q = \{a, \ldots, e\}$ is a set of issues. And, values of attributes are represented the attitude of agents to issues: 1 means that the agent is favorable to the issue, -1 means that the agent is against the issue and 0 means neutral.

In order to express the relation between agents, the follwing auxiliary function on U^2 [3] is defined as

$$\phi_q(x,y) = \begin{cases} 1 & if\ q(x)q(y) = 1\ \ or\ \ x = y \\ 0 & if\ q(x)q(y) = 0\ \ and\ \ x \neq y \\ -1 & if\ q(x)q(y) = -1 \end{cases} \qquad (1)$$

where $q(x)$ is the attitude of the agent x to the issue q. This means that, if $\phi_q(x,y) = 1$, the agents x and y have the same opinion about the issue q, if $\phi_q(x,y) = 0$, at least one agent has neutral approach to q and if $\phi_q(x,y) = -1$, they have different opinions about q.

We need the distance between x and y to evaluate the relation between x and y. Therefore we use Pawlak's definition as follows:

$$\rho_Q^*(x,y) = \frac{\sum_{q \in Q} \phi_q^*(x,y)}{|Q|} \qquad (2)$$

where

$$\phi_q^*(x,y) = \frac{1 - \phi_q(x,y)}{2} = \begin{cases} 0 & if\ q(x)q(y) = 1\ \ or\ \ x = y \\ 0.5 & if\ q(x)q(y) = 0\ \ and\ \ x \neq y \\ 1 & if\ q(x)q(y) = -1 \end{cases} \qquad (3)$$

Applying $\rho_Q^*(x,y)$ to the data in Table 1, we obtained Table 2.

U	1	2	3	4	5	6
1						
2	0.9					
3	0.9	0.2				
4	0.8	0.3	0.3			
5	1	0.1	0.1	0.2		
6	0.4	0.5	0.5	0.6	0.6	

Table 2. Distance functions between objects in Table 1

4 Interval importances to conflict analysis

In this section, we will add subjective evaluations for issues to conflict analysis. It is assumed that non-negative relative weights are given for all issues. Using a non-negative weight $w(q)$ for each issue q, a new distance function ρ_Q' is defined as follows:

$$\rho_Q'(x,y) = \frac{\sum\limits_{q \in Q} \phi_q^*(x,y)w(q)}{\sum\limits_{q \in Q} w(q)} \qquad (4)$$

where $\sum_{q \in Q} w(q) \neq 0$. Let $w'(q) = \dfrac{w(q)}{\sum_{q \in Q} w(q)}$. We can rewrite ρ'_Q under the normality condition, $\sum_{q \in Q} w'(q) = 1$, as follows:

$$\rho'_Q(x, y) = \sum_{q \in Q} \phi_q^*(x, y) w'(q) \tag{5}$$

When an expert's knowledge is given as the following weights, then the distance function with weights is calculated in Table 3 using (4).

$$w(a) = 0.2; w(b) = 0.8; w(c) = 0.5; w(d) = 1.0; w(e) = 0.6.$$

U	1	2	3	4	5	6
1						
2	0.87					
3	0.90	0.23				
4	0.81	0.32	0.29			
5	1.00	0.13	0.10	0.19		
6	0.35	0.52	0.55	0.65	0.65	

Table 3. Distance Functions with weights

When we can use many experts' knowledge, they are formulated as interval density functions [1] as shown in Section 2. It is assumed that an expert gives normal weights, that is, the sum of them becomes 1. When there are plural experts, then the following functions (w_*, w^*) becomes an interval density function.

Proposition: When there exist plural normal weights, w_i $(i = 1, \ldots, n)$, over the disjoint space Q, two functions are defined as

$$w_*(q) = \min_{i \in \{1, \cdots, n\}} w_i(q)$$

$$w^*(q) = \max_{i \in \{1, \cdots, n\}} w_i(q)$$

Then, (w_*, w^*) becomes an interval density function.

Proof: It is clear that

$$\sum_{q \in Q} w_*(q) \leq 1 \text{ and } \sum_{q \in Q} w^*(q) \geq 1$$

holds. If there exists some $q' \in Q$ such as $\sum_{q \in Q} w_*(q) + (w^*(q') - w_*(q')) > 1$, then there is no set of normal weights which $w^*(q')$ belongs to. Therefore, for all $q' \in Q$,

$$\sum_{q \in Q} w_*(q) + (w^*(q') - w_*(q')) \leq 1$$

holds. Similarly, for all $q' \in Q$,

$$\sum_{q \in Q} w^*(q) - (w^*(q') - w_*(q')) \geq 1$$

holds. Consequently, (w_*, w^*) becomes an interval density function.   Q.E.D.

Using functions (w_*, w^*), instead of ρ'_Q we can write a distance function $\rho_Q^w = (\rho_{Q*}^w, \rho_Q^{w*})$ as follows;

$$\rho_{Q*}^w(x, y) = \frac{\sum_{q \in Q} \phi_q^*(x, y) w_*(q)}{\sum_{q \in Q} w_*(q)} \tag{6}$$

and

$$\rho_Q^{w*}(x, y) = \frac{\sum_{q \in Q} \phi_q^*(x, y) w^*(q)}{\sum_{q \in Q} w^*(q)} \tag{7}$$

When many experts' knowledge are given as Table 4, then the distance function with weights is calculated in Table 5 using (6) and (7).

Q	importance
a	[0.15, 0.30]
b	[0.20, 0.35]
c	[0.10, 0.25]
d	[0.25, 0.40]
e	[0.10, 0.20]

Table 4. Experts' knowledge with interval density functions

5 Conclusion

A new approach of conflict analysis with interval importance representing experts' knowledge is proposed under the assumption that an expert's knowledge is given as a relative importance for each attribute. Importance degrees from experts are formulated as interval importance degrees using interval density functions and then, conflict degrees between two agents are obtained as an interval value.

The presented approach for conflict analysis depends on experts' knowledges which lead to interval conflicts. In order to judge some relationship between two agents as one of conflict, neutral, and alliance, the judgement measure proposed by Pawlak can be used.

	1	2	3	4	5	6
1						
2	[0.875, 0.886]					
3	[0.933, 0.938]	[0.187, 0.188]				
4	[0.750, 0.753]	[0.353, 0.375]	[0.300, 0.313]			
5	[1.000, 1.000]	[0.120, 0.125]	[0.063, 0.067]	[0.233, 0.250]		
6	[0.375, 0.400]	[0.487, 0.500]	[0.533, 0.563]	[0.600, 0.625]	[0.600, 0.625]	

Table 5. Distance functions by interval density functions

References

1. Y.Maeda and H. Tanaka: Non-additive probability measures with interval density functions. J. Japan Society for Fuzzy Theory and Systems, to appear (in Japanese).
2. Z.Pawlak: On conflicts. Int. J. Man-Machine Studies. **21**(1984)127–134.
3. Z.Pawlak: An inquiry into anatomy of conflicts. Int. J. Infomation Sciences. **109**(1998)65–78.
4. Z.Pawlak: Rough Sets-Theoretical Aspects of Reasoning about Data. Kluwer Academic Publishers. Dordrecht.(1991)

Incorporating Personal Databases by Granular Computing

Yoshitsugu Kakemoto

Information and Mathematical Dept.,
The Japan ResearchInstitute, Limited.
Ichibancyo 16, Chiyoda-ku Tokyo,
102-0082 Japan
E-mail:kakemoto@tyo.sci.jri.co.jp

Abstract. As the end-user computing grows up, the volume of information defined by users is increasing. Therefore, incorporating the information defined by users is a core component of the knowledge management. In this paper, the author proposes a method for incorporating personal databases, which is based on granular computing and the relational database theory.

1 Introduction

As the end-user computing grows up, the volume of information defined by users is increasing. Using the databases defined by users is very convenient for our daily work. At the same time the personal databases are also an important source of knowledge of an organization. It is necessary to incorporate personal databases for using them as primary knowledge sources.

A possible way for the database incorporation is the relation transformation based on the normalization theory of relational databases[1]. However, the normalization theory focuses on the formal aspect of relational databases only. To incorporate the personal databases defined by users, a method that reflects the meaning of a domain is required.

Data mining based on granular computing is essentially a "reverse" engineering of database processing. The latter organizes and stores data according to the given semantics, while the former is "discovering" the semantics from stored data[5]. This assertion suggests that data mining based on granular computing is an efficient way for incorporating personal databases.

In this paper, the author proposes a method for incorporating personal data resources. This method is based on granular computing and the relational database theory. At first, anomalies on personal databases are discussed, and then the proposed incorporating method and its theoretical background are described.

2 Properties of Personal Data

At the first step of our study, we made an inquiry about personal databases. As a result of this inquiry, the following derivations can be found in the databases defined by users.

391

- Derivation on data definition.
  In Japanese language environment, there are many ways to express some words with same meaning. However, the words with same meaning are processed as different words.
- Derivation on database schema definition.
  - Derivation on attribute definition.
    For example, when defining a field about a customer's name, a user may use two attributes: the family name and the first name, but the other user may define only one attribute: the name.
  - Functional dependencies.
    Functional dependencies define a relation between key values and dependent values. There may be various key definitions on same relation. Even though same relation definition is provided for users, each user probably defines different keys on the relation.
  - Relation.
    If a relation on a database satisfies only the first normal form in the normalization theory, sometimes, abnormal results are obtained by an ordinary SQL operation. These abnormal results are caused by structure of relation that is intermixed several different relations. Usually, a translation into the second normal form is considered when a database application developer meets these abnormal results.

In the relational database theory, the normalization theory has become a core one for database normalization[1]. Many personal databases do not clear all levels of relational database normal forms. From the practical view, it is sufficient if every personal database satisfies the third normal form in the relational database theory.

Formal translating methods from a first normal form relation to the second normal form relation have been developed. Most methods translated relations based on their syntactical aspects, but the main task that translates a first normal form relation has semantic issues. When we translate a first form relation, we should pay an attention to its semantic aspects. So we need to develop a semantic based translation method for translating a first normal form relation into a second normal form relation.

3 Schema Discovery by Granular Computing

3.1 Functional Dependencies and Normalization

In this section, we give a brief overview about the functional dependencies and the relational database normalization theory according to the literature[2].

Let sets X and Y be two attribute sets of a relation $R(A_1, A_2, \ldots, A_n)$, where $X \cup Y = \Omega_R$ and $X \cap Y = Z(\neq \emptyset)$ where Ω_R is the set of all attributes on a relation R. The functional dependency $X \to Y$ is defined as the definition 1.

Definition 1. We say the functional dependency from X to Y in R exits, if the condition (1) is satisfied in all instances r of R.

$$(\forall t, t' \in R)(t[X] = t'[X] \Rightarrow t[Y] = t'[Y]) \tag{1}$$

The multi-valued dependency $X \longrightarrow\longrightarrow Y$ is a generalization of the functional dependency defined as Definition 2.

Definition 2. X and Y are an attribute on a relation R. We say X decides Y in multi-value or Y depends on X in multi-value when the condition 2 is satisfied all instances r in R.

$$(\forall t, t' \in R)(t[X] = t'[X] \Rightarrow t[X \cup Y] = t'[Z]) \in R \tag{2}$$
$$\wedge (t'[X \cup Y], t[Z] \in R)$$

where $Z = \Omega_R - (X - Y)$.

The definition says that the new tuples $(t[X \cup Y], t[Z])$ and $(t'[X \cup Y], t[Z])$ that are made from tuples t and t' are also tuples in the R where t and t' satisfy the condition $t[X] = t'[X]$.

We say two projections $R[X]$ and $R[Y]$ on the relation R is information lossless decomposition if $R = R[X] \times R[Y]$. The necessary and sufficient condition for information lossless decomposition is guaranteed by the following propositions and theorem.

Proposition 3. *If X and Y are information lossless decomposition of R, $R \subseteq R[X] \bowtie R[Y]$.*

Proposition 4. *The necessary and sufficient condition for $R \subseteq R[X] \bowtie R[Y]$ is the multi-valued dependency $X \cap Y \longrightarrow\longrightarrow X|Y$ that is satisfied on R.*

By the proposition 3 and 4, the following theorem can be obtained.

Theorem 5. *The necessary and sufficient condition for $R \subseteq R[X] \bowtie R[Y]$ is the multi-valued dependency $X \cap Y \longrightarrow\longrightarrow X|Y$ is satisfied on R.*

3.2 Granular Computing

We introduce the notations and theorems on granular computing followed by Lin[5]. An equivalence relation divides the universe into disjoint elementary sets. A binary relation decomposes a universe into elementary neighborhoods that are not necessarily disjoint. The decomposition is called a binary granulation and the collection of the granules a binary neighborhood system[3, 4].

Let $(U; B^i; C^i, i = 0, 1, 2, \ldots)$ be a collection of granular structure where U is a set of entities or an NS-space imposed by B, B^i is elementary neighborhoods and C^i is elementary concepts $(= NAME(B^i_j))$ and each C^i is an NS-space. The relationships among attribute values of a table are defined by the elementary sets of a multiple partition. Inclusion of an elementary set in another elementary set is an inference on the corresponding elementary concepts. A functional dependency of two columns is a refinement of two corresponding partitions.

On the granular structure, the following rules are defined[5].

1. Continuous inference rules:
   A formula $C_j \rightarrow D_h$ is a continuous inference rule, if $NEIGH(P_j) \subseteq NEIGH(Q_h)$. Here, $NEIGH(P_j)$ means a neighborhoods of P_j.
2. Softly robust continuous inference rules: A formula $C_j \rightarrow D_h$ is a softly continuous inference rule, if $NEIGH(P_j) \subseteq NEIGH(Q_h)$ and $|NEIGH(P_j) \cap NEIGH(Q_h)| \geq threshold$.
3. High level continuous inference rules: Suppose $PN = B^i, QN = B^j, QN = B^j, QN = B^j$, and $j \neq i$ are two nested granular structures, that is, $PN^i \prec PN^{(i+k)}$ and $QN^i \prec QN^{(i+k)}$. Write $P = PN^m$ and $Q = QN^n$, where $m \geq m$ and $k > 0$.. A formula $C_j \rightarrow D_h$ is a high level continuous inference rule, if $NEIGH(P_j) \subseteq NEIGH(Q_h)$ and $|NEIGH(P_j) \cap NEIGH(Q_h)| \geq threshold$.

The above rules can be regarded as the generalization for the definition about functional dependencies in relational database. Furthermore, we can extend these concepts to the method for discovering functional dependencies.

4 Incorporating Personal Databases

The main procedure of incorporating personal databases is described as follows.

1. Data normalization.
   The data normalization procedure is consist of following two sub-procedures.
   (a) Continuous data quantization.
   (b) Word correction.
   The first sub-procedure is the continuous data quantization if a data set is continuous values. The second sub-procedure is the data correction. For the word correction, we use a simple structured thesaurus. In the thesaurus, the following Japanese specific word correct relation are stored.
   - Conversion rules between han-kaku kana and zen-kaku kana.
   - Special characters with same meaning.
   - Rules about okuri-gana.
   According to these rules, different characters with same meaning is corrected automatically.
2. Obtaining elementary concepts and elementary neighborhoods for each attribute on relations.
3. Detection of originally identical attributes.
   If $|C^i \cap C^j| \geq threshold$, the attributes i and j seem to be identical attributes where C^i is the elementary concepts of the attribute i, and C^j is the elementary concepts of the attribute j.

4. Detection of functional dependencies.

Functional dependencies in a relation are found according to the following procedure.

(a) For each attribute in a relation, level of the derivable rule on the granular structure is determined.

(b) According to the determined level, the following schema transformations are possible.

- If the continuous inference rule is established between C^i and C^j, a functional dependency between attributes i and j is established.

- If the continuous inference rule is established from C^i to C^j and C^k at the same time, the attributes j and k seem to be the attributes in which a functional dependency between attributes i and $j \cap k$ is established.

- If the softly robust continuous inference rules is established between C^i and C^j, a multi-valued dependency between attributes i and j is established. Moreover, the attribute j can be another relation. If the attribute j is decomposed into another relation, this decomposition is the information lossless decomposition. It is evident from Definition 2 and the properties of the rules on the granular structure.

5 Conclusion

In this paper, a method for incorporating personal databases was proposed. We described that incorporating information defined by users is a core component of the knowledge management. Some kinds of deviations of data and schema were argued. Another type of data deviation that was not argued in this paper is how to handle null values. How to handle null values is depend on different users. We are also developing a more sophisticated method for handling this issue.

References

1. E.F. Codd. *Further Normalization of the Database Relational Model*, Prentice-Hall (1972) 33-64.

2. Y. Masunaga. *The Foundation of Relational Database*, OHM Publisher (1990).

3. Lin, T.Y. "Granular Computing on Binary Relations 1: Data Mining and Neighborhood Systems", L. Polkowski and A. Skowron (eds.) *Rough Sets in Knowledge Discovery 1*, Physica-Verlag (1998) 107-121.

4. Lin, T.Y. "Granular Computing on Binary Relations 2: Rough Set Representation and Belif Function", L. Polkowski and A. Skowron (eds.) *Rough Sets in Knowledge Discovery 1*, Physica-Verlag (1998) 122-140.

5. Lin, T.Y. "Data Mining: Granular Computing Approach", Zhong, N. and Zhou, L. (eds.) *Methodologies for Knowledge Discovery and Data Mining*, LNAI 1574, Springer-Verlag (1999) 24-33.

Knowledge-Driven Discovery of Operational Definitions

Jan M. Żytkow

Computer Science Department, UNC Charlotte, Charlotte, N.C. 28223
and Institute of Computer Science, Polish Academy of Sciences
zytkow@uncc.edu

Abstract. Knowledge representation which is internal to a computer lacks empirical meaning so that it is insufficient for the investigation of the external world. All intelligent systems, including robot-discoverers must interact with the physical world in complex, yet purposeful and accurate ways. We argue that operational definitions are necessary to provide empirical meaning of concepts, but they have been largely ignored by the research on automation of discovery and in AI. Individual operational definitions can be viewed as algorithms that operate in the real world. We explain why many operational definitions are needed for each concept and how different operational definitions of the same concept can be empirically and theoretically equivalent. We argue that all operational definitions of the same concept must form a coherent set and we define the meaning of coherence. No set of operational definitions is complete so that expanding the operational definitions is one of the key tasks in science. Among many possible expansions, only a very special few lead to a satisfactory growth of scientific knowledge. While our examples come from natural sciences, where the use of operational definitions is especially clear, operational definitions are needed for all empirical concepts. We briefly argue their role in database applications.

1 Operational definitions provide empirical meaning

Data about external world are obtained by observation and experiment. Sophisticated procedures and instruments are commonly used to reach data of scientific value. Yet we rarely think systematically about methods by which data have been procured, until problems occur. When a set of data is inconsistent with our expectations, we start asking: "How was this particular measurement obtained?", "What method has been used?", "How is this method justified?". Often it turns out that a method must be changed. Because data can be wrong in so many ways, sophisticated knowledge is required in order to examine and improve measurement methods.

It is critical to the growth of scientific knowledge to study new situations, for which no known method can measure a particular quantity. For instance, we may wish to measure temperatures lower than the capabilities of all existing instruments. Or we want to measure temperature change inside a living cell, as the cell undergoes a specific process.

When no known method applies, new methods must be discovered. New measurement methods must expand the existing concepts. For instance, a new thermometer must produce measurements on a publicly shared scale of temperature.

Discovery of new measurement methods, which we also call operational definitions, is the central problem in this paper. We provide an algorithm that demonstrates how empirical knowledge is used to construct new operational definitions, how new methods can be empirically verified and how choices can be made among competing methods.

We end each section with a few basic claims about operational definitions.

Claim 1: For each empirical concept, measurements must be obtained by repeatable methods that can be explained in detail and used in different laboratories.

Claim 2: The actual verification in empirical science is limited to empirical facts. Operational definitions determine facts; thus they determine the scope of scientific verification.

Claim 3: In contrast, scientific theories often make claims beyond the facts that can be empirically verified at a given time. Theoretical claims often apply to all physical situations, whether we can observe them or not.

In this paper we use examples of numerical properties of objects and their pairs. The numbers that result from measurements, for instance temperature or distance, we call *values* of empirical concepts.

Claim 4: Operational definitions can be classified in several dimensions: (a) they apply to objects, states, events, locations and other empirical entities; (b) they may define predicates of different arity, for instance, properties of individual objects, object pairs (distance) or triples (chemical affinity); (c) some operational definitions provide data while others prepare states that possess specific properties, such as the triple point of water.

2 The AI research has neglected operational definitions

Operational semantics links the terms used in scientific theories with direct observations and manipulations (Bridgman, 1927; Carnap, 1936). While important in empirical science, the mechanisms that produce high quality experiments have been neglected not only in the existing discovery systems but in the entire domain of artificial intelligence.

The distinction between formalism and its interpretation, also called semantics, has been applied to the study of science since 1920's and 1930's. Scientific theories have been analyzed as formal systems whose language is empirically interpreted by operational definitions.

A similar distinction applies to discovery systems and to knowledge they create. A discovery mechanism such as BACON (Langley, Simon, Bradshaw & Zytkow, 1987) can be treated as (1) a formal system that builds equations from data that are formally tuples in the space of the values of independent and dependent variables plus (2) a mechanism that procures data.

Similarly to scientists, BACON and other discovery systems use plans to propose experiments. Each experiment consists in selecting a list of values $x_1, ..., x_k$ of empirical variables $X_1, ..., X_k$, and in obtaining the value y of a dependent variable Y which provides the "world response" to the empirical situation characterized by $x_1, ..., x_k$. But instead of real experiments, the values of dependent variables are either typed by the user or computed in simulation, in response to the list of values of independent variables.

This treatment bypasses real experimentation and measurements. Other papers and collections that consider many components of the scientific methods (Kulkarni & Simon, 1987; Sleeman, Stacey, Edwards & Gray, 1989; Shrager & Langley, 1990; Valdes-Perez, 1995) neglect operational definitions of concepts.

In the wake of robotic discovery systems, operational semantics must, at the minimum, provide realistic methods to acquire data. Żytkow, Zhu & Hussam (1990) used a robotic mechanisms which conducted automatically experiments under the control of FAHRENHEIT. In another robotic experiment, Żytkow, Zhu & Zembowicz (1992) used a discovery process to refine an operational definition of mass transfer. Huang & Zytkow (1997) developed a robotic system that repeats Galileo's experiment with objects rolling down an inclined plane. One operational definition controlled the robot arm so that it deposited a cylinder on the top of an inclined plane, while another measured the time in which the cylinder rolled to the bottom of the plane.

While operational semantics must accompany any formalism that applies to the real world, it has been unnoticed in AI. Jackson's claim (1990) is typical: "a well-defined semantics ...reveals the meaning of ...expressions by virtue of their form." But this simply passes on the same problem to a broader formalism, that includes all the terms used in formal semantics. Those terms also require real-world interpretation that must be provided by operational definitions.

Plenty of further research must be conducted to capture the mechanisms in which operational definitions are used in science and to make them applicable on intelligent robots.

Claim 5: Formal semantics are insufficient to provide empirical meaning.

Claim 6: Robotic discoverers must be equipped in operational definitions.

3 Operational definitions interact with the real world

Early analyses of operational definitions used the language of logic. For instance, a dispositional property "soluble in water" has been defined as

If x is in water then (x is soluble in water if and only if x dissolves)

But a more adequate account is algorithmic rather than descriptive:

```
Soluble (x)
    Put x in water!
    Does x dissolve?
```

As an algorithm, operational definition consists of instructions that prescribe manipulations, measurements and computations on the results of measurements. Iteration can enforce the requirements such as temperature stability, which can be preconditions for measurements. Iteration can be also used in making measurements. The loop exit condition such as the equilibrium of the balance, or a coincidence of a mark on a measuring rod with a given object, triggers the completion of a step in the measurement process.

Procedures that interpret independent and dependent variables can be contrasted as manipulation and measurement mechanisms. Each independent variable requires a manipulation mechanism which sets it to a specific value, while a response value of an dependent variable is obtained by a measurement mechanism. In this paper we focus on measurement procedures.

It happens that an instruction within procedure P does not work in a specific situation. In those cases P cannot be used. Each procedure may fail for many reasons. Some of these reasons may be systematic. For instance, a given thermometer cannot measure temperatures below -40C because the thermometric liquid freezes or above certain temperature, when it boils. Let us name the range of physical situations to which P applies by R_P.

Often, a property is measured indirectly. Consider distance measurement by sonar or laser. The time interval is measured between the emitted and the returned signal. Then the distance is calculated as a product of time and velocity. Let $C(x)$ be the quantity measured by procedure P. When P terminates, the returned value of C is $f(m_1, ..., m_k)$, where $m_1, ..., m_k$ are the values of different quantities of x or the empirical situation around x, measured or generated by instructions within P, and f is a computable function on those values.

Claim 7: Each operational definition should be treated as an algorithm.

Claim 8: The range of each procedure P is limited in many ways, thus each is merely a partial definition applicable in the range R_P.

Claim 9: An operational definition of concept C can measure different quantities and use empirical laws to determine the value of C: $C(x) = f(m_1, ..., m_k)$

Claim 10: An operational definition of a concept $C(x)$ can be represented by a descriptive statement: "If x is in R_P then $C(x) = f(m_1, ..., m_k)$"

4 Each concept requires many operational definitions

In everyday situations distance can be measured by a yard-stick or a tape. But a triangulation method may be needed for objects divided by a river. It can be extended to distance measurement from the Earth to the Sun and the Moon. Then, after we have measured the diameter of the Earth orbit around the Sun, we can use triangulation to measure distances to many stars.

But there are stars for which the difference between the "winter angle" and the "summer angle" measured on the Earth, is non-measurably small, so another method of distance measurement is needed. Cefeids are some of the stars within the range of triangulation. They pulsate and their maximum brightness varies according to the logarithm of periodicity. Another law, determined on Earth and applied to stars claims that the perceived brightness of a constant light source diminishes with distance as $1/d^2$. This law jointly with the law for cefeids allows us to determine the distance to galaxies in which individual cefeids are visible.

For such galaxies the Hubble Law was empirically discovered. It claims proportionality between the distance and red shift in the lines of hydrogen spectrum. The Hubble Law is used to determine the distance of the galaxies so distant that cefeids cannot be distinguished.

Similarly, while a gas thermometer applies to a large range of states, in very low temperatures any gas freezes or gas pressure becomes non-measurably small. A thermometer applied in those situations measures magnetic susceptibility of paramagnetic salts and uses Curie-Weiss Law to compute temperature. There are high temperatures in which no vessel can hold a gas, or states in which the inertia of gas thermometer has unacceptable influence on the measured temperature. Measurements of thermal radiation and other methods can be used in such cases.

Claim 11: Empirical meaning of a concept is defined by a set of operational definitions.

Claim 12: Each concrete set is limited and new methods must be constructed for objects beyond those limits.

5 Methods should be linked by equivalence

Consider two operational definitions P_1 and P_2 that measure the same quantity C. When applied to the same objects their results should be empirically equivalent within the accuracy of measurement. If P_1 and P_2 provide different results, one or both must be adjusted until the empirical equivalence is regained.

From the antiquity it has been known that triangulation provides the same results, within the limits of measurement error, as a direct use of measuring rod or tape. But in addition to the empirical study of equivalence, procedures can be compared with the use of empirical theories and equality of their results may be proven.

Triangulation uses a basic theorem of Euclidean geometry that justifies theoretically the consistency of two methods: by the use of yard-stick and by triangulation. To the extent in which Euclidean geometry is valid in the physical world, whenever we make two measurements of the same distance, one using a tape while the other using triangulation, the results are consistent.

Claim 13: Methods can differ by their accuracy and by degree to which they influence the measured quantity.

Claim 14: When two operational definitions define the same property and apply to the same objects, their results should be empirically equivalent. If they are not, additional data are collected and methods are adjusted in order to restore their equivalence.

Claim 15: When two operational definitions define the same concept $C(x)$, it is possible to prove their equivalence. The prove consists in deducing from a verified

empirical theory that the statements that represent them are equivalent, that is,
$f_1(m_1, ..., m_k) = f_2(n_1, ..., n_l)$

Claim 16: When the statements that represent two procedures use empirical
laws $C(x) = f_1(m_1, ..., m_k)$, $C(x) = f_2(n_1, ..., n_l)$, theoretical equivalence of both
procedures follows from those laws.

Claim 17: The more general and better verified are the theories that justify the
equivalence of two procedures P_1 and P_2, the stronger are our reasons to believe
in the equivalence of P_1 and P_2.

Claim 18: Proving the equivalence of two procedures is desired, because the
empirical verification of equivalence is limited.

6 Operational definitions of a concept form a coherent set

We have considered several procedures that measure distance. But distance can
be measured in many other ways. Even the same method, when applied in different laboratories, varies in details. How can we determine that different measurements define the same physical concept? Procedures can be coordinated by
the requirements of empirical and theoretical equivalence in the areas of common
application. However, we must also require that each method overlaps with some
other methods and further, that each two methods are connected by a chain of
overlapping methods.

Definition: A set $\Phi = \{\phi_1, ..., \phi_n\}$ of operational definitions is coherent **iff** for
each i, j = 1,...,n

(1) ϕ_i is empirically equivalent with ϕ_j. Notice that this condition is trivially
satisfied when the ranges of both operational definitions do not overlap;

(2) there is a sequence of definitions $\phi\text{-}i_1, ..., \phi\text{-}i_k$, such that $\phi\text{-}i_1 = \phi_i$, $\phi\text{-}i_k = \phi_j$, and for each $m = 2, ..., k$ the ranges of $\phi\text{-}i_m$ and $\phi\text{-}i_{m+1}$ intersect.

The measurements of distance in our examples form such a coherent set. Rod
measurements overlap with measurements by triangulation. Different versions
of triangulation overlap with one another. The triangulation applied to stars
overlaps with the method that uses cefeids, which in turn overlaps with the
method that uses Hubble Law.

Similarly, the measurements with gas thermometer have been used to calibrate the alcohol and mercury thermometers in their areas of joint application.
For high temperatures, measurements based on the Planck Law of black body
radiation overlap with the measurements based on gas thermometers. For very
low temperatures, the measurements based on magnetic susceptibility of paramagnetic salts overlap with measurements with the use of gas thermometer.

Claim 19: Each empirical concept should be defined by a coherent set of operational definitions. When the coherence is missing, the discovery of a missing
link becomes a challenge.

For instance, the experiment of Millikan provided a link between the charge
of electron and electric charges measured by macroscopic methods.

Claim 20: By examining theoretical equivalence in a coherent set Φ of operational definitions we can demonstrate that the values measured by all procedure in Φ are on the same scale.

Claim 21: Operational definitions provide means to expand to new areas the range of the laws they use.

7 Laws can be used to form new operational definitions

Operational definitions can expand each concept in several obvious directions, towards smaller values, larger values, and values that are more precise. But the directions are far more numerous. Within the range of "room" temperatures, consider the temperature inside a cell, temperature of a state that is fast varying and must be measured every second, or temperature on the surface of Mars. Each of these cases requires different methods. A scientist may examine the shift of tectonic plates by comparing the distances on the order of tens of kilometers over the time period of a year, when the accuracy is below a millimeter.

Whenever we consider expansion of operational definitions for an empirical concept C to a new range R, the situation is similar:

(1) we can observe objects in R for which C cannot be measured with the needed accuracy;

(2) some other attributes $A_1, ..., A_n$ of objects in R can be measured, or else those objects would not be empirically available;

(3) some of $A_1, ..., A_n$ are linked to C by empirical laws or theories. We can use one or more of those laws in a new method: measure some of $A_1, ..., A_n$ and then use laws to compute the value of C.

Consider the task: determine distance D from Earth to each in a set R of galaxies, given some of the measured properties of R: $A_1, A_2, ..., A_n$. Operational definitions for $A_1, ..., A_n$ are available in the range R. For instance, let A_2 measure the redshift of hydrogen spectrum. Let $D = h(A_2)$ be Hubble Law. The new method is:

```
For a galaxy g, when no individual cefeids can be distinguished:
Measure A2 of the light coming from g by a method of spectral analysis
Compute the distance D(Earth, g) as h(A2(g))
```

The same schema can yield other operational definitions that determine distance by properties measurable in a new range, such as yearly parallax, perceived brightness or electromagnetic spectrum.

Some laws cannot be used even though they apply to galaxies. Consider $D = a/\sqrt{B}$ (B is brightness). It applies even to the most remote sources of light. But B used in the law is the absolute brightness at the source, not the brightness perceived by an observer. Only when we could determine the absolute brightness, we could determine the distance to galaxies by $D = a/\sqrt{B}$.

The following algorithm can be used in many applications:

```
Algorithm:
Input: set of objects observed in range R
       attribute C that cannot be measured in R
       set of attributes A1,...,Ak that can be measured in R
       set {F1,...,Fp} of known operational definitions for C
       set LAWS of known empirical laws
Output: a method by which the values of C can be determined in R

Find in LAWS a law L in which C occurs
    Let B1,...,Bm be the remaining attributes that occur in L
    Verify that C can be computed from L, and the values of B1,...,Bm
    Verify that {B1,...,Bm} is subset of {A1,...,Ak},
        that is, B1,...,Bm can be measured in at least some situations in R
    Use L and B1,...,Bm to create new procedure F for C
    Make F consistent with procedures in {F1,...,Fp}
```

After the first such procedure has been found, the search may continue for each law that involves C.

In set-theoretic terms, each expansion of concept C to a new range R can be viewed as a mapping from the set of distinguishable classes of equivalence with respect to C for objects in R to a set of possible new values of C, for instance, the values larger than those that have been observed with the use of the previous methods. But possible expansions are unlimited. The use of an existing law narrows down the scope of possible concept expansions to the number of laws for which the above algorithm succeeds. But the use of an existing law does not merely reduce the choices, it also justifies them. Which of the many values that can be assigned to a given state corresponds to its temperature? If laws reveal the real properties of physical objects, then the new values which fit a law indicate concept expansion which has a potential for the right choice.

Claim 22: Whenever the empirical methods expands to new territories, new discoveries follow. New procedures are instrumental to that growth.

Claim 23: Each new procedure expands the law it uses to a new range. If procedures P_1 and P_2 use laws L_1 and L_2 respectively, and produce empirically inconsistent results for new objects in range R, the choice of P_1 will make L_2 false in R.

If a number of procedures provide alternative concept expansions, various selection criteria can be used, depending on the goal of research.

Claim 24: Among two methods, prefer the one which has a broader range, for it justifies concept expansion by a broader expansion of an existing law.

Claim 25: Among two methods, prefer the one which has a higher accuracy, since it provides more accurate data for the expansion of empirical theories.

Claim 26: Methods must and can be verified in their new area of application or else, the empirical laws they apply would be mere definitions.

8 Operational definitions apply to all empirical concepts

While explicit operational definitions are rarely formed by experimental scientists, they become necessary in autonomous robots. A robot explorer can also benefit from mechanisms for generation of new procedures.

Operational meaning applies to databases. They are repositories of facts that should be shared as a major resource for knowledge discovery and verification. But data and knowledge can be only useful for those who understand their meaning. Operational definitions describe how the values of all fields were produced.

Similarly to our science examples, operational definitions can be generated from data and applied in different databases. Consider a regularity L, discovered in a data table D, which provides accurate predictions of attribute C from known values of $A_1, ..., A_n$. L can be used as a method that determines values of C.

Consider now another table D_1, that covers situations similar to D, but differs in some attributes. Instead of test C, tests $B_1, ..., B_m$ are provided, which may or may not be compatible with C. Suppose that a doctor who has been familiar with test C at his previous workplace, issues a query against D_1 that includes attribute C which is not in D_1. A regular query answering mechanism would fail, but a mechanism that can expand operational meaning of concepts may handle such a query (Ras, 1997). A quest Q for operational definition of concept C with the use of $B_1, ..., B_m$ will be send to other databases. If an operational definition is found, it is used to compute the values of C in the doctor's query.

References

Bridgman, P.W. 1927. *The Logic of Modern Physics*. The Macmillan Company.

Carnap, R. 1936. Testability and Meaning, *Philosophy of Science, 3*.

Huang, K. & Zytkow, J. 1997. Discovering Empirical Equations from Robot-Collected Data, Ras Z. & Skowron A eds. *Foundations of Intelligent Systems*, Springer, 287-97.

Jackson, P. 1990. *Introduction to Expert Systems*, Addison-Wesley.

Kulkarni, D., & Simon, H.A. 1987. The Process of Scientific Discovery: The Strategy of Experimentation, *Cognitive Science, 12*, 139-175.

Langley, P.W., Simon, H.A., Bradshaw, G., & Zytkow J.M. 1987. *Scientific Discovery; An Account of the Creative Processes*. Boston, MA: MIT Press.

Ras, Z. 1997. Resolving queries through cooperation in multi-agent systems, in eds. T.Y. Lin & N. Cercone, *Rough Sets and Data Mining*, Kluwer Acad. Publ. 239-258.

Shrager, J. & Langley, P. eds. 1990. *Computational Models of Scientific Discovery and Theory Formation*, Morgan Kaufmann Publ.

Sleeman, D.H., Stacey, M.K., Edwards, P., & Gray, N.A.B., 1989. An Architecture for Theory-Driven Scientific Discovery, *Proceedings of EWSL-89*.

Valdes-Perez, R. 1995. Generic Tasks of scientific discovery, *Working notes of the Spring Symposium on Systematic Methods of Scientific Discovery*, AAAI Technical Reports.

Żytkow, J.M., Zhu, J. & Hussam A. 1990. Automated Discovery in a Chemistry Laboratory, in: *Proc. Eight National Conf. on Artificial Intelligence*, AAAI Press, 889-894.

Żytkow, J., Zhu, J. & Zembowicz, R. 1992. Operational Definition Refinement: a Discovery Process, *Proc. 10th Nat'l Conf. on Artificial Intelligence*, AAAI Press, 76-81.

A Closest Fit Approach to Missing Attribute Values in Preterm Birth Data

Jerzy W. Grzymala-Busse[1], Witold J. Grzymala-Busse[2], and Linda K. Goodwin[3]

[1] Department of Electrical Engineering and Computer Science,
University of Kansas, Lawrence, KS 66045, USA
[2] RS Systems, Inc., Lawrence, KS 66047, USA
[3] Department of Information Services and the School of Nursing,
Duke University, Durham, NC 27710, USA

ABSTRACT: In real-life data, in general, many attribute values are missing. Therefore, rule induction requires preprocessing, where missing attribute values are replaced by appropriate values. The rule induction method used in our research is based on rough set theory.

In this paper we present our results on a new approach to missing attribute values called a closest fit. The main idea of the closest fit is based on searching through the set of all cases, considered as vectors of attribute values, for a case that is the most similar to the given case with missing attribute values. There are two possible ways to look for the closest case: we may restrict our attention to the given concept or to the set of all cases. These methods are compared with a special case of the closest fit principle: replacing missing attribute values by the most common value from the concept. All algorithms were implemented in system OOMIS. Our experiments were performed on preterm birth data sets collected at the Duke University Medical Center.

KEYWORDS: Missing attribute values, closest fit, data mining, rule induction, classification of unseen cases, system OOMIS, rough set theory.

1 Introduction

Recently data mining, i.e., discovering knowledge from raw data, is receiving a lot of attention. Such data are, as a rule, imperfect. In this paper our main focus is on missing attribute values, a special kind of imperfection. Another form of imperfection is inconsistency—the data set may contain conflicting cases (examples), having the same values of all attributes yet belonging to different concepts (classes).

Knowledge considered in this paper is expressed in the form of rules, also called production rules. Rules are induced from given input data sets by algorithms based on rough set theory. For each concept lower and upper approximations are computed, as defined in rough set theory [4, 6, 12, 13].

Often in real-life data some attribute values are *missing* (or *unknown*). There are many approaches to handle missing attribute values [3, 5, 7]. In this paper we will discuss an approach based on the closest fit idea. The closest fit algorithm for missing attribute values is based on replacing a missing attribute value by existing values of the same attribute in another case that resembles as much as possible the case with the missing attribute values. In searching for the closest fit case, we need to compare two vectors of attribute values of the given case with missing attribute values and of a searched case.

There are many possible variations of the idea of the closest fit. First, for a given case with a missing attribute value, we may look for the closest fitting cases within the same concept, as defined by the case with missing attribute value, or in all concepts, i.e., among all cases. The former algorithm is called *concept closest fit*, the latter is called *global closest fit*.

Secondly, we may look at the closest fitting case that has all the same values, including missing attribute values, as the case with a missing attribute value, or we may restrict the search to cases with no missing attribute values. In other words, the search is performed on cases with missing attribute values or among cases without missing attribute values.

During the search, the entire training set is scanned, for each case a proximity measure is computed, the case for which the proximity measure is the largest is the closest fitting case that is used to determine the missing attribute values. The proximity measure between two cases e and e' is the Manhattan distance between e and e', i.e.,

$$\sum_{i=1}^{n} \text{distance } (e_i, e_i'),$$

where

$$\text{distance } (e_i, e_i') = \begin{cases} 0 & \text{if } e_i \text{ and } e_i' \text{ are symbolic and } e_i \neq e_i', \\ 1 & \text{if } e_i = e_i', \\ 1 - \dfrac{|e_i - e_i'|}{|a_i - b_i|} & \text{if } e_i \text{ and } e_i' \text{ are numbers and } e_i \neq e_i', \end{cases}$$

where a_i is the maximum of values of A_i, b_i is the minimum of values of A_i, and A_i is an attribute.

In a special case of the closest fit algorithm, called the most common value algorithm, instead of comparing entire vectors of attribute values, the search is reduced to just one attribute, the attribute for which the case has a missing value. The missing value is replaced by the most frequent value within the same concept to which belongs the case with a missing attribute value.

2 Rule Induction and Classification of Unseen Cases

In our experiments we used LERS (Learning from Examples based on Rough Set theory) for rule induction. LERS has four options for rule induction; only one, called LEM2 [4, 6] was used for our experiments. Rules induced from the lower approximation of the class *certainly* describe the class, so they are called *certain*. On the other hand, rules induced from the upper approximation of the class describe only *possibly* (or *plausibly*) cases, so they are called *possible* [8]. Examples of other data mining systems based on rough sets are presented in [14, 16].

For classification of unseen cases system LERS uses a modified "bucket brigade algorithm" [2, 10]. The decision to which class a case belongs is made on the basis of two parameters: strength and support. They are defined as follows: *Strength* is the total number of cases correctly classified by the rule during training. The second parameter, *support*, is defined as the sum of scores of all matching rules from the class. The class C for which the support, i.e., the value of the following expression

$$\sum_{\text{matching rules } R \text{ describing } C} \text{Strength}(R)$$

is the largest is a winner and the case is classified as being a member of C. The above scheme reminds non-democratic voting in which voters vote with their strengths.

If a case is not completely matched by any rule, some classification systems use *partial matching*. During partial matching, system AQ15 uses the probabilistic sum of all measures of fit for rules [11]. Another approach to partial matching is presented in [14]. Holland *et al.* [10] do not consider partial matching as a viable alternative of complete matching and thus rely on a default hierarchy instead. In LERS partial matching does not rely on the input of the user. If complete matching is impossible, all partially matching rules are identified. These are rules with at least one attribute-value pair matching the corresponding attribute-value pair of a case.

For any partially matching rule R, the additional factor, called *Matching_factor* (R), is computed. Matching_factor is defined as the ratio of the number of matched attribute-value pairs of a rule with a case to the total number of attribute-value pairs of the rule. In partial matching, the class C for which the value of the following expression

$$\sum_{\text{partially matching rules } R \text{ describing } C} \text{Matching\_factor}(R) * \text{Strength }(R)$$

is the largest is the winner and the case is classified as being a member of C.

During classification of unseen (testing) cases with missing attribute values, missing attribute values do not participate in any attempt to match a rule during complete or partial matching. A case can match rules using only actual attribute values.

3 Description of Data Sets and Experiments

Data sets used for our experiments come from the Duke University Medical Center. First, a large data set, with 1,229 attributes and 19,970 cases was partitioned into two parts: training (with 14,977 cases) and testing (with 4,993 cases). We selected two mutually disjoint subsets of the set of all 1,229 attributes, the first set containing 52 attributes and the second with 54 attributes and called the new data sets Duke-1 and Duke-2, respectively. The Duke-1 data set contains laboratory test results. The Duke-2 test represents the most essential remaining attributes that, according to experts, should be used in diagnosis of preterm birth. Both data sets were unbalanced because only 3,103 cases were preterm, all remaining 11,874 cases were fullterm.

Table 1. Missing attribute values

Number of missing attribute values in data sets processed by

	Global closest fit	Concept closest fit	Most common value
Duke-1	1,1641	505,329	0
Duke-2	615	1,449	0

Similarly, in the testing data set, there were only 1,023 preterm cases while the number of fullterm cases was 3,970.

Both data sets, Duke-1 and Duke-2, have many missing attribute values (Duke-1 has 505,329 missing attribute values, i.e., 64.9% of the total number of attribute values; Duke-2 has 291,796 missing attribute values, i.e., 36.1% of the total number of attribute values).

First, missing attribute values were replaced by actual values. Both data sets were processed by the previously described five algorithms of the OOMIS system: global closest fit and concept closest fit, among all cases with and without missing attribute values, and most common value.

Since the number of missing attribute values in Duke-1 or Duke-2 is so large, we were successful in using only three algorithms. The version of looking for the closest fit *among all cases without missing attribute values* returned the unchanged, original data sets. Therefore, in the sequel we will use names *global closest fit* and *concept closest fit* for algorithms that search among all cases with missing attribute values. For Duke-1 the concept closest fit algorithm was too restrictive: All missing attribute values were unchanged, so we ignored the Duke-1 data set processed by the concept closest fit algorithm. Moreover, global closest fit or concept closest fit algorithms returned data sets with only reduced number of missing attribute values. The results are presented in Table 1.

Since using both closest fit options result in some remaining missing attribute values, for the output files the option most common value was used to replace all remaining missing attribute values by the actual attribute values. Thus, finally we

Table 2. Training data sets

		Global closest fit	Concept closest fit	Most common value
Duke-1	Number of conflicting cases	8,691	–	10,028
	Number of unique cases	6,314	–	4,994
Duke-2	Number of conflicting cases	7,839	0	8,687
	Number of unique cases	7,511	9,489	6,295

obtained five pre-processed data sets without any missing attribute values.

To reduce error rate during classification we used a very special discretization. First, in the training data set, for any numerical attribute, values were sorted. Every value v was replaced by the interval $[v, w)$, where w was the next bigger values than v in the sorted list. Our approach to discretization is the most cautious since, in the training data set, we put only one attribute value in each interval. For testing data sets, values were replaced by the corresponding intervals taken from the training data set. It could happen that a few values come into the same interval.

Surprisingly, four out of five training data sets, after replacing missing attribute values by actual attribute values and by applying our cautious discretization, were inconsistent. The training data sets are described by Table 2.

For inconsistent training data sets only possible rule sets were used for classification. Certain rules, as follows from [8], usually provide a greater error rate. Rule induction was a time-consuming process. On a DEC Alpha 21164 computer, with 512 MB of RAM, 533 MHz clock speed, rule sets were induced in elapsed real time between 21 (for Duke-2 processed by the concept closest fit algorithm) and 133 hours (for Duke-2 processed by the global concept fit algorithm). Some statistics about rule sets are presented in Table 3.

As follows from Table 3, as a result of unbalanced data sets, the average rule strength for rules describing fullterm birth is much greater than the corresponding rule strength for preterm birth. Consequently, the error rate on the original rule sets is not a good indicator of the quality of a rule set, as follows from [9].

Our basic concept is the class of preterm cases. Hence the set of all correctly predicted preterm cases are called true-positives, incorrectly predicted preterm cases (i.e., predicted as fullterm) are called false-negatives, correctly predicted fullterm cases are called true-negatives, and incorrectly predicted fullterm cases are called false-positives.

Sensitivity is the conditional probability of true-positives given actual preterm birth, i.e., the ratio of the number of true-positives to the sum of the number of true-

Table 3. Rule sets

			Global closest fit	Concept closest fit	Most common value
Duke-1	Number of rules	Preterm	734	–	618
		Fullterm	710	–	775
	Average strength of rule set	Preterm	4.87	–	8.97
		Fullterm	39.08	–	44.73
Duke-2	Number of rules	Preterm	1,202	483	1,022
		Fullterm	1,250	583	1,642
	Average strength of rule set	Preterm	2.71	9.69	4.60
		Fullterm	15.8	43.99	11.37

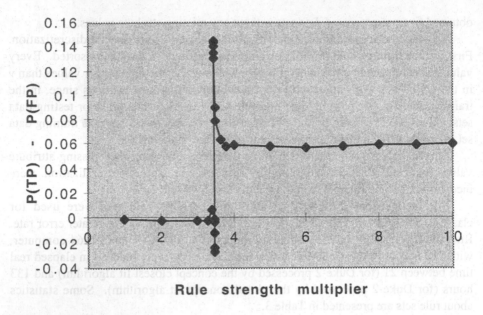

Fig. 1. P(TP) – P(FP) versus rule strength multiplier for Duke-2 data set and most common value method used for replacing missing attribute values

positives and false-negatives. It will be denoted by P(TP), following notation from [15]. Specificity is the conditional probability of true-negatives given fullterm birth, i.e., the ratio of the number of true-negatives to the sum of the number of true-negatives and false-positives. It will be denoted by P(TN). Similarly, the conditional probability of false-negatives, given actual preterm birth, and equal to 1 – P(TP), will be denoted by P(FN) and the conditional probability of false-positives, given actual fullterm birth, and equal to 1 – P(TN), will be denoted by P(FP). Obviously,

$$\text{Sensitivity} + \text{Specificity} = P(TP) - P(FP) + 1,$$

so all conclusions drawn from the observations of the sum of sensitivity and specificity can be drawn from observations of P(TP) – P(FP). Another study of the sum of sensitivity and specificity was presented in [1].

Following [9], we computed the maximum of the difference between the conditional probabilities for true-positives given actual preterm birth and false-positives given actual fullterm birth as a function of the rule strength multiplier for the preterm rule set. A representative chart is presented in Fig. 1. For completeness, a typical chart (Fig. 2) shows how the true-positive, true-negative and total error rate change as a function of the rule strength multiplier. The total error rate is defined as the ratio of the number of true-positives and true-negatives to the total number of testing cases.

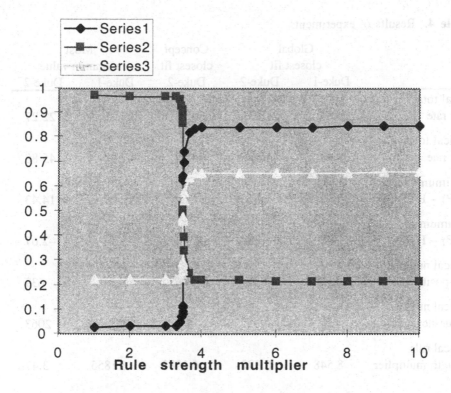

Fig. 2. Sensitivity (series 1), specificity (series 2), and total error rate (series 3) versus rule strength multiplier for Duke-2 data set and most common value method used for replacing missing attribute values

Again, following the idea from [9], in our experiments we were increasing the strength multiplier for each five rules describing preterm birth and observed P(TP) – P(FP). For each rule set, there exists some value of the rule strength multiplier, called *critical*, for which the values of P(TP) – P(FP) jumps from the minimal value to the maximal value. The respective values of true positives, true negatives, etc., and the total error rate, are also called *critical*. The results are summarized in Table 4. The total error rate, corresponding to the rule strength multiplier equal to one, is called *initial*.

The corresponding values of P(TP) – P(FP) are presented in Table 4. The critical total error rate from Table 4 is computed as the total error rate for the maximum of P(TP) – P(FP).

4 Conclusions

In our experiments the only difference between the five rule sets used for diagnosis of preterm birth is handling the missing attribute values. The maximum of the sum of sensitivity and specificity (or the maximum of P(TP) – P(FP)) is a good

Table 4. Results of experiments

	Global closest fit		Concept closest fit	Most common value	
	Duke-1	Duke-2	Duke-2	Duke-1	Duke-2
Initial total error rate	21.67	21.93	20.75	22.15	22.27
Critical total error rate	68.48	64.09	54.30	42.40	45.88
Maximum of P(TP) – P(FP)	3.65	5.97	11.69	17.07	14.43
Minimum of P(TP) – P(FP)	−15.96	−11.28	−5.37	−3.52	−2.67
Critical number of true-positives	882	838	747	615	639
Critical number of true-negatives	692	955	1535	2261	2063
Critical rule strength multiplier	8.548	6.982	6.1983	6.1855	3.478

indicator of usefulness of the rule set for diagnosis of preterm birth. It is the most important criterion of quality of the rule set. In terms of the maximum of the sum of sensitivity and specificity (or, equivalently, the maximum of $P(TP) - P(FP)$), the best data sets were processed by the most common value algorithm for missing attribute values. Note that the name of the algorithm is somewhat misleading because, in our experiments, we used this algorithm to compute the most common attribute value for each concept separately. The next best method is the concept closest fit algorithm. The worst results were obtained by the global closest fit.

The above ranking could be discovered not only by using the criterion of the maximum of the sum of sensitivity and specificity but also by using other criteria, for example, the minimum of the sum of sensitivity and specificity, the number of critical true-positive cases, critical false-positive cases, etc.

The initial total error rate is a poor indicator of the performance of an algorithm for handling missing attribute values. Similarly, the number of conflicting cases in the input data is a poor indicator.

Finally, it can be observed that the smaller values of the minimum of $P(TP) - P(FP)$ correspond to the smaller values of the maximum of $P(TP) - P(FP)$, so that the sum of the absolute values of these two numbers is roughly speaking constant.

References

[1]   Bairagi, R. and Suchindran C.M.: An estimator of the cutoff point maximizing sum of sensitivity and specificity. Sankhya, Series B, Indian Journal of Statistics **51** (1989) 263–269.

[2]   Booker, L. B., Goldberg, D. E., and Holland, J. F.: Classifier systems and genetic algorithms. In Machine Learning. Paradigms and Methods. Carbonell, J. G. (ed.), The MIT Press, 1990, 235–282.

[3]   Grzymala-Busse, J. W.: On the unknown attribute values in learning from examples. Proc. of the ISMIS-91, 6th International Symposium on Methodologies for Intelligent Systems, Charlotte, North Carolina, October 16–19, 1991, 368–377, Lecture Notes in Artificial Intelligence, vol. 542, 1991, Springer-Verlag.

[4]   Grzymala-Busse, J. W.: LERS—A system for learning from examples based on rough sets. In Intelligent Decision Support. Handbook of Applications and Advances of the Rough Sets Theory. Slowinski, R. (ed.), Kluwer Academic Publishers, 1992, 3–18.

[5]   Grzymala-Busse, J.W. and Goodwin, L.K.:   Predicting preterm birth risk using machine learning from data with missing values.   Bull. of Internat. Rough Set Society 1 (1997) 17–21.

[6]   Grzymala-Busse, J. W.: LERS—A knowledge discovery system. In Rough Sets in Knowledge Discovery 2, Applications, Case Studies and Software Systems, ed. by L. Polkowski and A. Skowron, Physica-Verlag, 1998, 562–565.

[7]   Grzymala-Busse, J.W. and Wang A.Y.: Modified algorithms LEM1 and LEM2 for rule induction from data with missing attribute values.   Proc. of the Fifth International Workshop on Rough Sets and Soft Computing (RSSC'97) at the Third Joint Conference on Information Sciences (JCIS'97), Research Triangle Park, NC, March 2–5, 1997, 69–72.

[8]   Grzymala-Busse, J.W. and Zou X.:   Classification strategies using certain and possible rules. Proc. of the First International Conference on Rough Sets and Current Trends in Computing, Warsaw, Poland, June 22–26, 1998. Lecture Notes in Artificial Intelligence, No. 1424, Springer Verlag, 1998, 37–44.

[9]   Grzymala-Busse, J. W., Goodwin, L.K., and Zhang, X.: Increasing sensitivity of preterm birth by changing rule strengths. Submitted for the 8th Workshop on Intelligent Information Systems (IIS'99), Ustronie, Poland, June 14–18, 1999.

[10]   Holland, J. H., Holyoak K. J., and Nisbett, R. E.:   Induction. Processes of Inference, Learning, and Discovery. The MIT Press, 1986.

[11]   Michalski, R. S., Mozetic, I., Hong, J. and Lavrac, N.: The AQ15 inductive learning system: An overview and experiments. Department of Computer Science, University of Illinois, Rep. UIUCDCD-R-86-1260, 1986.

[12]   Pawlak, Z.: Rough sets. International Journal Computer and Information Sciences 11 (1982) 341–356.

[13]   Pawlak, Z.: Rough Sets. Theoretical Aspects of Reasoning about Data. Kluwer Academic Publishers, 1991.

[14]   Stefanowski, J.: On rough set based approaches to induction of decision rules. In Polkowski L., Skowron A. (eds.) Rough Sets in Data Mining and Knowledge Discovery, Physica-Verlag, 1998, 500–529.

[15]   Swets, J.A. and Pickett, R.M.: Evaluation of Diagnostic Systems. Methods from Signal Detection Theory. Academic Press, 1982.

[16]   Ziarko, W.: Systems: DataQuest, DataLogic and KDDR. Proc. of the Fourth Int. Workshop on Rough Sets, Fuzzy Sets and Machine Discovery RSFD'96, Tokyo, Japan, November 6–8, 1996, 441–442.

Visualizing Discovered Rule Sets with Visual Graphs Based on Compressed Entropy Density

Einoshin Suzuki and Hiroki Ishihara

Division of Electrical and Computer Engineering,
Faculty of Engineering,
Yokohama National University
79-5, Tokiwadai, Hodogaya, Yokohama, 240-8501, Japan
suzuki@dnj.ynu.ac.jp

Abstract. This paper presents a post-processing algorithm of rule discovery for augmenting the readability of a discovered rule set. Rule discovery, in spite of its usefulness as a fundamental data-mining technique, outputs a huge number of rules. Since usefulness of a discovered rule is judged by human inspection, augmenting the readability of a discovered rule set is an important issue. We formalize this problem as a transformation of a rule set into a tree structure called a visual graph. A novel information-based criterion which represents compressed entropy of a data set per description length of the graph is employed in order to evaluate the readability quantitatively. Experiments with an agricultural data set in cooperation with domain experts confirmed the effectiveness of our method in terms of readability and validness.

1 Introduction

Knowledge Discovery in Databases (KDD) [4] represents a novel research area for discovering useful knowledge from large-scale data. With the rapid proliferation of large-scale databases, increasing attention has been paid to KDD. In KDD, rule discovery [1, 7, 9] represents induction of local constraints in a data set. Rule discovery is, due to its applicability, one of the most fundamental and important methods in KDD.

In general, a huge number of rules are discovered from a data set. In order to evaluate interestingness of a discovered rule set precisely, it is desirable to decrease the number of uninteresting rules and to output the rule set in a readable representation. However, conventional rule-discovery methods [1, 7, 9] consider mainly generality and accuracy of a rule, and readability[1] of a discovered rule set has been curiously ignored. Usefulness of a rule can be only revealed through human inspection. Therefore, visualization of a discovered rule set is considered to be highly important since it augments their readability.

Rule discovery can be classified into two approaches: one is to discover strong rules each of which explains many examples, and the other is to discover weak

---

[1] In this paper, we define readability as simplicity and informativeness.

rules each of which explains a small number of examples [8, 10]. This paper belongs to the first approach, and presents a method which transforms a set of strong rules with the same conclusion into a readable representation. As a representation, we consider a visual graph which explains the conclusion with premises agglomerated with respect to their frequencies. There exist methods for discovering graph-structured knowledge, such as Bayesian network [6] and EDAG [5]. However, our method is different from these methods since readability is our main goal. We propose, as a novel criterion for evaluating readability of a visual graph, compressed entropy density which is given as compressed entropy of the data set per description length of the graph. We demonstrate the effectiveness of our method by experiments using an agricultural data set in cooperation with domain experts.

2 Problem Description

In this paper, we consider transforming a data set D and a rule set R into a visual graph $G(D, S)$, where S is a subset of R and represents the rule set contained in $G(D, S)$. We assume that the number $|S|$ of rules in the rule set S is specified as a threshold by the user prior to the transformation.

The data set D consists of several examples each of which is described with a set of propositional attributes. Here, a continuous attribute is supposed to be discretized with an existing method such as [3], and is coverted to a nominal attribute. An event representing that an attribute has one of its values is called an atom. Proportion of examples each of which satisfies an atom a is represented by $\Pr(a)$.

The rule set R consists of $|R|$ rules $r_1, r_2, \cdots, r_{|R|}$, which are discovered with an existing method [7, 9] from the data set D.

$$R = \{r_1, r_2, \cdots, r_{|R|}\} \quad (1)$$

In KDD, important classes of rules include an association rule [1] and a conjunction rule [7, 9]. In an association rule, every attribute is assumed to be binary, and a value in an atom is restricted to "true". An association rule represents a rule of which premise and conclusion are either a single atom or a conjunction of atoms. In a conjunction rule, every attribute is assumed to be nominal. A conjunction rule represents a rule of which premise is either a single atom or a conjunction of atoms, and conclusion is a single atom. In this paper, we consider conjunction rules since they assume a more general class of attributes than association rules. For simplification, we assume that each rule r_i has the same conclusion x.

$$r_i = y_{i1} \wedge y_{i2} \wedge \cdots \wedge y_{i\nu(i)} \rightarrow x \quad (2)$$

where $y_{i1}, y_{i2}, \cdots, y_{i\nu(i)}$, x represent a single atom with different attributes respectively.

A visual graph $G(D, T)$ represents, in a graph format, a rule set T which consists of $|T|$ rules $t_1, t_2, \cdots, t_{|T|}$. As mentioned above, this rule set T is a subset of the input rule set R. A visual graph $G(D, T)$ is a tree structure in which $n(D, T)$ premise nodes $b_1(D, T), b_2(D, T), \cdots, b_{n(D,T)}(D, T)$ has their respective arc to a conclusion node $b_0(D, T)$. Here, the conclusion node $b_0(D, T)$ represents the atom x of conclusions in the rule set T. A premise node $b_i(D, T)$ represents the premises of rules each of which has the i-th most frequent atom d_i in the rule set T. Our method constructs a premise node $b_i(D, T)$ with an ascending order of i, and a rule represented in a premise node is no longer represented in the successive premise nodes. When more than two atoms have the same number of occurrence, the atom with the smallest subscript is selected first. Figure 1 shows an example of a rule set and its corresponding visual graph. In the visual

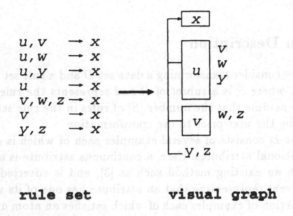

rule set   visual graph

Fig. 1. Example of a rule set and its corresponding visual graph

graph, the upmost atom x represents a conclusion node, and the other nodes are premise nodes. In the figure, the most frequent atom u is first agglomerated as a node, and the premises of the four rules each of which contains u represent the premise node 1. Although three rules contain atom v, the premise node 2 represents two rules since one of the three rules is employed in the premise node 1.

While a visual graph is uniquely determined by a rule set S, there are $_{|R|}C_{|S|}$ ways of selecting a subset S from the rule set R. In the next section, we describe how to choose a subset S from the rule set R in order to obtain a visual graph $G(D, S)$ with high readability.

3   Transformation of a Rule Set into a Visual Graph

3.1   Compressed Entropy Density Criterion

In order to obtain a visual graph with high readability, an appropriate subset S should be selected from the rule set R. In this paper, we consider an evaluation criterion for the readability of a visual graph, and propose a novel method which does not necessarily require user interaction.

The readability of a visual graph depends on two main factors. One factor is graph complexity which can be represented by the number of nodes and arcs in the graph. A complex graph is considered to have low readability. For example, if we consider intuitively, a graph with 300 nodes has lower readability than a graph with 30 nodes. The other factor is graph description-power which can be represented by the information content of the data set D in the graph. For example, if a visual graph A represents a subset of a rule set represented by another visual graph B and these two graphs have the same complexity, A has lower readability than B.

As explained in the previous section, a visual graph represents a tree structure of depth one in which each premise node has an arc to a conclusion node. Since the atom in the conclusion node is fixed and the depth is one, visual graphs vary with respect to the atoms in the premise nodes. Assuming that every atom has the same readability, graph complexity can be approximated by the number of atoms in the premise nodes. We can also consider the branching factor of the conclusion node, but we ignore it since it is equal to the number of premise nodes and can be approximately estimated by the number of atoms.

In order to provide an intuitive interpretation to the evaluation criterion, we represent graph complexity by its description length. If there are A kinds of atoms, the description length of an atom is $\log_2 A$ bit. Therefore, complexity $U(D, T)$ of a visual graph $G(D, T)$ is given as follows.

$$U(D, T) \equiv |G(D, T)| \log_2 A \qquad (3)$$

where $|G(D, T)|$ represents the number of atoms in the visual graph $G(D, T)$. Since A is fixed, $U(D, T)$ is a linear function of $|G(D, T)|$.

Since a visual graph and a rule set has one-to-one correspondence, the information content of a data set D represented by a visual graph $G(D, T)$ is equivalent to the information content of the data set D represented by the rule set T. The information content is calculated with respect to either the whole rule set or each rule. In rule discovery, although readability should be considered with respect to the whole rule set, usefulness is considered for each rule. Therefore, we take the latter approach. We first obtain the information content of a data set D represented by each rule in the rule set T, and then regard their add-sum as the graph-description power $V(D, T)$ for the data set D of the visual graph $G(D, T)$. Note that, this formalization ignores dependency among rules. We have also pursued another formalization in which premises of rules are mutually exclusive. However, this approach has turned out to be less effective by experiments with an agricultural data set.

In ITRULE rule discovery system [7], Smyth employed compressed entropy of a data set D by a rule $t : y \rightarrow x$ as an evaluation criterion J-measure $J(t, D)$ of the rule.

$$J(t, D) \equiv \Pr(y) \left[\Pr(x|y) \log_2 \frac{\Pr(x|y)}{\Pr(x)} + \Pr(\overline{x}|y) \log_2 \frac{\Pr(\overline{x}|y)}{\Pr(\overline{x})} \right] \qquad (4)$$

where $\bar{x}$ represents the negation of the atom x. J-measure is a single quantitative criterion which simultaneously evaluates the generality $\Pr(y)$, the accuracy $\Pr(x|y)$ and the unexpectedness $\Pr(x|y)/\Pr(x)$ of a rule, and is reported to be effective in rule discovery [7]. Interested readers can consult [7] for theoretical foundation and empirical behavior of J-measure. In this paper, we represent information content of a data set D by each rule t with J-measure $J(t, D)$. Therefore, graph description-power $V(D, T)$ for a data set D of a visual graph $G(D, T)$ is given as follows.

$$V(D, T) \equiv \sum_{t \in T} J(t, D) \tag{5}$$

Note that readability of a visual graph $G(D, T)$ decreases with respect to graph complexity $U(D, T)$, and increases with respect to graph description-power $V(D, T)$. The former is represented by the description length of the graph, and the latter is represented by the compressed entropy of the data set by the graph. Here, the quotient of graph description-power by graph complexity represents compressed entropy of the data set per description length of the graph, and can be regarded as density of compressed entropy. If this quotient of a graph is large, we can regard the graph as representing information of the data set with high density. We propose, as the evaluation criterion of readability of a visual graph, compressed entropy density $W(D, T)$ which is given as follows.

$$W(D, T) \equiv \frac{V(D, T)}{U(D, T)} \tag{6}$$

Behavior of $W(D, T)$ cannot be analyzed exactly since it is highly dependent on the nature of the input data. Probabilistic analysis, based on average performance over all possible input data sets, is too difficult to carry out directly without invoking unrealistic assumptions concerning the nature of the inputs. We leave more rigorous analysis of the problem for further research.

3.2 Search Method

Our algorithm obtains a rule set S by deleting, one by one, rules in the input rule set R until the number of rules becomes $|S|$. In a KDD process, we cannot overemphasize the importance of user interaction [2]. In rule visualization, users may iterate visualization procedure by inspecting the output and specifying new conditions. Therefore, our algorithm employs hill climbing since its computation time is relatively short. Our algorithm is given as follows.

1. (Set) $T \leftarrow R$
2. (Delete rules)
   (a) while($|T| > |S|$)
   (b) $T \leftarrow \arg \max_{T-\{t\}} W(D, T - \{t\})$

3. (Return) Return $G(D, T)$

4 Application to an Agriculture Data Set

In this section, we demonstrate the effectiveness of our method by applying it to "Agriculture" data sets. "Agriculture" is a series of data sets which describes agricultural statistics such as various crops for approximately 3200 municipalities in Japan. We have followed suggestions of domain experts and analyzed "semi-mountainous municipalities". Japanese ministry of agriculture specified approximately 1700 municipalities as semi-mountainous for conservation of agriculture in mountainous regions, and analysis on these municipalities is highly demanded. We have used the 1992 version of "Agriculture", and there are 1748 semi-mountainous municipalities as examples in the data set.

Since Japan has diverse climates, there are many crops each of which is cultivated in a restricted region. An atom representing the absence of such a crop is frequent in discovered rules. However, such an atom is uninteresting to domain experts since it represents another view of climatic conditions. In order to ignore such atoms, we employed 148 attributes each of which has a positive value in at least one-third of municipalities. These attributes represent, for instance, summaries of municipalities, shipments of crops and acreages of crops. In discretizing a continuous attribute, we first regarded "0" and missing values as a new value, then employed equal-frequency method [3] of three bins.

According to domain experts, conditions on high income are their main interests. First, we settled the atom of the conclusion "agricultural income per farmhouse = high". We obtained a rule set which consists of 333 rules with a rule discovery method [9]. For the rule set S in the output visual graph, we settled as $|S| = 15$. Figure 2 shows the result of this experiment.

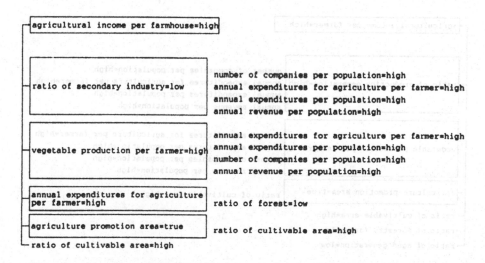

Fig. 2. Visual graph for conclusion "agricultural income per farmhouse = high" with $|S| = 15$

Atoms in this figure can be classified into four groups. The first group represents that a considerable amount of subsidies are granted by the administration. Atoms which belong to this group are "agriculture promotion area=true", "annual expenditures for agriculture per farmer=high", "annual expenditures per population=high" and "annual revenue per population=high". These atoms represent that agriculture is highly-promoted by administrations, and their financial status are excellent. The second group represents that vegetables are well-cultivated. Atoms which belong to this group are "vegetable production per farmer=high", "ratio of cultivable area=high" and "ratio of forest = low". These atoms represent that high income is gained with vegetables, and acreage for vegetables is large. According to domain experts, difference in cultivation technique of vegetables has considerable influence on income. The third group represents that companies are highly-active, and "number of companies per population=high" belongs to this group. This atom represents that a municipality is located close to cities, each of which gives opportunity of shipment and side income. The fourth group represents that a municipality depends mainly on agriculture, and "ratio of secondary industry=low" belongs to this group. This atom represents that, for instance, each farmer has large acreage. This analysis shows that each atom in the premise nodes in figure 2 is appropriate as a reason of "agricultural income per farmhouse = high".

In the next experiment, the atom in the conclusion is settled to "agricultural income per farmer = high", and a rule set which consists of 335 rules is obtained with the same procedure. Figure 3 shows the visual graph obtained by our method with the same conditions.

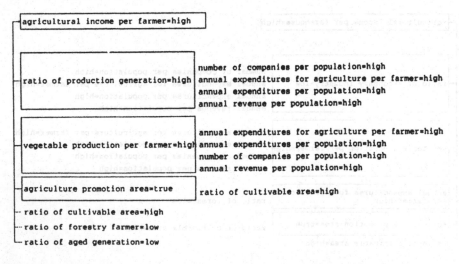

Fig. 3. Visual graph for conclusion "agricultural income per farmer = high" with $|S| =$ 15

In Japan, municipalities of "agricultural income per farmer = high" are almost equivalent to municipalities of "agricultural income per farmhouse = high". Large-scale farmhouses are dominant in these municipalities. Since atoms in the premise nodes in figure 3 are similar to those in figure 2, this visual graph can be validated with similar discussions as above.

In the last experiment, the atom in the conclusion is settled to "agricultural income per 10A = high", and a rule set which consists of 319 rules is obtained with the same procedure. Figure 3 shows the visual graph obtained by our method with the same conditions.

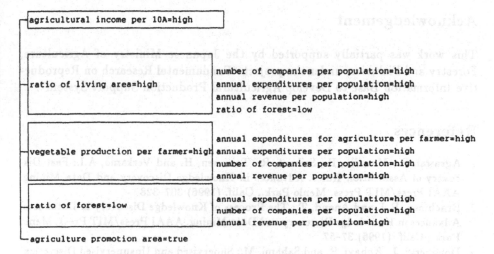

Fig. 4. Visual graph for conclusion "agricultural income per 10A = high" with $|S| = 15$

Unlike the other two visual graphs, visual graph in figure 4 has "ratio of living area = high", and considers "ratio of forest = low" as more important. It should be also noted that atoms "ratio of secondary industry=low" and "ratio of production generation=high" have disappeared. These results can be explained that some of municipalities in which large-scale farmhouses are dominant are excluded in "agricultural income per 10A = high", and cultivation techniques are more important for this conclusion.

From figure 2 to 4, each obtained visual graph has a simple structure and contains valid rules. Domain experts evaluated these three results, and claimed that each visual graph has a simple structure and thus has high readability. They also concluded that each visual graph contains accurate and valid rules in explaining the conclusion.

5   Conclusion

Existing rule discovery methods induce a huge number of rules, and inspection of these rules for judging their usefulness requires considerable efforts for humans.

In order to circumvent this problem, we proposed a novel method for transforming a discovered rule set into a visual graph which has a simple structure for representing information of a data set. For this transformation, we presented a novel criterion: compressed entropy density which is given by the quotient of compressed entropy by the description length of the graph. Our method has been applied to an agricultural data set for 1748 municipalities in Japan, and the results were evaluated by domain experts. Obtained visual graphs have high readability and contain valid rules even for these experts. We consider that this fact demonstrates the effectiveness of our method.

Acknowledgement

This work was partially supported by the Japanese Ministry of Agriculture, Forestry and Fisheries, under the project "Fundamental Research on Reproductive Information Base to Develop Agricultural Production Support System".

References

1. Agrawal, R., Mannila, H., Srikant, R., Toivonen, H. and Verkamo, A.I.: Fast Discovery of Association Rules. Advances in Knowledge Discovery and Data Mining. AAAI Press/MIT Press, Menlo Park., Calif. (1996) 307–328
2. Brachman, R.J. and Anand, T.: The Process of Knowledge Discovery in Databases. Advances in Knowledge Discovery and Data Mining. AAAI Press/MIT Press, Menlo Park., Calif. (1996) 37–57
3. Dougherty, J., Kohavi, R. and Sahami, M.: Supervised and Unsupervised Discretization of Continuous Features. Proc. Twelfth Int'l Conf. Machine Learning (ICML-95), Morgan Kaufmann, San Francisco, (1995) 194–202
4. Fayyad, U.M., Piatetsky-Shapiro, G. and Smyth, P.: From Data Mining to Knowledge Discovery: An Overview. Advances in Knowledge Discovery and Data Mining. AAAI Press/MIT Press, Menlo Park., Calif. (1996) 1–34
5. Gaines, B.R.: Transforming Rules and Trees. Advances in Knowledge Discovery and Data Mining. AAAI Press/MIT Press, Menlo Park., Calif. (1996) 205–226
6. Jensen, F.V.: An Introduction to Bayesian Networks. Springer-Verlag, New York (1996)
7. Smyth, P. and Goodman, R.M.: An Information Theoretic Approach to Rule Induction from Databases. IEEE Trans. Knowledge and Data Eng. 4 (4) (1992) 301–316
8. Suzuki, E.: Autonomous Discovery of Reliable Exception Rules. Proc. Third Int'l Conf. Knowledge Discovery & Data Mining (KDD-97), AAAI Press, Menlo Park, Calif. (1997) 259–262
9. Suzuki, E.: Simultaneous Reliability Evaluation of Generality and Accuracy for Rule Discovery in Databases. Proc. Fourth Int'l Conf. Knowledge Discovery & Data Mining (KDD-98), AAAI Press, Menlo Park, Calif. (1998) 339–343
10. Suzuki, E. and Kodratoff, Y.: Discovery of Surprising Exception Rules based on Intensity of Implication. Principles of Data Mining and Knowledge Discovery, Lecture Notes in Artificial Intelligence 1510. Springer-Verlag, Berlin (1998) 10–18

A Distance-Based Clustering and Selection of Association Rules on Numeric Attributes

Xiaoyong Du[1,2], Sachiko Suzuki[3], Naohiro Ishii[1]

1. Department of Intelligence and Computer Science,
Nagoya Institute of Technology, Nagoya, Japan
E-mail:{duyong, ishii}@egg.ics.nitech.ac.jp
2. School of Information, Renmin University of China, Beijing, China
duyong@mail.ruc.edu.cn
3. Department of Computational Intelligence and Systems Science,
Tokyo Institute of Technology, Yokohama, Japan
sachi@ntt.dis.titech.ac.jp

Abstract. Association rule is a kind of important knowledge extracted from databases. However, a large number of association rules may be extracted. It is difficult for a user to understand them. How to select some "representative" rules is thus an important and interesting topic. In this paper, we proposed a distance-based approach as a post-processing for association rules on numeric attributes. Our approach consists of two phases. First, a heuristic algorithm is used to cluster rules based on a matrix of which element is the distance of two rules. Second, after clustering, we select a representative rule for each cluster based on an objective measure. We applied our approach to a real database. As the result, three representative rules are selected, instead of more than 300 original association rules.

Keywords: Association rules, Rule clustering, Rule selection, Numeric attributes, Objective Measures, Discretization.

1 Introduction

Data mining has been recognized as an important area of database research. It discovers patterns of interest or knowledge from large databases. As a kind of important pattern of knowledge, association rule has been introduced. An association rule is an implication expression: $C_1 \Rightarrow C_2$, where C_1 and C_2 are two conditions. It means that when the condition C_1 is true, the conclusion C_2 is almost always true.

Association rule is first introduced in Agrawal et al.'s papers [AIS93, AS94]. They considered only bucket type data, like supermarket databases where the set of items purchased by a single customer is recorded as a transaction. When we focus on data in relational databases, however, we have to consider various types of data, especially continuous numeric data. For example, $(age \in [40,60]) \Rightarrow (own\_house = yes)$. In this case, we may find hundreds or thousands of association rules corresponding to a specific attribute. Fig. 1 shows all rules (about 300) that we extracted from an adult database. The rules have the form "$fnlwgt \in [a, b] \Rightarrow (income < 50K)$", where $fnlwgt$ is a numeric attribute and income is a decision attribute. We order the rules by the ranges in the LHS. It is not accteptable to show all rules to users. To tackle this problem, Fukuda et. al. [FMMT96a, FMMT96b] proposed so-called optimized association rule. It extracts a single association rule from all candidates which maximizes some index of the rules, for example, support. In many cases, however, it is just a common sense rule and has no value at all.

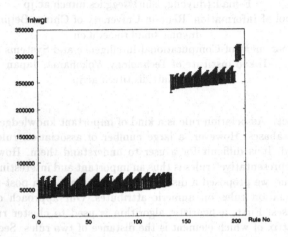

Fig. 1. Many similar rules are extracted

To overcome this shortcoming, in our opinion, it is reasonable to divide the process of discovering association rules into two steps: one is to find all candidates of which support and confidence are greater than the thresholds given by users; the other is to select some representative rules from all candidates. Although most of existing papers contributed to the first step, an incremental interesting has been paid on the second step [KMR+94, MM95, Fre98, GB98, Kry98a, Kry98b, WTL98]. Various measures for interestingness of association rules have been proposed.

In general, the evaluation of the interestingness of discovered rules has both an objective and a subjective aspect. Kiemettinen et al.[KMR+94] proposed a simple formalism of rule templates to describe the structure of interesting rules, like what attributes occur in the antecedent and what attribute is the consequent. Liu et al. [LHC97] proposed a user-defined impression to analyze discovered rules. Other authors choose to look for objective measures for rule selection. Gago et al.[GB98] defined a distance between two rules, and select n rules such that they are the most distinguished. Major et al.[MM95] proposed a set of measures, like simplicity, novelty, statistical significant, and a stepwise selection process. Kryszkiewicz [Kry98a, Kry98b] defined a cover operator for association rule on bucket data, and found a least set of rules that covers all association rule by the cover operator. However, since downward closure property is not true for association rules on numeric attribute, Cover operation is not appropriate for rule selection.

In this paper, we focus on selection of association rules on numeric attributes. We assume that a set R of association rules have been extracted. We then select a subset of R as representative rules of R. Our approach is first to cluster association rules according to the distance between rules, and then to select a representative rule for each class. In this paper, we also focus on objective measures for association rules. We observe from Fig. 1 that many similar rules exist. It is because a rule candidate which is close to a rule with high support and confidence is most possibly an association rule too. Hence, it is reasonable to define a representative rule for a set of similar rules. Two objective measures are proposed for clustering and selection of rules, respectively.

The paper is organized as follows: In Section 2, we present basic terminology and an overview of the work. Section 3 defines a distance between rules which is used for grouping similar rules. In Section 4, we propose a coverage measure for selection of representative rules. In Section 5, we present some experimental results. Section 6 concludes and presents our future work.

2 Overview of Our Work

In this section we present basic terminology for mining association rules on numeric attributes, and then give an overview of our approach.

Assume there is a relation $D(A_1, A_2, \cdots, A_n, C)$, where A_i is an attribute name, and C is a decision attribute. For a tuple $t \in D$, $t.A_i$ denotes the value of A_i at t. An association rule is an expression of the form $C_1 \Rightarrow C_2$, where C_1 and C_2 are two expressions, called left-hand side (LHS) and right-hand side (RHS) of the rule, respectively. In this paper, we consider association rules on numeric attributes with the form:

$$R : (a_1 \leq A_1 < b_1) \wedge \cdots \wedge (a_n \leq A_n < b_n) \Rightarrow (C = yes)$$

where A_i is a numeric attribute and C is a Boolean attribute. Without confusion, we usually denote a rule by an area P in the n dimension space. $t \in P$ means $(a_1 \leq t.A_1 < b_1) \wedge \cdots \wedge (a_n \leq t.A_n < b_n)\}$.

Two measures, *support* and *confidence*, are commonly used to rank association rules. The support of an association R, denoted by $supp(R)$, is defined by $|\{t|t \in P\}|/|D|$ [1]. It means how often the value of A occurs in the area P as a fraction of the total number of tuples. The confidence of an association rule, denoted by $conf(R)$, is defined by $|\{t|t \in P \wedge t.C = yes\}|/|\{t|t \in P\}|$. It is the strength of the rule.

For a pair of *minsup* and *minconf* specified by the user as the thresholds of support and confidence, respectively, an association rule is called "interesting" if both its support and confidence are over the minimal thresholds. Let Ω denote the set of all interesting rules. That is $\Omega = \{R|supp(R) \geq minsup \wedge conf(R) \geq$

---

[1] or $|\{t|t \in P \wedge t.C = yes\}|/|D|$.

$minconf\}$. Our purpose is to extract a set of representative rules from Ω. Our approach consists of the following two steps:

(1) Clustering. We define a distance between two rules, and a diameter of a set of rules based on distance of rule pairs. Intuitively, the rules in Fig. 1 should be clustered into three groups.
(2) Selection. For each cluster, we select exactly one rule as its representative rule. We define a coverage for each rule. It measures the degree of a certain rule to "cover" all others.

In the following two sections, we discuss these two aspects respectively.

3 Clustering Association Rules

Let $\Omega = \{r_1, \cdots, r_n\}$ be a set of association rules. Each rule r_i contains an area in LHS. We denote also the area as r_i without confusion. In the followings, we use the word "rule" and "area" in the same meaning.

Definition 1. Let r_1 and r_2 be two rules. The distance of r_1 and r_2 is defined by

$$dist(r_1, r_2) = \sqrt{\Sigma_{i=1}^n((a_i^{(1)} - a_i^{(2)})^2 + (b_i^{(1)} - b_i^{(2)})^2)} \tag{1}$$

where $r_i = \{a_1^{(i)} \le A_1 < b_1^{(i)}, \cdots, a_n^{(i)} \le A_n < b_n^{(i)}\}$ for $i = 1, 2$.

In this definition, we view the left and right terminals of a range on a numeric attribute as two independent parameters. Thus a rule can be represented as a point in a $2n$ dimension space. The distance of two rules is defined as the distance of the two points in the space.

Definition 2. Let $C = \{r_1, \cdots, r_m\}$ be a set of rules, $r \in C$ be a rule. A (average) distance of r to C is defined by

$$dist(r, C) = \Sigma_{r_i \in C} dist(r, r_i)/m \tag{2}$$

Definition 3. Let C_1 and C_2 be two sets of rules. The (average) distance between C_1 and C_2 is defined by

$$dist(C_1, C_2) = \Sigma_{r_i \in C_1, r_j \in C_2} dist(r_i, r_j)/(|C_1| \cdot |C_2|) \tag{3}$$

where $|C_1|$ and $|C_2|$ are the numbers of rules in C_1 and C_2, respectively.

The diameter of a cluster is the average distance of all pairs of rules in the cluster.

Definition 4. Let $C = \{r_1, \cdots, r_m\}$ be a set of rules. A diameter of C is defined by

$$d(C) = \Sigma_{r_i, r_j \in C} dist(r_i, r_j)/(m(m-1)) \tag{4}$$

Definition 5 . Let $C = \{C_1, \cdots, C_k\}$, where $C_i \subseteq \Omega$. C is called a clustering of Ω if for a given threshold d_0, the followings are satisfied.

1. $C_i \cap C_j = \phi, (i \neq j)$
2. $d(C_i) \leq d_0$
3. $dist(C_i, C_j) \geq d_0, (i \neq j)$

This definition gives a basic requirement for clustering. Obviously, the further the distance between clusters, the better the clustering. In other words, we expect to maximize the sum of the distance of all pairs of clusters. However, there are $O((n!)^2/2^n)$ number of candidates for clusterings. It is impossible to obtain an optimized clustering by a native aproach.

In this section, we propose a heuristic approach to construct a clustering. It is a hill-climbing algorithm working on a matrix of which cell represents the distance of two rules. That is

$$D = (dist(r_i, r_j))_{n \times n}$$

We always select two rules (or two sets of rules) between which the distance is the minimal. Hence, our algorithm consists of a loop, each of which combines two lines/columns of the matrix of which crosspoint cell has the minimal value.

While combining two rules (or two sets of rules), we have to recompute the distance between the combined cell and the other rules. The following properties can be used for this increamental recomputing. They can be derived from the definitions of diameter and distance, and Fig. 2.

Property 6 . Let $C_1 = \{r_1, \cdots, r_m\}$, $C_2 = \{s_1, \cdots, s_n\}$ be two sets of rules. Assume $d(C_1) = d_1$, and $d(C_2) = d_2$, and $dist(C_1, C_2) = dist$. The diameter of $C_1 \cup C_2$ can be evaluated by the following formula.

$$
\begin{aligned}
d(C_1 \cup C_2) &= \sum_{r,s \in C_1 \cup C_2} dist(r,s)/((m+n)(m+n-1)) \\
&= \frac{(\sum_{r,s \in C_1} + \sum_{r,s \in C_2} + \sum_{r \in C_1, s \in C_2} + \sum_{s \in C_1, r \in C_2}) dist(r,s)}{(m+n)(m+n-1)} \\
&= \frac{m(m-1)d(C_1) + n(n-1)d(C_2) + (2mn)dist(C_1, C_2)}{(m+n)(m+n-1)} \\
&= \frac{m(m-1)d_1 + n(n-1)d_2 + (2mn)dist}{(m+n)(m+n-1)}
\end{aligned}
$$

Property 7 . Let $C_1 = \{r_1, \cdots, r_m\}$, $C_2 = \{s_1, \cdots, s_n\}$ be two clusters. C_3 be another cluster. Assume C_1 and C_2 are combined to a new cluster $C_1 \cup C_2$, then the distance between C_3 and $C_1 \cup C_2$ can be evaluated by the following formula.

$$
\begin{aligned}
dist(C_3, C_1 \cup C_2) &= (\sum_{r \in C_3, s \in C_1 \cup C_2} dist(r,s))/(|C_3| \cdot |C_1 \cup C_2|) \\
&= \frac{\sum_{r \in C_3, s \in C_1} dist(r,s) + \sum_{r \in C_3, s \in C_2} dist(r,s)}{|C_3| \cdot |C_1 \cup C_2|} \\
&= \frac{|C_3| \cdot |C_1| \cdot dist(C_3, C_1) + |C_3| \cdot |C_2| \cdot dist(C_3, C_2)}{|C_3| \cdot |C_1 \cup C_2|} \\
&= \frac{md_1 + nd_2}{m+n}
\end{aligned}
$$

428

where $d_1 = dist(C_3, C_1)$, and $d_2 = dist(C_3, C_2)$.

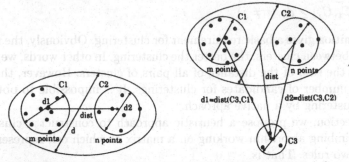

(1) Diameter of $C_1 \cup C_2$ (2) Distance between C_3 and $C_1 \cup C_2$

Fig. 2. Diameter and distance of clusters

The algorithm consists of a loop of two steps. The first step is to select the minimal distance from the upper triangle of the matrix. If the value is less than the threshold d_0, the corresponding two rules (clusters) should be combined. The next step is to generate a new matrix which has smaller size.

Algorithm 8. Clustering
Input: a matrix $D(i,j)$
Output: clustering $C = \{C_1, \cdots, C_k\}$
Method:

1. $C_i = \{i\}$ for $i = 1, \cdots, k$; $d = min_{i \neq j}\{D(i,j)\}$; Assume $D(s,t)$ is the minimal distance element is D.
2. While $(d \leq d_0)$ Do {
   2-1. combine C_s and C_t, and let the new C_s be $C_s \cup C_t$,
   2-2. delete C_t from C.
   2-3. generate a new matrix $D' = (e_{i,j})_{(n-1) \times (n-1)}$, where

$$\begin{cases} e_{s,s} = \dfrac{n_s(n_s - 1)d_{s,s} + n_t(n_t - 1)d_{t,t} + 2n_s n_t d_{s,t}}{(n_s + n_t)(n_s + n_t - 1)} \\ e_{s,j} = (n_s * d_{s,j} + n_t * d_{t,j})/(n_s + n_t), & j \neq s,t \\ e_{i,j} = d_{i,j}, & i,j \neq s,t \end{cases}$$

   where n_s and n_t are the size of the s-th and t-th clusters, $d_{i,j}$ is the distance between C_i and C_j.
   2-4. find the minimal distance from D': Let $D'(s,t) = min_{i \neq j}\{D'(i,j)\} = d$.
   }
3. Output C. Assume the final matrix is $D'_{m \times m}$. Then the diameter of cluster C_i is $e_{i,i}$.

The complex of this algorithm is $O(n^3)$. This is because that the most expensive step is finding the minimal element of the matrix in each loop.

Example 1. Let us consider a simple example. The rules contain only one attribute in its LHS. That is, all rules can be represeted as a range in this case. Let $\Omega = \{[1,3],[3,5],[2,4],[6,7],[7,9]\}$. The distance matrix is

$$D_1 = \begin{pmatrix} 0 & 2\sqrt{2} & \sqrt{2} & \sqrt{41} & 6\sqrt{2} \\ & 0 & \sqrt{2} & \sqrt{13} & 4\sqrt{2} \\ & & 0 & 5 & 5\sqrt{2} \\ & & & 0 & \sqrt{5} \\ & & & & 0 \end{pmatrix}$$

Assume that the threshold $d_0 = 2$. The algorithm runs as follows.

1. Find $D_1(1,3)$ which value is the minimal in D_1. Since the value $D_1(1,3) < d_0$, we combine the first and the third line/column at first. The new matrix a 4×4 one.

$$D_2 = \begin{pmatrix} \sqrt{2} & (3/2)\sqrt{2} & (\sqrt{41}+5)/2 & (11/2)\sqrt{2} \\ 0 & & \sqrt{13} & 4\sqrt{2} \\ & & 0 & \sqrt{5} \\ & & & 0 \end{pmatrix}$$

2. In the new matrix, the minimal value except the elements in the diagonal line is $D_2(1,2) = (3/2)\sqrt{(2)} < d_0$ Hence, we need to combine of the first and second line/column of D_2. The reduced new matrix D_3 is,

$$D_3 = \begin{pmatrix} (4/3)\sqrt{2} & (\sqrt{41}+5+\sqrt{13})/3 & 5\sqrt{2} \\ & 0 & \sqrt{5} \\ & & 0 \end{pmatrix}$$

3. Finally, since the minimal value cell $D_3(2,3) > d_0$, the algorithm stops.

Ω is thus divided to three clusters. One is $\{[1,3],[3,5],[2,4]\}$, and the others are $\{[6,7]\}$ and $\{[7,9]\}$.

4 Selecting Representative Rules

The next phase of our approach is to select a representative rule for each cluster. Since all rules in the same cluster are similar, it is reasonable to select only one as a representative rule.

Definition 9. Let $C = \{r_1, \cdots, r_n\}$ be a cluster of rules, and $R \in C$. The coverage of R to C is defined as

$$\alpha(R) = (\Sigma_{r \in C} \|r \cap R\| / \|r \cup R\|)/|C| \tag{5}$$

where $\|X\|$ is the volume of the area X. $r \cup R$ and $r \cap R$ are defined in an ordinary way. A rule R is called representative rule of C if $\alpha(R)$ is the maximal.

The measure $\alpha(R)$ reflects the degree of one certain rule to cover all others. It can be used as an objective measure for selection. In the following section, we can see from an example that this measure is better than the others like support.

Example 2. Let us consider Example 1 once again. For cluster $\{[1,3],[3,5],[2,4]\}$, we can evaluate that $\alpha([1,3]) = 4/9$, $\alpha([3,5]) = 4/9$, and $\alpha([2,4]) = 5/9$. Hence, [2,4] should be selected as the representative rule of the cluster. The other two clusters are single element clusters. The rule itself is thus the representative rule of the cluster. Hence, we finally obtain a set of representative rules for Ω. It is $\{[2,4],[6,7],[7,9]\}$.

It is easy to develop an algorithm with $O(n^2)$ complexity to select a representative rule from the cluster C.

5   Experiments

The first experiment is to apply our approach to analyse a set of association rules extracted from an adult database. The association rule has the form "$fnlwgt \in [a,b] \Rightarrow (income < 50K)$". The RHS of the rule can be viewed as a Boolean attribute. The database contains 32560 tuples. When we set $minconf = 0.8$ and $minsup = 0.03$, we obtained 310 rules.

In the first step, we represent these rules as points in a 2D space. By our algorithm, they formed three clusters (Fig 3(a)). Furthermore, three rules are selected from three clusters, respectively. The representative rule of the cluster 1 is showed in Fig. 3(b).

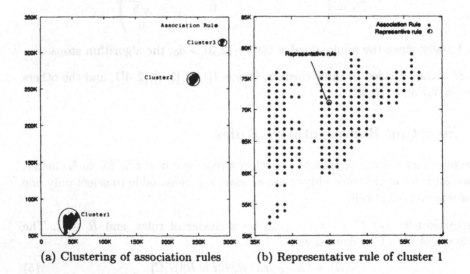

(a) Clustering of association rules   (b) Representative rule of cluster 1

Fig. 3. The X-axis and Y-axis represent the left and right terminal of the range in the LHS of a rule, respectively.

The second experiment is to compare our coverage measure with the support measure as selection metric. We consider another attribute "age" in the adult database to see the association relation between "age" and "income", that is, pattern of rule "$Age \in [a, b] \Rightarrow Income < 50K$". Let the threshold of confidence θ_c be 0.8. Fig. 4 (a) shows the range which support is the maximal and confidence is greater than θ_c. From the figure we can see that the selected range covers a large part of which confidence is less than θ_c. It is because that the left part of the range is with a confidence which is much higher than the θ_c. To be opposite, Figure 4 (b) shows the range of which coverage is the maximal and its confidence and support are greater than the given thresholds.

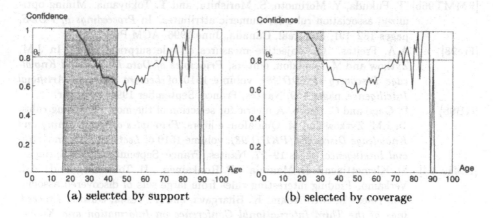

(a) selected by support   (b) selected by coverage

Fig. 4. Comparison of the measure of coverage and support

6   Conclusions and Further Work

Selection of representative and useful association rules from all candidates is a hard problem. Although it depends on user's interests in nature, we believe that some objective measures are helpful for users to select. For association rules on numeric attributes, we observed that there exist many similar rules. We thus propose a distance-based clustering algorithm to cluster them. The clustering algorithm is a heuristic hill-climbing and matrix-reducing procedure. The complexity is $O(n^3)$, where n is the number of association rules. We also propose an objective measure called coverage for selection of representative rule for each cluster.

Some further work is needed. How to deal with attributes with different types and/or scales in the LHS of the rules is interesting. Further evaluation of the effectiveness of our approach in real applications is also necessary.

Acknowledgments The authors would like to thank the anonymous reviewer who provided critical and detail comments.

References

[AIS93]   R. Agrawal, T. Imielinski, and R. Srikant. Mining association rules between sets of items in large databases. In *Proceedings of SIGMOD*, pages 207–216, Washington, USA, May 1993.

[AS94]   R. Agrawal and R. Srikant. Fast algorithms for mining association rules. In *Proceedings of the 20th VLDB Conference*, pages 487–499, Santiago, Chile, 1994.

[FMMT96a]   T. Fukuda, Y. Morimoto, S. Morishita, and T. Tokuyama. Data mining using two-dimentional optimized association rules: Scheme, algorithms and visualization. In *Proceedings of SIGMOD*, pages 13–23, Montreal, Canada, June 1996. ACM Press.

[FMMT96b]   T. Fukuda, Y. Morimoto, S. Morishita, and T. Tokuyama. Mining optimized association rules for numeric attributes. In *Proceedings of PODS*, pages 182–191, Montreal, Canada, June 1996. ACM Press.

[Fre98]   A.A. Freitas. On objective measures of rule surprisingness. In J.M. Zytkow and M. Quafafou, editors, *Principles of Data Mining and Knowledge Discovery (PKDD'98)*, volume 1510 of *Lecture Notes in Artificial Intelligence*, pages 1–9, Nantes, France, September 1998. Springer.

[GB98]   P. Gago and C. Bento. A metric for selection of the most promising rules. In J.M. Zytkow and M. Quafafou, editors, *Principles of Data Mining and Knowledge Discovery (PKDD'98)*, volume 1510 of *Lecture Notes in Artificial Intelligence*, pages 19–27, Nantes, France, September 1998. Springer.

[KMR+94]   M. Klemettinen, H. Mannila, P. Ronkainen, H. Toivonen, and A.I. Verkamo. Finding interesting rules from large sets of discovered association rules. In N.R. Adam, K. Bhargava, and Y. Yesha, editors, *Proceedings of the Third International Conference on Information and Knowledge Management*, pages 401–407, Maryland, USA, November 1994. ACM Press.

[Kry98a]   M. Kryszkiewicz. Representative association rules. In X. Wu, R. Kotagiri, and K.B. Korb, editors, *Research and Developement in Knowledge Discovery and Data Mining (PAKDD'98)*, volume 1394 of *Lecture Notes in Artificial Intelligence*, pages 198–209, Melbourne, Australia, April 1998. Springer.

[Kry98b]   M. Kryszkiewicz. Representative association rules and mining condition maximum consequence association rules. In J.M. Zytkow and M. Quafafou, editors, *Principles of Data Mining and Knowledge Discovery (PKDD'98)*, volume 1510 of *Lecture Notes in Artificial Intelligence*, pages 361–369, Nantes, France, September 1998. Springer.

[LHC97]   B. Liu, W. Hsu, and S. Chen. Using general impressions to analyze discivered classification rules. In *Proceedings of KDD-97*, pages 31–36, California, USA, 1997.

[MM95]   J.A. Major and J.J. Mangano. Selecting among rules induced from a hurricane database. *Journal of Intelligent Information Systems*, (4):39–52, 1995.

[WTL98]   K. Wang, S.H.W. Tay, and B. Liu. Interestingness-based interval merger for numeric association rules. In *Proceedings of KDD-98*, 1998.

Knowledge Discovery for Protein Tertiary Substructures

Chao-wen Kevin Chen[1] and David Y. Y. Yun[2]

Laboratory of Intelligent and Parallel Systems, College of Engineering, Univ. of Hawaii

cwchen@spectra.eng.hawaii.edu[1], dyun@spectra.eng.hawaii.edu[2]

Abstract. Mining for common motifs in protein tertiary structures holds the key to the understanding of protein functions. However, due to the formidable problem size, existing techniques for finding common substructures are computationally feasible only under certain artificially imposed constraints, such as using super-secondary structures and fixed-length segmentation. This paper presents the first, pure tertiary-level algorithm that discovers the common protein substructures without such limitations. Modeling this as a maximal common subgraph (MCS) problem, the solution is found by further mapping into the domain of maximum clique (MC). Coupling a MC solver with a graph coloring (GC) solver, the iterative algorithm, CRP-GM, is developed to narrow down towards the desired solution by feeding results from one solver into the other. The solution quality of CRP-GM amply demonstrates its potential as a new and practical data-mining tool for molecular biologists, as well as several other similar problems requiring identification of common substructures.

1. Introduction

This paper describes a new algorithm capable of discovering maximal common substructures from large, complex graph representations of given structures of interest. The ability to produce high-quality solutions in reasonable time has been a long standing challenge, since the maximal common subgraph (MCS) problem is known to be NP-hard. Overcoming the size limitation of current pattern discovery techniques based on conventional graph theory turns out to be even more significant. Finally, the algorithm is demonstrated to be not only a general, useful data-mining tool but also an effective method for analysis of protein structure, and function.

In recent years, molecular biologists have been devoting their efforts on the analysis of protein structure commonality. It is of great interest for a number of reasons. The detection of common structural patterns (or, motifs) between proteins may reveal the functional relationships. Moreover, the results of Jones and Thirup [1] have indicated that the three-dimensional structure of proteins can often be built from substructures of known proteins. In other words, the mining of protein motifs may in fact hold the key to the question of how proteins fold into unique and complicated 3D structures. The understanding of the 'protein folding' problem will further contribute to the design of new and more effective drugs with specific 3D structures.

A number of automated techniques have been developed for this purpose. Rosmann et al. pioneered the technique of superimposing two proteins. Approaches using variations of structure representation, similarity definition, and optimization techniques have been deployed [2,3,4]. Most representative among these techniques include those of Grindley et al.[3], and Holm and Sander [4]. Grindley et al. pre-processed the protein tertiary structures into a collection of coarser representations, the secondary structures, then performed maximal common subgraph matching on the resultant representations. Holm and Sander discarded the notion of secondary structure, and, instead, pre-segmented the proteins into fixed-length patterns. Then a Monte Carlo random walk algorithm is used to locate large common segment sets.

All the aforementioned techniques are subject to artificially imposed constraints, such as using super-secondary structures and fixed length segmentation, which could damage the optimality of the solution. This paper presents a new maximal common sub-graph algorithm that overcomes those limitations.

2. Protein Common Substructure Discovery by MCS

Similar 3D protein structures have similar inter-residue distances. The most often used inter-residue distance is the distance between residue centers, i.e. C^α atoms. By using the inter-C^α distance, the similarity can be measured independent of the coordinates of the atoms.

The similarity of two proteins P_1 and P_2 tertiary structures can be defined as,

$$S = \sum_{i=1}^{M} \sum_{j=1}^{M} \phi(i, j)$$

(1)

where M is the number of matched C^α atom pairs from P_1 and P_2, and $\phi(i, j)$ is a similarity measure between the matched pair i and j, which is defined as a threshold step function that outputs 1 when $d_{threshold} - |d_{P_1}(i, j) - d_{P_2}(i, j)| \geq 0$, otherwise 0. This removes any contribution of unmatched residues to the overall similarity.

Definition 1: The **Protein Common Tertiary Substructure (PCTS) Problem** is defined as that of maximizing similarity measure of eq. (1), seeking the maximum number of matched C^α atom pairs satisfying the distance measure.

2.1 Maximal Common Subgraph Approach

In recent years, graph matching algorithms have been liberally used to perform protein structure analysis (such as the work of Grindley et al. [3]).

Definition 2: A *graph* $G(V,E)$ is defined as a set of *vertices* (nodes), V, together with a set of *edges*, E, connecting pairs of vertices in V ($E \subseteq V \times V$). A *labeled graph* is one in which labels are associated with the vertices and/or edges.

The protein structures can be easily represented as labeled graphs. For the purpose of PCTS problem, proteins are considered labeled graphs with vertices being the

C^α atoms, and edges labeled with the C^α-to-C^α distances between the vertices. Then the largest common substructures between two proteins is simply the *maximal common sub-graph* (MCS) *isomorphism* problem:

Definition 3: Two graphs, G_1 and G_2, are said to be *isomorphic* if they have the same structure, i.e. if there is a one-to-one correspondence or match between the vertices and their (induced) edges. A *common sub-graph* of G_1 and G_2, consists of a sub-graph H_1 of G_1, and a subgraph H_2 of G_2 such that H_1 is isomorphic to H_2.

The flexibility allowed by the similarity measure can be easily incorporated into this graph theoretical approach for solving the PCTS problem. For example, the angle or bond rigidity in the protein geometry could be relaxed. The similarity measure, then, only needs to allow a looser edge label and the distance.

2.2 Transforming to Maximum Clique Problem

Brint and Willett [5] performed extensive experiments in the 80's and concluded that the MCS problem can be solved more effectively in the maximum clique domain, which can be done by using the following transformation.

Definition 4: A *clique* is a complete graph. The *Maximum Clique* (MC) Problem is to find the clique with the maximum number of nodes in a given graph.

[Transforming from MCS to MC] Barrow et al. [6] gave a transform to convert MCS into the MC problem by the following procedures:

Given a pair of labeled graphs G_1 and G_2, create a *correspondence graph* C by,
1) Create the set of all pairs of same labeled nodes, one from each of the two graphs.
2) Form the graph C whose nodes are the pairs from (1). Connect any two node pairs $N_1(A_i, B_x)$, $N_2(A_j, B_y)$ in C if the labels of the edges from A_i to A_j in G_1 and B_x to B_y in G_2 are the same.

Solving the MCS problem becomes that of finding the maximum clique of C and then map the solution back into a MCS solution by the inverse transformation.

3. Algorithms

3.1 Exploiting the Relations Between MC and Graph Coloring

Both problems of maximal common subgraph and maximum clique are NP-hard. Numerous MC algorithms have been developed over the years. However, their solution quality tends to vary significantly from test case to test case, mainly because they are mostly heuristic algorithms trying to solve a multi-dimensional optimization problem with local optima "traps". Another NP-hard problem of graph coloring (GC) is tightly coupled with MC in an iterative loop aiming to converge to the optimal solution of either problem, or in many cases both. In this section, only the most relevant parts to the MC-GC solver are included, leaving other details in [7]. The algorithmic framework of the MC-GC solver is shown in Figure 1.

Given a graph $G(V,E)$, the relation between MC and GC is fundamentally expressed by the following well-known theorem:

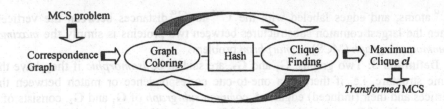

Fig. 1. Algorithmic Framework for CRP-MCS

***Theorem 1*:** Given the size of the maximum clique, $\omega(G)$, and the chromatic number, $\lambda(G)$, then $\omega(G) \leq \lambda(G) \leq (\Delta+1)$, where Δ is the maximum degree of G.[8]

With the chromatic number and maximum clique size bounding each other, it provides a perfect termination condition for the loop process shown in Figure 1. If such situation occurs, then the optimal solutions for both problems are solved simultaneously.

In order to devise a set of heuristics for clique-finding and graph coloring, the following definitions and theorems are utilized.

Definition 5: Given a coloring for **G**, the *color-degree* of vertex v_i, cdeg(v_i), is defined as the number of different colors of the adjacent nodes, the *color-vector* of vertex v_i, cv(v_i), is defined as the set of colors that v_i can use not conflicting with the colors assigned to its adjacent nodes.

Lemma 1: For any set V of vertices, let the size of the maximum clique that includes S be $\omega(G \mid V)$. Then, $\omega(G \mid V) \leq (\text{cdeg}(V) + |V|)$. (Proof omitted here.)

Definition 6: A set $D \subset V$ is defined to be dominant if $\forall v \in (V \setminus D), \exists u \in D \to (u, v) \in E$. Given a complete coloring C for **G**, for any color c, $\{v \mid c \in cv(v)\}$ forms a set of dominant S_c vertices. The color that corresponds to the smallest S_c is called the (***minimal***) *dominant color*.

Assuming that the graph has uniform probability for the edge connection, then the probability of a vertex in any dominant set can be derived as follows,

***Theorem 2*:** Given a random graph G_n^p (V,E), where the graph size is n, the edge probability for each pair of vertices $e(u,v) = p$, and a specific maximum clique is ω, for a complete coloring C for **G**, if the *minimal dominant vertex set* is S_c, then $\forall v \in S_c$, the probability that v belongs to a clique of size ω is

$$\left(\frac{p}{1-(1-p)^{|S_c|}}\right)^{|\omega|-1} \cdot p^{\binom{\omega-1}{2}}.$$

Therefore, selecting a vertex from the smallest dominant vertex set means a higher probability for it to be in the maximum clique. This then underlies the strategy of using a GC solution as an initializing "seed" for the MC computing process.

Definition 7: When coloring a graph, the *color reduction* of node v_i is defined as the process of removing colors from $cv(v_i)$ in conflict with the colors of all of its neighbors.

Graph coloring is generally accomplished by sequentially assigning colors to uncolored vertices. The risk of such sequential decision process is that once a vertex is colored with color c when there is more than one choice, due to color reduction, the adjacent vertices are forced to use the other available colors. Consequently, the coloring solution could be misdirected away from the optimal due to premature color decisions. The *color reduction* process is used in this work precisely to prevent premature commitments in the effort of achieving minimal coloring.

Definition 8: A *Saturated Clique (SC)* is defined as a clique cl whose size is equal to the union of all node color vectors, i.e., $| \underset{v \in cl}{\cup} cv(v) | = | cl |$.

3.2 Solving MCS via MC (and GC)

Based on the observation of the close relations between graph coloring and maximum clique problems, a complementary algorithm, CRP-MCS, that combines graph coloring and clique-finding algorithms is designed to solve the maximum clique problem. A resource management methodology [9], called *Constrained Resource Planning* (CRP), provides the guiding principles and motivates the solution strategies for both the coloring and clique-finding processes of the iterative loop. Solution from one, and its derived information, is used to initialize the counterpart process, and execute alternatingly until a solution is found. Such an initialization process is called '*seeding*' in this work. Each sub-algorithm terminates upon completion of its targeted goal and then hands over the result to the other. The entire iterative process terminates when certain criteria are met, and the maximal clique solution is transformed into a MCS solution.

3.2.1 Clique-Finding Algorithm

Each graph coloring process produces different color distribution. Since our clique-finding algorithm relies on the coloring information, the coloring result C comes from previous coloring process naturally becomes the seed for clique-finding. The set of nodes that use the dominant color is set to be the seed, or pivot vertices, for the clique-finding process, and large cliques are sought in $Nbr(v)$ for each pivot v.

In addition, for any clique in the graph, each color contributes at most one vertex. Moreover, once a vertex is chosen to add into a temporary clique, vertices that do not connect to it have to be disregarded, thus may result in some colors being disregarded without contributing any vertex. Therefore, it is highly desirable to preserve as many colors as possible during the process of searching for a clique. Similar to the principle of selecting the pivot vertices above, the color that contains fewest vertices is chosen. Then within the selected color, the vertex v that has highest color degree is selected and added into the temporary clique.

The clique-finding algorithm is summarized as follows.

Algorithm *CLIQUE-FINDING* (input: coloring C, largest clique found cl^0)

1: Let B_U = UpperBound(G, C). If $| cl^0 |$ = B_U, terminate.
2: Locate dominant color c, set Pivot node set P

$= \{v \mid cv(v) \geq (B_U - 1), v \in \text{color } c', |c'|=|c|\}$

3: For each p in P,
   Set $G' = \{v \mid v \in \text{Nbr}(p)\}$. cl=NULL, set tmp-cl= [p]
   While |tmp-cl| > 0,
      bcdeg = MaxCDEG(G')

   select c that $MIN\{v \mid v \in G \land cv(v) = c \land cdeg(v) \geq (bcdeg-1)\}$

   If ties, select c that $MAX \sum_{v,cv(v)=c} \deg(v)$

   Pick node v from c that MAX(cdeg(v))

   If ties, pick v that $MAX |\{e(u,w) \mid u, w \in Nbr(v)\}|$

   If InHash([v, tmp-cl]), select another node
   Set tmp-cl = [v, tmp-cl], add tmp-cl into HASH

   Set $G' = G' - \overline{Nbr(v)}$
   If |G'| = 0, Call BackTrack()

3.2.2 Coloring Algorithm

As discussed earlier, color reduction (CR) plays an active role in the GC algorithm. It not only helps to reduce the solution space by removing conflicted colors from the color vectors, but also assists to reach the chromatic number and decide the convergence of the solution.

Theorem 4: **[Coloring Lower Bound B_L]** Given a graph G, assume that a coloring initializes with k colors. If during performing pure color reduction, there is any, (1) node with zero size cv, or (2) clique cl and $m = \left| \bigcup_{1 \text{ to } |cl|} v_i \ (v_i \in cl) \right| < |cl|$, then G needs at least $(k+1)$ colors $(k+m-|cl|)$ if (2) is the case.

Moreover, since $\omega(G) \leq \lambda(G)$, the lower bound for any graph coloring process would be the maximum of the B_L derived from previous coloring process and the largest clique that was found in earlier clique-finding process.

Because that the largest clique found in earlier clique-finding processes may in fact be the new lower bound for graph coloring, it is treated as the 'seed' for a new coloring process. Specifically, for the largest clique cl with size k, each vertex in cl is assigned with a unique color from 1 through k.

In order to perform color reduction by using the concept of saturated cliques (SC), a set of cliques needs to be identified. Since the proposed algorithm is an iterative process, all the cliques found by clique-finding process and stored in the hash can be utilized. The more cliques collected, the higher chance of more SC's for color reduction, thus postpones unnecessary forced coloring. This could lead to the use of fewer colors for coloring the entire graph. A supplement algorithm designed for this purpose is described in [11].

For any state of the coloring, vertices that have smaller color vectors tend to have less chance of being assigned colors on them. In order to avoid overuse too many colors, it is critical to process these vertices as earlier as possible. Therefore, these vertices are regarded as the most constrained tasks to be accomplished. Meanwhile, although there is already fewer choices than the others for coloring these vertices, careless assigning color would still result in overuse of colors. Thus, each color from the color vector needs to be examined to determine which would have least impact on the rest. The impact is evaluated by the reduction of color vector sizes of uncolored neighbor set, then the color with least impact would be assigned to the vertex. The complete algorithm is shown below.

Algorithm *COLORING* (input: graph **G**, hash memory: **HASH**)

1. Set B_L =MAX(B_L, MAX(HASH))
2. $\forall v \in G$, assign v a color vector cv(v)={1 ... B_L}
3. Color the largest untried clique cl in with 1~|cl|
4. Perform color reduction, and update B_L by Theorem 4
5. while there is a node uncolored,
   Select a vertex v with MIN(|cv(v)|)

   If ties, select v that $MIN\sum_{c \in cv(v)}|\{u\,|\,u \in Nbr(v) \wedge c \in cv(u)\}|$

   Select $c \in cv(v)$ that $MIN\,|\{u\,|\,u \in Nbr(v) \wedge c \in cv(v)\}|$

   If ties, select c that $MAX\sum_{u}|cv(v)|$ after CR.

   Perform color reduction.

3.3 Generic Pre-Processing/Dynamic Accessing Strategy for Large Problems

Although the MC solution provides an effective means for solving MCS, the $O(m^2n^2)$ space requirement for storing the adjacent matrix for the connection information is simply too large for applications like the protein common tertiary substructure problem. Discarding the adjacent matrix and re-compute the adjacency between vertices could alleviate the space consumption, however, it still requires $O(m^2n^2)$ computation time for visiting all the possible connections, which is now taking more time since it needs to be recalculated.

A generic pre-processing/dynamic accessing technique is developed in this work to handle such situation. For the convenience of discussion, it is described for the protein substructure problem, but it can be extended to a more general context easily.

Assume that P_1 of size m and P_2 of size n are the two proteins to be explored. The dominant subroutine and needs to be repeatedly performed during the computation is to finding the adjacent vertex pairs. To be more specific, given that v_1 in P_1 and v_2 in P_2 is paired (matched), it is crucial to determine which vertex pair in the correspondence graph is compatible with vertex pair (v_1, v_2). Namely, to find out all vertex pairs (u_i, u_j), $u_i \in P_1, u_j \in P_2 \ni |d(v_1, u_i) - d(v_2, u_j)| \le d_{threshold}$. The complexity for a specific vertex pair alone is $O(mn)$, and grows to $O(m^2n^2)$ if the connections for all vertex pairs need to be re-computed. When the problem is small enough to fit in the primary memory space, this can be done by simply a table look-up at the adjacency matrix for the correspondence graph. However, the space consumption is too expensive for the protein common tertiary problems.

Instead of searching through the *mn* vertex pairs repeatedly, the complexity can be reduced by pre-sorting all vertices in P_2 with respect to each vertex in P_2 in ascending order in terms of edge labels (distances). The complexity of the sorting can be done with $O(n^2 \log n)$, which needs to be done only once, with the cost of additional $O(n^2)$ space to the storage for the protein itself, rather than the expensive $O(m^2 n^2)$. Each time to find the adjacent vertex pairs for vertex pair (v_1, v_2), it can be simply done dynamically with the following algorithm:

Algorithm *DynamicFindAdjacentPairs* (input: v_1, v_2)

1. Set $L = \varnothing$, S=sorted_list(v_1)
2. For each vertex v ($v \neq v_1$) in P_1,
   d=distance(v_1, v), $l = d - d_{threshold}$, $u = d + d_{threshold}$
   (s, e)=RangeSearch(l, u, S)
   For $i = s$ to e, $L = \{L \mid \text{index}(v_2, I)\}$. Return L

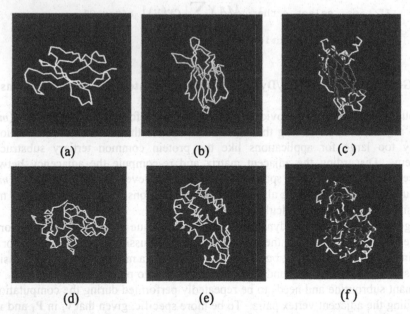

(a)   (b)   (c)

(d)   (e)   (f)

Fig. 1. (a) The backbone view of Synaptotagmin (1rsy), and **(b)** Fibronectin Type III domain (1fna) **(c)** The similar structures between (a) 1rsy and (b) 1fna after alignment. (71 pairs matched with r.m.s.d. = 1.742632 Angstrom) Red : Similarity from (a) 1rsy, Blue : Similarity from (b) 1fna . **(d)** The backbone view of Hen egg-white lysozyme (1lyz), and **(e)** T4 phage lysozyme (2lzm) **(f)** The similar structure structures between (d) 1lyz and (e) 2lzm after alignment. (106 pairs matched with r.m.s.d. = 3.923571 Angstrom) Red: Similarity from (e) 1lyz, Blue : Similarity from (f) 2lzm

4. Experiments

A set of protein files [6] frequently referenced in the molecular biology literature is selected to test the CRP-MCS algorithm. The protein sizes range from 108 to 497 C^α atoms. They are Hen egg-white lysozyme (1lyz), T4 phage lysozyme (2lzm), actinoanthin (1acx), superoxide dismutase (1cob), tumor necrosis factor (1tnf), methylamine dehydrogenase

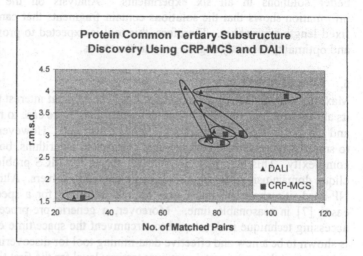

Fig. 3. Comparison of the protein common tertiary substructure discovery between using DALI and proposed CRP-MCS. Each circled area represents one of the six experiments.

(2mad), defensin (1dfn), neurotoxin (1sh1), fibronectin cell-adhesion module type III-10 (1fna), and synaptotagmin I (1rsy). The program is implemented in C and run on an SGI Onyx(R10000) machine using single processor. Each experiment on a given pair of proteins is set with 10 min. time limit, and the best result produced is used to align the two proteins and derive the error measurement.

The solution quality is measured with (a) number (N) of matched C^α atom pairs,

and, (b) the root-mean-square deviation (r.m.s.d.) which is defined as $\sqrt{\dfrac{\sum_{i=1}^{N} d_i^2}{N}}$,

where d_i is the distance between the *i-th* pair of C^α atoms. The results are compared against those from the DALI web-server (http://www2.embl-ebi.ac.uk/dali/), which is an implementation of Holm and Sander's work (4). The time cut-off of 10 minutes for structures with a homolog in the representative set seems to be sufficient.

Two typical alignment results are shown in Figure 2, where the optimally aligned protiens using the discovered largest common structures are shown. The number of matched C^α atom pairs and the r.m.s.d. values are also included.

There are totally six experiments conducted. They are (1) 1dfn vs. 1sh1, (2) 1fna vs.1rsy, (3) 1acx vs.1cob, (4) 1acx vs.1tnf, (5) 1acx vs. 2mad, and (6) 1lyz vs. 2lzm. The results obtained by submitting to DALI server and by the CRP-MCS algorithm are plotted in Figure 3, where each experiment is shown in a circled area. Since the DALI server usually provides only one solution for each submission, the comparison is made by setting corresponding threshold parameters in the CRP-MCS solution such that either the no. of matched pairs or the r.m.s.d. as close to the DLAI one as possible. As shown in the comparison chart, the CRP-MCS algorithm resulted in better solutions in all six experiments. Analysis on the corresponding pair information shows that the solutions contain fragments that can not be detected by fixed-length approaches. Therefore, this tool is expected to provide more flexibility and optimality for analyzing the protein structures.

5. Discussions

Maximal common sub-graph approach has been of great interest to many areas due to its ability to extract the largest common substructure, the ease to mapping the problem and the flexibility to incorporating various constraints. However, it has been limited to small-size problems due to the lack of efficient algorithms, both in space and time complexity. This algorithm, CRP-MCS, solves the MCS problem in the maximum clique domain using MC and GC as complementary solvers. Although the problem is NP-hard, it's shown to reach near optimal solutions for a spectrum of benchmark cases [7] in reasonable time. Moreover, a generic pre-processing and dynamic-accessing technique is developed to circumvent the space/time overhead. The result is shown to be a new and effective data-mining tool for discovering the large common substructures between proteins at pure tertiary level for the first time.

The tool allows fully free matching among the C^α atoms for the protein problem. As demonstrated in the experiments, it has the capability to find near-optimal common substructures that are not possibly to be detected by conventional techniques that use pre-defined patterns or segments. Thus it provides the molecular biologists more flexibility to discover the common tertiary substructures.

Structural similarity is an important yet difficult data-mining problem. The CRP-MCS algorithm presented here is shown to successfully bring the graph-based approach to an important real-world problem and is expected have more applications in assisting the discovery of new knowledge in other related areas also [7].

References

1. T.A.Jones and S. Thirup, "Using Known Substructures in Protein Model Building and Crystallography," *EMBO J.*, vol. 5, no. 4, pp. 819-822, 1986.
2. G. Vriend and C. Sander, "Detection of common three-dimensional substructures in proteins," *Proteins*, vol. 11, pp.52-58, 1991.
3. H. M. Grindley, P. J. Artymiuk, D. W. Rice, and P. Willett, "Identification of Tertiary Structure Resemblance in Proteins Using a Maximal Common Subgraph Isomorphism Algorithm," *J. Mol. Biol.* Vol. 229, pp. 707-721, 1993.
4. L. Holm and C. Sander, "Protein Structure Comparison by Alignment of Distance Matrices," *J. Mol. Biol.*, vol. 233, pp. 123-138, 1993.
5. A. T. Brint and P. Willett, "Algorithms for the Identification of Three-dimensional Maximal Common Substructures." *J. Chem. Inform. Comput. Sci.* v27, pp. 152-158, 1987.
6. H. G. Barrow and R. M. Burstall, "Subgraph Isomorphism, Matching Relational Structures and Maximal Cliques," *Information Processing Letters*, vol. 4, pp. 83-94, 1976.
7. Chao-wen Chen, Algorithms for Maximal Common Subgraph Problem Using Resource Planning, Ph.D. Dissertation, University of Hawaii, 1999.
8. B. Bollbas, *Extermal Graph Theory*, Academic Press, 1978.
9. Kent, N. P. and D. Y. Y. Yun, "A Planning/Scheduling Methodology for the Constrained Resource Problem," *Proceedings of the Eleventh International Joint Conference on Artificial Intelligence*, (1989), 20-25.

Integrating Classification and Association Rule Mining : A Concept Lattice Framework

Keyun Hu   Yuchang Lu   Lizhu Zhou   Chunyi Shi
Computer Science Department, Tsinghua University,
Beijing 100084, P.R.China
hky@s1000e.cs.tsinghua.edu.cn, {lyc, dcszlz, scy}@tsinghua.edu.cn

Abstract Concept lattice is an efficient tool for data analysis. In this paper we show how classification and association rule mining can be unified under concept lattice framework. We present a fast algorithm to extract association and classification rules from concept lattice.

1. Introduction

Concept lattice, also called Galois lattice, was first proposed by Wille[1]. A node of concept lattice is a formal concept, consisting of two parts: the extension (examples the concept covers) and intension (descriptions of the concept). Concept lattice gives a vivid and concise account of relations (generalization /specialization) among those concepts through Hasse Diagram.

Classification rule mining and association rule mining are two important data mining techniques. There are already some classification systems based on concept lattice. Empirical evaluation shows that concept lattice based systems have comparable performance with those typical systems such as C4.5 [5]. Association rule mining is a hot research topic in data mining recently. Some authors have shown that concept lattice is a nature framework for association rule mining [4]. In this paper we would show that concept lattice is an appropriate tool for integrating association and classification rule mining. Some author also discussed the topic [2]. But we argue that concept lattice embodies the relationships between concepts in a more understandable way. Therefore it is very interesting dealing with the task under the context of concept lattice.

2 Basic Notions of Concept Lattice

In this section we recall necessary basic notions of concept lattice briefly. the detail description can be found in [1].

Suppose given the context (O, D, R) describing a set O of objects, a set D of descriptors and a binary relation R, there is a unique corresponding lattice structure, which is known as *concept lattice*. Each node in lattice L is a pair, noted (X, Y), where $X \in P(O)$ is called *extension* of the concept, $Y \in P(D)$ is called *intension* of concept. Each pair must be complete with respect to R. i.e.:

(1) $X = \{x \in O \mid \forall y \in Y, yRx\}$; (2) $Y = \{y \in D \mid \forall x \in X, yRx\}$。

A partial order relation can be built on all concept lattice nodes. Given $H_1 = (X_1, Y_1)$ and $H_2 = (X_2, Y_2)$, let $H_1 < H_2 \Leftrightarrow Y_1 \subset Y_2$, the precedent order means H_1 is a direct parent of H_2. The Hasse diagram of the lattice can be generated using the partial order relation. If $H_1 < H_2$ and there is no other node H_3 such that $H_1 < H_3 < H_2$ there is an edge from H_1 to H_2.

Below is an example of context and corresponding lattice and Hasse diagram.

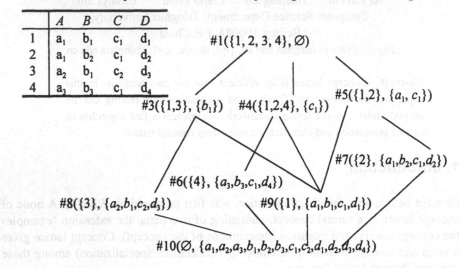

	A	B	C	D
1	a_1	b_1	c_1	d_1
2	a_1	b_2	c_1	d_2
3	a_2	b_1	c_2	d_3
4	a_3	b_3	c_1	d_4

$\#1(\{1,2,3,4\}, \varnothing)$

$\#5(\{1,2\}, \{a_1, c_1\})$

$\#3(\{1,3\}, \{b_1\})$ $\#4(\{1,2,4\}, \{c_1\})$

$\#7(\{2\}, \{a_1,b_2,c_1,d_2\})$

$\#6(\{4\}, \{a_3,b_3,c_1,d_4\})$

$\#8(\{3\}, \{a_2,b_1,c_2,d_3\})$ $\#9(\{1\}, \{a_1,b_1,c_1,d_1\})$

$\#10(\varnothing, \{a_1,a_2,a_3,b_1,b_2,b_3,c_1,c_2,d_1,d_2,d_3,d_4\})$

In our algorithm a node in lattice is denoted by $(C=|X|, Y)$, as content of X does not matter. Now it's easy to see if C is bigger than some threshold t, X is a maximal large item set.

Implication rules can be derived from concept lattice. Rule $Q \Rightarrow R$ holds if and only if the smallest concept (intent) containing Q also contains R[6].

3 Building the Lattice

In order to reduce the number of nodes in lattice, it is necessary to introduce a support threshold. We adapt Bordat's algorithm [3] by introducing a support threshold ε and making other minor improvement. Because lattice-constructing algorithm only find only maximal itemsets, hence they're much faster than Apriori algorithm.

The lattice L is initialized with topmost node $(|O|, \varnothing)$ and expanded by constructing its subnode recursively. In the algorithm we use an array of pointer PX to keep track of first appearance of all single attr-val pair in the lattice. This structure will be used later in the association rule mining. Once the support of a node is found lower than ε, the node will no longer be expanded. We improved the original algorithm by utilizing counting information. That is, instead of checking whether extensions of two attr-val pair set are identical, we check whether the count of either extension is equal to the count of extension of the union of the two attr-val pair set. Experiments show that it is about five time faster than original algorithm (when ε is set to 0). The lattice built by this algorithm is in fact a "frequent" lattice, i.e. it contains only those nodes whose support is greater than ε. Thus the algorithm reduces the complexity of building the complete lattice.

4 Rule extraction from the lattice

In this section we present an algorithm which generates all non-redundant rules for a given item set (set of attr-val pair) as right hand side (RHS). We first find the smallest node containing the item set, then launch a breadth-first traverse to its sub-lattice. From each node we generate all non-redundant rule. Because the way we build the lattice, the support of all the rule generated are greater than minimal support. We first produce all rule whose confidence is 1 (i.e. implication rules) and then appropriate lower confidence rules. The way we generate (implication) rule relies on following observations.

Observation 1 If a node $H=(C, X)$ has only one parent node $P=(C', X')$, then
(1) The left hand side(LHS) of rules generated from H consists of a single item.
(2) For each attr-val $p \in \{X-X'\}$, there is a rule $p \Rightarrow X-p$.

Suppose the LHS of a rule generated from H consists of more than one item (attr-val). If all these item are also included in X', they must have been already treated in parent P because of our top down traverse fashion; if there exists an item $p \in \{X-X'\}$ in LHS, the rule is redundant with respect to $p \Rightarrow X'-p$, the latter is simpler.

Observation 2 If a node $H=(C, X)$ has d parent nodes $P_1(C_1, X_1)$, $P_2(C_2, X_2)$, $\cdots$, $P_d(C_d, X_d)$, there is a rule $p \Rightarrow X-p$ for each item $p \in \{X-(X_1 \cup Y_2 \cup ... \cup X_d\}$.

Because any item $p \in \{X-(X_1 \cup X_2 \cup ... \cup X_d\}$ is the first time appearing in the lattice, it is obvious its confidence is 1. Any rules whose LHS strictly include p is redundant with respect to $p \Rightarrow X-p$.

Observation 3 If a node $H=(C, X)$ has two parent nodes $P_1(C_1, X_1)$, $P_2(C_2, X_2)$, $\forall p_1 \in \{X_1-X_1 \cap X_2\}$ and $\forall p_2 \in \{X_2-X_1 \cap X_2\}$, there is a rule $p_1 p_2 \Rightarrow X-p_1 p_2$

That is because if there are two items coming from the same parent, their relationship must have been described before. So Any rules whose LHS strictly include $p_1 p_2$ would be redundant with respect to $p_1 p_2 \Rightarrow X-p_1 p_2$.

Observation 3 can be generalized to the case of any number of parent nodes.

In the algorithm, we adopt a heuristic search strategy. If an item set can not form a implication rule, it is saved in a candidate set. In the next loop, all items in the candidate set are joined in an Apriori-like manner. Then new candidates are tested against whether they can form an implication rule.

As to rules whose confidences are below 1, we use a data structure PX (see previous section) to aid computing confidence. PX points to first appearance of every single element of LHS. Function PX(lhs) finds the first appearance of LHS and thus its support. If LHS of such rule is included in another rule, it will be discarded since longer LHS rule have higher confidence.

The computation depends heavily on judging whether several elements are included in a common parent. We introduce a bit vector V to do the judgement efficiently. Every element in the node has a bit vector. If the element also appears in a parent, the corresponding bit will be set. Thus any combination of those elements can be judged by simple and fast AND operations.

Rule Extraction Algorithm for specific RHS
1.   **GenRule**(itemset rhs)

2.   Find first node H containing rhs by breadth-first traverse
3.   queue←H, ruleset←∅, singleset←∅
4.   while queue not empty
5.   Remove H from queue head. push all children of H into queue tail
6.   if H not visited   then { GenRuleFromNode(H); mark H as visited}
7.   endwhile
8.   **GenRuleFromNode(H=(C,X))**
9.   d←number of parents of H: $(C_1,X_1)...(C_d,X_d)$
10.   if (d= =1)   {ruleset=ruleset∪{p→rhs | p∈{X-X$_1$}};return;}
11.   for every parent of H compute their union S and generate array V.
12.   ruleset=ruleset∪{p→rhs, | p∈{X-S}};
13.   ruleset=ruleset∪{p→rhs, conf=‖PX(p)‖/‖H‖ | p∈S, not sameparent(p, rhs), p not generated before}
14.   L←{S1∪S2| S1∈S, S2∈S}
15.   while L not empty
16.   S←∅
17.   for every element K in L
18.   if each item in K aren't included in same parent and not R∈ruleset that R⊆K
19.   ruleset=ruleset∪{K→rhs, sup=C/‖O‖, conf=1}
20.   else
21.   S←S∪K
22.   if not sameparent(K,rhs) then ruleset=ruleset∪{K→rhs, conf=‖PX(K)‖/‖H‖ }
23.   endif
24.   endfor
25.   L←{S1∪S2| S1∈S, S2∈S, ‖S1∪S2‖=‖S1‖+1}
26.   endwhile
27.   return ruleset;

Line 14-27 generates rules whose LHS contain more than one item in an Apriori-like manner. The algorithm is written according to above observations. The rules generated are sorted by confidence (larger to smaller). If confidence is same, higher support would be first. The classification is done by matching the new instance against every rule from begin to end. If no rule fires, then the majority class is used. When building the lattice, class attribute values are treated as an ordinary attribute and are added to the lattice. The rule extraction algorithm is run a number of times by assigning every class attribute value as parameter value. Then all rules are collected together performing the classification.

5 Experiments and Conclusions

We implement our algorithm and do the comparison using MLC++. First we did some preliminary test on lattice constructing algorithm and found it much faster than Apriori. This is because the algorithm produces only maximal large item sets. Thus the comparison between them doesn't seem to have much meaning.

In this section we mainly present the result of comparing C4.5 and our algorithm. We use 10 datasets form UCI Repository for the comparison. In our experiment, minimum confidence is set to 0.6 and minimum support is set to 0.01. Our experiment is done on a PC with 64Mb PII 233 running windows 98. Our algorithm is referred as CLACF (Concept Lattice based Association and Classification rules mining Framework). The discretization is done using entropy method in MLC++.

Datasets	C4.5	CLACF	Rule time	No. of rules	Lattice time
Breast	5.0	3.8	20.1	307	24.6
Diabetes	25.8	27.8	2.14	35	0.9
Glass	31.3	18.9	0.88	49	0.82
Heart	19.2	16.6	3.51	528	11.1
Iris	4.7	4.0	0.0	19	0.0
Led7	26.5	23.7	0.83	278	8.0
Monk1	19.0	9.0	0.22	127	2.5
Monk2	30.1	19.9	0.38	169	4.3
Monk3	8.3	5.1	0.28	113	2.4
Pima	24.5	27.1	1.0	25	0.4
Average	18.5	15.6	2.93	165	5.43

Column 2 and column 3 show error rates of C4.5 and CLACF. CLACF outperforms C4.5 in 8 out of 10 datasets, and has an average error rate of 15.6, which is lower than 18.5 of C4.5. Column 4 to 6 give the execution time of the two algorithms and the number of rule generated respectively. We can see the algorithm produces relatively smaller set of rules comparing with [2] while retaining accuracy.

In this paper we propose a framework to integrate classification and association rule mining based on concept lattice. We adapt an existing lattice constructing algorithm to generating a 'frequent' lattice and present an efficient algorithm to produce association/classification rules from the lattice. In our future work, we will focus on developing faster algorithm by further exploring the relationship stored in the concept lattice.

Acknowledgments This research has been supported by Natural Science Foundation of China and National 973 Fundamental Research Program.

References

1. Wille, R. Reconstructing Lattice Theory: an Approach Based on Hierarchies of concepts. in *Ordered sets*, Reidel(1982).
2. Liu, B. Hsu, H. Ma, Y. Integrating classification and association rule mining. *kdd-98*, New York, Aug 27-31 (1998)
3. Njiwoua, P., Nguifo E.M.: Forwarding the choice of bias LEGAL-F: using feature selection to reduce the complexity of LEGAL. in *Proceedings of BENELEARN-97,ILK and INFOLAB*, Tiburg University, The Netherlands (1997) 89-98
4. Hu,K., Lu, Y., Shi, C.: Incremental discovering association rules: a concept lattice approach. In *Proceedings of PAKDD-99*, Beijing (1999) 109-113
5. Sahami, M.: Learning classification rules using lattices. In *Machine learning:ECML-95*. Grete, Greece (1995)343-346
6. Godin, R., Missaoui, R.: An incremental concept formation approach for learning from databases. *Theoretical Computer science*, 133(1994)387-419

Using Rough Genetic and Kohonen's Neural Network for Conceptual Cluster Discovery in Data Mining

Hoang Kiem, Do Phuc

University of Natural Sciences, HCM city
Department of Information Technology
227 Nguyen Van Cu St., District 5, HCM city, Vietnam
Email: hkiem@htco.com.vn

Abstract. We consider the problem of discovering the conceptual clusters from a large database. From Z. Pawlak's information system in rough set theory, we define an information matrix, information mappings and some concepts in data mining literature such as large sets, association rules and conceptual cluster. We propose a combined method of information matrix, Kohonen's neural network for large set discovery and genetic algorithm for conceptual cluster validity. We present an application of our method to a student database for discovering the rules contributing to the training of the gifted students.

1 Introduction

Data Mining (DM) is to discover the interesting patterns present implicitly in large database [7]. In this paper, we study the problem of conceptual cluster discovery from a large database. This problem is stated as: given a set of objects, conceptual clustering discovery is to find clusters of objects based on a conceptual closeness among objects [1],[2],[3],[4]. We proposed a method for solving and expanding this problem. Based on Z. Pawlak's information system [9], we define an information matrix and some concepts then we employ a combined Kohonen's self-organizing algorithm (SOA) and Genetic algorithm for conceptual cluster discovery and building rules from these discovered concepts. We build an information matrix in the computer memory for improving the speed of mining process. The paper is organized as follows. Section 1: Introduction. Section 2: Formal definitions. Section 3: Problem statement. Section 4: Using SOA for discovering large descriptor sets. Section 5: Using GA for cluster validity. Section 6: An application to a student database. Section 7: Conclusions and future works.

2 Formal definitions

In this section, we define an information matrix and some concepts related to our proposed method. Based on these definitions, we implement a set of functions for processing the mining tasks in the computer memory instead of scanning the whole database in disk. Therefore, we can improve significantly the speed of mining process.

2.1 Definition 1: Information matrix

Information matrix is defined as $B=(O,D)$ where $O=\{o_1,...,o_n\}$ is a finite set of n objects and $D=\{d_1,...,d_m\}$ is a finite set of m descriptors. Let b_{ij} (i=1,...,n and j=1,...,m) be the element of matrix B, $b_{ij}=1$ if o_i has d_j, otherwise $b_{ij}=0$.

2.2 Definition 2: Information mappings

Given a finite set O of n objects and a finite set D of m descriptors [5]. Let P(D) be a power set of D, P(O) be a power set of O. Information mapping χ is defined as: $\chi: D \rightarrow \{0,1\}$. Given $o \in O$ and $d \in D$, $\chi(o,d) = 1$ if o has d, otherwise $\chi(o,d)=0$. Mappings ρ and λ are defined as: $\rho: P(D) \rightarrow P(O)$ and $\lambda: P(O) \rightarrow P(D)$ where:

Given $S \subseteq D$ then $\rho(S) = \{o \in O: \forall d \in S, \chi(o,d)=1\}$
Given $X \subseteq O$ then $\lambda(X) = \{d \in D: \forall o \in X, \chi(o,d) = 1\}$

2.3 Definition 3: Large descriptor set

Given an information matrix $B=(O,D)$ and a threshold τ which is the MINSUP of the large item set in data mining literature[7]. A large descriptor set S is a subset of D that satisfy condition: $Card(\rho(S))/Card(O)>=\tau$, where Card is the cardinality of set.

2.4 Definition 4: Binary association rule

Given an information matrix $B=(O,D)$ and a threshold τ. Let S be a large descriptor set of B. Let L_i, L_j be the subsets of S. A binary association rule with threshold τ is a mapping from L_i to L_j and is denoted as $L_i \rightarrow L_j$.

2.5 Definition 5: Confidence factor of a binary association rule

Let S be a large descriptor set of B, L_i, L_j be the subsets of S, $L_i \rightarrow L_j$ be a binary association rule with a threshold τ. The confidence factor $CF(L_i \rightarrow L_j)$ of this rule is calculated by $Card(\rho(L_i) \cap \rho(L_j)) / Card(\rho(L_i))$.

2.6 Definition 6: Concept

A concept is a pair $C=(X,S)$ where $X \subseteq O$ and $S \subseteq D$. X and S satisfy following conditions:
a) $X \subseteq \rho(S)$ and $\lambda(X) = S$
b) $\forall L_i, L_j \subseteq S$ and $Card(L_i) = Card(L_j) = 1$ then $\rho(L_i) \subseteq \rho(L_j)$.

3 Problem statement

Problem 1: Given an information matrix B and a threshold τ, find all large descriptor sets of B. The large descriptor set determines the popular descriptors of data objects. The threshold τ determines a measure of popularity [7].

Problem 2: Given an information matrix B and a threshold τ, find k conceptual clusters $C_1,...,C_k$ where $C_j = (X_j, S_j)$. These conceptual clusters satisfy: a) $\cap X_i = \varnothing$ for i=1,...,k ; b) $\cap S_i = \varnothing$ for i=1,...,k ; c) $Card(X_i)/Card(O) >= \tau$; d) Maximize the ratio $Card(X_1 \cup ... \cup X_k)/Card(O)$ e) C_i is a concept. Conceptual cluster determines an object set that has the same set of descriptors. Based on the concept $C=(X,S)$, we build rule $L_i \rightarrow L_j$ where $L_i \cup L_j = S$ and $L_i \cap L_j = \varnothing$. It means that if object has all the descriptors of L_i (rule antecedent) then object has all the descriptors of L_j (rule consequent).

4 Using SOA for discovering large descriptor sets

In this section, we employ SOA for discovering the potential large descriptor sets [6]. SOA can be summarized as follows:

Step 1. Initialize all weight vectors of Kohonen's neural network
Step 2. Select the node with minimum distance d_v to the input vector $v(t)$.
Step 3. Update weight vectors of nodes that lie within a nearest neighbor set
of the node (i_o, j_o): $w_{ij}(t+1) = w_{ij}(t) + \alpha(t)(v(t)-w_{ij}(t))$
for $i_o-N_c(t) <= i <= i_o+N_c(t)$ and $j_o-N_c(t) <= j <= j_o+N_c(t)$
Step 4. Update time $t = t+1$, add new input vector and go to (Step 2)

In the above algorithm, d_v is Euclidean distance, $\alpha(t)$ is a gain ratio $(0<=\alpha(t)<=1)$ and $N_c(t)$ is the radius of neighbor set. $N_c(t)$ and $\alpha(t)$ are decreased monotonically with time. The algorithm finishes when $\alpha(t) =0$ or $N_c(t)=0$.

Given an information matrix in table 1, each row of this matrix corresponds to an input vector of Kohonen's neural network.

Table 1. An information matrix for large descriptor set discovery.

	d1	d2	d3	d4	d5	d6
o1	1	1	1	0	0	0
o2	1	1	1	0	0	0
o3	1	1	1	1	0	0
o4	0	0	1	1	1	1
o5	0	0	0	1	1	1
o6	0	0	0	1	1	1

After running SOA, we have the potential large descriptor sets:

$\{d_1, d_2, d_3\}$, $\{d_4, d_5, d_6\}$, $\{d_1, d_2, d_3, d_4\}$. With $\tau=50\%$, $\{d_1, d_2, d_3\}$, $\{d_4, d_5, d_6\}$ are large descriptor set; $\{d_1, d_2, d_3, d_4\}$ is not a large descriptor set because $Card(\rho(\{d_1, d_2, d_3, d_4\}))/Card(O)=33.3\% < \tau$.

5 Using GA for cluster validity

Large descriptor sets discovered by SOA are used for building the initial GA population. We hold that the subset of a large descriptor set is also a large descriptor set [7]. Let $L=\{L_1,...,L_k\}$ be a set of k large descriptor sets, we employ GA[8] for finding a set $\{S_1,...,S_k\}$ where $S_i \subseteq L_i$ (i=1,...,k) and $(S_i,\rho(s_i))$ is a concept. A chromosome is a set of BS_i, each BS_i is a bit string corresponding to a large descriptor set. With two large descriptor sets $\{d_1, d_2, d_3\}$ and $\{d_4, d_6\}$, we have chromosome $\{d_1:1, d_2:1, d_3:1, d_4:1, d_5:0, d_6:1\}$. The genetic representation of population P is a set of chromosomes. A typical population P with 3 chromosomes is as follows: $P(t)= \{111111, 100111,001100\}$. The genetic operations are defined as:

5.1 Crossover operator

Given two parental chromosomes: $\{a_1, a_2, a_3, a_4, a_5, a_6\}$ and $\{b_1, b_2, b_3, b_4, b_5, b_6\}$ where $a_i, b_i \in \{0,1\}$(i=1,...,6). The crossover will swap a portion of two parental chromosomes and yield the offspring: $\{a_1, a_2, a_3, b_4, b_5, b_6\}$ and $\{b_1, b_2, b_3, a_4, a_5, a_6\}$.

5.2 Mutation operator

Given a chromosome $\{a_1, a_2, a_3, a_4, a_5, a_6\}$. Select a random position $h \in [1..6]$. Let h be the selected position, if $a_h = 1$ then a_h is changed to 0 and vice versa.

5.3 Fitness factor and fitness value

Fitness factor S_{ij}: Let S_{ij} be a subset of chromosome BS_i, we build set Q containing all two-element subsets of S_{ij}. Let $\{a, b\}$ be an element of Q. From $\{a, b\}$, we build two rules $\{a\} \rightarrow \{b\}$ and $\{b\} \rightarrow \{a\}$ and calculate the CFs of these rules. The Fitness factor of S_{ij} is the average of CF_s of $2 \times Card(Q)$ rules which are built up from Q.
Fitness value of a chromosome BS_i is the average of fitness factor of all S_{ij} in chromosome BS_i.

6 An application to a student database

We employ our proposed method for discovering the conceptual clusters from a student database. An information matrix with 1000 rows and 100 columns is built up from this database. In this matrix, each row corresponds to a record and each column corresponds to a descriptor. Some descriptors of the information matrix are "parent of student are teachers"; "student is ranked in good level of learning"; "student wins a prize of computer science competition".

The size of Kohonen's output layer is 100x100. With the threshold $\tau=0.7$ (70%), we discover some large descriptor as {student wins a prize of a math competition; student is interested in math}; {student is ranked in good level of learning,; parents of student are teachers}; {student is interested in math; student is interested in foreign language; student is interested in computer science}.

We employ the following values for GA parameters: number of chromosomes is 50; number of generations is 300; crossover probability is 0.1; mutation probability is 0.1. The GA give us some discovered conceptual clusters as {student is ranked in good level of learning; student has good behavior; parents of student are teachers; Student has the self-learning time greater than 6 hours every day}; {student is interested in math; student is interested in foreign language; student is interested in computer science}; {student lives in country; income of student family is lower than $100 every month; student is ranked in fair level of learning}.

7   Conclusions and future works

We gathered some preliminary result in using a combined information matrix, GA and SOA for cluster discovery in data mining. The experiment shows very encourage in large data set. A matrix expressed in bit is also used for keeping the whole information matrix in main memory to increase the efficiency of conceptual cluster discovery. We continue to study how to change binary information matrix to fuzzy information matrix and use fuzzy cluster discovery for the fuzzy database.

References
1. Bezdek, J.C.: Cluster validity with fuzzy sets. In J. Cybernetics, Vol. 3, No 3, (1974)
2. Eui-Hong Han: Hyper-graph based clustering in high dimensional data sets, Data Engineering, IEEE, March (1998), 15-22
3. Hoang Kiem, Do Phuc: A combined multi-dimensional data model, self-learning algorithm and genetic algorithm for cluster discovery in data mining- Proc of the PAKDD'99 conference, Bejing, China, (1999), 54-59
4. Hoang Kiem, Do Phuc: Using data mining in education, Magazine of science and technology development, Vol. 1, No. 4, VNU-HCM, Vietnam, (1999), 28-37
5. Ho Tu Bao: Automatic unsupervised learning on Galois lattice, Proceedings of IOIT conference, Hanoi, Vietnam, (1996), 27-43
6. L.P.J Veelenturf: Analysis and application of Artificial Neural Networks, Prentice hall, (1995), 182-214
7. Pieter Adrians, Dolf Zantige: Data Mining, Addison Wesley, Longman, 1996
8. Zbigniew Michalewicz: Genetic Algorithms + Data Structures = Evolution Programs, Springer Verlag, (1992), 13-67
9. Z.Pawlak: Data Mining-A Rough Set Perspective, Methodologies for Knowledge Discovery and Data Mining, Springer Verlag, (1999), 3-12

Towards Automated Optimal Equity Portfolios Discovery in a Knowledge Sharing Financial Data Warehouse

Yi-Chuan Lu[1] and Hilary Cheng[2]

[1] Department of Information Management, Financial Data Mining Laboratory,
Yuan Ze University, Chung-Li 320, Taiwan, R.O.C.
imylu@saturn.yzu.edu.tw
[2] Department of Business Administration, Financial Data Mining Laboratory,
Yuan Ze University, Chung-Li 320, Taiwan, R.O.C.
hilary@saturn.yzu.edu.tw

Abstract. We propose a knowledge discovery process for multi-factor portfolio management on a financial decision support system. We first construct an OPen Intelligent Computing System (OPICS) to support time series management and knowledge management. A system, Cyclone, which efficiently supports financial applications, is developed under the OPICS. We then introduce a data mining solution for equity portfolio construction using the simulated annealing algorithm. Two data sets consist of small stocks ranging from 11/86 to 10/91 and from 6/93 to 5/96 are used. The corresponding rates of return of Russell 2000 index are collected as benchmarks for evaluation based on the Sharpe ratios and • the turnover ratios. The result shows that the simulated annealing algorithm outperforms both the market index and the gradient maximization method.

1 Introduction

Competition in the investment business is intense and increasing. Historically, investment managers who actively select stocks employ a team of analysts who understand various industries, visit companies, and utilize quantitative techniques to learn important information to help them recommend which stocks to own. While the Internet technology and data availability are rapidly changing, quantitative techniques also become more sophisticated. The data warehousing and data mining techniques are thus acquired across the financial services as well as the banking industry.

We propose a knowledge discovery process for multi-factor portfolio management on a financial decision support system. The process consists of two construction phases: the knowledge management functionality, and the data mining solution. We first construct an OPen Intelligent Computing System (OPICS) to support time series management and knowledge management. A system, Cyclone, is developed under the OPICS. The Cyclone is designed to allow power users to adjust parameters, simulate portfolios easily and efficiently, and eventually, to share knowledge with other power users. We then introduce a data mining solution for equity portfolio construction using the simulated annealing algorithm. Two data sets consist of small stocks ranging from 11/86 to 10/91 and from 6/93 to 5/96 are used. The corresponding rates of return of Russell 2000 index are collected as benchmarks. The evaluation is based on the Sharpe ratios as well as the turnover ratios. The result shows that the simulated annealing algorithm outperforms the market index as well as the gradient maximization method.

2 A Motivating Example

The process of constructing a portfolio involves defining a universe of stocks, dealing with data integrity issues, selecting an appropriate construction technique, determining the way of using training and testing data, setting up the construction rules and constraints, and selecting a publicly available benchmark for evaluating the test results. The whole process is known as the back-testing model in the financial investment society. Schock and Brush in [1] show an example in managing a small-stock portfolio, which exploits the above process. The first step is to construct a sequence of monthly universe, by ranking a database on market capitalization, then eliminating the largest 1,000 stocks and using the next 600 excluding limited partnerships, investment trusts and stocks with very low trading volumes.

The second step is to include values and other proven measures to the universe. The common factors are earnings to price ratio, book value to price ratio, cash flow to price ratio, volatility adjusted price momentum, etc. The rest of work is to apply a decision model to construct monthly portfolios through time and compute the corresponding returns. The resulting returns of portfolios are then used to compare with those of the selected benchmarks. To optimize the flow of information and "knowledge-worker-to-knowledge-worker" interaction so that companies can make better trading decisions, the specific data and modeling results should then be shared and managed. This is the essence of knowledge management.

3 Knowledge Management

A successful knowledge management means that our data processes enhance the way people work together, enable knowledge workers and partners to share information easily so that they can build on each other's ideas and work more effectively and efficiently. Though data for financial applications are simple data, the data typically includes time series. The empirical research based on time series thus is a data intensive activity that needs a knowledge management system with data and time modeling capabilities, computational intelligence and performance functionalities (see Schmidt and Marti [2], Dreyer, Dittrich, and Schmidt [3]).

455

We develope a time-series management system, OPICS, specifically for security investment research. The OPICS is a component-based system enhanced with distributed processing capability. To effectively manage the data sets derived by security firms' knowledge workers, a time series data set is organized into two parts, the header and the data. The header is a meta-data about the time series basis and its derived data sets. We also develop a time series management system (TSMS), Cyclone, under the OPICS platform to enable knowledge workers in a security research firm to share their idea and models with others. See Lu and Cheng in [4] and [5] for the inter-modules, the system architecture, as well as other functionalities.

4 Data Mining for Multi-factor Portfolio Construction

Building a multi-factor excess return model to select a portfolio stocks becomes a widely used tool for portfolio management. We consider the portfolio construction model as a global optimization problem. To control risk throughout a portfolio construction process, we consider the model's objective function as to maximize the return over risk. The optimization algorithms to be examined in this paper are the gradient maximization method (see Brush [6]) and the simulated annealing technique (see kirkpatrick [7]).

We use the Sharpe excess return (see Sharpe [8]) to illustrate the concerns of portfolio managers who seek for high returns with low risks. We apply the simulated annealing algorithm in conjunction with the above integrated return/risk portfolio model. Since the values of the Sharpe excess return can be either positive or negative, we write the dynamic rule in a symmetrical form $E = 1/(1 + e^{SharpeExcessReturn})$, which is referred as an energy measure. For each given temperature, the thermal equilibrium can be described by a probability distribution function with respect to the occurrence of a state with energy E, that is,

$$P(E) = (1/Z(T))e^{(-E/(K_B*T))}, \tag{1}$$

where $Z(T)$ is a normalized constant, E is the energy of the state, $P(E)$ is the probability of finding a unit in that state, T is the temperature, and K_B is the Boltzmann's constant. Let ΔE be the change of the energy E. The traditional hill-climbing algorithm only accepts ΔE when it is less than zero, while the simulated annealing algorithm allows positive ΔE being accepted with probability greater than 0. The process of the simulated annealing involves three steps: the generation of a new state, the acceptance criterion of the new state, and the condition of the cooling schedule, which will be briefly discussed as follows.

Generating a new state for the purpose of portfolio optimization is to explore the current state region around the current weighting combination. By sequentially changing each factor's weight up and down a bit from the original weighting combination, local search determines a combination of factor changes that identifies a best direction of improving the portfolio return. Having established a direction of local search, we next make a bigger step to weight changes

Table 1. Portfolio construction using data during 11/86 - 10/91

Data Set (11/86 - 10/91)	Year(s)	Portfolio Return	Becnmark Return	Excess Return	Sharpe Ratio	Turnover Ratio
GM	1	-11.31%	-18.04%	6.73%	0.24	8.14%
	2	6.45%	3.21%	3.24%	0.18	8.01%
	3	11.73%	8.64%	3.09%	0.18	7.48%
	4	1.61%	-0.08%	1.69%	0.09	6.75%
	5	8.02%	6.97%	1.05%	0.05	6.16%
SA	1	-9.95%	-18.04%	8.09%	0.32	8.97%
	2	7.21%	3.21%	4.00%	0.26	9.20%
	3	13.06%	8.64%	4.42%	0.23	10.43%
	4	2.01%	-0.08%	2.09%	0.11	11.31%
	5	8.64%	6.97%	1.67%	0.09	12.06%

in the portfolio. When ΔE is less than zero, we accept the new state with probability 1. If ΔE is greater than zero, then we accept the new state with the probability $e^{-\Delta E/T}$, where T represents the current temperature.

The annealing schedule is also critical to the performance of the algorithm. Without any prior knowledge of energy landscapes, one can only hope to derive an appropriate cooling schedule for a specific random process. As being a cooling schedule of the temperature $T(t)$, it must be able to decrease from a given sufficiently high temperature T_0 down to a zero degree. To experiment an initial temperature, we first let the system be free running with a 100% acceptance rate for a certain number of iterations. By sampling all energy states, we can calculate the standard deviation to estimate an initial temperature. We then implement the cooling schedule using the formula, $T(t) = T_0/(1 + t)$, where T_0 is the initial temperature. The equilibrium detection during a particular temperature is measured by the formula, $|T_{est} - T_c| < 0.01 * T_c$, where T_{est} is the estimated temperature from N samples and T_c is the current temperature (see Szu [9]).

5 Implementation and Results

We focus on the universe, which is designed to capture the attractive long-term return potential associated with small stocks. We purchase a stock when it ranks above 10% in any economic sectors. A stock will be sold when it ranks below the 30% by economic sector. The round-trip trading cost is 3.6%. The total number of stocks in our portfolio is between 50 and 60. We also choose two widely used value measures and two widely used growth measures to form an integrated return/risk portfolio. The value measures are earning to price ratio and book value to price ratio, whereas the growth measures are short-term (four quarters) earnings change to price ratio and long-term (three-year) earnings change to price ratio.

We collect two data sets: one from 11/86 to 10/91 and the other from 6/93 to 5/96. The corresponding rates of return of Russell 2000 index are collected as

Table 2. Portfolio construction using data during 6/93 - 5/96

Data Set (6/93 - 5/96)	Year(s)	Portfolio Return	Becnmark Return	Excess Return	Sharpe Ratio	Turnover Ratio
GM	1	36.51%	4.41%	32.10%	1.06	8.67%
	2	19.31%	11.97%	7.34%	0.26	7.74%
	3	20.62%	15.81%	4.81%	0.23	6.88%
SA	1	36.11%	4.41%	31.70%	1.06	7.96%
	2	20.92%	11.97%	8.95%	0.35	8.41%
	3	24.03%	15.81%	8.22%	0.32	9.10%

benchmarks. The results in Tables 1 and 2 show that the simulated annealing algorithm (SA) outperforms both the gradient maximization (GM) and market index in all time periods. Also, the longer the time period, the lower the Sharpe ratio, that indicates the risk is proportional to the time period. Although the simulated annealing has superior rates of returns, it has higher turnover ratios.

References

1. Schock, V.K., Brush, J.S.: Capturing Returns and Controlling Risk in Managing a Small-Stock Portfolio. Small Cap Stocks Investment and Portfolio Strategies for the Institution Investor, Chapter 13 (1993) 295-326
2. Schmidt, D., Marti, R.: Time Series, A Neglected Issue in Temporal Database Research? Recent Advances in Temporal Databases, Workshops in Computing Series, Springer (1995) 214-232
3. Dreyer, W., Dittrich, A., Schmidt, D.: An Object-oriented Data Model for A Time Series Management System. Proceedings of the 7th International Working Conference on Scientific and Statistical Database Management (SSDBMS'94), Charlottesville, Virginia. (Sept. 1994) 186-195
4. Lu, Y.-C., Cheng, H., Hsu, C., Jung, M.: Financial Decision Support System: A Distributed and Parallel Approach. Proceedings of 1999 Workshop on Distributed System Technologies and Applications, Taiwan (1999) 425-431
5. Lu, Y.-C., Cheng, H.: A Time Series Management System for Financial Decision Support: Models, Techniques, and Implementations. Technical Report, OPICS - Financial Data Mining Lab., Yuan Ze University, Taiwan (1999)
6. Brush, J.S., Schock, V.K.: Gradient Maximization: An Integrated Return/Risk Portfolio Construction Procedure. The Journal of Portfolio Management (summer 1995) 89-98
7. Kirkpatrick, S., Gelatt, C.D., Vecchi, M.P.: Optimization by Simulated Annealing. Science **220** (May 1983) 671-680
8. Sharpe, W.F.: Mutual fund performance. Journal of Business **39:1, Part 2** (January 1966) 119-138
9. Szu, H., Hartley, R.: Nonconvex Optimization by Fast Simulated Annealing. Proceeding of IEEE **75:11** (November 1987)

Rule-Evolver: An Evolutionary Approach for Data Mining

Carlos Lopes*, Marco Pacheco**, Marley Vellasco** Emmanuel Passos*

*Departamento de Engenharia Elétrica, Pontifícia Universidade Católica Rio de Janeiro
Rua Marquês de São Vicente, 225, Gávea, Rio de Janeiro, RJ, CEP. 22453-900, BRASIL

{ch|marco|marley|emmanuel}@ele.puc-rio.br

*Dep. de Engenharia e Sistemas e Computação, Universidade do Estado do Rio de Janeiro
Rua São Francisco Xavier, 524, Maracanã, Rio de Janeiro, RJ, CEP. 20550-013, BRASIL

This paper presents a genetic model and a software tool (*Rule-Evolver*) for the classification of records in Databases (DB). The model is based on the evolution of association rules of the IF-THEN type, which provide a high level of accuracy and coverage. The modeling of the Genetic Algorithm consists of the definition of chromosomes representation, the evaluation function, and the genetic operators. The *Rule-Evolver* is a tool that provides an environment for the evaluation of the genetic model and implements the interface with DBs. The case studies evaluate the performance of the model in several benchmark DBs. The results obtained are compared with those of other models, such as Artificial Neural Nets, Neuro-Fuzzy Systems and Statistical Models.

1. Evolutionary Data Mining Systems

Genetic Algorithms (GAs) have been successfully used in optimization problems [1] and some data mining models can be found in the literature [2]. In the context of GAs, classification consists of the evolution of association rules.

The quality of the rules evolved is measured through their Accuracy and Coverage. The accuracy of an association rule, IF C THEN P, measures the rule's degree of confidence (Equation 1).

$$Acuracy = \frac{|C \cap P|}{|C \cap P| + |C \cap \overline{P}|} \tag{1}$$

Rule's coverage may be interpreted as the comprehensive inclusion of all the records that satisfy the rule. Equation 2 represents the definition of rule's coverage.

$$Coverage = \frac{|C \cap P|}{|C \cap P| + |\overline{C} \cap P|} \tag{2}$$

2. Genetic Algorithm Modeling for Data Mining

The genetic algorithm consists of 4 main components: Chromosome Representation, Evaluation Function, Genetic Operators and Initialization of the Population.

In chromosome representation, categorical or discrete attributes represent a finite value-set, or mapped values, within a set of integers. Quantitative or continuous attributes represent value ranges in the attribute domain.

Thus, a chromosome must represent an association rule by means of the value range of the predictive, quantitative and categorical attributes (Fig. 1).

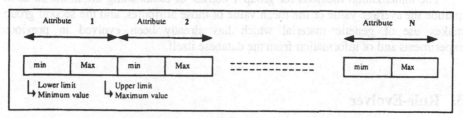

Fig. 1. Representation of the chromosome for the classification task by a vector of 2N values.

In Fig. 1, each gene has two real numbers: the minimum and the maximum values. This representation makes it possible to formulate rules such as:

IF {((Attr. **1** ∈ [mim **1**, max **1**]) *and* (Attr. **2** ∈ [mim **2**, max **2**]) *and* ... *and* (Attr N ∈ [mim **N**, max **N**])}*THEN* **Target Attribute = P**

where min. **X** and max. **X** indicate the minimum and maximum values of each predictive attribute that has been defined as quantitative (for categorical attributes, only min. X is used), and P is the value of the Target Attribute, which has been identified as an objective in the initial phase of the process.

The advantage of this representation lies in its high level of comprehensibility and in the fact that the domain of the attributes values are evolved as real numbers.

Several evaluation functions were implemented and tested with three types of rewards: accuracy, coverage, and/or both [3], according to Equation 3.

$$if\ ((accuracy\ (i) \neq 0)\ and/or\ (coverage\ (i) \neq 0))\ then \tag{3}$$
$$fitness(i) = f(i)*accuracy\ (i)*coverage\ (i);$$

Among the functions (ten) that were implemented, the Cbayesian [3] function is worthy of note. The Cbayesian function is inspired by the Bayesian classifiers [4] and represents the product of the probabilities that the values of a rule's attributes pertain to an interval, given that the class of the current record is the one that has been specified as the objective. Equation 4 presents this function, where A_i is an attribute, a_i is a value interval, C is the target attribute, and c is the value of the specified class.

$$P(A_1 = a_1| C = c) * P(A_2 = a_2| C = c) * ... P(A_k = a_k| C = c) \tag{4}$$

Several genetic operators have been tested in this project: one-point, two-point, average, and uniform crossovers, simple mutation, and a new mutation operator called "don't care". This operator eliminates a given attribute in the composition of the rule, i.e., the entire domain of the attribute is considered valid for the rule.

Methods to constrain the Genetic Algorithm's search space or to introduce promising solutions were implemented to initialize the population, thus increasing its performance:

1. Random seedless initialization methods: a) including the average value of this attribute; b) including the median value of this attribute;
2. Random initialization methods with seeds: a) Seeds from previous evolutions; b) Seeds from random Database records.

The initialization methods for group 1 consist of establishing the limits so as to include the average value or the mean value of those attributes, and the second group makes use of genetic material which has already been evolved in previous experiments and of information from the database itself.

3. Rule-Evolver

The Rule-Evolver is a data mining environment which incorporates a dedicated genetic algorithm model to evolve classification rules.

The *Rule-Evolver* is capable of extracting all the association rules which differentiate a specific record cluster from the other records in a database. It comprises 4 dedicated modules:

- *Selection of database attributes* → Allows the user to choose attributes of interest based on each attribute's average, variance, and variation coefficient;
- *Interpretation* → Presents the best rules found in the *IF (A1 and A2 and A3 and ... An) THEN P* format;
- *Graphic Previewing* → Plots the accuracy, coverage, and fitness value (evaluation function) graphs of the individuals in the course of genetic evolution;
- *Parameterization of the environment* → Allows the user to specify rates, parameters, evaluation functions, operators, and evolutionary techniques.

4. Case Studies

The benchmark databases used in this study were obtained from "ftp://ftp.ics.uci.edu/pub/machine-learning-databases/" repository. Two case studies are summarized in this article: Iris Plants Database and Tic-Tac-Toe Endgame Database. These databases were divided into 2 sets: a training set and a test set.

The **"Iris Plants Database"** comprises 150 records divided into 3 classes of 50 records each (Iris Setosa, Versicolour, and Virginica), with 4 attributes: the plant's petal and sepal width and length.

In Table 1, the classification results obtained by the *Rule-Evolver* are compared with those obtained by means of other techniques: the NEFCLASS Neuro-Fuzzy System; the Hierarchical Neuro-Fuzzy System (NFHQ), the Bayesian Neural Net with the gaussian approximation method, and a Bayesian Neural Net with the Markov Chain Monte Carlo method (MCMC).

Table 1. Iris and Tic-Tac-Toe database as benchmark

IRIS DATABASE (Error In The Training And Test Set - %)			TIC-TAC-TOE DATABASE (Error In The Test Set - %)	
MODEL	TRAINING	TEST	MODEL	TEST
NEFCLASS [5]	2,67	4	NewId [6]	84
NFIIQ [7]	2,67	2,67	CN2 [6]	98,1
RNB (BNN) [8]	0	2,67	MBRTalk [6]	88,4
RNB (MCMC) [8]	0	2,67	IB3-CI [6]	99,1
RULE-EVOLVER (GA)	0	13,34	RULE-EVOLVER (GA)	97,6

The results indicate that, though the **Rule-Evolver** obtained good results in the training set, it presented a low performance level in the test set. Of the 10 records reported as errors (13,34%), 3 records were wrongly classified between the *Iris Versicolour* and *Virginica* classes, and 7 records were **not** classified, i.e., they were not covered by any of the rules obtained during training. This problem is typical of models that use training and test sets: the model does not succeed in finding rules whose attribute values are not in the training set.

The second database tested was the **"Tic-Tac-Toe Endgame database"**, which encodes the complete set of possible final configurations of the game board under the assumption that 'x' has made the first move. It is composed of 958 records, of which 626 are "x wins" with 9 attributes, each corresponding to a position on the game board.

Table 2 presents the decoding of the rules found by the **Rule-Evolver** which lead "x to win", where symbol # (don't care) means that "x wins" regardless of the values contained in that position of the board. Symbol "b" represents a blank field, and symbol "o" represents the other player.

Table 2. Decoding of the rules found by the *Rule-Evolver* for *Tic-Tac-Toe*

Rule 1			Rule 2			Rule 3			Rule 4		
bx	#	#	#	#	bx	x	#	#	#	x	#
#	bx	#	#	bx	#	x	xo	#	#	x	#
#	#	bx	bx	#	#	x	#	#	#	x	#
216 Records			153 Records			41 Records			39 Records		

Rule 5			Rule 6			Rule 7			Rule 8		
#	#	x	x	x	x	#	#	#	#	#	#
#	#	x	#	#	#	x	x	x	#	#	#
#	#	x	#	#	#	#	#	#	x	x	x
46 Records			43 Records			39 Records			44 Records		

This example shows that 621 of the 626 (99,2%) records were covered by the rules evolved by the *Rule-Evolver* and 5 records were not covered on account of the Rule 3, which specialized in position 5 on the board (xo).

The results indicate that 4 records have been classified as being characteristics of "x's" victory, though they are not, and 3 records that lead to "x wins" have not been classified. The model therefore fits 280 of the 287 records of the test set, thus presenting a success rate of 97,6 %.

5. Conclusions

Record classification through the evolution of associative rules with the use of Genetic Algorithms has proved to be a promising procedure for characterizing the record clusters in a database. When compared with other methods (Artificial Neural Nets, Statistical Models), the advantage of rule discovery by means of GAs is that the rules evolved are self-explanatory.

The current model is incapable of evolving correctly the interval of a rule's attribute values for values which are not exemplified in the training set. However, the genetic model is capable of generating highly accurate rules with a high level of coverage without any conflicts for the intervals present in training.

To evaluate one generation of chromosome rules, the *Rule-Evolver* makes a single pass over the data, which may be all available in the memory, thus reducing processing time. The *Rule-Evolver* automatically tries to load the whole database into memory; if not possible, it will access the data through the DBMS.

For the databases tested (hundred of records of less than ten attributes), the *Rule-Evolver*'s run time was satisfactory - about 6 minutes in average on a Pentium II 350 Mhz for the evaluation of 80 generations of populations of 100 individuals each.

The scalability of the genetic model in applications with large databases (say with millions of records) has not been accessed yet.

References

1. D. Dasgupta, Z. Michalewicz, Evolutionary Algoritms in Engineering Applications, (Eds.) Springer, 1997.
2. N. J. Radcliffe (Project Manager) – GA-MINER: Parallel Data Mining with Hierarchical Genetic Algorithms Final Report, University of Edinburgh, 1995.
3. Carlos Henrique Pereira Lopes, "Classificação de Registros em Banco de Dados por Evolução de Regras de Associação utilizando Algoritmos Genéticos", Dissertação de Mestrado, Departamento de Engenharia Elétrica, PUC-Rio, Abril 1999, "www.ica.ele.puc-rio.br".
4. Shen W. M. Bayesian probability theory – A general method for machine learning MCC-Carnot-101-93. Microelectronics and Computer Technology Corporation, Austin, TX, 1993.
5. Rudolf Kruse, Detlef Nauck. "NEFCLASS-A Neuro-Fuzzy Approach for the Classification of Data", Proc. of the 1995 ACM Symposium on Applied Computing, Nashville, 1995.
6. Aha, D. W. Incremental constructive induction: An in stance-based aproach. In Proceedings of 8th International Workshop on Machine Learning, 117-121. Evanston, ILL: Morgan Kaufmann, 1991.
7. Flávio Joaquim de Souza. "Modelos Neuro-Fuzzy Hierárquicos", Tese de Doutorado, DEE, PUC - Rio. Abril 1999, "www.ica.ele.puc-rio.br".
8. Edison Américo Huarsaya Tito. "Aprendizado Bayesiano para Redes Neurais". Dissertação de Mestrado, DEE, PUC – Rio. Março 1999, "www.ica.ele.puc-rio.br".

Decision Making with Probabilistic Decision Tables

Wojciech Ziarko

Computer Science Department

University of Regina

Regina, Saskatchewan, S4S 0A2, Canada

Abstract. The paper is concerned with the decision making with predictive models acquired from data called probabilistic decision tables. The methodology of probabilistic decision tables presented in this article is derived from the theory of rough sets. In this methodology, the probabilistic extension of the original rough set theory, called variable precision model of rough sets, is used. The theory of rough sets is applied to identify dependencies of interest occurring in data. The identified dependencies are represented in the form of a decision table which subsequently is analyzed and optimized using rough sets-based methods. The original model of rough sets is restricted to the analysis of functional, or partial functional dependencies. The variable precision model of rough sets can also be used to identify probabilistic dependencies, allowing for construction of probabilistic predictive models. The main focus of the paper is on decision making aspect of the presented approach, in particular on setting the parameters of the model and on decision strategies to maximize the expected gain from the decisions.

1 Introduction

Standard decision tables are tabular models of the functional dependencies between input conditions and decisions or actions taken in response to the occurrence of some combinations of conditions. They have been used in software engineering, circuit design and other application areas for years [6]. The dependency is encoded by the table designer in the form of a set a disjoint decision rules covering all possible input situations. However, in many problems related to decision making with uncertainty, machine learning, pattern recognition and data mining, the condition-decision dependency is typically unknown and almost always non-deterministic. Often, it is hidden in empirical data. A number of analytical methodologies have been developed in recent years to approximate this kind of the dependency for the purpose of prediction or better understanding of the nature of the relationship, for example, by using decision trees, neural networks or rough sets [3,5, 9-12].

In this paper, we will focus on using decision tables extracted from data for that purpose. The research into decision tables acquisition from data was initiated by Pawlak in the context of rough sets theory[1,2]. His original works were concerned with the acquisition of deterministic, or partially deterministic tables.

We demonstrate how an extended approach, called variable precision rough sets model (VPRS), can be applied to acquistion of non-deterministic decision tables with probabilistic characterization of their decision accuracy [8]. In what follows, the review of the methods of rough sets for the above mentioned data-based modeling problem is presented and illustrated with simple examples. A comprehensive discussion of the optimal decision making strategies and parameter setting for the model is also included. Generally, the objective is not to construct a predictive system which would guarantee always correct predictions (which is typically impossible) , but to have a system which would support decisions with sufficient success rate in the longer run, or sufficient expected gain or profit from the decision making.

The paper is organized as follows. We first discuss the basics of the formal model of decision tables acquired from data. Then, the main definitions of the variable precision model of rough sets are introduced. In the next sections, they are used to define extended notions of the dependency between attributes, of the extended reduct and core attributes. A separate section is devoted to the discussion of the optimal decision making with the probabilistic decision tables.

2   Decision Tables Acquired from Data

Generally, the decision table is defined here as a tabular representation of a relation discovered in data. The relation is identified through a classification process in which data objects having the same values of selected attributes, or having the same values of properly selected functions of the attributes (for example, using some attribute value discretization technique), are considered to be identical. It should be noted, however, that this kind of the decision table does not necessarily represent functional relationship as it is the case with "classical" decision tables known in software engineering and other areas. More precisely, the data-extracted decision table is defined as follows:

Let U be the universe of objects $e \in U$ and $a \in A$ be the attributes of the objects, that is functions $a : e \rightarrow a(e)$ assigning some features (attribute values) to objects. We assume that every attribute maps into a finite set of values, $v_a \in range(a)$. The attributes are divided into two categories, condition attributes $C = \{a_1, a_2, ..., a_m\}$ and the decision attributes D. Typically, the condition attributes represent measurable properties of objects whereas decision attributes are the "predictive" attributes (variables) whose values are normally predicted based on known values of condition attributes. We will assume here, without loss of generality, that there is only one binary-valued decision attribute $d \in D$ and one value v_d^i ($i = 0$ or 1) of this attribute has been selected as a prediction or modeling "target". With all these assumptions, the decision table can be expressed as a quadruple $< U, C, d, v_d^i >$. Each of the two values v_d^0, v_d^1 of the decision attribute d corresponds to a set of objects matching that particular value. We will denote these sets as X^0 and X^1 respectively. Clearly, $X^0 = \neg X^1$ and $X^0 \cup X^1 = U$. Our objective in the construction and analysis of the decision tables is to develop a simple predictive model for the target set which would

CLASS	S	H	E	C	$P(T = 1\|E_i)$	$P(T = 0\|E_i)$
E_1	0	0	1	0	0.10	0.90
E_2	1	0	2	1	0.85	0.15
E_3	1	1	1	0	0.01	0.99
E_4	0	2	1	1	1.00	0.00
E_5	1	2	1	0	0.82	0.18
E_6	1	0	1	0	0.12	0.88
E_7	1	2	2	1	0.92	0.08
E_8	0	0	2	1	0.91	0.09

Table 1. Classification by condition attributes only

enable us to predict, with an acceptable confidence, whether an object matching a combination of attribute values occurring in the decision table belongs to the target set, or to its complement.

3 VPRS Model of Rough Sets

In data mining and predictive modeling applications the variable precision model of rough sets (VPRS) was used for analysis of decision tables extracted from data. The VPRS model extends the capabilities of the original model of rough sets to handle probabilistic information. The main aspects of the VPRS model are presented below.

Let R be an equivalence relation (called the indiscernibility relation) and let R^* be the set of equivalence classes of R. Typically, the relation R represents the partitioning of the universe U in terms of the values of condition attributes as defined in Section 2. Also, let $E \in R^*$ be an equivalence class (elementary set) of the relation R. With each class E we can associate the estimate of the conditional probability $P(X|E)$ by the formula: $P(X|E) = card(X \cap E)/card(E)$ assuming that sets X and E are finite. This situation is illustrated in Table 1 which represents the classification of raw data in terms of condition attributes S,H,E,C, with each class E_i being assigned probabilities $P(T = 1|E_i)$ and $P(T = 0|E_i)$.

Let $0 \leq l < u \leq 1$ be real-valued approximation *precision control* parameters called lower and upper limits respectively. For any subset $X \subseteq U$ we define the *u-positive* region of X, $POS_u(X)$ as a union of those elementary sets whose conditional probability $P(X|E)$ is not lower than the upper limit , that is

$$POS_u(X) = \bigcup \{E \in R^* : P(X|E) \geq u\}$$

The *u-positive* region of X represents an area in the universe which contains objects with relatively high probability of belonging to the set X.

The *(l,u)-boundary* region $BNR_{l,u}(X)$ of the set X with respect to the lower and upper limits ℓ and u is a union of those elementary sets E for which the conditional probability $P(X|E)$ is higher than the lower limit ℓ and lower than the upper limit u. Formally,

CLASS	S	H	E	C	E_i REGION
E_1	0	0	1	0	NEG
E_2	1	0	2	1	POS
E_3	1	1	1	0	NEG
E_4	0	2	1	1	POS
E_5	1	2	1	0	POS
E_6	1	0	1	0	NEG
E_7	1	2	2	1	POS
E_8	0	0	2	1	POS

Table 2. The probabilistic decision table with $u = 0.8$ and $l = 0.2$

$$BNR_{l,u}(X) = \bigcup \{E \in R^* : l < P(X|E) < u\}$$

The boundary area represents objects which cannot be classified with sufficiently high confidence (represented by u) into set X and which also cannot be excluded from X with the sufficiently high confidence (represented by $1 - l$).

The *l-negative* region $NEG_l(X)$ of the subset X, is a collection of objects which can be excluded from X with the confidence not lower than $1 - l$, that is,

$$NEG_l(X) = \bigcup \{E \in R^* : P(U - X|E) \geq 1 - l\}$$

The *l-negative* region represents objects of the universe for which it is known that it is relatively unlikely that they would belong to X.

In the Table 2, each of the classes E_i of the Table 1 is assigned to one of the *rough approximation regions*, according to the above definitions, with $u = 0.8$ and $l = 0.2$. The decision table in which each combination of condition attributes is assigned its approximation region with respect to the target value of the decision attribute (in this example, $T = 1$) is called *probabilistic decision table* [8]. The probabilistic decision table can be used to predict the target value of the decision attribute, or its complement, with probabilities not lower than u and $1 - l$, respectively.

4 (l, u)-Dependency in Decision Tables

The analysis of decision tables extracted from data involves inter-attribute dependency analysis, identification, elimination of redundant condition attributes and attribute significance analysis [2]. The original rough sets model-based analysis involves detection of functional, or partial functional dependencies and subsequent dependency-preserving reduction of condition attributes. In this paper, we extend this idea by using *(l,u)-probabilistic dependency* as a reference rather than functional or partial functional dependency. To define (l, u)-probabilistic dependency we will assume that the relation R corresponds to the partitioning

CLASS	S	H	E	C	E_i REGION
E_1	0	0	1	0	NEG
E_2	1	0	2	1	POS
E_3	1	1	1	0	NEG
E_4	0	2	1	1	POS
E_5	1	2	1	0	BND
E_6	1	0	1	0	BND
E_7	1	2	2	1	POS
E_8	0	0	2	1	POS

Table 3. The probabilistic decision table with $u = 0.83$ and $l = 0.11$

of the universe U in terms of values of condition attributes C in the decision table $< U, C, d, v_d^i >$, ($i = 0$ or 1). In other words, we assume that objects having identical values of the attributes are considered to be equivalent.

The *(l,u)-probabilistic dependency* $\gamma_{l,u}(C, d, i)$ between condition attributes C and the decision attribute d in the decision table $< U, C, d, v_d^i >$ is defined as the total relative size of (l, u)-approximation regions of the subset $X^i \subseteq U$ corresponding to target value of the decision attribute . In other words, we have

$$\gamma_{l,u}(C, d, i) = (card(POS_u(X^i)) + card(NEG_l(X^i)))/card(U)$$

The dependency degree can be interpreted as a measure of the probability that a randomly occurring object will be represented by such a combination of condition attribute values that the prediction of the corresponding value of the decision attribute could be done with the acceptable confidence, as represented by (l, u) pair of parameters.

To illustrate the notion of (l, u)-dependency let us consider the classification given in Table 1 again. When $u = 0.80$ and $l = 0.15$ the dependency equals to 1.0. This means that every object e from the universe U can be classified either as the member of the target set with the probability not less than 0.8, or the member of the complement of the target set, with the probability not less than 0.85. The lower and upper limits define acceptable probability bounds for predicting whether an object is, or is not the member of the target set.

If (l, u)—dependency is less than one it means that the information contained in the table is not sufficient to make either positive, or negative prediction in some cases. For instance, if we take $u = 0.83$ and $l = 0.11$ then the probabilistic decision table will appear as shown in Table 3. As we see, when objects are classified into boundary classes, neither positive nor negative prediction with acceptable confidence is possible. This situation is reflected in the $(0.11, 0.83)$—dependency being 0.7 (assuming even distribution of atomic classes $E_1, E_2, ..., E_8$ in the universe U).

5 Optimization of Precision Control Parameters

An interesting question, inspired by practical applications of the variable precision rough set model, is how to set the values of the precision control parameters l

and u to achieve desired quality of prediction. It is, in fact, an optimization problem, strongly connected to the external knowledge of possible gains and losses associated with correct, or incorrect predictions, respectively. It also depends on the quality of the information encoded in data used to create the probabilistic decision table. In general, setting lower values of l and higher values of u results in increasing the size of the boundary area on the expense of positive and negative regions. In practical terms, this means that we my not be always able to make decisions with the confidence level we would like it to be. If nothing is known about the potential gains or losses associated with the decisions, the reasonable goal is to increase the likelihood of positive correct prediction about the target value of the decision attribute, i.e. above random guess probability of success (by positive correct prediction we mean correctly predicting that the selected value will occur). Similarly, we are interested in increasing the probability of negative correct prediction, i.e. predicting correctly that a particular target value will not occur. We would like this probability to be above random guess probability of success as well. That is, given the distribution of the target value of the decision attribute to be $(p, 1-p)$, where p is the probability that an object has the target value of the decision attribute, and $1-p$ is the probability that it does not, the reasonable settings of the parameters are $0 \leq l < p$ and $1 \geq u > p$. With the settings falling into these limits, in the negative region the prediction that object does not belong to the target set would be made with the confidence higher than random guess, i.e. with the probability not less than $1-l > 1-p$ and, in the positive region, the prediction that an object belongs to the target set would be made with the probability not less than u,

Clearly, other factors can affect the selection of the precision control parameters. In particular, an interesting question is how to set those parameters in a game playing situation, where each decision making act is carrying a cost (*bet cost* $b > 0$) and incorrect decision results in a *loss* whereas correct decision results in a *win*. Because there are two possible outcomes of the decision, and one can pick any of these outcomes, there are two kinds of losses and two kinds of wins:

- positive win, when positive outcome is bet (that is, that the target value will occur) and that outcome really occurred; the win is denoted here as $q^{++} > 0$ and the cost of this betting is denoted as b^{+};

- positive loss, when positive outcome is bet but that outcome did not occur; the loss is denoted here as $q^{+-} < 0$;

- negative win, when the negative outcome is bet (that is, that the target value will not occur) and that outcome really occurred; the win is denoted here as $q^{--} > 0$ and the cost of this betting is denoted as b^{-};

- negative loss, when the negative outcome is bet but that outcome did not occur; the loss is denoted here as $q^{-+} < 0$;

In addition to the assumptions listed above we will assume that both positive and negative wins are not smaller than the cost of betting, that is $q^{--} \geq b^{-} > 0$

and $q^{++} \geq b^+ > 0$, and that the absolute values of both negative and positive losses are not smaller than the bet, that is $|q^{-+}| \geq b^-$ and $|q^{+-}| \geq b^+$.

Also, with each approximation region we will associate an *expected gain* function, which is the weighted average of wins and losses in the respective region. Our decision making strategy assumes that in the positive region the positive outcome is bet, and that in the negative region, the negative outcome is bet. The bet in the boundary region will depend on the value of the expected gain function, and we will assume that the bet which maximizes the gain function is selected. The gain functions $Q(approximation\ region)$ are defined as follows:

- $Q(POS) = p(+|POS)*q^{++}+p(-|POS)*q^{+-}$ where $p(+|POS)$ and $p(-|POS)$ are conditional probabilities of positive and negative outcomes respectively within the positive region;

- $Q(NEG) = p(+|NEG) * q^{-+} + p(-|NEG) * q^{--}$ where $p(+|NEG)$ and $p(-|NEG)$ are conditional probabilities of positive and negative outcomes respectively within the negative region;

- $Q(BND) = p(+|BND)*q^{++}+p(-|BND)*q^{+-}$ or $Q(BND) = p(+|BND)*q^{-+} + p(-|BND) * q^{--}$, depending on the bet, whichever value is higher with the positive, or negative bet, where $p(+|BND)$ and $p(-|BND)$ are conditional probabilities of positive and negative outcomes respectively within the boundary region.

Let us note that:

1. $Q(POS) \geq u * q^{++} + (1-u) * q^{+-}$ and
2. $Q(NEG) \geq l * q^{-+} + (1-l) * q^{--}$.

The uncertain decision is considered advantageous and justified if the expected gain is not lower than the cost of the bet, i.e. if $Q(POS) \geq b$ and $Q(NEG) \geq b$, assuming that positive outcome is bet in the positive region and negative outcome is bet in the negative region. By focusing on these two regions we can determine from (1) and (2) the bounds for parameters l and u to maximize the size of positive and negative regions while guaranting that $Q(POS) \geq b$ and $Q(NEG) \geq b$. From conditions $u * q^{++} + (1 - u) * q^{+-} \geq b$ and $l * q^{-+} + (1 - l) * q^{--} \geq b$ we get the following bounds for the precision control parameters:

$$1 \geq u \geq \frac{b^+ - q^{+-}}{q^{++} - q^{+-}} \text{ and } 0 \leq l \leq \frac{b^- - q^{--}}{q^{-+} - q^{--}} .$$

To maximize the sizes of both positive and negative areas the upper limit should assume the minimal range value and the lower limit should assume the maximal range value, that is:

$$u = \frac{b^+ - q^{+-}}{q^{++} - q^{+-}} \text{ and } l = \frac{b^- - q^{--}}{q^{-+} - q^{--}} .$$

We should be aware however that these bounds set only the requirements how the rough approximation regions should be defined in order to obtain desired expected results of decision making processes. The actual data set may not support these bounds in the sense that the positive, negative or both regions may be empty resulting in the boundary area covering the whole universe. In general, it can be demonstrated that in the boundary area, regardless whether positive or negative bet is made, the expected gain is always less than the respective bet, that is $Q(BND) < b^+$, if the positive bet is taken, and $Q(BND) < b^-$, if the negative bet is taken. Consequently, if the decision has to be made in the boundary area, one should take the one which maximizes $Q(BND)$, but in the longer run the "player" is in the loosing position anyway in the boundary area. The expected gain G from making decisions based on the whole decision table, according with the assumptions and decision strategy described above, is given by:

$$G = p(POS) * Q(POS) + p(NEG) * Q(NEG) + p(BND) * Q(BND)$$

where $p(POS)$, $p(NEG)$ and $p(BND)$ are the probabilities of respective approximation regions (the probabilities mentioned here can be approximated based on frequency distribution of data records belonging to the respective regions). Only if the overall expected gain G is higher than the expected cost of betting the "player" is winning in the longer run. This clearly sets the limit on the applicability of the probabilistic decision tables to support decision making.

6 Summary

We briefly review in this section the main points of the described approach to predictive modeling and decision making.

The main distinguishing feature of this approach is that it is primarily concerned with the acquisition of decision tables from data and with their analysis and simplification using notions of attribute dependency, reduct, core and attribute significance. The decision tables represent "discovered" inter-data dependencies which implies that, in general, a number of decision tables can be extracted from a given data collection. An important issue in the whole process of decision table acquisition from data is a choice of the mapping from original attributes, in which raw data are expressed, to finite-valued attributes used in the decision table. This is an application domain-specific task, often requiring deep knowledge of the domain. One popular technique is discretization of continous attributes. However, the discretization of continuous attributes [7], is a comprehensive research topic in itself whose discussion goes beyond the scope of this article. The decision making with probabilistic decision tables is typically uncertain. The decision strategy involves making positive prediction in the positive region, negative prediction in the negative region, and positive or negative prediction in the boundary region, depending on the value of the gain function. The techniques described in this article are aimed at constructing probabilistic decision tables which would support uncertain decision making leading to long range gains rather than to correct decisions in each case. They seem to be applicable to practical problems involving making guesses based on past data, such as stock market price movements prediction or market research.

7 Acknowledgment

The research reported in this article was partially supported by a research grant awarded by Natural Sciences and Engineering Research Council of Canada.

References

1. Pawlak, Z. Grzymala-Busse, J. Słowiński, R. and Ziarko, W. (1995). Rough sets. *Communications of the ACM*, 38, 88–95.
2. Pawlak, Z. (1991). *Rough Sets - Theoretical Aspects of Reasoning about Data.* Kluwer Academic.
3. Ziarko, W. (ed.) (1994). *Rough Sets, Fuzzy Sets and Knowledge Discovery.* Springer Verlag.
4. Ziarko, W. (1993). Variable precision rough sets model. *Journal of Computer and Systems Sciences*, vol. 46, no. 1, 39-59.
5. Polkowski, L., Skowron, A. (eds.) (1998). *Rough Sets in Knowledge Discovery.* Physica Verlag, vol. 1-2.
6. Hurley, R. (1983). *Decision Tables in Software Engineering.* Van Nostrand Reinhold.
7. Son, N. (1997). Rule induction from continuous data. In: Wang, P.(ed.), *Joint Conference of Information Sciences.* Duke University, Vol. 3, 81–84.
8. Ziarko, W. (1998). Approximation region-based decision tables. In: Polkowski, L., Skowron, A. (eds.). *Rough Sets and Current Trends in Computing.* Lecture Notes in AI 1424, Springer Verlag, 178-185.
9. Zhong, N., Dang, J., Ohsuga, S. (1998). Soft techniques to data mining. In: Polkowski, L., Skowron, A. (eds.). *Rough Sets and Current Trends in Computing.* Lecture Notes in AI 1424, Springer Verlag, 231-238.
10. Munakata, T. (1998). *Fundamentals of the New Artificial Intelligence.* Springer Verlag.
11. Lenarcik, A., Piasta, Z. (1998). Rough classifiers sensitive to costs varying from object to object. In: Polkowski, L., Skowron, A. (eds.). *Rough Sets and Current Trends in Computing.* Lecture Notes in AI 1424, Springer Verlag, 221-230.
12. Tsumoto, S. (1998). Formalization and induction of medical expert system rules based on rough set theory. In: Polkowski, L., Skowron, A. (eds.) (1998). *Rough Sets in Knowledge Discovery.* Physica Verlag, vol. 2,.307-323.

The Iterated Version Space Learning

Jianna Jian Zhang[1] and Nick J. Cercone[2]

[1] Brock University, St. Catharines
Ontario, Canada, L2S 3A1
jianna@brocku.ca
[2] University of Waterloo, Waterloo
Ontario, Canada, N2L 3G1
ncercone@uwaterloo.ca

Abstract. Inspired with Version Space learning, the Iterated Version Space Algorithm (IVSA) has been designed and implemented to learn disjunctive concepts. IVSA dynamically partitions its search space of potential hypotheses of the target concept into contour-shaped regions until all training instances are maximally correctly classified.

1 Introduction

Since mid 1950s, many AI researchers have developed learning systems that automatically improve their performance. Vere's Multiple Convergence Algorithm [12] and Mitchell's Candidate Elimination Algorithm [6] introduced a novel approach to concept learning known as the Version Space Algorithm (VSA). Unlike other learning algorithms, which used either generalization or specialization alone, VSA employed both. VSA has advantages – no back tracking for any seen training instances and a unique concept description that is consistent with all seen instances. VSA has weaknesses – training instances must be noise free and the target concept must be simple. These problems have prevented VSA from practical use outside the laboratories.

During the last few years, many improved algorithms based on VSA have been designed and/or implemented. In section 2, we first introduce VSA and then highlight two improved methods compare with the IVSA approach. The discussion in section 2 focuses on learning a disjunctive concept from six training instances, which are noise free so that problems caused by learning disjunctive concepts can be isolated from problems caused by noise training instances. Section 3 presents the overall approach of IVSA. Preliminary experimental results on several ML databases [10] and English pronunciation databases [13] are presented in Section 4. Discussions on each specific test and sample rules are also given in Section 4. In Section 5, we summarize current research on IVSA and give suggestions for future research.

2 Version Space Related Research

2.1 The Version Space Algorithm

A *version space* is a representation that contains two sets of hypotheses, the general hypotheses (G set) and the specific hypotheses (S set). Both G and S must be consistent with all examined instances. Positive instances make the S set more general to include all positive instances seen, while negative instances make the G set more specific to exclude all negative instances seen. If the training instances are consistent and complete, G and S sets eventually merge into one hypothesis set. This unique hypothesis is the learned concept description.

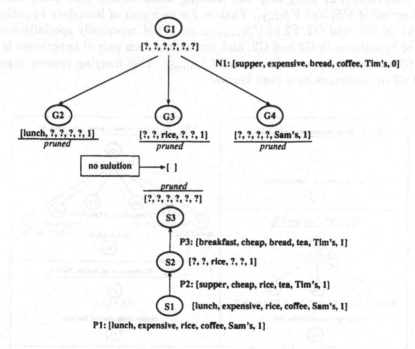

Fig. 1. The Version Space after the Fourth Instance (P3)

The following six noise-free training instances have been selected to illustrate problems with VSA. The value '1' or '0' in each instance indicates that the patient had a positive or negative allergic reaction respectively.

P_1 = (lunch, expensive, rice, coffee, Sam's, 1)
N_1 = (supper, expensive, bread, coffee, Tim's, 0)
P_2 = (supper, cheap, rice, tea, Tim's, 1)
P_3 = (breakfast, cheap, bread, tea, Tim's, 1)
P_4 = (supper, expensive, rice, tea, Bob's, 1)
P_5 = (supper, cheap, rice, coffee, Sam's, 1)

Figure 1 shows that as soon as instance P_3 is processed, the new specific hypothesis S3 must be discarded due to over-generalization. When either G or

S set becomes empty, the version space is collapsed, and thus "No legal concept description is consistent with this new instance as well as all previous training instances" [6], although a concept description of "tea or rice" can be easily derived by hand. To improve VSA learning, Hirsh has designed a new algorithm, the Incremental Version Space Merging (IVSM) [3].

2.2 The Incremental Version Space Merging

Instead of building one version space, IVSM constructs many version spaces $VS_{1...n}$ where n is the number of training instances. For each $i \in n$, IVSM first constructs VS_i using only one training instance, and then computes the intersection of VS_i and $VS_{(i-1)}$. That is, for each pair of boundary hypotheses G1, S1 in VS_i and G2, S2 in $VS_{(i-1) \cap (i-2)}$, IVSM repeatedly specializes each pair of hypotheses in G1 and G2, and generalizes each pair of hypotheses in S1 and S2 to form a new version space $VS_{i \cap (i-1)}$. This merging process repeats until all the instances have been learned.

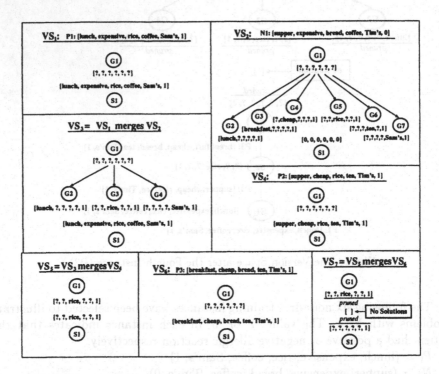

Fig. 2. The IVSM Approach after Processing the Fourth Instance (P_3)

The same six training instances are used to demonstrate IVSM learning. In Figure 2, after IVSM has computed the intersection for VS_5 and VS_6, the resulting specific hypothesis [?, ?, ?, ?, ?, ?, 1] is overly generalized. According to the IVSM merging algorithm [3], the current specific hypothesis must be pruned. IVSM, therefore, does not offer a solution for this particular exercise.

2.3 The Parallel Based Version Space Learning

Another recent research into VSA is Parallel Based Version Space (PBVS) learning [4]. Like the IVSM approach, PBVS also uses a version space merging algorithm, except that PBVS divides the entire set of training instances into two groups and constructs two version spaces simultaneously from each group, and then merges these two version spaces as IVSM does. Figure 3 shows the PBVS learning process using the the same six instances. Again when PBVS merges the VS_1 into VS_2, the resulting boundary sets are empty. Therefore, PBVS learning fails to learn this set of training instances due to the same reason that causes the IVSM learning fails.

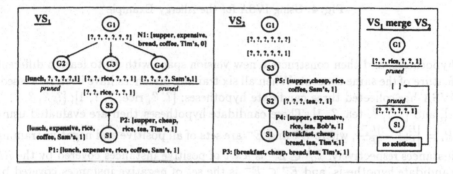

Fig. 3. The PBVS Approach after Processing the Fourth Instance

3 The Iterated Version Space Learning

The allergy example is simple and can be described with two hypotheses. But when the number of training instances, attributes, and classes are getting larger and larger, it becomes more and more difficult to detect which attribute value would be a true feature that distinguishes instances of different classes. However, VSA has already provided a natural way of separating different features. That is, whenever VSA collapses, the search has encountered a new feature. This is one of the new idea behind IVSA.

3.1 Learning the Allergy Example with IVSA

Before showing the detailed algorithm and approach, let us apply the same six allergy instances to IVSA. As Figure 1 shows when the version space is collapsed by processing P_3, instead of failing, IVSA first collects G3 and S2 as candidate

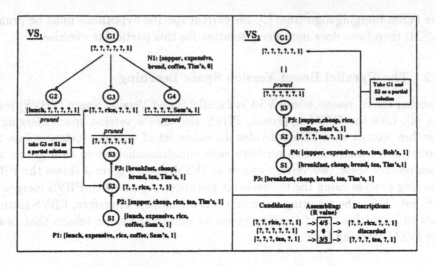

Fig. 4. Using IVSA for the Allergy Example

hypotheses, and then constructs a new version space with P_3 to learn a different feature of the same concept. When all six training instances have been processed, IVSA has collected three candidate hypotheses: [?, ?, rice, ?, ?, 1]; [?, ?, ?, ?, ?, 1]; and [?, ?, ?, tea, ?, 1]. These candidate hypotheses then are evaluated using $R_i = \frac{|E_i^+|}{|E^+|} - \frac{|E_i^-|}{|E^-|}$, where E^+ and E^- are sets of all positive and negative training instances respectively. $E_i^+ \subset E^+$ is a set of positive instances covered by the ith candidate hypothesis, and $E_i^- \subset E^-$ is the set of negative instances covered by the same candidate hypothesis. For the allergy example, $R_1 = \frac{4}{5}$, $R_2 = 0$, and $R_3 = \frac{3}{5}$. Therefore, [?, ?, rice, ?, ?, 1] and [?, ?, ?, tea, ?, 1] are selected as the concept description: $((A_3 = rice) \vee (A_4 = tea)) \rightarrow$ allergy.

3.2 Learning from Noisy Training Instances

When training instances contain noise, the noise interferes or even stops the learning. With IVSA, noisy training instances are simply ignored. Here we use the same allergy example in Section 2.1 plus a noise instance $N_2 =$ (supper, cheap, rice, tea, Tim's, 0). Figure 5 shows this learning process. In the first version space, IVSA simply ignores N_2 just like it ignores instances representing different features such as P_3 in Figure 4 in the second version space. Because N_2 is negative, IVSA amalgamates the second version space with P_3. But if the incorrect instances was classified as possitive, IVSA would start with this instance and later the hypothesis generated from this noisy instance would be discarded. The learned concept description does not interfered by N_2 because IVSA recognizes that N_2 does not represent the feature of the concept.

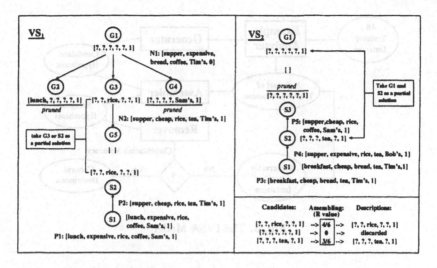

Fig. 5. Learning Noisy Training Instances with IVSA

3.3 The IVSA Model

Learning a concept is similar to assembling a multi-dimensional jigsaw puzzle from a large selection of possible pieces. The target concept can be viewed as the puzzle and an ordered list of disjunctive hypotheses can be viewed as groups of puzzle pieces. One method of solving this problem is to repeatedly generate any possible missing pieces and add them to the puzzle until it is complete. IVSA is based on this puzzle assembling method.

As shown in Figure 6, IVSA contains the Example Analyser, Hypothesis Generator, Assembler, and Remover. The Example Analyser provides statistical evaluation for each attribute value provided by the instance space to determine the order of input trining instances. The Hypothesis Generator produces a set of candidate hypotheses from the given set of training instances. The Hypothesis Assembler repeatedly selects the most promising hypothesis from a large number of candidate hypotheses according to the statistical evaluation provided by the Example Analyser, and then tests this hypothesis in each position in a list of accepted hypotheses. If adding a new hypothesis increases concept coverage, it is placed in the position that causes the greatest increase; otherwise this hypothesis is discarded. After the candidate hypotheses have been processed, the list of accepted hypotheses is examined by the Hypothesis Remover to see if any of the hypotheses can be removed without reducing accuracy. If the learning accuracy is satisfactory, the accepted hypothesis set becomes the learned concept description. Otherwise, the set of incorrectly translated instances are fed back to the generator, and a new learning cycle starts.

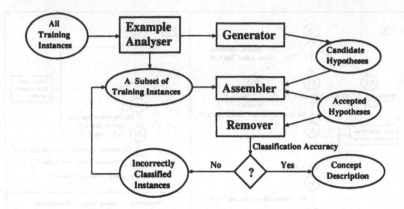

Fig. 6. The IVSA Model

4 Experimental Results on UCI Databases

IVSA is tested on some machine learning databases [10]. To demonstrate the consistency of IVSA, a ten-fold cross validation test is used. the cross validation test is defined as follows:

> **Definition 1.** *Let I be the set of positive and negative instances given, i be the index for 10 ten-fold tests, and j be the index for test instances,*
> *then* $\left\{ Test_i = \{x_j | x_j \in I\}_{j=i}^{(j=j+10)\leq|I|} \right\}_{i=1}^{10}$ *and* $\{Train_i = \{I - Test_i\}\}_{i=1}^{10}$.

That is, for each fold of the test, use 90% of instances to train the system and then with the rules learned from the 90% instances, testing on 10% unseen instances.

4.1 Learning the Mushroom Database

The mushroom database [7] has a total of 8,124 entries (tuples or instances). Each tuple has 22 feature attributes and one decision attribute. The 22 feature attributes have 2–5 values and the decision attribute has two values (or classes) 'p' (poison) or 'e' (eatable). Because the mushroom database is noise-free, any machine learning program should be able to learn it accurately. For example, STAGGER "asymptoted to 95% classification accuracy after reviewing 1,000 instances" [8], HILLARY has learned 1,000 instances and reported an average accuracy about 90% on ten runs [5], a back propagation network developed in [2] has generated 'crisp logical rules' that give correct classification of 99.41%, and variant decision tree methods used in [11] have 100% accuracy by a ten-fold cross validation test [11]. With IVSA, the predictive accuracy shown in Figure 1 on the mushroom database has reached 100% with 9 rules.

Table 1. Ten-fold Tests on Mushroom Data (CPU: MIPS R4400)

Run Number	Number of Instances 90%	Number of Instances 10%	Accuracies Training	Accuracies Testing	Number of Rules	CPU Time (h/m/s)
1	7,311	813	100.00%	100.00%	9	01/42/14
2	7,311	813	100.00%	100.00%	9	02/09/42
3	7,311	813	100.00%	100.00%	9	01/45/41
4	7,311	813	100.00%	100.00%	9	01/53/12
5	7,312	812	100.00%	100.00%	9	01/40/58
6	7,312	812	100.00%	100.00%	9	02/30/08
7	7,312	812	100.00%	100.00%	9	01/46/51
8	7,312	812	100.00%	100.00%	9	01/59/00
9	7,312	812	100.00%	100.00%	8	01/46/40
10	7,312	812	100.00%	100.00%	9	01/56/16
Ave.	7,312	812	100.00%	100.00%	9	01/55/04
S.D.	0.49	0.49	0.00	00.00	0.30	859.94

4.2 Learning the Monk's Databases

The Monk's Databases contains three sets: Monk-1, Monk-2, and Monk-3. Each of the three sets is originally partitioned into training and testing sets [10] [9]. IVSA is trained and tested on Monk-1, Monk-2, and Monk-3. In Table 2, the experiment shows that 5, 61, and 12 rules learned from Monk-1, Monk-2, and Monk-3 databases gives 100%, 81.02%, and 96.30% classification accuracies on three sets of 432 previously unseen instances.

Table 2. Tests on Monk's Databases (CPU: 296 MHz SUNW, UltraSPARC-II)

Data Base	Instances Training	Instances Testing	Accuracy Training	Accuracy Testing	# of Rules	CPU Time (seconds)
Monk-1	124	432	100.00%	100%	5	3
Monk-2	169	432	100.00%	81.02%	61	38
Monk-3	122	432	100.00%	96.30%	12	5

Rules learned from Monk-1, (2 2 ? ? ? ? ? 1), (3 3 ? ? ? ? ? 1), (1 1 ? ? ? ? ? 1), (? ? ? ? 1 ? 1), (? ? ? ? ? ? 0), show exactly the desired concept description with minimum number of rule allowed by the concept language, which can be rewritten as: $(head\_shape = body\_shape) \vee (jacket = red) \rightarrow monk$. For the Monk-2 database, 61 rules learned which is relatively large compared with the other two sets (Monk-1 and Monk-3) due to a highly disjunctive (or irregular) concept. However, it can be improved with more statistical analysis or some improved instance space (or representation space) shown in [1] the predictive accuracy can be as high as 100% [Bloedorm et al., 1996, p.109]), although this method is highly specified for only Monk-2 database. Twelve rules are learned from the Monk-3 database with 96.3% classification accuracy despite 5% noise added to the Monk-3 training instances: (1 1 1 1 3 1 0), (1 2 1 2 3 1 0), (2 2 1 2 2 1 0), (2 2 1 3 3 1 0), (2 2 1 3 3 2 0), (2 3 1 1 3 1 1), (3 3 1 1 3 2 1), (3 3 1 1 4 1 1), (? ? ? ? 4 ? 0), (? 1 ? ? ? ? 1), (? 2 ? ? ? ? 1), (? ? ? ? ? ? 0)

4.3 Learning English Pronunciation Databases

IVSA has been applied to learn English pronunciation rules [13]. The task is to provide a set of rules that transform input English words into sound symbols using four steps: (1) decompose words into graphemes, (2) form syllables from graphemes, (3) stress marking on syllables, and (4) transform them into a sequence of sound symbols. Learning and testing results are shown in Table 3.

Table 3. Learning and Testing Results for Individual Steps

Step	Learning Accuracy				Testing Accuracy				# of Rules
	Inst.	Words	Inst.	Words	Inst.	Words	Inst.	Words	
(1)	118,236	17,951	99.58%	99.19%	13,050	1,995	98.18%	94.89%	1,030
(2)	56,325	23,684	97.23%	96.34%	6,241	2,656	96.36%	95.41%	248
(3)	56,325	23,684	78.30%	72.26%	6,241	2,656	77.95%	72.78%	2,080
(4)	118,236	17,951	98.14%	95.31%	16,418	2,656	96.93%	92.23%	1,971

5 Conclusions

We have presented a new concept learning method IVSA, its approach, and test results. Our analysis of previous research shows that the empty version space signals a new feature of the same target concept presented by a particular instance. The hypotheses generated by previous version spaces belong to one region of the target concept while the current hypotheses generated by a new version space belong to another region of the same concept. IVSA takes the advantage of an empty version space, using it to divide the regions of a concept, and correctly handles noisy training instances.

A concept description can be divided into regions, and each region can be represented by a subset of training instances. These subsets can be collected according to the statistical analysis on each attribute value provided by the Example Analyser. The technique of re-arranging the order of training instances according to the importance of a particular attribute value provides a practical method to overcome order bias dependency of the training instances.

The demonstration on learning noisy training instances shows that IVSA has strong immunity to noisy data, and has the ability to learn disjunctive concept. The preliminary experimental results show that rules learned by IVSA obtain high accuracy when applied to previously unseen instances. In the future, we will intensively test IVSA with additional databases and improve the Example Analyser to obtain higher learning speed and smaller numbers of rules.

Acknowledgments

The authors are members of the Institute for Robotics and Intelligent Systems (IRIS) and wish to acknowledge the support of the Networks of Centers of Excellence program of the Government of Canada, the Natural Sciences and Engineering Research Council, and the participation of PRECARN Associates Inc.

References

1. E. Bloedorm, S. Ryszard, and S. Michalski. The AQ17-DCI system for data-driven constructive induction and its application to the analysis of world economics. In *Proc. of Nineth International Symposium on Foundations of Intelligent Systems*, pages 108–117, Zakopane, Poland, June 1996.
2. W. Duch, R. Adamczak, and K. Grabczewski. Extraction of logical rules from training data using backpropagation networks. *Proceedings of the The First Online Workshop on Soft Computing*, pages 19–30, 1996.
3. Haym Hirsh. Generalizing Version Spaces. *Machine Learning*, 17:5–46, 1994.
4. T. Hong and S. Tseng. Learning concepts in parallel based upon the strategy of Version Space. *IEE Transactions on Knowledge and Data Engineering*, 6:857–867, 1994.
5. W. Iba, J. Wogulis, and P. Langley. Trading off simplicity and coverage in incremental concept learning. *Proceedings of the 5th International Conference on Machine Learning*, pages 73–79, 1988.
6. T.M. Mitchell. *Version Spaces: An Approach to Concept Learning*. PhD thesis, Stanford University, CA, 1979.
7. J. Schlimmer, editor. *The Audubon Society Field Guide to North American Mushrooms*. Alfred A. Knopf, New York, 1981.
8. J.C. Schlimmer. Incremental adjustment of representations for learning. In P. Langley, editor, *Proc. of the Fourth International Workshop on Machine Learning*, pages 79–90, Irvine, CA, June 1987.
9. S.B. Thurn and et al. The monk's problem: A performance comparison of different learning algorithms. Technical Report CMU-CS-91-197, Carnegie Mellon University, Pittsburgh, December 1991.
10. UCI. UCI repository of machine learning databases and domain theories. on ftp://ftp.ics.uci.edu/pub/machine-learning-databases/, 1996.
11. P.E. Utgoff, N.C. Berkman, and J.A. Clouse. Decision tree induction based on efficient tree resturcturing. *Machine Learning*, 29:5–44, 1997.
12. S.A. Vere. Induction of concepts in predicate calculus. In *Proc. IJCAI-75, Advance Papers of the Fourth International Joint Conference on Artificial Intelligence*, volume 1, pages 281–287, Tbilisi, USSR, 1975.
13. Jianna Jian Zhang. *The LEP Learning System: An IVSA Approach*. PhD thesis, University of Regina, Regina, Canada, 1998.

An Empirical Study on Rule Quality Measures

Aijun An and Nick Cercone

Department of Computer Science, University of Waterloo
Waterloo, Ontario N2L 3G1 Canada
Email: {aan, ncercone}@uwaterloo.ca

Abstract. We describe statistical and empirical rule quality formulas and present an empirical comparison of them on standard machine learning datasets. From the experimental results, a set of formula-behavior rules are generated which show relationships between a formula's performance and dataset characteristics. These formula-behavior rules are combined into formula-selection rules which can be used in a rule induction system to select a rule quality formula before rule induction.

1 Introduction

A rule induction system generates decision rules from a set of data. The decision rules determine the performance of a classifier that exploits the rules to classify unseen objects. It is thus important for a rule induction system to generate decision rules with high predictability or reliability. These properties are commonly measured by a function called rule quality. A rule quality measure is needed in both rule induction and classification. A rule induction process is usually considered as a search over a hypothesis space of possible rules for a decision rule that satisfies some criterion. In the rule induction process that employs general-to-specific search, a rule quality measure can be used as a search heuristic to select attribute-value pairs in the rule specialization process; and/or it can be employed as a significance measure to stop further specialization. The main reason to focus special attention on the stopping criterion can be found in the studies on *small disjunct problems* [9]. The studies indicated that small disjuncts, which cover a small number of training examples, are much more error prone than large disjuncts. To prevent small disjuncts, a stopping criterion based on rule consistency (i.e., the rule is consistent with the training examples) is not suggested for use in rule induction. Other criteria, such as the G2 likelihood ratio statistic as used in CN2 [7] and the degree of logical sufficiency as used in HYDRA [1], have been proposed to "pre-prune" a rule to avoid overspecialization. Some rule induction systems, such as C4.5 [12] and ELEM2 [2], use an alternative strategy to prevent the small disjunct problem. In these systems, the rule specialization process is allowed to run to completion (i.e., it forms a rule that is consistent with the training data or as consistent as possible) and "post-prunes" overfitted rules by removing components that are deemed unreliable. Similar to pre-pruning, a criterion is needed in post-pruning to determine when to stop this generalization process. A rule quality measure is also needed in classification. It is possible that an unseen example satisfies multiple decision rules that indicate different classes. In this situation, some conflict resolution scheme must be applied to assign the unseen object to the most appropriate class. It is therefore useful for each rule to be associated with a numerical factor representing its classification power, its reliability, etc. We survey and evaluate statistical and empirical rule quality measures, some of which have been discussed by Bruha [5]. In our evaluation, ELEM2 [2] is used as the basic learning and classification algorithms. We report the experimental results from using these formulas in ELEM2 and compare the results by indicating the significance level of the difference between each pair of the formulas. In addition, the relationship between the performance of a formula and a dataset is obtained by automatically generating formula-behavior rules from the experimental results. The formula-behavior rules are further combined into formula-selection rules which can be employed by ELEM2 to select a rule quality formula before rule induction. We report the experimental results showing the effects of formula-selection on ELEM2's predictive performance.

2 Rule Quality Measures

Many rule quality measures are derived by analysing the relationship between a decision rule R and a class C. The relationship can be depicted by a 2×2 *contingency table* [5], which consists of a cross-tabulation of categories of observations with the frequency for each cross-classification shown:

	Class C	Not class C	
Covered by rule R	n_{rc}	$n_{r\bar{c}}$	n_r
Not covered by R	$n_{\bar{r}c}$	$n_{\bar{r}\bar{c}}$	$n_{\bar{r}}$
	n_c	$n_{\bar{c}}$	N

where n_{rc} is the number of training examples covered by rule R and belonging to class C; $n_{r\bar{c}}$ is the number of training examples covered by R but not belonging to C, etc; N is the total number of training examples; n_r, $n_{\bar{r}}$, n_c and $n_{\bar{c}}$ are marginal totals, e.g., $n_r = n_{rc} + n_{r\bar{c}}$, which is the number of examples covered by R. The contingency table can also be presented using relative rather than absolute frequencies as follows:

	Class C	Not class C	
Covered by rule R	f_{rc}	$f_{r\bar{c}}$	f_r
Not covered by R	$f_{\bar{r}c}$	$f_{\bar{r}\bar{c}}$	$f_{\bar{r}}$
	f_c	$f_{\bar{c}}$	1

where $f_{rc} = \frac{n_{rc}}{N}$, $f_{r\bar{c}} = \frac{n_{r\bar{c}}}{N}$, and so on.

2.1 Measures of Association

A measure of association indicates a relationship between the classification for the columns and the classification for the rows in the 2×2 contingency table.

Pearson χ^2 Statistic assumes contingency table cell frequencies are proportional to the marginal totals if column and row classifications are independent, and is given by

$$\chi^2 = \sum \frac{(n_o - n_e)^2}{n_e}$$

where n_o is the observed absolute frequency of examples in a cell, and n_e is the expected absolute frequency of examples for the cell. A computational formula for χ^2 can be obtained using only the values in the contingency table with absolute frequencies [6]: $\chi^2 = \frac{N(n_{rc}n_{\bar{r}\bar{c}} - n_{\bar{r}c}n_{r\bar{c}})^2}{n_c n_{\bar{c}} n_r n_{\bar{r}}}$. This value measures whether the classification of examples by rule R and one by class C are related. The lower the χ^2 value, the more likely the correlation between R and C is due to chance.

G2 Likelihood Ratio Statistic measures the distance between the observed frequency distribution of examples among classes satisfying rule R and the expected frequency distribution of the same number of examples where rule R selects examples randomly. The value of this statistic can be computed as

$$G2 = 2(\frac{n_{rc}}{n_r}log_e\frac{n_{rc}N}{n_r n_c} + \frac{n_{r\bar{c}}}{n_r}log_e\frac{n_{r\bar{c}}N}{n_r n_{\bar{c}}}).$$

The lower the G2 value, the more likely the apparent association between the two distributions is due to chance.

2.2 Measures of Agreement

A measure of agreement concerns the main diagonal contingency table cells.

Cohen's Formula Cohen [8] suggests comparing the actual agreement on the main diagonal $(f_{rc} + f_{\bar{r}\bar{c}})$ with the chance agreement $(f_r f_c + f_{\bar{r}} f_{\bar{c}})$ by using the normalized difference of the two:

$$Q_{Cohen} = \frac{f_{rc} + f_{\bar{r}\bar{c}} - (f_r f_c + f_{\bar{r}} f_{\bar{c}})}{1 - (f_r f_c + f_{\bar{r}} f_{\bar{c}})}$$

When both elements f_{rc} and $f_{\bar{r}\bar{c}}$ are reasonably large, Cohen's statistic gives a higher value which indicates the agreement on the main diagonal.

Coleman's Formula Coleman [3, 5] defines a measure of agreement between the first column and any particular row in the contingency table. Bruha [5] modifies Coleman's measure to define rule quality, which actually corresponds to the agreement on the upper-left element of the contingency table. The formula normalizes the difference between actual and chance agreement:

$$Q_{Coleman} = \frac{f_{rc} - f_r f_c}{f_r - f_r f_c}.$$

2.3 Measure of Information

Given class C, the amount of information necessary to correctly classify an instance into class C whose prior probability is $P(C)$ is defined as $-log_2 P(C)$. Given rule R, the amount of information we need to correctly classify an instance into class C is $-log_2 P(C|R)$, where $P(C|R)$ is the posterior probability of C given R. Thus, the amount of information obtained by rule R is $-log_2 P(C) + log_2 P(C|R)$. This value is called *information score* [10]. It measures the amount of information R contributes and can be expressed as

$$Q_{IS} = -log_2\frac{n_c}{N} + log_2\frac{n_{rc}}{n_r}.$$

2.4 Measure of Logical sufficiency

The logical sufficiency measure is a standard likelihood ratio statistic, which has been applied to measure rule quality [1]. Given a rule R and a class C, the degree of logical sufficiency of R with respect to C is defined by

$$Q_{LS} = \frac{P(R|C)}{P(R|\bar{C})}$$

where P denotes probability. A rule for which Q_{LS} is large means that the observation of R is encouraging for the class C – in the extreme case of Q_{LS} approaching infinity, R is sufficient to establish C in a strict logical sense. On the other hand, if Q_{LS} is much less than unity, the observation of R is discouraging for C. Using frequencies to estimate the probabilities, we have $Q_{LS} = \frac{\frac{n_{rc}}{n_c}}{\frac{n_{r\bar{c}}}{n_{\bar{c}}}}$.

2.5 Measure of Discrimination

Another statistical rule quality formula is the measure of discrimination, which is applied in ELEM2 [2]. The formula was inspired by a query term weighting formula used in the probability-based information retrieval. The formula measures the extent to which a query term can discriminate relevant and non-relevant documents [13]. If we consider a rule R as a query term in an IR setting, positive examples of class C as relevant documents, and negative examples as non-relevant documents, then the following formula can be used to measure the extent to which rule R discriminates positive and negative examples of class C:

$$Q_{MD} = log \frac{P(R|C)(1 - P(R|\bar{C}))}{P(R|\bar{C})(1 - P(R|C))}$$

where P denotes probability. The formula represents the ratio between the rule's positive and negative odds and can be estimated as $Q_{MD} = \frac{\frac{n_{rc}}{n_{r\bar{c}}}}{\frac{n_{\bar{r}c}}{n_{\bar{r}\bar{c}}}}$.

2.6 Empirical Formulas

Some rule quality formulas are not based on statistical or information theories, but from intuitive logic. Bruha [5] refers to these as *empirical* formulas. We describe two empirical formulas that combine two characteristics of a rule: consistency and coverage. Using the elements of the contingency table, the consistency of a rule R can be defined as $cons(R) = \frac{n_{rc}}{n_r}$ and its coverage as $cover(R) = \frac{n_{rc}}{n_c}$.

Weighted Sum of Consistency and Coverage Michalski [11] proposes to use the weighted sum of consistency and coverage as a measure of rule quality:

$$Q_{WS} = w_1 \times cons(R) + w_2 \times cover(R)$$

where w_1 and w_2 are user-defined weights with their values belonging to $(0, 1)$ and summed to 1. This formula is applied in an incremental learning system YAILS [14]. The weights in YAILS are specified automatically as: $w_1 = 0.5 + \frac{1}{4}cons(R)$ and $w_2 = 0.5 - \frac{1}{4}cons(R)$. These weights are dependent on consistency. The larger the consistency, the more influence consistency has on rule quality.

Product of Consistency and Coverage Brazdil and Torgo [4] propose to use a product of consistency and coverage as rule quality:

$$Q_{Prod} = cons(R) \times f(cover(R))$$

where f is an increasing function. The authors conducted a large number of experiments and chose to use the following form of f: $f(x) = e^{x-1}$. This setting of f makes the difference in coverage have smaller influence on rule quality, which results in the rule quality formula to prefer consistency.

3 Experiments with Rule Quality Measures

3.1 The Learning System

ELEM2 uses a *sequential covering* learning strategy; it reduces the problem of learning a disjunctive set of rules to a sequence of learning a single conjunctive rule that covers a subset of positive examples. Learning a conjunctive rule begins by considering the most general rule precondition, then greedily searches for an attribute-value pair that is most relevant to class C according to the following function: $SIG_C(av) = P(av)(P(C|av) - P(C))$, where av is an attribute-value pair and P denotes probability. The selected attribute-value pair is then added to the rule precondition as a conjunct. The process is repeated until the rule is as consistent with the training data as possible. Since a "consistent" rule may be a small disjunct that overfits the training data, ELEM2 "post-prunes" the rule after the initial search for the rule is complete. To post-prune a rule, ELEM2 first computes a rule quality value according to the formula of measure of discrimination Q_{MD} (Section 2.5). It then checks each attribute-value pair in the rule in the reverse order in which they were selected to see if removal of a pair will decrease the rule quality value. If not, the pair is removed.

After rules are induced for all classes, the rules can be used to classify new examples. The classification procedure in ELEM2 considers three possible cases: (1) *Single match*. The new example satisfies one or more rules of the same class. In this case, the example is classified to that class. (2) *Multiple match*. The new example satisfies more than one rules of different classes. In this case, ELEM2 computes a decision score for each of the matched classes as: $DS(C) = \sum_{i=1}^{k} Q_{MD}(r_i)$, where r_i is a matched rule that indicates class C, k is the number of this kind of rules, and $Q_{MD}(r_i)$ is the rule quality of r_i. The new example is then classified into the class with the highest decision score. (3) *No match*. The new example is not covered by any rule. Partial matching is conducted. If the partially-matched rules do not agree on classes, a partial matching score between new example e and a partially-matched rule r_i with n attribute-value pairs, m of which match the corresponding attributes of e, is computed as $PMS(r_i) = \frac{m}{n} \times Q_{MD}(r_i)$. A decision score for a class C is computed as $DS(C) = \sum_{i=1}^{k} PMS(r_i)$, where k is the number of partially-matched rules indicating class C. The new example is classified into the class with the highest decision score.

3.2 Experimental Design

We evaluate the rule quality formulas described in Section 2 by determining how rule quality formulas affect the predictive performance of ELEM2. In our experiments, we run versions of ELEM2, each of which uses a different rule quality formula. The formulas: $Q_{MD}, Q_{Cohen}, Q_{Coleman}, Q_{IS}, Q_{LS}, Q_{WS}$, and Q_{Prod} are used exactly as described in Section 2. The χ^2 statistic is used in two ways: (1) $Q_{\chi^2_{.05}}$: In post-pruning, the removal of an attribute-value pair depends on whether the rule quality value after removing an attribute-value pair is greater than $\chi^2_{.05}$, i.e., the tabular χ_2 value for the significance level of 0.05 with one degree of freedom. If the calculated value is greater than $\chi^2_{.05}$, then remove the attribute-value pair; otherwise check other pairs or stop post-pruning if all pairs have been checked. (2) $Q_{\chi^2_{.05+}}$: In post-pruning, an attribute-value pair is removed if and only if the rule quality value after removing the pair is greater than $\chi^2_{.05}$ and no less than the rule quality value before removing the pair. The G2 statistic, denoted as $Q_{G2_{.05+}}$, is used in the same way as $Q_{\chi^2_{.05+}}$.

Our experiments are conducted using 22 benchmark datasets from the UCI Repository of Machine Learning database. The datasets represent a mixture of characteristics shown in Table 1. ELEM2 removes all the examples containing missing values before rule induction. For datasets with missing values (such as "crx"), the number of examples shown in Table 1 is the number after removal.

Datasets	Number of classes	attributes	examples	Class Distribution	Domain
1 abalone	3	8	4177	Even	Predicting the age of abalone from physical measurements
2 australia	2	14	690	Even	Credit card application approval
3 balance-scale	3	4	625	Uneven	Balance scale classification
4 breast-cancer	2	9	683	Uneven	Medical diagnosis
5 bupa	2	6	345	Uneven	Liver disorder database
6 crx	2	15	653	Uneven	Credit card applications
7 diabetes	2	8	768	Uneven	Medical diagnosis
8 ecoli	8	7	336	Uneven	Predicting protein localization sites
9 german	2	20	1000	Uneven	Credit database to classify people as good or bad credit risks
10 glass	6	9	214	Uneven	Glass identification for criminological investigation
11 heart	2	13	270	Uneven	Heart disease diagnosis
12 ionosphere	2	33	351	Uneven	Classification of radar returns
13 iris	3	4	150	Even	Iris plant classification
14 lenses	3	4	24	Uneven	Database for fitting contact lenses
15 optdigits	10	64	3823	Even	Optical recognition of handwritten digits
16 pendigits	10	16	7494	Even	Pen-based recognition of handwritten digits
17 post-operative	3	8	87	Uneven	Postoperative Patient Data
18 segment	7	18	2310	Even	image segmentation
19 tic-tac-toe	2	9	958	Uneven	Tic-Tac-Toe Endgame database
20 wine	3	13	178	Uneven	Wine recognition data
21 yeast	10	8	1484	Uneven	Predicting protein localization sites
22 zoo	7	16	101	Uneven	Animal classification

Table 1. Description of Datasets.

3.3 Results

On each dataset, we conduct a ten-fold cross-validation of a rule quality measure using ELEM2. The results in terms of predictive accuracy mean over the 10 runs on each dataset for each formula are shown in Figure 1. The average of the

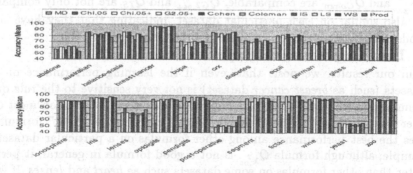

Fig. 1. Results on the 22 datasets

accuracy means for each formula over the 22 datasets is shown in Table 2, where

	Q_{WS}	Q_{MD}	Q_{LS}	$Q_{Coleman}$	Q_{Prod}	$Q_{G2.05+}$	Q_{IS}	$Q_{\chi^2_{.05+}}$	Q_{Cohen}	$Q_{\chi^2_{.05}}$
Average	82.09	81.34	81.33	80.66	80.33	79.85	79.63	79.51	79.05	72.95

Table 2. Average of accuracy means for each formula over the datasets.

the formulas are listed in decreasing order of average accuracy means. Whether a formula with a higher average is significantly better than a formula with a lower average is determined by paired t-tests using the S-Plus statistics software. The t-test results in terms of p-values are reported in Table 3. A small p-value indicates that the null hypothesis (the difference between the two formulas is due to chance) should be rejected in favor of the alternative at any significance level above the calculated value. The p-values that are smaller than 0.05 are shown in bold-type to indicate that the formula with higher average is significantly better than the formula with the lower average at the 5% significance level. For example, Q_{WS} is significantly better than $Q_{Coleman}$, $Q_{G2.05+}$, Q_{IS}, $Q_{\chi^2_{.05+}}$,

	Q_{WS}	Q_{MD}	Q_{LS}	$Q_{Coleman}$	Q_{Prod}	$Q_{G2.05+}$	Q_{IS}	$Q_{\chi^2_{.05+}}$	Q_{Cohen}	$Q_{\chi^2_{.05}}$
Q_{WS}	NA	0.0819	0.1421	**0.0119**	0.0692	**0.002**	**0.002**	**0.0073**	**0.0463**	**0.0026**
Q_{MD}	-	NA	0.9719	0.1323	0.4032	**0.0183**	**0.01**	0.0545	0.1328	**0.0069**
Q_{LS}	-	-	NA	0.0539	0.4389	**0.0026**	**0.0046**	0.0694	0.149	**0.0076**
$Q_{Coleman}$	-	-	-	NA	0.7858	0.0526	**0.035**	0.256	0.3187	**0.0137**
Q_{Prod}	-	-	-	-	NA	0.6947	0.5621	0.4325	0.3962	**0.0111**
$Q_{G2.05+}$	-	-	-	-	-	NA	0.5088	0.7512	0.6316	**0.0282**
Q_{IS}	-	-	-	-	-	-	NA	0.9117	0.733	**0.0316**
$Q_{\chi^2_{.05+}}$	-	-	-	-	-	-	-	NA	0.6067	**0.0144**
Q_{Cohen}	-	-	-	-	-	-	-	-	NA	**0.0179**
$Q_{\chi^2_{.05}}$	-	-	-	-	-	-	-	-	-	NA

Table 3. Significance levels (p-values from paired t-test) of improvement.

Q_{Cohen} and $Q_{\chi^2_{.05}}$; Q_{MD} and Q_{LS} are significantly better than $Q_{G2.05+}$, Q_{IS} and $Q_{\chi^2_{.05}}$; and all formulas are significantly better than $Q_{\chi^2_{.05}}$ at the 5% significance level. Generally speaking, Q_{WS}, Q_{MD} and Q_{LS} are comparable even if their performance does not agree on a particular dataset. $Q_{Coleman}$ and Q_{Prod}, and $Q_{\chi^2_{.05+}}$ and Q_{Cohen} are comparable. $Q_{G2.05+}$ and Q_{IS} are not only comparable, but also similar on each particular dataset, indicating that they have similar trends with regard to n_{rc}, n_r, n_c and N in the contingency table.

4 Learning from the Experimental Results

From our results, we posit that, even if the learning performance on some datasets (such as *breast cancer* dataset) is not very sensitive to the rule quality formula used, the performance greatly depends on the formula on most of the other datasets. It would be desirable that we can apply a "right" formula that gives the best performance among other formulas on a particular dataset. For example, although formula $Q_{\chi^2_{.05}}$ is not a good formula in general, it performs better than other formulas on some datasets such as *heart* and *lenses*. If we can find conditions under which each formula leads to good learning performance, we can select "right formulas" for different datasets and can improve the predictive performance of the learning system further.

To find out this regularity, we use our learning system, i.e., ELEM2, to learn the formula selection rules from the experimental results shown in the last section. The learning problem is divided into (1) learning formula-behavior rules for each rule quality formula that describe the conditions under which the formula produces "good", "medium" or "bad" results, and (2) combining the rules for all the formulas that describe the conditions under which the formulas give the "good" results. The resulting set of rules is the formula-selection rules that can be used by the ELEM2 classification procedure to perform formula selection.

4.1 Data Representation

To learn formula-behavior rules we construct training examples from Figure 1 and Table 1. First, on each dataset, we decide the relative performance of each formula as "good", "medium", or "bad". For example, on the *abalone* dataset, we say that the formulas whose accuracy mean is above 60% produce "good" results; the formulas whose accuracy mean is between 56 and 60 produce "medium" results; and other formulas give "bad" results. Then, for each formula, we construct a training dataset in which an training example describes the characteristics of a dataset and the performance of the formula on the dataset. Thus, to learn behavior rules for each formula, we have 22 training examples. The dataset characteristics are described in terms of number of examples, number of attributes, number of classes and the class distribution. Samples of training examples for learning behavior rules of Q_{IS} are shown in Table 4.

Number of			Class	
Examples	Attributes	Classes	Distribution	Performance
4177	8	3	Even	Good
690	14	2	Even	Medium
625	4	3	Uneven	Bad
683	9	2	Uneven	Medium

Table 4. Sample of training examples for learning the behavior of a formula

4.2 The Learning Results

ELEM2 with its default rule quality formula (Q_{MD}) is used to learn the "behavior" rules from the training dataset constructed for each formula. Table 5 shows samples of generated rules for each formula, where N stands for the number of examples, NofA is the number of attributes, NofC is the number of classes, and the column "No. of Support Datasets" means the number of the datasets that support the corresponding rule. These rules serve two purposes. We summarize predictive performance of each formula in terms of dataset characteristics. We build a set of formula-selection rules by combining all "good" rules, i.e., the rules that predicts "good" performance for each formula, and use them to select a "right" rule quality formula for a (new) dataset. For formula selection, we can use the ELEM2 classification procedure that takes formula-selection rules to classify a dataset into a class of using a particular formula.

4.3 ELEM2 with Multiple Rule Quality Formulas

With formula-selection rules, we can apply ELEM2's classification procedure to select a rule quality formula before using ELEM2 to induce rules from a dataset.

Formula	Condition	Decision	Rule Quality	No. of Support Datasets
Q_{WS}	(NofA≤20)and(NofC=2)	Good	1.23	8
	(N<3823)and(classDistr=Even)	Good	0.77	4
Q_{MD}	(N>625)and(8<NofA≤18)	Good	1.56	6
	(NofC>8)	Good	1.07	3
Q_{LS}	(ClassDistr=Even)	Good	1.22	6
	(N≤24)	Bad	1.61	1
$Q_{Coleman}$	(N>768)and(NofA>8)	Good	1.52	5
	(N>1484)	Good	1.34	4
	(351<N≤683)and(NofA≤9)	Bad	1.57	2
Q_{Prod}	(N≤214)and(NofA≤13)	Good	1.66	5
	(NofA>20)	Good	1.05	2
	(351<N≤653)	Bad	1.40	2
$Q_{G2.05+}$	(N>1484)	Good	1.77	4
	(NofA>20)	Good	1.26	2
	(NofA≤7)and(NofC>2)and(ClassDistr=Uneven)	Bad	1.38	3
Q_{IS}	(N>1484)	Good	1.77	4
	(NofA>20)	Good	1.26	2
	(NofA≤7)and(NofC>2)and(ClassDistr=Uneven)	Bad	1.38	3
$Q_{\chi^2.05+}$	(87<N≤178)	Good	1.27	3
	(13<NofA<15)	Bad	1.57	2
Q_{Cohen}	(101<N≤1484)and(NofA≤8)and(NofC>2)	Good	1.34	4
	(768<N≤2310)and(8<NofA≤18)	Bad	1.40	2
$Q_{\chi^2.05}$	(N≤87)	Good	1.57	2
	(9<NofA<14)and(NofC≤2)	Good	1.57	2
	(N>87)	Bad	1.15	15

Table 5. Formula Behavior Rules

Thus, ELEM2 can use different formulas on different datasets. To show this strategy, we conduct ten-fold evaluation of ELEM2 on the 22 datasets we used. The result is shown in Figure 2, in which the average accuracy mean from the "flexible" ELEM2 (labeled "Combine" in the graph) is compared with ones using individual formulas. We also conduct paired t-tests to see how much the flexible

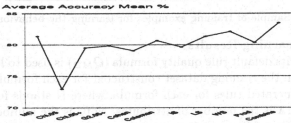

Fig. 2. Average of accuracy means of each formula on the 22 datasets

ELEM2 improves over ELEM2 with a single rule quality formula. The p-values from the t-test are shown in Table 6. "Combine" improves Q_{WS}, Q_{MD} and Q_{LS} at the 2.5% significance level; and it improves other formulas more significantly at the 0.5% significance level.

	Q_{WS}	Q_{MD}	Q_{LS}	$Q_{Coleman}$	Q_{Prod}	$Q_{G2.05+}$	Q_{IS}	$Q_{\chi^2.05+}$	Q_{Cohen}	$Q_{\chi^2.05}$
Combine	0.0182	0.0217	0.0228	0.0031	0.0002	0.0009	0.0007	0.0005	0.0083	0.0006

Table 6. Significance levels of the improvement of "Combine" over individual formulas

5 Conclusions

We have described and evaluated various statistical and empirical formulas for defining rule quality measures. The performance of these formulas varies among datasets. The empirical formulas, especially Q_{WS}, work very well. Among statistical formulas, Q_{MD} and Q_{LS} work the best on the tested datasets and are comparable with Q_{WS}. To determine the regularity of a rule quality formula's performance in terms of dataset characteristics, we used ELEM2 to induce rules from a dataset constructed from the experimental results. These rules provided ideas about the situations in which a formula leads to good, medium or bad performance. These rules can also be used to automatically select a rule quality formula before rule induction. Our experiment showed that this selection can lead to significant improvement over the rule induction system using a single formula. Future work includes testing our conclusions on more datasets to obtain more reliable formula-selection rules.

Acknowledgment

The authors are members of the Institute for Robotics and Intelligent Systems (IRIS) and wish to acknowledge the support of the NCE program of the Government of Canada, NSERC, and PRECARN Associates Inc.

References

1. Ali, K. and Pazzani, M. 1993. "HYDRA: A noise-tolerant relational concept learning algorithm". *Proceedings of IJCAI'93*, Chambery, France. Morgan Kaufmann.
2. An, A. and Cercone, N. 1998. "ELEM2: A Learning System for More Accurate Classifications." *Lecture Notes in Artificial Intelligence 1418.*
3. Bishop, Y.M.M, Fienberg, S.E. and Holand, P.W. 1991. *Discrete Multivariate Analysis: Theory and Practice*. The MIT Press.
4. Brazdil, P. and Torgo, L. 1990. "Knowledge Acquisition via Knowledge Integration". In: *Current Trends in Knowledge Acquisition*, IOS Press.
5. Bruha, I. 1993. "Quality of Decision Rules: Empirical and Statistical Approaches". *Informatica*, 17, pp.233-243.
6. Bruning, J.L. and Kintz, B.L. 1997. *Computational Handbook of Statistics*, Addison-Wesley Educational Publishers Inc.
7. Clark, P. and Niblett, T. 1989. "The CN2 Induction Algorithm". *Machine Learning*, 3, pp.261-283.
8. Cohen, J. 1960. "A Coefficient of Agreement for Nominal Scales". *Educational and Psych. Meas.* 22, pp.37-46.
9. Holte, R., Acker, L. and Porter, B. 1989. "Concept Learning and the Problem of Small Disjuncts". *Proceedings of IJCAI'89*, Detroit, Michigan.
10. Kononenko, I. and Bratko, I. 1991. "Information-Based Evaluation Criterion for Classifier's Performance". *Machine Learning*, 6, 67-80.
11. Michalski, R.S. 1990. "Pattern Recognition as Rule-Guided Inductive Inference". *IEEE Transactions on Pattern Analysis and Machine Intelligence*, PAMI-2, 4.
12. Quinlan, J.R. 1993. *C4.5: Programs for Machine Learning*. Morgan Kaufmann Publishers. San Mateo, CA.
13. Robertson, S.E. and Sparck Jones, K. 1976. "Relevance Weighting of Search Terms". *J. of the American Society for Information Science*. Vol.27. pp.129-146.
14. Torgo, L. 1993. "Controlled Redundancy in Incremental Rule Learning". *ECML-93*, pp.185-195.

Rules as Attributes in Classifier Construction

Marcin S. Szczuka

Institute of Mathematics, Warsaw University
Banacha Str. 2, 02-097, Warsaw, Poland
e-mail: szczuka@mimuw.edu.pl

Abstract. A method for constructing classification (decision) systems is presented. The use of decision rules derived using rough set methods as new attributes is considered. Neural networks are applied as a tool for construction of classifier over reconstructed dataset. Possible profits of such an approach are briefly presented together with results of preliminary experiments.

1 Introduction

In the process of constructing classification (decision) sytems we have several objectives in mind. Among others, we concern robustness, versatility, adaptiveness, compactness and intuitive understanding of produced solution. Of course it is tough job to fulfill all the expectations, especially if our system is based only on the information contained in the data. In this paper we are trying to address the issue of compactness and adaptiveness of a classifier. We propose a method of treating decision rules as a source for new features. Using those rules we construct new set of data that is easier to classify. A simple artificial neural network is used for this purpose

The classifier constructed in such a way shows, according to preliminary experiments, some nice features. It is usually smaller and simpler than rough set classifier having comparable quality. It is also easier to explain intuitively as it has less components.

The paper begins with introduction of basic notions. Then some foundational features of rule based rough set classifiers are presented. Next sections contain short description of proposed solution and the initial experimental results.

2 Basic notions

The structure of data that is subject of our study is represented in the form of *information system* [9] or, more precisely, the special case of information system called *decision table*.

Information system is a pair of the form $\mathbf{A} = (U, A)$ where U is a *universe* of *objects* and $A = (a_1, ..., a_m)$ is a set of *attributes* i.e. mappings of the form $a_i : U \rightarrow V_a$, where V_a is called *value set* of the attribute a_i. The decision table is also a pair of the form $\mathbf{A} = (U, A \cup \{d\})$ where the major feature that is different from the information system is the distinguished attribute d. In case of decision table the attributes belonging to A are called *conditional attributes* or simply *conditions* while d is called *decision* (sometimes *decision attribute*). We will further assume that the set of decision values is finite and by $rank(d)$ we will refer to its cardinality. The i–th *decision class* is a set of objects $C_i = \{o \in U : d(o) = d_i\}$, where d_i is the i–th decision value taken from decision value set $V_d = \{d_1, ..., d_{rank(d)}\}$

For any subset of attributes $B \subset A$ *indiscernibility relation* IND(B) is defined as follows:

$$xIND(B)y \Leftrightarrow \forall_{a \in B} a(x) = a(y) \tag{1}$$

where $x, y \in U$.

Having indiscernibility relation we may define the notion of reduct. $B \subset A$ is a *reduct* of information system if $IND(B) = IND(A)$ and no proper subset of B has this property.

Decision rule is a formula of the form

$$(a_{i_1} = v_1) \wedge ... \wedge (a_{i_k} = v_k) \Rightarrow d = v_d \tag{2}$$

where $1 \leq i_1 < ... < i_k \leq m$, $v_i \in V_{a_i}$. Atomic subformulae $(a_{i_1} = v_1)$ are called *conditions*. We say that rule r is *applicable* to object, or alternatively, the object *matches* rule, if its attribute values satisfy the premise of the rule. With the rule we can connect some characteristics. *Support* denoted as $Supp_\mathbf{A}(r)$ is equal to the number of objects from $\mathbf{A}$ for which rule r applies correctly i.e. premise of rule is satisfied and the decision given by rule is similar to the one preset in decision table. $Match_\mathbf{A}(r)$ is the number of objects in $\mathbf{A}$ for which rule r applies in general. Analogously the notion of matching set for a collection of rules may be introduced. By $Match_\mathbf{A}(R, o)$ we denote the subset M of rule set R such that rules in M are applicable to the object $o \in U$.

3 Rule based decision systems

Among the others, we may use the decision (classification) support systems based on rules derived from data. There are several approaches to generate such rules. They differ in the way the rules are generated as well as in the form of rule representation and use. Nevertheless, all the approaches have some common, basic questions to answer. One of them, probably most important one while classifying new, unseen objects is this about trustworthness of a rule or group of rules. Depending on approach, there may be several issues to solve while deciding what the decision for new-coming object should be.

Given a set of decision rules $R = (r_1, ..., r_m)$ derived from data by some method and the new object o_i, we may face several problems while trying to make decision. Namely:

1. They may be no rule in R that is applicable to o_i. In other words the values of conditional attributes of o_i do not satisfy conditions of any rule in R. In

such a case we cannot make decision since there is no knowledge within our rule set that covers the case of o_i.

2. There are several rules in R that are applicable to o_i but they give contradictory outputs. This situation, known as *conflict* between rules must be resolved by applying procedures to measure the confidence of particular rules (or groups of them).

There is a number of possible solutions to above two problems. Usually to resolve the problem of non-applicability of rules one of three methods may be applied:

- The object is assigned the *default* value of decision according to preset assumptions.
- The rule that has best (according to a given criterion) applicability is chosen and the decision is determined by this rule. The applicability criterion may be based e.g. on number of conditions in the rule that are satisfied by object. Other such criterion may be induced by preferences about decision value like in case of ordered decision domain.
- The "don't know" signal is returned to the user.

Of course, rule based decision systems are usually build in the manner to avoid the situation of not recognizing new object. But still, the actual accuracy depends on the quality of derived rules.

The matter of resolving conflicts between rules may be even more complicated, especially in case when we have bunch of them and no external, additional information about their applicability and importance. To cope with that problem, several techniques may be applied (refer to [7]). Bringing all of them here is rather impossible but we discuss some below.

The most popular way to establish final decision is based on comparison of the number of rules form different decision classes that are applicable to a given object. The object is assigned to the class determined by majority of the rules (in comparison with other classes). This method, however, causes unification of rule importance. This may be a serious weakness and in order to avoid it weights may be assigned to rules (or groups of them). The method we exploit in our experiments is based on the following formula describing weight for set of rules:

$$
W_{BSS}(M,o) = \begin{cases} \dfrac{\sum\limits_{r \in Match(M,o)} card(Supp_A(r) \cdot SC_A(r))}{\sum\limits_{r \in M} card(Supp_A(r) \cdot SC_A(r))} & \text{if } \sum\limits_{r \in M} card(Supp_A(r) \cdot SC_A(r)) > 0 \\ 0 \text{ otherwise} \end{cases} \quad (3)
$$

where $SC_A(r)$ is called stability coefficient and it is determined during the process of rule calculation using dynamic reducts (see [3], [4] for detailed explanation). To give some intuition about $SC_A(r)$ is is worth knowing that it mainly depends on frequency of occurrence of rule r in the set of optimal rules at subsequent steps of the dynamic algorithm for rule generation (see [4]). We use this method because numerous experiments (see [4], [3], [11]) prove that it is, on the average, better than other.

4 Rough set rule induction

The process of creating rules with use of rough set techniques is essential for our ideas of classifier construction. Therefore some basic information about methods for rule induction is needed. The base for deriving rules is reduct calculation. Numerous practical experiments show that usually there is a need for calculation of several reducts in order to get satisfiable quality of classification. Most of the cases involving larger set of data require calculation of reducts and rules with use of dynamic techniques. From technical point of view the process of calculating the reducts and rules is computationally exhaustive and for real-world solutions some approximate techniques like heuristics or genetic algorithms are engaged (see e.g.[12]).

The derived set of rules R may be for some reason unsatisfactory. The major concerns are:

- The number of rules is excessive so the cost of storing, checking against and explaining the rules is not acceptable.
- The rules are too general, so they do not really contain any valid knowledge, or too specific, so they describe very small part of the universe in too much detail.

To avoid at least part of the problems mentioned above we may apply shortening procedures. Those procedures allow to shorten the rules and, in consequence, reduce the number of them. The process of rule shortening comprises of several steps that, in consequence, lead to removing some descriptors from a particular rule. Usually, after shortening, the number of rules decreases as repetitions occur in the set of shortened rules. There are several methods leading to this goal, for details review e.g. [1], [4], [13].

5 Rules as attributes

In the classical approach, once we have decision rules we are at the end of classifier construction. But there is also other way of treating the rules since they describe relations existing in our data. Therefore we may treat them as features of objects. In this view, the process of rule extraction becomes the one of new feature extraction. These features are of higher "order" since they are taking into account specific configurations of attribute values with respect to decision.

Let us consider set of rules $R = (r_1, ..., r_m)$. We may construct the new decision table based on them.

With every rule r_i in R we connect a new attribute ar_i. The decision attribute remains unchanged as well as the universe of objects. The values of attributes over objects may be defined in different ways depending on the nature of data. For the purposes of this research we use the following three possibilities:

- $ar_i(o_j) = d_k$ where d_k is the value of decision returned by rule r_i if it is applicable to object o_j, 0 (or any other constant) otherwise.

- $ar_i(o_j) = const$ (usually $const$ equal 1 or -1) if the rule r_i applies to the object o_j, 0 (or any other constant) otherwise.
- In case of tables with binary decision $ar_i(o_j) = 1$ if the rule r_i applies to the object o_j and the output of this rule points at decision value 1. $ar_i(o_j) = -1$ if the rule r_i applies to the object o_j and the output of this rule points at decision value 0. When the rule is not applicable $ar_i(o_j) = 0$.

Due to technical restrictions in further steps of classifier construction it is sometimes necessary to modify above methods by e.g. encoding the decision values in first of the approaches in order to use neural network as it is in our case.

It can be easily seen how important is to keep the rule set within reasonable size. Otherwise the newly produced decision table may become practically unmanageable due to the number of attributes.

6 The making of classifier

Equipped with the decision table extracted with use of the set of rules we may now proceed with construction of final classification (decision) system. In order to keep computation within reasonable size with respect to time and spatial complexity we apply very simple and straight methods. Namely we use a simple sigmoidal neural network with no hidden layers (see [6]). The overall process of classifier construction is illustrated in Figure 1.

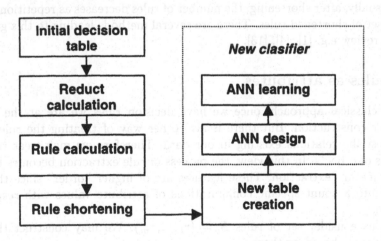

Fig. 1. The layout of new classifier

We start with initial, training decision table for which we calculate reducts and the set of possibly best rules. We may derive rules in dynamic or non-dynamic way depending on the particular situation (data). These rules are then used to construct new decision table in the manner described in previous section.

Over such constructed new data table we build neural network based classifier to classify newly formed objects. Then classifier is checked against quality on testing set.

Of course with the proposed scheme we may construct various classifiers as some parameters may be adjusted on any step of this process. In the process of reduct and rule calculation we may establish restrictions for number and size of reducts (rules) as well as on rule specificity, generality, coverage and so on. During neural network construction we may apply different learning algorithms. The learning coefficients of those algorithms may vary as well.

To complete the picture of classifier it is important to add a handful of technical details. For the purpose of the research presented in this paper we used dynamic calculation of rules based on genetic algorithm and incorporating some discretisation techniques for attributes continuous in their nature. For details consult [5]. On the side of neural network we used simple architecture with neurons having classical sigmoid or hyperbolic tangent as the activation function. Usually, the network is equipped with bias and trained using gradient descent with regularisation, momentum and adaptive learning rate (see [6],[2]).

The simple architecture of neural network has one additional advantage. From its weights we may decipher the importance of particular attributes (rules) for decision making. It is usually not the case of more complicated neural architecture for which such an interpretation is difficult and the role of single inputs is not transparent.

7 Experimental results

The proposed methods have been tested against real data tables. For testing we used two benchmark datasets taken from repository [14] and one dataset received from medical sources. Table 1 below describes basic parameters of decision tables used in experiments. The EEG data was originally represented as matrix of

Dataset	Objects	Attributes	Attribute type	$rank(d)$
Monk1	432	6	binary	2
Monk2	432	6	binary	2
Monk3	432	6	binary	2
Lymphography	148	18	symbolic	4
EEG	550	105	binary	2

Table 1. Datasets used for experiments.

signals that was further converted to binary form by applying wavelet analysis and discretisation techniques as originally proposed in [11], [5] and developed in [10]. The MONK datasets have preset partition into training and testing set, rest of the data sets were tested using cross-validation method.

The rules were calculated using dynamic techniques. Then we performed several experiments using different rule shortening ratio. The table 2 below shows best results. Columns in this table describe number of rules used for new table

Data sets	Number of rules	Shortening ratio	Method	Error Proposed	Other
Monk1	31	0.6	TT	0/0.03	0/0
Monk2	26	0.6	TT	0/0.06	0/0.049
Monk3	44	0.6	TT	0/0.051	0/0.046
Lymphography	78	0.8	CV-10	0.03/0.19	0/0.15
EEG	13	0.3	CV-5	0/0.01	0.11/0.16

Table 2. The results of experiments.

(*Number of rules*), shortening ratio of rules (between 0 and 1), method of training/testing (TT=train & test, CV-n = n-fold cross-validation), average error on training/testing set as a fraction of the number of cases and best results got from other rough set methods for comparison. The experiments were performed several times in order to get averaged (representaive) results. The comparison is made with best result got from application of combined rough set methods. However, it is important to mention that those best classifiers are usually based on much larger sets of rules.

The results are comparable to those published in [8] and [3] but they usually use much less rules and simpler setting of classifier than in case of best results in [4] and [3]. The most significant boost is visible if we compare the outcome of classification using only the calculated rules with classical weight setting. Especially, in the case of small shortening ratio which corresponds to significant reduction of rules, the impact of methods proposed is clearly visible.

8 Conclusions

The proposed approach allows to construct classifier with combination of rule based systems and neural networks. The rough set rules derived with respect to the discernibility of object seem to posses extended importance, if used as new feature generators. Application of neural network in last stage of classifier construction allows better fitting to particular set of data and makes further addition of new knowledge to th system easier due to its adaptiveness.

Initial experiments show promising results, especially in cases of binary decision. Reduction of the number of rules used makes system obtained in this way closer to natural intuitions.

As the work on this issue is on its beginning, there is still a lot to do in many directions. Most interesting from our point of view is further investigation of relationships between process of rule induction with rough sets and their further quality as new attributes.

499

Acknowledgement : I want to thank Jan Bazan and Piotr Wojdyłło for sharing
their expertise and allowing to use some of their solutions and tools. This work
was supported by grant #8T11C02412 of Polish State Committee for Scientific
Research and ESPRIT project 20288 - CRIT 2.

References

1. Agrawal R., Manilla H., Srikant R., Toivonen H., Verkamo I., Fast Discovery of
   Association Rules, In: Proceedings of the Advances in Knowledge Discovery and
   Data Mining, AAAI-Press/MIT Press, 1996, pp. 307-328
2. Arbib M. A.(ed.), The Handbook of Brain Theory and Neural Networks, MIT
   Press, Cambridge MA, 1995
3. Bazan J., A Comparison of Dynamic and non-Dynamic Rough Set Methods for
   Extracting Laws from Decision Tables, In: Skowron A., Polkowski L.(ed.), Rough
   Sets in Knowledge Discovery 1, Physica Verlag, Heidelberg, 1998, pp. 321-365
4. Bazan J., Approximate reasoning methods for synthesis of decision algorithms (in
   Polish), Ph. D. Thesis, Department of Math., Comp. Sci. and Mechanics, Warsaw
   University, Warsaw 1998
5. Nguyen Sinh Hoa, Nguyen Hung Son, Discretization Methods in Data Mining, In:
   Skowron A., Polkowski L.(ed.), Rough Sets in Knowledge Discovery 1, Physica
   Verlag, Heidelberg, 1998, pp. 451-482
6. Karayannis N. B., Venetsanopoulos A. N., Artificial Neural Networks: Learning
   algorithms, Performance Evaluation and Applications, Kluwer, Dordrecht, 1993
7. Michalski R., Tecuci G., Machine Learning IV: A Multistrategy Approach, Morgan-
   Kaufmann, San Francisco, 1994
8. Michie D., Spiegelhalter D. J., Taylor C. C., Machine Learning, Neural and Statistical Classification, Ellis Horwood, London, 1994
9. Pawlak Z., Rough Sets: Theoretical Aspects of Reasoning about Data, Kluwer,
   Dordrecht, 1991
10. Szczuka M., Wojdyłło P., Neuro-Wavelet Classifiers for EEG Signals Based on
    Rough Set Methods, submitted (June '99) to Neurocomputing
11. Wojdyłło P., Wavelets, Rough Sets and Artificial Neural Networks in EEG Analysis,
    Proceeding of RSCTC'98, Lecture Notes in Artificial Intelligence 1424, Springer
    Verlag, Berlin, 1998, pp. 444-449
12. Wróblewski J., Covering with Reducts - A Fast Algorithm for Rule Generation,
    Proceeding of RSCTC'98, Lecture Notes in Artificial Intelligence 1424, Springer
    Verlag, Berlin, 1998, pp. 402-407
13. Ziarko W., Variable Precision Rough Set Model, Journal of Computer and System
    Sciences, 40 (1993), pp. 39-59
14. The Machine Learning Repository, University of California at Irvine,
    http://www.ics.uci.edu/ mlearn/MLRepository.html

An Algorithm to Find the Optimized Network Structure in an Incremental Learning

Jong Chan Lee[1], Won Don Lee[2] and Mun-Sung Han[3]

[1] Dept. of AI, ChungWoon Univ., Hongsung, ChungNam, KOREA
jclee@aisun.cwunet.ac.kr
[2] Dept. of Computer Science, ChungNam Nat'l Univ., Taejun, KOREA
wdlee@chungnam.ac.kr
[3] Speech Understanding Lab., ETRI, Taejun, KOREA
msh@etri.re.kr

Abstract. In this paper [1] we show a new learning algorithm for pattern classification. A scheme to find a solution to the problem of incremental learning algorithm is proposed when the structure becomes too complex by noise patterns included in the learning data set. Our approach for this problem uses a pruning method which terminates the learning process with a predefined criterion. Then an iterative model with a 3 layer feedforward structure is derived from the incremental model by appropriate manipulation. Note that this network is not fully connected between the upper and lower layers. To verify the effectiveness of the pruning method, the network is retrained by EBP. We test this algorithm by comparing the number of nodes in the network with the system performance, and the system is shown to be effective.

1 Introduction

Conventional iterative models such as EBP usually have a fixed feedforward network structure and use an algorithm to gradually modify the weights of networks as learning proceeds. So this approach does not allow to expand the network during training. This approach sometimes has a critical limitation, depending on the trial and error method or ad hoc schemes to obtain an appropriate architecture for learning patterns.

Therefore another approach is devised to solve this problem by adding nodes to the network when necessary. This type of learning is referred to as incremental learning as the network grows as training occurs. As a procedure, Lee et al.[2] have proposed an incremental algorithm using Fisher's Linear Discriminant Function(FLDF)[1]. This model searches an optimal projection plane based on the statistical method for pattern classification. And then, after projecting patterns on this projection plane, this model starts a search procedure for an optimal hyperplane based on an entropy measure and thus determines the neuron in the structure.

---

[1] This research is supported by Brain Science and Engineering Research Program in Korea

Lee et al.[3] introduced a neural network learning algorithm which transforms a structure of an incremental model into that of an iterative model. This model showed that the weights and thresholds as well as the structure of the 3 layer feedforward neural network can be found systematically by examining the instances statistically. It is well known that a major part of the learning capability is in the architecture of its models.

In iterative models the approaches to solving this problem are as follows. Kung et al.[5] proposed a method which is learning with a network structure with a predefined node number. But this method had a problem in that it converges less than the theoretical bases. Sietsma and Dow[7] devised an algorithm which assigns many nodes in prior learning and then removes nodes by making an observation of inactive nodes in learning. Though this algorithm can be applied to simple problems, when a problem is more complex one encounters many difficulties. Hanson and Pratt[9] proposed an algorithm which removes hidden nodes with a constraint term in EBP function. But this algorithm has a side effect which reduces the probability of the convergence. Hagiwara[8] proposed an algorithm which considers a proper node number and weight values concurrently. But this algorithm needs much time to converge. Moody and Rognvaldsson[10] proposed adding the complexity-penalty term, but it has much more complicated form and also demands much more computational complexity. Wangchao et al.[11] introduced the sparselized pruning algorithm for a higher-order neural network, but it is applied after all the higher-order weights are trained.

The incremental model uses an algorithm trying to produce a neural network with near-optimal architecture intelligently. But this model has a drawback in that it can be extremely extended by noises included in patterns. In this paper we propose a method to solve this problem.

2 Background

2.1 The Incremental Network Model

In this paper we present a pattern by a vector of n components and describe a pattern classifier as a mapping of the input pattern space, a subset of n dimensional real space, to the set of classes 1,2,...,k. In order to make the output decisions we develop the constructs which build internal representations of a class description. For doing this, we represent one unit as a hyperplane specified by elements (weight vector, threshold value) and then partition the space as follows.

$$Hyperplane(P) = \{X | X \in R^n \text{ and } W^T X = T\} \quad (1)$$

where, X : Input pattern, R^n : Real space, W : Weight, T : Threshold
The hyperplane P separates the space H^n into two sets, P^L and P^R. Thus input X belongs to P^R or P^L by P.

$$P^R = \{X | X \in H^n \text{ and } W^T X \geq T\}$$
$$P^L = \{X | X \in H^n \text{ and } W^T X < T\}$$

The network structure of the model consists of one input unit, a number of hidden units and output units as many as the number of classes. Each unit has a weight vector and a threshold value. The input X is broadcasted to all units and for each unit at most one path is activated. Thus, in the whole network one path at most is activated for each input vector.

2.2 The Training Process

During the training phase, a collection of classified patterns describing a desired class is presented to an incrementally formed network of neurons. And the weight vector and threshold values of units of the network are determined by an adaptation process. Each unit is assigned to represent a certain hyperplane which is part of the discriminant hypersurface represented by the network. The training set is presented a number of times and at each presentation the network is expanded by adding a number of new units. The adaptation process is carried out at each unit independently of others.

In the adaptation process, the Fisher's linear discriminant function[2] is used in order to determine the optimal hyperplane. Fisher's linear discriminant function provides the optimal weight for the input pattern data for an arbitrary distribution. Fisher's formula for n classes is shown in (2). The optimality is characterized by the overall measure representing the mutual distances between a set of projected points of a class and that of another class and is achieved by maximizing the overall measure, B, standardized by V. We use Cauchy-Schwartz inequality for obtaining maximum value of $V^{-1}B$.

$$\frac{W^T BW}{W^T VW} = \frac{W^T[\sum_i (\overline{X_i} - \overline{X})(\overline{X_i} - \overline{X})^T]W}{W^T[\sum_i \sum_j (X_{ij} - \overline{X_i})(X_{ij} - \overline{X_i})^T]W} \quad (2)$$

Let $\beta = V^{1/2}W$, then (2) becomes $\frac{\beta^T V^{-1/2}BV^{-1/2}\beta}{\beta^T \beta}$. This formulus attains the highest value when vector β becomes the eigenvector e_1 which is associated with the highest eigenvalue λ of the matrix $V^{-1/2}BV^{-1/2}$. Thus weight vector(W) is obtained as $V^{-1/2}e_1$.

The threshold value determining the position of the hyperplane is obtained based on the following entropy function.

$$H(C|(\delta^d)) = PL^{(*)}H(p_1) + PR^{(*)}H(p_2) \quad d : cursor\ position \quad (3)$$

After projection(W^TX), the projected points are divided into two parts by a dividing plane placed on a cursor position. After entropy is measured, the plane is moved one at a time from $d_1(W^TX_1)$ to $d_{n-1}(W^TX_{n-1})$. The optimal position is where the smallest value of entropy is found. Let $n_1(n_2)$ be the number of left (right) region, then $PL^* = n_1/(n_1 + n_2)$ and $PR^* = n_2/(n_1 + n_2)$. Let x_{ij} be the number of class j events in each region i(left, right). The probability p_{ij} whose class will be j is x_{ij}/n_i. Then, $H(p_i) = -\sum p_{ij}log_2 p_{ij}$ i=1,2 j=1,...,class number.

503

2.3 Translation into an iterative model structure

In this section we present a transforming procedure which converts the incremental network topology into an iterative one.

Step 1 : The input layer consists of as many nodes as the number of variables (dimensions) in learning patterns.

Step 2 : The first layer has the same number of nodes as that of the hidden nodes except for leaf nodes in the incremental model. The weight and the threshold between the input layer and the first layer are fully connected and have the same values as those of hidden nodes in the incremental model.

Step 3 : The second layer has as many nodes as the number of leaf nodes representing the region of each class in the incremental model. Each region is made by the intersection of hyperplanes in the first layer, thus this layer is characterized by AND : when all the inputs are active, the output is active. Weight and threshold between the first layer(j) and the second layer(i) is determined by each path, from the input node to the leaf node in the incremental model : that is,

$$\sigma_{iL} = -1, \sigma_{iR} = +1$$
$$W_{ij}(weight) = \sigma_{iD}, T_i = \Sigma|\sigma_{iD}| - 0.5$$

Here, i=0,...,(the number of discriminated region-1) and σ_{iD}(D=L(left space), D=R(right space)) denotes i_{th} path.

Step 4 : In the third layer there are as many nodes as the number of classes. The intersected regions from the second layer are unioned in each class region. Weight and threshold between the second(j) and the third(i) layer is determined by the class of each region.

$$\text{If the region from } i_{th} \text{ path contains class j} : R_{ij} = 1$$
$$\text{Otherwise} : R_{ij} = 0$$

$$W_{ij}(\text{Weight}) = R_{ij}$$
$$T_i(\text{Threshold}) = \Sigma R_{ij} - 0.5$$

For further information, please refer to [3].

3 Pruning in the incremental learning Model

3.1 Pruning algorithm for the proposed incremental model

Most of the network structure of the incremental learning algorithm have the shape of the binary trees and hence the number of the nodes does not need to be predetermined as the structure is determined. But an incremental model has to solve a new problem, as there can be too many nodes to be added. This is because a learning pattern set can contain many noisy data. The algorithm above also has this problem as it iterates until there is only one class of data contained at a divided space. As the algorithm proceeds recursively, not only the computational time but also the memory wastes are increased. Moreover, the network performance can be degraded because of the noise effect. Therefore, an algorithm must be devised to overcome these problems. The following formulas are added to the algorithm introduced in the section 2.2 to solve those problems:

$$IF(N(\bullet) > PruneRate) \; Continue \; the \; learning \; procedure.$$
$$ELSE \; Terminate \; the \; learning \; procedure.$$

where Prune Rate is a criterion for the percentage of the noise. And $N(\bullet)$ for the learning pattern P at node i is determined as follows:

$$N_i(P) = \frac{(\#TE - \#MCE)}{\#MCE} \times 100.0 \tag{4}$$

- #TE : Total number of patterns at node i.

- #MCE : The number of patterns of a class with the most number among the #TE patterns.

3.2 Selective-learning algorithm

In the section of 2.3 we introduced a method which transforms the structure of an incremental model into that of a 3 layer iterative model. This transformed network structure is not a fully connected but partially connected one except for between the input layer and the hidden layer. In this section we propose a method to reduce the network structure using an iterative learning algorithm. This algorithm is based on the observation that, in the algorithm explained in the section 3.1, the pruning procedure corresponds to a node reduction, and the partial connection between layers corresponds to a reduction of weights set size. The network structure of this model consists of nodes with the threshold function and the sigmoid function. This structure is shown in Fig. 1. We use the EBP algorithm to train the structure. As an initial structure, the first layer is constructed with the nodes with the same weights and threshold values determined in incremental learning. And the bias nodes are added to the second hidden layer and the output layer.

The learning procedure of this model is described below. The first hidden layer(O) is activated as follows:

$$IF((\sum w_{ij}X_j) > T_i) \; O_i = 1$$
$$ELSE \; O_i = 0$$

An EBP learning is performed on the upper layer using the output from the first layer. As our model is partially connected, the EBP is done according to the following:

$$\Delta_p w_{ji}(n+1) = C_{ji}(\eta \delta_{p_j} O_{p_i} + \alpha \Delta_p w_{ji}(n)) \tag{5}$$

where C_{ji} is 1 if there is a connection between node j in the upper layer and node i in the lower layer, and 0 otherwise.

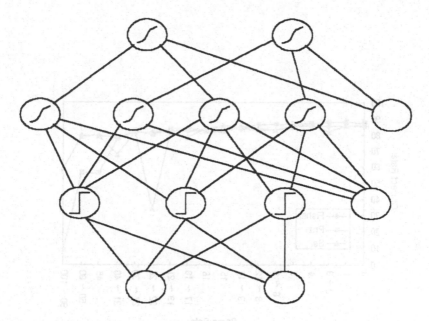

Fig. 1. The network structure of the selective learning model

4 Implementation

To test our system, we use sleep stage scoring data sets and the speech data used by Peterson and Barney. And we implemented the following and measured the performance.

1. Fisher : Weight vector and threshold values are determined using the algorithm in section 2.2. Prun Rate(PR, $0 \leq PR \leq 100$) is varied and the performance is measured.
2. Prun : After transforming into the network topology explained in section 2.3, we train the fully connected network by EBP with the weights and the thresholds initialized from the procedure 1.
3. Sel : After transforming into the network topology explained in section 2.3, we train the partially connected network by EBP with weights and thresholds initialized from the procedure 1. The performance is measured as the number of connections varies.

PR is increased by 1% from 0 to 30 %, and by 5% after 30%. The learning rate η and the momentum rate α are set to 0.2 and 0.7, respectively. From the experiment, we observe that the performance of the "Sel" upto PR equals 21% and the "Prun" upto PR equals 40% is similar or even better than that of the "Fisher". The result is shown in Fig. 2. Fig. 3 shows that the number of the nodes decreases as PR increases. Fig. 4 shows that the number of connections decreases as the PR increases. As "Prun" has fully connected network structure while "Sel" has a partially connected one, the number of the connections of the "Prun" is more than that of the "Sel".

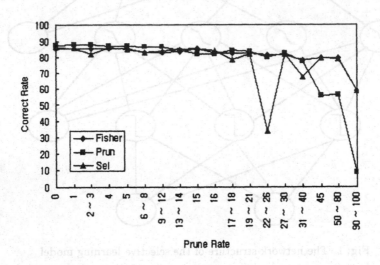

Fig. 2. The performance comparison of 3 learning methods

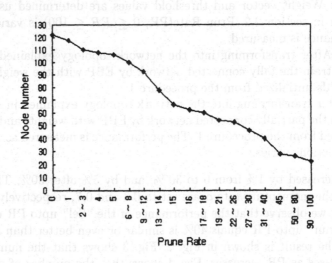

Fig. 3. The number of nodes vs. PR.

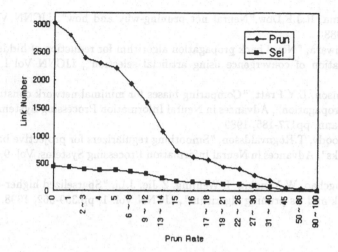

Fig. 4. The number of connections vs. PR.

5 Conclusion

In this paper a solution is proposed to solve the problem that the network structure of an incremental model can be extended excessively when the learning pattern contains many noisy patterns. The proposed method uses a predefined parameter, PR, to stop the recursive process in making the network structure. After this binary tree network structure is transformed into the three layer feedforward structure, the EBP is employed to train the structure further. An appropriate number of nodes and the corresponding weights between the nodes are determined, which is the aim of the pruning process.

References

[1] R.A.Johnson, D.W.Wichern, "Applied Multivariate Statistical Analysis. 2ED", pp470-542, Prentice-Hall International Editioned, 1988.

[2] J.C.Lee, Y.H.Kim, W.D.Lee, S.H.Lee, "Pattern Classifying Neural Network using Fisher 's linear discriminant function", Vol1, IJCNN June, 1992.

[3] J.C.Lee, Y.H.Kim, W.D.Lee, S.H.Lee, "A method to find the structure and weights of layered neural networks, WCNN, pp552-555, June 1993.

[4] M. Hagiwara, "A simple and effective method for removal of hidden units and weights", Neurocomputing 6, pp207-218, 1994

[5] S.Y.Kung, J.N.Hwang,"An algebric projection analysis for optimal hidden units size and learning rates in back-propagation learning", IJCNN Vol I, pp363-370, 1988

[6] A.S.Weigend, D.E.Rumelhart, B.A.Huberman, "Generalization by weight-elimination with application to forecasting", Advances in Neural Information Processing Systems 3, Morgan Kaufmann, pp875-882, 1991

[7] J.Sietsma, R.J.F.Dow,"Neural net pruning-why and how", IJCNN Vol I, pp325-333, 1988

[8] M.Hagiwara, "Novel back propagation algorithm for reduction of hidden units and acceleration of convergence using artificial selection", IJCNN Vol I, pp625-630, 1990

[9] S.J.Hanson, L.Y.Pratt, "Comparing biases for minimal network construction with back-propagation", Advances in Neural Information Processing Systems 1, Morgan Kaufmann, pp177-185, 1989.

[10] J.E.Moody, T.Rognvaldsson, "Smoothing regularizers for projective basis function networks", Advances in Neural Information Processing Systems, Vol. 9, pp.585-591, 1997.

[11] L.Wangchao, W.Yongbin, L.Wenjing, Z.Jie, J.Li, "Sparselized higher-order neural network and its pruning algorithm", IJCNN Vol. 1, pp.359-362, 1998.

Patterns in Numerical Data:
Practical Approximations to Kolmogorov
Complexity

T.Y. Lin[1,2]

[1] Department of Mathematics and Computer Science
San Jose State University, San Jose, California 95192
[2] Department of Electrical Engineering and Computer Science
University of California, Berkeley, California 94720
tylin@cs.sjsu.edu, tylin@cs.berkely.edu

Abstract. Intuitively, patterns of numerical sequences are often interpreted as formulas. However, we observed earlier that such an intuition
is too naive. Notions analogous to Kolmogorov complexity theory are introduced. Based on these new formulations, a formula is a pattern only
if its pattern complexity is simpler than the complexity of data.

1   Introduction

Mathematicians routinely write down the general term of a given sequence by
inspecting its initial terms. Such actions involve pattern discovery and sequence
prediction. The automation of the latter has been an important area in machine
learning [5].

Intuitively, the pattern of a numerical finite sequence is often interpreted as
a formula that generates the finite sequence. Earlier, we observed [3], somewhat
surprisingly, that such a simple minded formulation leads to no prediction phenomenon; see Section 3. Briefly, given any real number, there is a "pattern" that
predicts it. This phenomenon prompts us a more elaborated notion of patterns.

One possible fundamental approach is Kolmogorov complexity theory [4], in
which patterns are interpreted as algorithms. This approach is theoretical; there
is no practical way to determine the complexity of any explicitly given finite
sequence. Aiming toward practical applications, various notions of pattern complexities, analogous to that of Kolmogorov's, are proposed; see Section 5. Following Kolmogorov, we define the complexities of patterns and data, but based on
function theoretic views, instead of algorithmic views; as a conclusion, *a formula
is a pattern, only if its complexity is simpler than that of the numerical data* .

The proposed theories are probably overly simplified notions, but are practically manageable approximations to that of Kolmogorov. Finally, we may want
to point out that mathematicians use not only the numerical values but also
their "physical" expressions to predict the sequence. In this paper, we focus,
however, only on the numerical values.

2 Patterns of Numerical Data

Given a finite numerical sequence,

$$Seq: \ a_1, a_2, \ldots a_n.$$

What would be the most natural meaning of its *pattern*? Intuitively, a pattern is a formula $G(x)$ that generates the finite sequence. Mathematically, a formula is not a precisely defined term, roughly, it can be interpreted as a function that can be expressed by well-known functions, such as polynomials, trigonometric, radial basis or other special functions. Since the formula $G(x)$ is often valid beyond n, such a formula is also referred to as a *generalization*.

Next let us rephrase the problem in geometry; the finite sequence can be viewed as a set of points in Euclidean plan.

$$(1, a_1), (2, a_2), \ldots (n, a_n) \ldots$$

and the problem is to find a function

$$G: i \ \longrightarrow \ a_i, \ i = 1, 2, \ldots n$$

whose graph is a "nice" curve passing through these given points. It seems clear there would have many "nice" curves. Occam's razors are needed; so pattern complexity theories are developed.

3 Intuitive Solution - No Prediction Phenomenon

Let us recall the following arbitrary prediction phenomenon [3]: Suppose we are given a sequence

$$1, 3, 5, 7$$

What would be the next number? Commonly, one would say, according to the pattern of the initial four terms,

the next number would be 9.

However, we have a somewhat surprising observation. We got following examples from using Matlab:

1. $f_6(x)$ predicts: $1, 3, 5, 7, 6$

$$f_6(x) = -0.1250x^4 + 1.2500X^3 - 4.3750X^2 + 8.2500X - 4.0000$$

2. $f_8(x)$ predicts: $1, 3, 5, 7, 8$

$$f_8(x) = -0.0417x^4 + 0.4167X^3 - 1.4583X^2 + 4.0833X - 2.0000$$

Let us recall the folowing elementary algebra: If m values are assigned to m points in a Euclidean space (of dimension n), then there is a polynomial of n variables which assumes the given m values at m given points.

NO PREDICTION THEOREM
Given a finite numerical sequence

$$a_1, a_2, a_3, ...a_n$$

and any real number r, there is a pattern that would predict the following pattern

$$a_1, a_2, a_3, ...a_n, r.$$

It is clear that mere formula can not be the right notion for patterns. In the next few sections, we develop various complexity theories, analogous to Kolmogorov's, to formulate the notion of patterns.

4 An Attempt from Algorithmic Information Theory

A finite numerical sequence is, of course, expressible as a bit stream. So one might be able to apply Kolmogorov complexity theory here. Let us recall few notions. Let K denote the Kolmogorov complexity: Let length(p) be the length of the program p using a consistent method of counting, such as binary length.

$K(a) = Min\{length(p) \mid p$ is any conceivable program that generates the string $a\}$

Let $length(a)$ be the length of a string a. Then a is said to be Kolmogorov random, if $K(a) \geq length(a)$. A finite sequence is said to have a pattern if $K(a) < length(a)$. Intuitively $K(a)$ measures the complexity of a pattern and $length(a)$ the complexity of data. Next let us quote few interesting propositions:

1. Almost all finite sequences are random (have no patterns).
2. Gödel type incomplete theorem: It is impossible to effectively prove that they are random.

Due to the last assertion, algorithmic information theory can not be useful here. More practically approaches that could approximate the Kolmogorov compexity are needed.

5 Complexity Theories of Patterns

Instead of capturing the algorithm that defines $G(x)$, we are looking for a method to describe $G(x)$ in terms of a class of known functions.

5.1 Algebraic Information Theory

In this section, we will use the class of polynomial functions as our basis. Though in apriori, it is not known that $G(x), x = 1, 21, \ldots n$ is a polynomial, but Weistrass approximation theorem states that $G(x)$ can be approximated by a polynomial. Since the domain $\{x \mid x = 1, 2, \ldots n\}$ is a finite set of points, so it is.

It is our religious belief that a shortest algorithm defines a simplest polynomial, and the degree is the best measure of its simplicity. So we believe G is the least degree of the polynomials that generate the finite sequence $a = \{a_x \mid x = 1, 2, \ldots n\}$. Note that $length(a) = (n - 1)$. By mimic Kolmogorov, let D denote the algebraic complexity and define

$$D(a) = Min\{degree(p) \mid p \text{ is any conceivable polynomials that generates the finite sequence } a\}$$

From the well ordering principle ([1], pp 11), there is a polynomial H whose degree is $D(a)$. This polynomial H is the desirable G on the domain $\{x \mid x = 1, 2, \ldots n)\}$. However, the natural domain of H is the real numbers; it is well beyond the original domain of G.

We need few notions: Let $length(a)$ be the length of a finite sequence a. Then a is said to be algebraic *random*, if $D(a) \geq length(a)$. A finite sequence is said to have a *pattern* if $D(a) < length(a)$. Intuitively $D(a)$ measures the complexity of an algebraic pattern and $length(a)$ the complexity of data. So H is the algebraic pattern, if a is not algebraic random.

Let us apply the theory to answer the no-prediction phenomenon. Note that $deg(f_6(x)) = 4$ and $length(Seq) = 4$, so $deg(f_6(x)) \geq length(Seq)$. By our theory, the polynomials found in Section 3 are not patterns. We should point out that $f_6(x)$ is excluded out by its degree (algebraic complexity). Certainly, it is conceivable that $f_6(x)$ may not be excluded out from algorithmic view, but our religious belief will *not* admit that.

Finally, we would like to point out that both algorithmic and algebraic patterns do meet the requirements of the first and second razors of Pedro Domingos [2], except the simplicity for the first razor is measured by two different metrics. Even though our religious belief stating that they are the same, in reality, it may be different, and the results may be addressing some deep issues; they will be investigated in near future.

5.2 Functional Information Theory

Instead of polynomial functions, we can also consider any Schauder basis (as Banach space [6]) of the function space (under consideration), such as, trigonometric functions in L^2-space, radial basis functions in L^p-space, and many others. In these categories of functions, it is less clear what would be the simplest one. For trigonometric functions, such as $sin(nx)$ or $cos(mx)$, we believe the least positive n or m are the simplest. Roughly, the weights (as used in neural networks) are the measures of their simplicity. The exact meaning of weights is, of course, Schauder basis specific; we shall not be specific here.

513

To mimic Kolmogorov, we need some notations: Let B be a selected Schauder basis. A linear combination of functions in B will be denoted as B-combination. As before, G denotes the function that generates the sequence x. Now, let F denote the functional complexity and define

$$F(x) = Min\{weight(p) \mid p \text{ is any conceivable } B\text{-combination that}$$
$$\text{approximates } G \}$$

As in previous section, the domain is a finite set of points, so G is actually a B-combination. To illustrate the idea, we will use the terminology of this section to explain the results in previous section. Let B be the set of all monomials. B is a Schauder basis. Since the domain is a finite set of points, G is exactly a linear combination of monomials, in other words, a polynomial. We got the result of previous section using the reasoning in this section.

As before, a finite sequence is said to have a *pattern* if $F(x) < length(x)$.

6 Conclusion

This paper examines various notions of patterns in finite numerical sequences. By adopting the approach of classical algorithmic information theory (Kolmogorov complexity theory), two approximations, called algebraic and functional information theories are proposed. Base on these theories, we conclude that a formula is a pattern only if its *pattern complexity* (with respect to its proper theory) is simpler than that of data. We believe the theories should be useful in scientific discovery or financial data mining.

References

1. G. Birkhoff and S. Mac Lane, A Survey of Modern Algebra Forth Ed., Macmillan Publising Co, New York, 1977
2. P. Domingos, Occam's Two Razors: The Sharp and the Blunt, In: R. Agrawal, P Stolorz, and G. Piatetsky-Shapiro, the Proceedings of the Fourth International Conference on Knowledge Discovery & Data Mining, New York City, August 27-31, 1998.
3. T. Y. Lin, "Discovering Patterns in Numerical Sequences Using Rough Set Theory," In: the Proceeding of the Third World Multi-conference on Systemics, Cybernetics and Informatics, Volume 5, Computer Science and Engineering. Orlando, Florida, July 31-August 4, 1999
4. Ming Li and P. M. Vitanyi, Komogorov Complexity and its Applications, in Handbook of Theoretical Computer Science J van Leeuwen
5. T. G. Dietterich and R. S. Michalski, Learning to Predict Sequences, In R. Micgalski, J. Carbonell, T. Michell(eds),Machine Learning: An Artificial Intelligeve Approach (Vol II,63-106), San Franciscom Morgan Kaufmann.
6. A. Mukherjea and K. Pothoven, Real and Functional Analysis, Plenum Press, New York, 1978

Performance Prediction for Classification Systems

F. Sun*

Corporate Research Center, Alcatel, Richardson, Texas
fushing.sun@aud.alcatel.com

Abstract. Performance prediction for classification systems is important. We present new techniques for such predictions in settings where data items are to be classified into two categories. Our results can be integrated into existing classification systems and provide an accurate and predictable tool for data mining. In any given classification case, our approach uses all available training data for building the classification scheme and guarantees zero classification errors on the training data. We re-use the same training data to predict the performance of that scheme. The method proposed here enables control of errors over two types of error for the classification task.

Keywords: Classification, Data Mining, Decision Support, Error Prediction.

1 Introduction

Performance prediction is useful for evaluating the performance of a classification system and for comparing or combining such systems. Thus, it is an important part of data mining [2]. Traditional performance prediction methods [5] withhold a portion of the given data during training and estimate the errors after training from that portion. Typically, the same process is done iteratively, and the average of all error estimations is the final estimate.

We have developed a new approach for estimating performance of classification systems. We carry out training using all available data and estimate errors using the same data. In two-class classification, the predicted error distribution can be used to control classification errors. In the next section, we describe how to choose a reliable classification method and create a classification family as the classifier in our system via the provided training data. Section 3 introduces how to use the same training data to estimate the performance of the classification family and come out a decision scheme for the classifier based on the performance estimation. We give experimental results and conclusions in Section 4 and 5 respectively.

---

* This work was done when the author studied in the Computer Science Department of the University of Texas at Dallas. The author is currently working for Alcatel Network Systems.

2 Construction of a Classification Family

One can employ any existing classification method such as decision tree, Bayesian classifier, neural network, to construct the classification family in this module, providing that the method will generate a vote count to a record to indicate its classification preference. In our system, we use a method which is developed in our lab to generate the classification family C [3]. We emphasize the way of how to generate the classification family, and because of the way we choose, we are able to do a good estimation of the system performance.

There are two disjoint populations $\mathcal{A}$ and $\mathcal{B}$ of records. We are given subsets $A \subseteq \mathcal{A}$ and $B \subseteq \mathcal{B}$ as training data. Given the training sets A and B, we first select an integer $d \geq 5$ and partition A into d nonempty subsets $A^1, A^2, \ldots, A^d$ of essentially equal cardinality and view $A^1, A^2, \ldots, A^d$ as a circular list. We choose another variable c be the smallest integer that is larger than $d/2$; We take the union of A^i and of the $(c-1)$ subsequent A^j and call that union A_i; that is, $A_i = \bigcup_{j=i}^{i+c-1} A^j$. We obtain $A_1, A_2, \ldots, A_d$ accordingly. Applying the analogous process to B, we obtain the similar sets of B_i. Between each pair of A_i and B_i, we use the chosen classification method to generate a classification family member C_i which gives e votes depending on different criteria [3]. Overall, the classification family C will give $d \cdot e$ votes to any record. If we define one vote for $\mathcal{A}$ is $+1$ and vote for $\mathcal{B}$ is -1, the final total number is referred as *vote total*. Obviously, due to the cancellation effect, the vote total could be in the range between $-d \cdot e$ and $+d \cdot e$. We use z to denote the vote total. In our experiments, we use $d = 10$ and $e = 4$. Thus, z is within the range between -40 and $+40$. We use all the training data to generate a classification family C. And this C will be used to classify new data, it is the classifier in this classification system. The vote total that C will produce to $\mathcal{A}$(resp. $\mathcal{B}$) can be viewed as a random variable Z_A(resp. Z_B). The following section describes how to estimate the probability distributions of Z_A and Z_B.

3 Performance of the Classification Family

In this section, we introduce how to estimate the vote total distribution that C will give to the new data. We use $\mathcal{A}$ class as an explanatory example and the same methodology applies to $\mathcal{B}$ class as well. We need to estimate three parameters, namely the mean, variance, and distribution shape of the vote total.

From last section, we know that C is composed by $C_1, C_2, \cdots, C_d$. A record is *unseen* to C_i if it is not included in A_i or B_i. The way that we generated C_i leaves some records in training data are unseen to C_i. That is equal to say, if we analyze how C_i performs on these unseen data, we are able to predict the performance of C_i on new data. Applying the same argument to all C_i in C, the aggregated performance is exactly the vote total prediction of C.

The vote count of C_i given to an unseen record k is denoted as $v(i, k)$. Let X_i be the random variable to represent the vote count given by C_i for unseen records. The unseen records for C_i is indeed $\overline{A}_i = A - A_i$. Thus, the mean and variance of X_i can be estimated by

$$(3.1) \qquad \hat{\mu}_{X_i} = [1/|\overline{A}_i|] \sum_{k \in \overline{A}_i} v(i, k)$$

$$(3.2) \qquad \hat{\sigma}_{X_i}^2 = [1/(|\overline{A}_i| - 1)] \sum_{k \in \overline{A}_i} [v(i, k) - \hat{\mu}_{X_i}]^2$$

respectively.

Since Z_A is the vote total of C and thus the mean value for Z_A is estimated by

$$(3.3) \qquad \hat{\mu}_{Z_A} = \sum_{i=1}^{d} \hat{\mu}_{X_i}$$

For the covariance estimation between X_i and X_j, we have two situations. First, if the set $\overline{A}_{ij} = \overline{A}_i \cap \overline{A}_j$ is nonempty, then we can estimate the covariance of X_i and X_j by

$$(3.4) \qquad \hat{\sigma}_{X_i X_j} = [1/|\overline{A}_{ij}|] \sum_{k \in \overline{A}_{ij}} [v(i, k) - \hat{\mu}_{X_i}][v(j, k) - \hat{\mu}_{X_j}]$$

If $\overline{A}_{ij}$ is empty, we estimate the covariance for the X_i and X_j by a linear approximation function which is constructed by the known covariance values and the amount of intersection of $\overline{A}_{ij}$ [7].

We achieve estimating the distribution shape of Z_A by first estimating a smaller distribution of the total unseen vote of each record and scale up this distribution to the same mean and variance values of Z_A. The mathematical details are described in [7]. Applying the above described methods to both training data classes A and B, we have two estimated distributions of how the classification family will vote for new data.

Based on C, a family $\mathcal{D}$ of *decision schemes* D_z is generated, where z ranges over the possible vote totals of C. We use this decision scheme D_z to declare a record to be in A if the vote total produced by C for that record is greater than or equal to z and declares the record to be in B otherwise. The scheme D_z classifies a record correctly if the vote total is greater than or equal to z, and thus does so with probability $P(Z_A \geq z)$. Conversely, misclassification by D_z of a record of $A - A$ and thus a type A error occur with probability $\alpha = P(Z_A < z)$. Analogous results hold for B. Clearly, if we know the distributions of Z_A on $A - A$ and of Z_B on $B - B$, we will have the α and β for each D_z.

One can define a decision function of D_z based on the two error values, namely α and β. The function can be utilized to control the classification error according to different requirements [7].

4 Experiments

We have implemented the above approaches to a classification system with decision support as *Lsquare*. This system takes the input training data, constructs a classification family C, estimate the vote total distributions on two classes, and provide the decision function. The user can choose the decision threshold in the decision function to fulfill the needs of controlling the error of misclassification. Several well-known datasets from the Repository of Machine Learning Databases and Domain Theories of the University of California at Irvine [6] have been tested with *Lsquare*. We show the results of graphs for the *Australian Credit Card* problem in (5.1). The data were made available by J. R. Quinlan. They represent 690 MasterCard applicants of which 307 are declared as positive and 383 as negative. We declare $\mathcal{A}$ (resp. $\mathcal{B}$) to be the set of negative (resp. positive) records. We obtain from $\mathcal{A}$ and $\mathcal{B}$ randomly selected subsets A and B, each containing 50% of the respective source set. We apply *Lsquare* to A and B, obtain the family C of classification methods, and compute the estimated error probabilities $\hat{\alpha}$ and $\hat{\beta}$. Then we apply C to $\mathcal{A} - A$ and $\mathcal{B} - B$ to verify the error probabilities. The graphs below show the results. The curves plotted with diamonds are the estimated $\hat{\alpha}$ and $\hat{\beta}$, while the curves plotted with crosses are the verified values.

(5.1)

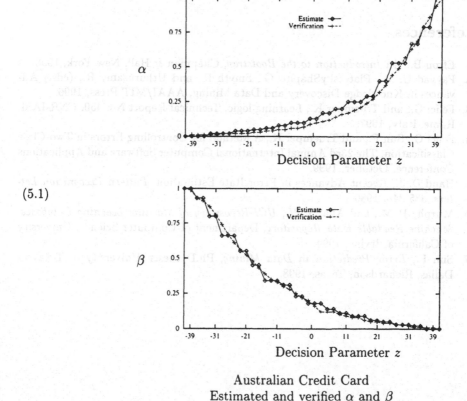

Australian Credit Card
Estimated and verified α and β

5 Conclusions and Future Research

We have developed new strategies and techniques to predict the performance of two-class classification systems. We predict its performance using the same data as in training the system. In two-class classification, the predicted error distributions can be used to control classification errors when new data are classified. Predicting performance by training data lets the learning system learn more without holding a portion of data for evaluation. The performance prediction is based on a system that has learned all the given data, hence it is representative of future performance.

The approaches have been implemented within a learning system *Lsquare*. It was tested by many well-known datasets in the machine learning community. It shows our performance prediction mechanism to be very reliable. We plan to test this scheme on using different classification methods such as Decision Trees, Neural Networks, and Nearest Neighbors Algorithm, etc. This will further trigger different analysis on performance prediction for different classifiers.

Acknowledgement

This project was supported in part by the Office of Naval Research under Grant N00014-93-1-0096. The author would also like to thank the support from University of Texas at Dallas during his study.

References

1. Efron B., *An Introduction to the Bootstrap*, Chapman & Hall, New York, 1993.
2. Fayyad U. M., PiatetskyShapiro G., Smyth P., and Uthurusamy R., (eds.) Advances in Knowledge Discovery and Data Mining, AAAI/MIT Press, 1996.
3. Felici G., and Truemper K., Learning logic, Technical Report No. 450, CNR-IASI, Rome, Italy, 1997.
4. Felici G., Sun F., and Truemper K., A Method for Controlling Errors in Two-Class Classification, The 23rd Annual International Computer Software and Applications Conference, October, 1999.
5. Hand D. J., Recent Advances in Error Rate Estimation, *Pattern Recognition Letters*, 335-346, 1986.
6. Murphy P. M., and Aha D. W., *UCI Repository of Machine Learning Databases: Machine Readable Data Repository*, Department of Computer Science, University of California, Irvine, 1994.
7. Sun F., *Error Prediction in Data Mining*, Ph.D thesis, University of Texas at Dallas, Richardson, Texas, 1998.

Flexible Optimization and Evolution of Underwater Autonomous Agents[1]

E. Eberbach[2], R. Brooks, S. Phoha

Applied Research Laboratory
The Pennsylvania State University
P. O. Box 30, State College, PA 16804
ejel@psu.edu, rrb@acm.org, sxp26@psu.edu

Abstract. The "Ocean SAmpling MObile Network" (SAMON) Project is a simulation testbed for Web-based interaction among oceanographers and simulation based design of Ocean Sampling missions. In this paper, the current implementation of SAMON is presented, along with a formal model based on process algebra. Flexible optimization handles planning, mobility, evolution, and learning. A generic behavior message-passing language is developed for communication and knowledge representation among heterogeneous Autonomous Undersea Vehicles (AUV's). The process algebra subsumed in this language expresses a generalized optimization framework that contains genetic algorithms, and neural networks as limiting cases.

1 Introduction

The global behavior of a group of interacting agents goes beyond juxtaposition of local behaviors. Wegner [15] indicates that interaction machines, formed by multiple agents, have richer behavior than Turing machines. Milner [7] indicates sequential processes cannot always represent concurrent interactive ones. Realistic applications of autonomous agents require new models and theories. Three fundamental questions remain open:
- How to produce intelligent global results from group local behaviors?
- How to decompose problems for solution by independent individual agents?
- How to integrate reactive and deliberative behaviors?

Our application uses process algebra and resource-bounded computation to solve these problems and plan mobile underwater robot group missions. In this paper, we describe a flexible optimization methodology for agent control and evolution. The

---

[1] Research supported by grant N00014-96-1-5026 from Office of Naval Research

[2] Support for this author was provided in part by NSERC under grant OGP0046501, on leave from Jodrey School of Computer Science, Acadia University, Wolfville, NS, Canada B0P 1X0, eugene.eberbach@acadiau.ca

optimization model unifies genetic algorithms and neural networks in a manner suited to reacting to changing dynamic environments with constrained resources.

2 SAMON Underwater Mobile Robot Testbed

ARL's SAMON testbed builds upon the ARL Information Science and Technology Division's existing AUV technology. The ONR SAMON project [11],[12] studies networks of Autonomous Underwater Vehicles (AUVs) for adaptive ocean sampling. It contains a Web-based testbed for distributed simulation of heterogeneous AUV missions, and advances adaptive autonomous agent design. A group of AUV's attempts missions in hazardous environments. The group is organized in a four level hierarchy (see Fig. 1).

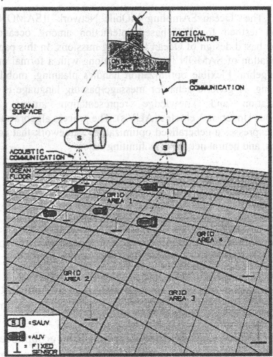

Fig. 1. SAMON hierarchy of Tactical Coordinator (TC), Supervisory Autonomous Underwater Vehicle (SAUV), Autonomous Underwater Vehicle (AUV) and Fixed Sensor Packages (FSP's)

A Tactical Coordinator initiates missions by transmitting orders to several Supervising Autonomous Underwater Vehicles (SAUVs). Each SAUV uses sonar to spontaneously form a group of subordinate AUV's. Each AUV collects data from Fixed Sensor Packages (FSPs) distributed throughout the region. This data is relayed to the commanding SAUV and Tactical Coordinator. SAUVs and AUVs all have identical controllers. Continuous sensor inputs are responded to by discrete decisions. It is a typical sense-plan-act system. ARL's AUV controller combines fuzzy logic with artificial neural networks as described in [13]. Signal processing routines use sensor inputs to estimate physical variables. Tasks are sequences of behaviors, which are sequences of atomic actions. Goal Achievement Functions (GAF) monitor system progress. The sequence of behaviors is flexible. New elements are inserted as required. The testbed allows remote access. It integrates remote heterogeneous simulators. A Geographic Information System (GIS) ARCINFO supports the Tactical Coordinator.

3. Process Algebra Model for Adaptive Autonomous Agents

Expressing and formulating emerging behavior requires a rigorous formal model, with the following characteristics:

- Agents are autonomous.
- Agents communicate asynchronously using message-passing.
- Agents are encapsulated.
- Agents can be heterogeneous.
- Agents communicate with a finite number of neighbors.
- Group reconfiguration, such as link and node migration, is possible.
- Groups exhibit complex behavior due to interaction among agents.
- Agents and groups adapt to bounded resources.

Appropriate formal models for autonomous agent design are π-calculus [7], interaction machines [15], cellular automata, and automata networks [4]. None adapt to bounded resources. Our model does, and it is as expressive as any other model.

Resource bounded computation is known under a variety of names, including anytime algorithms [16]. It trades off result quality for time or memory used to generate results. It is characterized by:

- Algorithm construction to search for bounded optimal answers.
- Performance measure and prediction.
- Composability.
- Meta-control.

We use a process algebra variant of resource-bounded computation to integrate deliberative and reactive approaches for action selection in real time. Our approach has been developed independently of anytime algorithms under the names modifiable algorithms [2], and $-calculus [3].

$-calculus proposes a general theory of algorithm construction. Everything is a $-expression: agents, behaviors, interactions, and environments. Elementary behaviors are $-expressions representing atomic process steps. Simple $-expressions consist of negation $\neg$, cost $\$$, send $\rightarrow$, receive $\leftarrow$, mutation $\lrcorner$, and user defined functions. More complex actions combine $-expressions using sequential composition $\circ$, parallel composition $\|$, general choice $\sqcup$, cost choice $\cup$, and recursive definition $:=$. $-expressions use prefix notation similar to Lisp. Each $-expression has an associated cost value. Data, functions, and meta-code are written as $(f\ \bar{x})$, where f is name and

$\vec{x} = (x_1, x_2, \dots)$ is a possibly countably infinite vector of parameters. $-expression syntax is summarized below. Let P denote compound $-expressions and α simple $-expressions:

$\alpha ::= (\neg\, \alpha)$	negation	$P ::= (^{\circ}_{i \in I} P_i)$	sequential composition
$\mid (\$\, P)$	cost	$\mid (\parallel_{i \in I} P_i)$	parallel composition
$\mid (\rightarrow (a\ \vec{Q}))$	send	$\mid (\cup_{i \in I} P_i)$	cost choice
$\mid (\leftarrow (a\ \vec{X}))$	receive	$\mid (\sqcup_{i \in I} P_i)$	general choice
$\mid (\lrcorner (a\ \vec{Q}))$	mutation	$\mid (f\ \vec{Q})$	user def. $-expression
$\mid (a\ \vec{Q}))$	user def. simple $-expr.	$\mid (:= (f\ \vec{X})\ P)$	recursive def.

I is a possibly countably infinite indexing set. We write empty parallel composition, general and cost choices as $\perp$, and empty sequential composition as ε. $\perp$ expresses logic false, and ε masks parts of $-expressions. Sequential composition is used when $-expressions run in order, and parallel composition when they are parallel. Cost choice expresses optimization, i.e. it selects the cheapest alternative. General choice is used when we are not interested in optimization. Call and definition, such as procedure and function definitions, specify recursion or iteration. This approach can describe all current heuristic methods, and provide a framework for choosing between heuristics.

Meta-control is a simple algorithm that attempts to minimize cost. Solution quality improves if time is available. Performance measures are cost functions, which represent uncertainty, time, or available resources. Crisp, probabilistic, and fuzzy-logic cost functions are part of the calculus. Users may define their own. Incomplete/uncertain information takes the form of invisible expressions whose cost is either unknown or estimated. Meta-control can choose between local search and global search. Global search methods, like genetic algorithms, process multiple points in the search space in parallel.

Scalability and composability is achieved by building expressions from subexpressions. Recursive definitions decompose expressions into atomic subexpressions. Composability of cost measures is assumed. Expression costs are functions of subexpression costs. Deliberation occurs in the form of select-examine-execute cycles corresponding to sense-deliberate-act. An empty examine phase produces a reactive algorithm.

Short (long) deliberation is natural for interruptible (contract) algorithms. Interruptible algorithms can be interrupted down to the level of atomic expressions. Interruptibility is controlled at two levels: choice of atomic expressions and the length of the deliberation phase. Contract algorithms, although capable of producing results whose quality varies with time allocation, must be given an agreed upon time allocation to produce results.

At the meta-level, execution is monitored and modified to minimize cost using "k-Ω optimization." Solutions are found incrementally. They may optimize any of several factors. Depending on problem complexity, cost function volatility and level of uncertainty, deliberation can be done for $k = 0, 1, 2, \dots$ steps or until termination.

Optimization is limited to alphabet Ω, a subset of the complete expression alphabet. This increases run-time flexibility.

We define an adaptive agent model, as parallel composition of component agents:

$$(\|_i A_i)$$

where $(:= (A_i) MA_i)$ defines agent i with meta-system control MA_i. Agent MA_0 is the environment. Each agent MA_i, $i>0$ has a finite neighborhood it communicates with, and $-expression:

$$(:= MA_i \, (^\circ \, (\text{init } P_{i0}) \, (\text{loop } P_i))),$$

where loop is the select-examine-execute cycle performing k-Ω optimization until the goal is satisfied. At which point, the agent re-initializes and works on a new goal:

$$(:= (\text{loop } P_i) \, (\sqcup \, (^\circ \, (\neg \text{ goal } P_i) \, (\text{sel } P_i) \, (\text{exam } P_i) \, (\text{exec } P_i) \, (\text{loop } P_i))$$
$$(^\circ \, (\text{goal } P_i) \, MA_i)))$$

This general model is an instance of resource-based computation, based on process algebras. It covers a wide class of autonomous agents, including SAMON AUV's. The graph and nodes can be arbitrary. We only require that the nodes "understand" the messages in the network. The environment is modeled as a $-expression, which can be a non-deterministic or stochastic (assuming incomplete knowledge of the environment). A distributed environment, if needed, can be modeled as a subnetwork, instead of a single node.

The SAMON network topology combines a 4 level tree (starting from the root: TC, SAUVs, AUVs, FSPs) and a star topology with the environment as a central node connected to all remaining nodes. All nodes communicate by message-passing through sensor and effectors. The input and output messages consist of orders (sonar or radio), reports (data or status), and sensory data from and to environment. In the distributed SAMON testbed, messages take the form of TCP/IP socket communication.

The hierarchical structure hides complexity, improves reliability, increases adaptation and execution speed. However, an optimal tree structure must be derived for a specific task. To find an acceptable tree, the architecture should evolve. For complicated tasks, a strict hierarchy may not suffice. For example, multiple robots pulling a heavy object must communicate with peer nodes to be successful. Temporary mobile links can do this. Cooperating agents need performance metrics with feedback to achieve their objectives.

4 Generic Behavior Message-Passing Language

SAMON allows vehicles from collaborating institutions to communicate and cooperate. However, existing AUVs (e.g., NPS Phoenix [1], FAU Ocean Explorer [12]) employ incompatible designs. It is too early to enforce a single standard AUV design. On the other hand, AUVs designed to perform similar missions should be able to cooperate. A unique aspect of SAMON is collaboration among heterogeneous AUVs. For this purpose, we propose a common communications language: *Generic Behavior Message-Passing Language*.

524

Our collaborative infrastructure (Fig.2) is a network of cooperating agents communicating by send and receive primitives. $-calculus provides the communications framework, and groups elementary behaviors into complex behaviors (missions), or scenarios (programs). Missions are programs of elementary behaviors. The amount of decomposition supported depends on the AUV implementation.

Each node is described by a cost expression, and implemented as an autonomous unit. Nodes can have heterogeneous architectures, but they share a generic behavior message-passing language to interact and cooperate. The nodes can be real or simulated vehicles, for instance FAU, NPS or ARL PSU AUV's, TC's (situated on land, air, or sea), SAUVs, environment, and base recovery vehicle. Nodes can be connected by an arbitrary topology, which we depict as a bus. Nodes communicate using their own message formats. Wrappers will convert messages to the generic language specification. Future controllers could produce messages directly in the standard form. Wrappers should be eliminated as a long term goal.

Generic behavior message-passing requirements language include:

- Modularization – each node should be independent, easy to replace and modify;
- Autonomy – objects communicate only by message-passing;
- Flexibility - allows arbitrary execution of behaviors;
- Extensible - for new types of vehicles, or new environments;
- Evolving - mechanisms for adaptation optimization;
- Research oriented - communication between real and simulated vehicles;
- Simple - but relatively complete;
- Programmable - messages can be added, but may not be accepted by all.

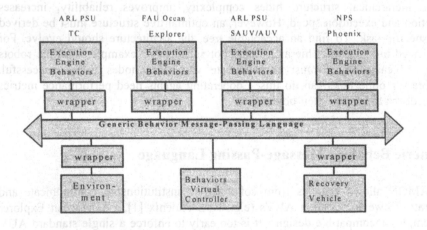

Fig. 2 SAMON testbed collaborative mission execution infrastructure

To satisfy these requirements, the language syntax is based on $-calculus. All behaviors are functions, transmitted between nodes using send and receive primitives. A set of predefined generic-behavior functions is a library on the node. Users can define new behaviors, which may not be understood by other controllers. Our definition does not specify function implementation. Functions are black boxes. The language has two behavior types:

- elementary behaviors, low level communication between entities, and
- agglomerate behaviors, group behaviors by scripting programs.

Send and receive primitives provide message-passing communication and synchronization between agents. They use the same communication channel name. Channels can be sonar, radio, satellite, etc. If a matching channel name is not found, the operation blocks. Parallel composition of send and the rest of program models asynchronous communication. A set of elementary behaviors has been formulated for undersea applications of hierarchical AUV networks.

5 Optimization and Adaptation

Adaptation occurs at all levels in the hierarchy. Optimization is limited. Individual AUV's, and the system as a whole, make decisions based on incomplete information about dynamic processes. Time is not available to compute strictly optimal solutions on-line. Instead, we compute a *satisficing* solution that provides a "good enough" answer; the best that can be found given current resources and constraints. The testbed uses multiple heterogeneous AUV's to collect data from an arbitrary undersea environment. Work takes place in hostile environments with noise-corrupted communication. Components are prone to destruction or failure. Under these conditions, efficient operation can not rely on static plans. Each AUV is self-contained and makes as many local decisions as possible. Operational information and sensor data travel in both directions in the hierarchy. The final two points imply informal activities that are difficult to implement in automated systems.

An AUV at any level in the hierarchy, when it receives a command, must choose from a number of strategies. Decisions are made by evaluating $-functions for the behaviors defining a strategy. Current network and AUV states are used as data by the $-function. $-functions can be derived in a number of ways. Much of computer science is based on deductively deriving asymptotic measures of algorithm computational complexity based on characteristics of the data input. They provide order of magnitude equations for best, worst, or possibly average algorithm performance based on input volume. Constants are irrelevant in asymptotic measures, since at some point the value of a higher order factor will be greater than a lower order one; no matter what constants are used. These measures are useful for determining algorithm scalability, but inadequate for deciding between two specific alternatives where constant factors are relevant. In addition, average complexity measures are generally based on questionable assumptions concerning the statistical distribution of input data, such as assuming all inputs are equally likely.

Computational complexity is almost irrelevant to NP-complete problems. However, computational complexity does provide a starting point for defining $-functions. In some cases, deduction alone can provide useful functions. In other cases, especially when noise is an important factor, deduction is insufficient. If that is so, empirical testing can be used. Testing is often simulation, where a number of runs are replicated with controllable factors set to fixed values and uncontrollable factors given random values. Results from a large number of tests provide data points, which can be used to derive functional approximations. Derivation of functional approximation can be done using a number of approaches, including statistical regression [8], rough sets [5], [14], and visualization [6], [9]. The $-functions found should then be tested to verify their ability to approximate the desired quality measures. Tests could involve either simulations or preferably physical experiments for AUV's.

The AUV's evaluate $-functions using values for the relevant factors, which express the current physical environment. Two natural limits exist to this approach: not all relevant factors can always be known with sufficient certainty, and the physical environment is subject to change. For that reason, we limit our optimization, performing what we call k-Ω optimization. The variable k refers to the limited horizon for optimization, necessary due to the unpredictable dynamic nature of the environment. The variable Ω refers to a reduced alphabet of information. No AUV ever has reliable information about all factors that influence all AUV's participating in a mission. To compensate for this, we mask factors where information is not available from consideration; reducing the alphabet of variables used by the $-function. This can be done by substituting a constant value, or a default function for the masked variables.

This approach allows each AUV to choose between strategies and accomplish its mission. $-functions provide a metric for comparing alternatives. By using k-Ω optimization to find the strategy with the lowest $-function, the AUV finds a satisficing solution. This avoids wasting time trying to optimize behavior beyond the foreseeable future. It also limits consideration to those issues where relevant information is available. This approach, using local optimizations to find globally acceptable satisficing solutions, can be generalized to other genetic approaches.

6 Conclusions

As part of an ONR program, a testbed is being established for combining heterogeneous AUV's for oceanographic sampling. A language describing generic AUV behaviors will be used to communicate between vehicles designed by independent research groups. Part of the language is a process algebra, which uses evolutionary primitives like mutation. The process algebra provides a framework for limited optimization. This optimization contains genetic algorithms and neural networks as limiting cases. Limiting the optimization allows it to be performed in real-time. Currently, the work is underway to implement the Generic Behavior Message-Passing Language, to experiment with cooperation and emerging behavior using resource-bounded optimization, and integrating Virtual Environment for

527

nowcasting and forecasting based on models and data form Harvard and Rutgers Universities.

Acknowledgments. We would like to recognize the contributions of multiple co-workers to this project. They include Dr. J. Stover, Dr. E. Peluso, Dr. P. Stadter, Marc Lattorela, Jen Ryland, and others too numerous to mention.

References

1. Brutzman D., et al: The Phoenix Autonomous Underwater Vehicle, in AI-Based Mobile Robots, MIT/AAAI Press, (1998)
2. Eberbach E.: SEMAL: A Cost Language Based on the Calculus of Self-Modifiable Algorithms, International Journal of Software Engineering and Knowledge Engineering, vol. 4, no. 3, (1994) 391-400.
3. Eberbach E.: Enhancing Genetic Programming by $-calculus, Proc. of the Second Annual Genetic Programming Conference GP-97, Morgan Kaufmann, (1997), 88 (a complete version in Proc. of the Tenth Australian Joint Conf. on AI AI'97, The ACS Nat. Committee on AI and ES, Perth, Australia, (1997), 77-83)
4. Garzon M.: Models of Massive Parallelism: Analysis of Cellular Automata and Neural Networks, Springer-Verlag, (1995)
5. Guan J.W., Bell D.A.: Rough computational methods for information systems, Artificial Intelligence, vol. 105, 77-103.
6. Keim D.A., Kriegel H.-P.: Visualization Techniques for Mining Large Databases: A Comparison, IEEE Transactions on Knowledge and Data Engineering, vol. 8, no. 6, (Dec. 1996) 923-938
7. Milner R., Parrow J., Walker D.: A Calculus of Mobile Processes, I & II, Information and Computation 100, (1992) 1-77.
8. Montgomery D.C.: Design and Analysis of Experiments, Wiley, New York, NY.
9. Nielson G.M., Hagen H., Mueller H.. Scientific Visualization, IEEE Computer Society, Washington, DC (1997)
10. Phoha S. et al: A Mobile Distributed Network of Autonomous Undersea Vehicles, Proc. of the 24th Annual Symposium and Exhibition of the Association for Unmanned Vehicle Systems International, Baltimore, MD, (1997)
11. Phoha S. et al: Ocean Sampling Mobile Network Controller, Sea Technology, (Dec. 1997) 53-58.
12. Smith S. et al, The Ocean Explorer AUV: A Modular Platform for Coastal Oceanography, Proc. of the 9th Intern. Symp. On Unmanned Untethered Submersible Technology, pp. 67-75, 1995.
13. Stover J.A., Hall D.L.: Fuzzy-Logic Architecture for Autonomous Multisensor Data Fusion, IEEE Transactions on Industrial Electronics. V. 43, no. 3. (Jun 1996) 403-410
14. Tsaptsinos D.: Rough Sets and ID3 Rule Learning: Tutorial and Application to Hepatitis Data, Journal of Intelligent Systems, vol. 8, no. 1-2, (1998) 203-223
15. Wegner P.: Why Interaction is More Powerful Than Algorithms, CACM, vol.40, no.5, (May 1997) 81-91
16. Zilberstein S.: Operational Rationality through Compilation of Anytime Algorithms, Ph.D. Dissertation, Dept. of Computer Science, Univ. of California at Berkeley (1993)

Ontology-Based Multi-agent Model of an Information Security System

V.I. Gorodetski[1], L.J. Popyack[2], I.V. Kotenko[3], V.A. Skormin[4]

1 - St.-Petersburg Institute for Informatics and Automation. E-mail: gor@mail.iias.spb.su
2 - USAF Research Laboratory, Information Technology Division, E-mail:popyack@rl.af.mil
3 - St.-Petersburg Signal University. E-mail: ivkote@robotek.ru
4 - Binghamton University, E-mail: vskormin@binghamton.edu

Abstract. The paper is focused on a distributed agent-based information security system of a computer network. A multi-agent model of an information security system is proposed. It is based on the established ontology of the information security system domain. Ontology is used as a means of structuring distributed knowledge, utilized by the information security system, as the common ground of interacting agents as well as for the agent behavior coordination.

Keywords: multi-agent system, information security, ontology.

1 Introduction

Existing computer security systems consist of a number of independent components that require an enormous amount of distributed and specialized knowledge facilitating the solution of their specific security sub-problems. Often, these systems constitute a bottleneck of the throughput, reliability, flexibility and modularity of the computational process. *A modern information security system* (ISS) should be considered as a number of independent, largely autonomous, network-based, specialized software agents operating in a coordinated and cooperative fashion designated to prevent particular kinds of threats and suppressing specific types of attacks. The modern multi-agent system technology presents a valuable approach for the development of an ISS that, when implemented in a distributed large scale multi-purpose information system, is expected to have important advantages over existing computer security technologies. An ontology-based multi-agent model of an ISS is considered herein. In Section 2, the conceptual level of an ISS model is outlined. In Section 3, we propose the *ontology of an information security domain*. The topology of a task-oriented distributed agent's knowledge and belief, providing a common ground for agent information exchange and utilized for agent behavior coordination and mutual understanding, is considered. In Section 4, we outline the ISS architecture and general principles of agents' negotiation and coordination within an agent-based ISS. In Section 5, modeling approach of an ISS is described. Section 6 contains brief analysis of relevant research associated with agent-based ISS. In conclusion, we outline the main results and future work aimed at utilizing agent-based technology for the ISS development.

2 Conceptual Agent-Based Model of ISS

Conceptually, a multi-agent ISS is viewed as a cooperative multitude of the following types of agents, distributed both across the network and on the host itself. (1) *Access control agents* that constrain access to the information according to the legal rights of particular users by realization of *discretionary access control rules* (ACR) specifying to each pair "subject - object" the authorized kinds of messages. Various access control agents cooperate for the purpose of maintaining the compliance with discretionary ACR on various sites of network. These agents supervise the flows of confidential information by realization of *mandatory ACR* not admitting an interception of confidential information. (2) *Audit and intrusion detection agents* detecting non-authorized access and alerting the responsible system (agent) about potential occurrence of a security violation. As a result of statistical processing of the messages formed in the information system, these agents can stop data processing, inform the security manager, and specify the discretionary ACR. A statistical learning process, crucial for the successful operation of these agents, is implemented. It utilizes available information about normal system operation, possible anomalies, non-authorized access channels and probable scripts of attacks. (3) *Anti-intrusion agents* responsible for pursuing, identifying and rendering harmless the attacker. (4) *Diagnostic and information recovery agents* accessing the damage of non-authorized access. (5) *Cryptographic, steganography and steganoanalysis agents* providing safe data exchange channels between the computer network sites. (6) *Authentication agents* responsible for the identification of the source of information, and whether its security was provided during the data transmission that provides the identity verification. They assure the conformity between the functional processes implemented and the subjects initiated by these processes. While receiving a message from a functional process, these agents determine the identifier of the subject for this process and transfer it to access control agents for realization of discretionary ACR. (7) *Meta-agents* that carry out the management of information security processes, provide coordinated and cooperated behavior of the above agents and assure the required level of general security according to a global criteria.

3 Ontology of Information Security Domain

Agents of a multi-agent ISS, performing the global information security task in a distributed and cooperative fashion, must communicate by exchanging messages. Message exchange requires that the agents are able to "understand", in some sense, each other. Mutual agent understanding implies that each agent "knows" (1) what kind of task it must and is able to execute, (2) what agent(s) it has to address when requesting help if its functionality and/or available information are not sufficient for dealing with a problem within its scope of responsibility, and (3) what are the forms and terms of message representation that are understood by the addressee. Therefore, each agent must possess its' own model and models of other agents.

One of the most promising approaches to model the distributed agents' knowledge, beliefs constituting the common ground of an entire multi-agent-system, is the utilization of domain *ontology* [3]. Like any other domain, ontology of the information security domain is a description of the partially ordered concepts of this domain and the relationships over them that should be used by the agents. This ontology describes, in a natural way, *ontological commitments* for a set of agents so that they might be able to communicate about a domain of discourse without a necessary operation of a globally shared theory. In such ontology, definitions associate the names of entities in the space of discourse with human-readable text describing the meanings of names, and formal axioms that constrain the interpretation and well-established use of these terms [3]. A part of the developed fragments of the information security domain ontology, that is associated with the tasks of agents responsible for auditing, detecting non-authorized access, and authentication, is depicted in Fig.1.

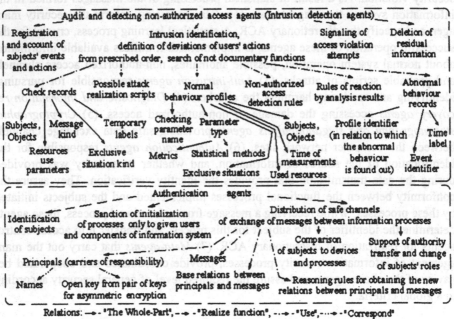

Relations: —→- "The Whole-Part", --→- "Realize function", ---→- "Use",--→- "Correspond"

Fig.1. Fragment showing ontology of information security domain

4 ISS Architecture and General Principles of Agents' Negotiation

Consider a number of basic ISS construction principles functioning as a community of integrated agents distributed in a network environment and allocated on several hosts. Each security agent should be host-based and operate on some segment of the computer network. In this case, we assume that each meta-agent is host-based as well. A meta-agent manages a set of the above-mentioned specialized agents, that, in turn, receive information from the agents-"demons" investigating the input traffic (login, password, etc.). The agents-demons perform monitoring of the input traffic to different servers located on the same host. In essence, they are software sensors that form various metrics of input traffic. All agents are expected to communicate that enables the ISS to detect attacks on the network when intrusion attempts are undertaken "locally" and (or) serially, even when each individual intrusion attempt cannot be interpreted as an intrusion. The offered set of agents can reside on any host and can cooperate through the meta-agent, which operates with the "top level" knowledge base and makes conclusions within the framework of one host. The information interchange between hosts is carried out either on a peer-basis, or by means of the meta-agent acting as the network layer manager.

The subsets of nodes and relations of ontology, used by particular agents for task solving, are determined by agents' functions. The nodes placed on the intersections of ontology fragments, reflecting the functions of two individual agents, constitute the shared knowledge jointly used by both agents in decision making (see Fig.2). Assume that in order to make a decision, agent 2 needs to access knowledge of nodes 1 and 2. But agents 1 and 3 have more detailed knowledge associated with these nodes. Therefore, agent 2

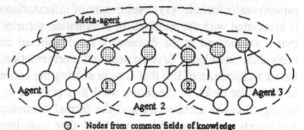

Ⓒ - Nodes from common fields of knowledge

Fig.2. Representation of agents' ontologies intersection

should receive this knowledge from them. Similar situations will take place during the interaction of other agents. It could be seen that agent 2 "is aware" only of nodes 1 and 2, it formulates its request in the terms understood by agents 1 and 3, receives from them knowledge, and is able to interpret it correctly.

5 ISS modeling approach

To demonstrate the validity of our approach we are in the process of developing a *modeling testbed of an ISS*. The hardware part of the testbed includes local computer network that has access to Internet. The software part is based on Unix and Windows OS and specialized agent-based program package that is being developed in Java and Visual C++. At the first step of the testbed realization, the attack intrusion detection environment was built. It includes facilities for network attack modeling, simple agents investigating the input traffic, intrusion detection agents and meta-agents.

6 Related works

Many existing and proposed ISSs use *a monolithic architecture*. Several approaches that exploit the idea of *distributed ISS* are given in [2, 4, 6, 7]. There exist few papers, for example, [1, 5, 8, 9] that consider *an agent-based approach* for an information security system design. Unfortunately, these papers (1) restrict themselves by solving only intrusion detection task, (2) do not pay needed attention to the agent cooperation problem and multi-agent system architecture, (3) ignore advantages of using intelligent agents. Nevertheless, even such a relatively simple agent-based approach as a model of ISS leads to a number of advantages such as efficiency, fault tolerance, resilience to subversion, scalability, etc. In our approach we have borrowed the idea to overcome all of these shortcomings.

7 Conclusion and Future Work

In this paper a multi-agent model of ISS is proposed based on *ontology*. *The main paper results* include: (1) development of information security domain ontology, that is associated with the multitude of information security tasks under consideration, and that is considered as the framework for distributed common knowledge and agent's individual knowledge development and representation; (2) development of an agent-based architecture of ISS that aims at solving the entire multitude of problems related to particular tasks. In the *future work* it is planned to develop the domain ontology, the agent-based architecture and the formal frameworks for distributed knowledge and beliefs representation in more detail. One more intention is to exploit "learning by feedback" methods to provide ISS by real-time adaptation properties.

Bibliography

1. J.Balasubramaniyan, J.Garcia-Fernandez, D.Isakoff, E.Spafford, D.Zamboni. An Architecture for Intrusion Detection using Autonomous Agents. In Proceedings of the 14th Annual Computer Security Applications Conference. Phoenix, Arizona. December 7-11, 1998.
2. S.Forrest, S.A.Hofmeyer, A.Somayaji. Computer Immunology. Communication of the ACM, vol.40, No.10, October 1997, pp.88-96.
3. T.R.Gruber. Toward principles for the design of ontologies used for knowledge sharing. In Proceedings of International Workshop on Formal Ontology, March 1993.
4. Hochberg, et al. "NADIR": An Automated System for Detecting Network Intrusion and Misuse. Computers and Security, vol.12, No.3, 1993, pp.235-248.
5. T.Lunt et al. Knowledge-based Intrusion Detection. In Proceedings of 1989 Governmental Conference Artificial Intelligence Systems. March, 1989.
6. P.A.Porras, P.G.Neumann. EMERALD: Event monitoring enabling responses to autonomous live disturbance. In Proceedings of 20-th National Information System Security Conference. National Institute of Standards and Technologies, 1997.
7. S.Stainford-Chen, et al. GrIDS: A Graph-based Intrusion Detection System for Large Networks. In Proceedings of the 19-th National Information System Security Conference. Vol.1, National Institute of Standards and Technology, October, 1996, pp.361-370.
8. S.J.Stolfo, A.L.Prodromidis, S.Tselepis, W.Lee, D.W.Fan, P.K.Chan. Jam: Java agents for meta-learning over distributed databases. In Proceedings of the 3rd International Conference on Knowledge Discovery and Data Mining, Newport Beach, CA, 1997, pp.74-81.
9. G.White, E.Fish, U.Pooch. Cooperating Security Managers: A Peer-Based Intrusion Detection System. IEEE Network, January/February 1996, pp.20-23.

Optimal Multi-scale Time Series Decomposition for Financial Forecasting Using Wavelet Thresholding Techniques

Taeksoo Shin[1], Ingoo Han[1]

[1]Graduate School of Management, Korea Advanced Institute of Science and Technology,
207-43 Cheongryangri-Dong, Dongdaemoon-Gu, Seoul, 130-012, Korea.
tsshin@msd.kaist.ac.kr, ighan@kgsm.kaist.ac.kr

Abstract. Wavelet analysis as a recently data filtering method (or multi-scale decomposition) is particularly useful for describing signals with sharp spiky, discontinuous or fractal structure in financial markets.

This study investigates the optimal several wavelet thresholding criteria or techniques to support the multi-signal decomposition methods of a daily Korean won / U.S. dollar currency market as a case study, specially for the financial forecasting with a neural network. The experimental results show that a cross-validation technique is the best thresholding criterion of all the existing thresholding techniques for an integrated model of the wavelet transformation and the neural network.

Key words: Discrete Wavelet Transform, Wavelet Packet Transform, Wavelet Thresholding Techniques, Neural Networks, Nonlinear Dynamic Analysis

1 Introduction

Traditionally, the fluctuation in financial market is treated as white noise. However, it is not true when trend is properly removed and we can clearly observe some business cycles, though they evolve with time. The goal of forecasting is to identify the pattern in the time series and use the pattern to predict its future path.

The issue of generalization in this interpretation becomes one of how to extract useful information from the noise-contaminated data, and to rebuild the pattern as closely as possible, while ignoring the useless noises.

Specially, the joint time-frequency filtering techniques such as wavelet transforms also have been shown to be useful in estimating coefficients of forecasting models. The principal advantage of applying the filtering methods is that the techniques make it possible to isolate relevant frequencies.

During the last decade a new and very versatile technique, the wavelet transform (WT), has been developed as a unifying framework of a number of independently developed methods (Mallat [34], Daubechies [18]). Recently, the literatures about the applications of wavelet analysis in financial markets were introduced (See Table 1).

One of the most important problems that has to be solved with the application of wavelet filters is the correct choice of the filter type and the filter parameters. The most difficult choice is that of the cut-off frequency of the filter which has to be specified either explicitly or implicitly (Mittermayr et al. [37]).

This study is intended to explore the wavelet universal thresholding algorithms to denoise data and compare its performance on the basis of the root mean squared error (RMSE) with that of other commonly used smoothing filters in financial forecasting.

We also evaluate the effectiveness of both these transform such as discrete wavelet transform and wavelet packet transform on daily Korean Won / US Dollar exchange rate market.

The remainder of this study is organized as follows. The next section reviews time-frequency decomposition, and then discrete wavelet transform (DWT) and wavelet packet transform (WPT). Section 3 introduces thresholding techniques for financial forecasting and describes best basis selection and best level criteria techniques (Tree Pruning Algorithm). Section 4 describes our model framework and Section 5 analyzes our experimental results. Finally Section 6 contains final comments.

Table 1.   Prior Case Studies Using Wavelet Transform Techniques Applied to Financial Markets

Author (Year)	Purpose	Data	Basis function	Methodology	Results
Pancham (1994)	Test the multi-fractal market hypothesis	Monthly, weekly, daily Index	-	-	Accepted the multi-fractal market hypo.
Cody (1994)	Present the concept of wavelets and the WT methods	General financial market data	DWT, WPT	Multi-scale linear prediction system	Suggested possible applications of the DWT to financial market analysis
Tak (1995)	Forecasting univariate time series	Standard & Poor's 500 index	Mexican-hat wavelet	ARIMA, detrending and AR, random walk, ANN	Outperformed than original data
Greenblatt (1996)	Analysis for structure in financial data	Foreign exchange rates	Coif-1,Coif-5	Best orthogonal basis, Matching pursuit, Method of frames, Basis pursuit	Found structure in financial data
McCabe and Weigend (1996)	Determine at which time-scale the series is most predictable	DM/US Dollar	Haar wavelet	Predictive linear models for multiresolution analysis	Rarely better than predicting the mean of the process
Høg (1996)	Estimate the fractional differencing parameter in Fractional Brownian Motion models for interest rate having the term structure	Monthly US 5-year yields on pure discount bonds (1965.11-1987.02)	Haar wavelet	ARFIMA(0,d+1,0) where H =d+1/2	$\tilde{d} = 0.900$ 95% confidence interval for d = [0.8711, 0.9289]
Høg (1997)	Analyze non-stationary but possibly mean-reverting processes	US interest rate	Haar wavelet	ARFIMA	Showed mean reversion of US interest rate
Aussem et al. (1998)	Predict the trend-up or down - 5 days ahead	S&P 500 closing prices	À trous wavelet	Dynamic recurrent NN & 1 nearest neighbors	86% correct prediction of the trend

2 Discrete Wavelet Transform and Wavelet Packet Transform

Recently, local atomic decompositions (wavelets, wavelet libraries) have become popular for the analysis of deterministic signals as an alternative to non-local Fourier representations. The Fourier transform is usually not to be used in case of non-stationary signals.

Each scale of wavelet coefficients provides a different dimension of the time series in the both time and frequency domains. Recently, due to the similarity between wavelet decomposition and the idea of combining both wavelet and NN has been proposed in various works (Bakshi and Stephanopoulos [4], [5]; Delyon et al. [19]; Geva [25]; Zhang [52]; Zhang and Benveniste [51]).

Recently, the wavelet transform was introduced as an alternatively technique for time-frequency decomposition (Daubechies [17], [18]). Wavelets are any of a set of special functions satisfying certain regularity conditions. Their support is finite; they are non-zero on a finite interval, and they are defined within finite frequency bands..

WT is a powerful method for multiresolution representation of signal data (Szu *et al.*, [44]). The discrete wavelet transform (DWT) expresses a time series as a linear combination of scaled and translated wavelets. Knowing which wavelets appear in a transform can provide information about the frequency content of the signal for a short time period.

DWT is generally calculated by the recursive decomposition algorithm known as the pyramid algorithm or tree algorithm (Mallat [34]), which offers the hierarchical, multiresolution representation of function (signal). As shown in Fig. 1(a), in the tree algorithm, the set of input data is passed through the scaling and the wavelet filters.

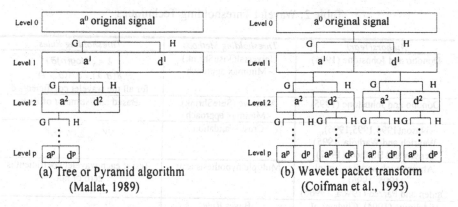

(a) Tree or Pyramid algorithm
(Mallat, 1989)

(b) Wavelet packet transform
(Coifman et al., 1993)

Fig. 1.. Discrete Wavelet Transform and Wavelet Packet Transform (G: The lowpass (or scaling) filter; H: The highpass (or wavelet) filter; $d^p := \{d_0^p, d_1^p, ..., d_{N/2-1}^p\}$, the detail coefficients (highpass filtered data) at the pth level of resolution; $a^p := \{a_0^p, a_1^p, ..., a_{N/2-1}^p\}$, the approximation coefficients (lowpass filtered data) at the pth level of resolution.)

Coifman and Meyer [12] develop wavelet packet functions as generalization of wavelets (DWT). In the pyramid algorithm the detail branches are not used for further calculations, i.e. only the approximations at each level of resolution are treated to yield approximation and detail obtained at level m+1. Application of the transform to both the detail and the approximation coefficients results in an expansion of the structure of the wavelet transform tree algorithm to the full binary tree (Coifman and Wickerhauer [15]; Coifman *et al.* [14]).

The main difference is that while in the DWT the detail coefficients are kept, and the approximation coefficients are further analyzed at each step, in the WPT both the approximation signal and the detail signal are analyzed at each step. This results in redundant information, as each level of the transform retains n samples. The process is illustrated in Fig. 1(b).

3 Wavelet Thresholding Techniques As Optimal Signal Decomposition for Financial Forecasting

Thresholding is a rule in which the coefficients whose absolute values (energies) are smaller than a fixed threshold are replaced by zeroes. The purpose of thresholding is to determine which are the good coefficients to keep, so as to minimize the error of approximation.

In this study, we define wavelet thresholding techniques as denoising, and smoothing techniques including best basis selection and best level algorithm to extract significant multi-scale information from the original time series.

Several approaches to thresholding have been introduced in the literature (See Table 2).

Table 2. Wavelet Thresholding Techniques

Authors(Year)	Thresholding Methods	Thresholding Rules
Donoho and Johnstone (1994)	Universal(VisuShrink) - Minimax approach	$\lambda = \sqrt{2\log(n)}\tilde{\sigma}$, $\delta(d,\lambda) = d \ 1(d > \lambda)$ for all the wavelet coefficients d
Donoho and Johnstone (1995)	Adaptive (SureShrink) - Minimax approach	Based on Estimator of Risk
Nason(1994,1995,1996), Weyrich and Warhola, 1995) Jensen and Bultheel (1997)	Cross-Validation	$CV = \frac{1}{n}\sum_{i=1}^{n}(y_i - \tilde{y}_i)^2$
Abramovich and Benjamini (1995, 1996), Ogden and Parzen (1996a, 1996b)	Multiple hypothesis tests	Test if each wavelet coefficient is zero or not.
Vidaković (1994), Clyde *et al.* (1995), Chipman *et al.* (1997)	Bayes Rule	
Goel and Vidaković (1995)	Lorentz curve	$p^0 = \frac{1}{n}\sum 1(d_i^2 \le d^2)$
Abramovich and Benjamini (1995)	The False Discovery Rate (FDR) approach to multiple hypo. Testing	
Johnstone and Silverman (1997)	Level-dependent Threshold	
Wang (1996), Johnstone and Silverman (1997)	Correlated errors	

Table 3. Wavelet Packet Basis Selection Algorithms

Authors (Year)	Basis Selection Algorithms	Contents
Daubechies (1988)	Method of Frames (MOF)	- Synthesis direction approach -A straight-forward linear algebra
Coifman and Wickerhauser (1992)	Best Orthogonal Basis	- Shannon entropy - Bottom-up tree searches
Mallat and Zhang (1993)	Matching Pursuit	-Synthesis direction approach
Chen (1995), Chen and Donoho (1995b), Chen et al. (1998)	Basis Pursuit	- Similar to MOF -A large-scale constrained opt.
Donoho (1995b)	CART	- Shannon entropy

In wavelet packet functions as generalization of wavelets (DWT), a best basis can explicitly contain the criterion of the coefficient selection. For stance, the best basis can be defined as the basis with the minimal number of coefficients, whose absolute value is higher than the predefined threshold.

Besides, best level algorithm (Coifman et al. [13]) computes the optimal complete sub-tree of an initial tree with respect to an entropy type criterion. The resulting complete tree may be of smaller depth than the initial one. The only difference from best basis selection algorithms is that the optimal tree is searched among the complete sub-tree of the initial tree.

4 Research Model Architecture

Our study is to analyze wavelet thresholding or filtering methods for extracting optimal multi-signal decomposed series (i.e. highpass and lowpass filters) as a key input variable fitting a neural network based forecasting model specially under chaotic financial markets (See Fig. 2).

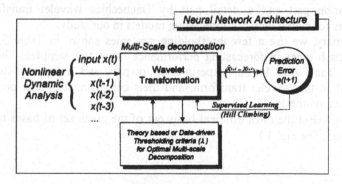

Fig. 2. Integration Framework of Wavelet Transformation and Neural Networks

4.1 Nonlinear Dynamic Analysis

In the chaos theory, it is proved that the original characteristics of the chaos can be reconstructed from a single time series by using a proper embedding dimension.

In this study, we use the dimension information specially to determine the size of time lagged input variables of neural network models. For example, the embedding dimension, 5 estimated in our study indicates that 4 time-lag data are matched to input factors of a neural network to predict the 5th data point of the time series.

4.2 Neural Networks

For time series predictions, the most popularly used neural networks are clearly time delay neural networks (TDNN; Weigend et al. [49]) and recurrent neural networks (RNN; Elman [24]). While in the dynamic context the recurrent neural networks can outperform the time delay neural networks, they occasionally are difficult to be trained optimally by a standard backpropagation algorithm due in part to the dependence of their network parameters (Kuan and Hornik [33]).

In this study, The basic model we experiment with is Backpropagation neural network (BPN) models which have a parsimonious 4 input nodes, 4 hidden nodes and 1 output node with single wavelet filter, i.e. highpass or lowpass filter within the network structure. The other model we experiment with is BPN models which have 8 input nodes, 8 hidden nodes and 1 output node with all the multiple filters.

5   Experimental Results

In this section, we evaluate prior methodology about wavelet thresholding using a case of the daily Korean Won / U.S. Dollar exchange rates are transformed to the returns using the logarithm and through standardization from January 10, 1990 to June 25, 1997. The learning phase involved observations from January 10, 1990 to August 4, 1995, while the testing phase ran from August 7, 1995 to June 25, 1997.

We transform the daily returns into the decomposed series such as an approximation part and a detail part by Daubechies wavelet transform with 4 coefficients for neural network forecasting models in our study.

In summary, we use a few thresholding strategies shown in Table 2, 3 and then compare each other in forecasting performance using test samples. The results are shown in Table 4-6. In our experiments, lowpass and highpass filters are both considered in the wavelet transform, and their complementary use provides signal analysis and synthesis.

First, we select the most efficient basis out of the given set of bases to represent a given signal (See Fig. 3.).

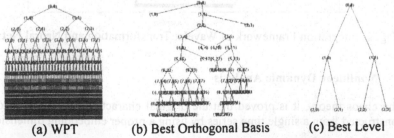

(a) WPT   (b) Best Orthogonal Basis   (c) Best Level

Fig. 3. WPT Analysis Using Daily Korean Won / US Dollar Returns Data [Parentheses contain a information about wavelet level index (left hand size) and wavelet coefficient index at the same level (right hand size)]

Table 4, 5, and 6 compare thresholding performances from different preprocessing methods in forecasting models.

Firstly, our experimental results (Table 4-6) show that WTs have proved to be very good methods for noise filtering and compressing data. This is doubtlessly due to the fact that varying resolution scales are treated, thus taking into account a range of superimposed phenomena.

Table 4 and 5 contain the comparison between hard and soft thresholding. Soft thresholding is hardly different from hard thresholding in the experimental results. Table 4-6 also show the results about the different performances among compression, denoising, best basis method, best level method, and cross-validation, etc.

But, except cross-validation method by DWT, any other method didn't significantly out-perform the others in viewpoint of neural network based forecasting performance. That is, only cross-validation method significantly has the best performance among their techniques and the other methods have almost the same results.

However, the data driven approach has some limitation as follows. That is, in fact, varying results can be obtained with different experimental conditions (signal classes, noise levels, sample sizes, wavelet transform parameters) and error measures, i.e. a cost function for global model optimization.

Ideally, the interplay between theory based and experimental (or data driven) approach to implement an optimal wavelet thresholding should provide the best performance of a model according to the above experimental conditions.

Table 4. A Discrete Wavelet Transform Thresholding Performance Using Test Samples

Threshold Techniques	Threshold Strategy	Filter Types	Network Structure	RMSE
-	-	-	Random Walks	2.939007
-	-	-	BPN(4-4-1)[c]	1.754525
Cross-validation	-	HP&LP[a]	BPN(8-8-1)	**1.676247**
Data Compression	Hard Thresholding	LP[b]	BPN(4-4-1)	1.766189
		HP&LP	BPN(8-8-1)	1.760744
Data Denoising	Soft Thresholding	LP	BPN(4-4-1)	1.767864
		HP&LP	BPN(8-8-1)	**1.751537**
	Hard Thresholding	LP	BPN(4-4-1)	1.766579
		HP&LP	BPN(8-8-1)	1.754131

a: Highpass and Lowpass filters. b: Lowpass filter.
c: BPN(I-H-O) = Backpropagation NN(I: Input Nodes; H: Hidden Nodes; O: Output Nodes).

Table 5. Wavelet Packet Transform Thresholding Performance Using Test Samples

Thresholding Techniques	Tresholding Strategy	Filter Types	Network Structure	RMSE
	-	-	BPN(4-4-1)	1.754525
Data Compression	Hard Thresholding	LP	BPN(4-4-1)	1.774456
		LP&HP	BPN(8-8-1)	1.759434
Data Denoising	Soft Thresholding	LP	BPN(4-4-1)	1.774456
		LP&HP	BPN(8-8-1)	1.759434

Table 6. Best Basis Selection and Best Level Technique Performance Using Test samples

Criteria	Contents	Filter Types	BPN Structure	RMSE
Best Orthogonal Basis	Coifman and Wickerhauser (1992)	LP	(4-4-1)	1.764243
		LP&HP	(8-8-1)	1.74329
Best Level	Coifman et al. (1994)	LP	(4-4-1)	1.767424
		LP&HP	(8-8-1)	1.748388

6 Concluding Remarks

Our research was motivated by a few problems central in time series analysis, i.e. how to extract non-stationary signals which may have abrupt changes, such as level shifts, in the presence of impulsive outlier noise under short-term financial time series. Our research indicates that a wavelet approach is basically an attractive alternative, offering a very fast algorithm with good theoretical properties and predictability in financial forecasting model design.

From our experimental results, wavelet shrinkage or denoising has also been theoretically proven to be nearly optimal from the following perspective: spatial adaptation, estimation when local smoothness is unknown, and estimation when global smoothness is unknown (Taswell [46]). In the future, the availability of these techniques will be promising more and more according to the domain features.

References

1. Abramovich, F., Benjamini, Y.: Thresholding of Wavelet Coefficients as Multiple Hypotheses Testing Procedure. Wavelets and Statistics, Antoniadis A., Oppenheim G. (eds.): Lecture Notes in Statistics 103, Springer-Verlag (1995) 5-14
2. Abramovich, F., Benjamini, Y.: Adaptive Thresholding of Wavelet Coefficients. Computational Statistics & Data Analysis 22 (1996) 351-361
3. Aussem, A., Compbell, J., Murtagh, F.: Wavelet-Based Feature Extraction and Decomposition Strategies for Financial Forecasting. Journal of Computational Intelligence in Finance, March/April (1998) 5-12
4. Bakshi, B. R., Stephanopoulos, G.: Wave-Net: A Multi-Resolution, Hierarchical Neural Network with Localized Learning. AIChE Journal, 39, 1, (1993) 57-81

5.   Bakshi, B. R., Stephanopoulos, G.: Reasoning in Time: Modeling, Analysis, and Pattern Recognition of Temporal Process Trends. Advances in Chemical Engineering, Vol. 22, (1995) 485-547

6.   Chen, S.S.: Basis Pursuit. Ph.D. Thesis, Stanford University, Department of Statistics, Nov. (1995)

7.   Chen, S.S., Donoho, D.L.: Basis Pursuit. Technical Report, Stanford University, Statistics Department. (1995)

8.   Chen, S.S., Donoho, D.L., Saunders, M.A.: Atomic Decomposition by Basis Pursuit. SIAM Journal on Scientific Computing, Vol. 20, No. 1, (1998) 33-61

9.   Chipman, H.A., Kolaczyk, E.D., McCulloch, R.E.: Adaptive Bayesian Wavelet Shrinkage. Journal of the American Statistical Association 92 (1997)

10.   Clyde, M., Parmigiani, G., Vidakovic, B.: Multiple Shrinkage and Subset Selection in Wavelets. Technical Report DP 95-37, Duke University (1995)

11.   Cody, M.A.: The Wavelet Packet Transform. Dr. Dobb's Journal, April (1994) 44-54

12.   Coifman, R., Meyer, Y.: Remarques sur lanalyse de Fourier a fenetre. C.R. Acad. Sci. Paris 312, (1992) 259-261

13.   Coifman, R.R., Meyer, Y., Quake, S., Wickerhauser, M.V.: Signal Processing and Compression with Wavelet Packets. In: Byrnes, J.S., Byrnes, J.L., Hargreaves, K.A., Berry, K. (eds.): Wavelets and Their Applications, (1994) 363-379

14.   Coifman, R.R., Y. Meyer, and M.V. Wickerhauser, Progress in Wavelet Analysis and Applications, in Y. Meyer and S. Roques (eds.), Editions Frontieres, France (1993) 77-93

15.   Coifman, R.R., Wickerhauser, M.V.: Entropy Based Methods for Best Basis Selection. IEEE Transactions on Information Theory Vol.38, No.2 (1992) 719-746

16.   Coifman, R.R., Wickerhauser, M.V.: Wavelets and Adapted Waveform Analysis. A Toolkit for Signal Processing and Numerical Analysis. Processing of Symposia in Applied Mathematics 47 (1993) 119-145

17.   Daubechies, I.: Time-Frequency Localization Operators: A Geometric Phase Space Approach. IEEE Transactions on Information Theory, 34, (1988) 605-612

18.   Daubechies, I.: Ten Lectures on Wavelets. SIAM, Philadelphia, PA, (1992)

19.   Delyon, B., Juditsky, A., Benveniste, A.: Accuracy Analysis for Wavelet Approximation. IEEE Transactions on Neural Networks, vol.6, (1995) 332-348

20.   Dohan, K., Whifield, P.H.: Identification and Chararterization of Water Quality Transients Using Wavelet Analysis. I. Wavelet Analysis Methodology. Wat. Sci. Tech. Vol. 36, No. 5 (1997) 325-335

21.   Donoho, D.L.: Denoising by Soft-Thresholding. IEEE Transactions on Information Theory, vol.41 (1995a) 6130-6270

22.   Donoho, D.L.: CART and Best-Ortho-Basis: A connection. Technical report, Department of Statistics, Stanford, CA (1995b)

23.   Donoho, D.L., Johnstone, I.M.: Ideal Spatial Adaptation via Wavelet Shrinkage. Bimetrika, vol. 81 (1994) 425-455

24.   Elman, J.L.: Finding Structure in Time. Cognitive Science 14, (1990) 179-211

25.   Geva, A.B.: ScaleNet-Muliscale Neural-Network Architecture for Time Series Prediction. IEEE Transactions on Neural Networks, Vol. 9, No.5, Sep. (1998)

26.   Goel, P., Vidakovic, B.: Wavelet Transformations as Diversity Enhancers. Discussion Paper 95-04, ISDS, Duke University, (1995)

27.   Greenblatt, S.A.: Atomic Decomposition of Financial Data. Second International Conference on Computing in Economics and Finance, Geneva, Switzerland, 26-28 June (1996)

28.   Høg, E.: A Fractional Differencing Analysis of Yield Curves by means of Wavelet Analysis. Second International Conference on Computing in Economics and Finance, Geneva, Switzerland, 26-28 June (1996)

29.   Høg, E.: Analyzing Continuous-time Long-memory Models with Wavelets. International Workshop on Wavelets and Statistics, Duke University, October 12-13, (1997)

30.   Jensen, M.J.: Using Wavelets to Obtain a Consistent Ordinary Least Squares Estimator of the Long Memory Parameter. Journal of Forecasting, (1998) (forthcoming).

31.   Jensen, M., Bultheel, A.: Wavelet Thresholding Using Generalized Cross Validation. The International Workshop on Wavelets and Statistics, Duke University, NC, U.S.A., 12-13 October (1997)

32.   Johnstone, I.M., Silverman, B.W.: Wavelet Threshold Estimators for Data with Correlated Noise. J. Roy. Statist. Soc. B., 59, (1997) 319-351

33.   Kuan, C.M., Hornik, K.: Convergence of Learning Algorithms with Constant Learning Rates. IEEE Transactions on Neural Networks. Vol. 2, No.5. (1991) 484-489

34. Mallat, A.: Theory for Multiresolution Signal Decomposition: The Wavelet Representation. IEEE Transactions on Pattern Analysis and Machine Intelligence, (1989) 674-693
35. Mallat, S.G., Zhang, Z.: Matching Pursuit with Time-Frequency Dictionaries. Technical Report 619, Courant Institute of Mathematical Sciences, (1993)
36. McCabe, T.M., Weigend, A.S.: Measuring Predictability Using Multiscale Embedding. Proceedings of the Ninth Yale Workshop on Adaptive and Learning Systems, Yale, June (1996) 13-18
37. Mittermayr, C.R., Nikolov, S.G., Hutter, H., Grasserbauer, M.: Wavelet Denoising of Gaussian Peaks: a Comparative Study. Chemometrics and Intelligent Laboratory Systems 34, (1996) 187-202
38. Nason, G.P.: Wavelet Regression by Cross-validation. Technical Report 447, Department of Statistics, Stanford University, (1994)
39. Nason, G.P.: Choice of the Threshold Parameter in Wavelet Function Estimation. In Wavelets and Statistics. Lecture Notes in Statistics 103, Antoniadis, A. and Oppenheim, G. (eds.), New York: Springer-Verlag, (1995) 261-280
40. Nason, G.P.: Wavelet Shrinkage Using Cross-Validation. Journal of the Royal Statistical Society, Series B, Vol. 58, (1996) 463-479
41. Ogden, T., Parzen, E.: Data Dependent Wavelet Thresholding in Nonparametric Regression with Change Points Applications. Computational Statistics and Data Analysis, 22 (1996a) 53-70
42. Ogden, T., Parzen, E.: Change-point Approach to Data Analytic Wavelet Thresholding. Statist. Comput., 6, (1996b) 93-99
43. Pancham, S.: Evidence of the Multifractal Market Hypothesis Using Wavelet Transforms. Ph.D. Thesis, Florida State Univ., (1994)
44. Szu, H., Telfer, B., Kadambe, S.: Neural Network Adaptive Wavelets for Signal Representation and Classification. Optical Engineering, Vol. 31, No.9, Sept. (1992) 1907-1916
45. Tak, B.: A New Method for Forecasting Stock Prices Using Artificial Neural Network and Wavelet Theory. Ph.D. Thesis, Univ. of Pennsylvania (1995)
46. Taswell, C.: The What, How, and Why of Wavelet Shrinkage Denoising. Technical Report CT-1998-09, Computational Toolsmiths, Stanford, (1998)
47. Vidakovic, B.: Nonlinear Wavelet Shrinkage With Bayes Rules and Bayes Factors. Discussion Paper 94-24, ISDS, Duke University, (1994)
48. Wang, Y.: Function Estimation via Wavelet Shrinkage for Long-memory Data. Annals of Statistics 24 (1996) 466-484
49. Weigend, A.S., Huberman, B.A., Rumelhart, D.E.: Predicting the Future: A Connectionist Approach. Intl. J. Neur. Sys. 1, (1990) 193-209
50. Weyrich, N., Warhola, G.T.: De-noising Using Wavelets and Cross-validation. In Approximation Theory, Wavelets and Applications, NATO ASI Series C, 454, Singh, S.P. (ed.), Dordrecht: Kluwer, (1995) 523-532
51. Zhang, Q., Benveniste, A.: Wavelet Networks. IEEE Transactions on Neural Networks, Vol. 3, No. 6, (1992) 889-898
52. Zhang, Q.: Using Wavelet Networks in Nonparametric Estimation. IEEE Transactions on Neural Networks, vol.8, (1997) 227-236

Computerized Spelling Recognition of Words Expressed in the Sound Approach

Michael Higgins
Yamaguchi University
Department of Perceptual Sciences and Design Engineering
1677-1 Yoshida
Yamaguchi, Japan 753-0841
and
Wojciech Ziarko
Computer Science Department
University of Regina
Regina, Saskatchewan, Canada S4S 0A2

Abstract. This article deals with the possible computer applications of the Sound Approach to English phonetic alphabet. The authors review their preliminary research into some of the more promising approaches to the application of this phonetic alphabet to the processes of machine learning, computer spell-checking, etc. Applying the mathematical approach of rough sets to the development of a data-based spelling recognizer, the authors delineate the parameters of the international cooperative research project with which they have been engaged since 1997, and point the direction of both the continuation of the current project and of future studies, as well.

1 Introduction

In 1993-1994, the first author developed and did initial testing on a new system of phonetic spelling of the sounds in English as an aid to learning better English pronunciation and improving listening and spelling skills in English for Japanese students of English. The method, subsequently entitled *Sound Approach* was tested initially on Japanese high school and university students. The results of the testing indicated that the creation of a *sound map* of English was very helpful in overcoming several common pronunciation difficulties faced by Japanese learners of English as well as improving their English listening, sight reading, and spelling skills [1]. It was further tested on Japanese kindergarten children (ages 3-6), primary school pupils (ages 6-11), and Russian primary school pupils (ages 9-10) and secondary school students (ages 11-13) with similar results [2-3]. It was further tested on a wide range of international ESL (English as a Second Language) students at the University of Regina. These latest results, while still preliminary, indicate that it is an effective and useful tool for helping any non-native speaker of English to overcome pronunciation and orthographic barriers to the effective use of English. The current stage of development for ESL/EFL

(English as a Second Language/ English as a Foreign Language) includes lesson plans for teachers, flip-cards and a workbook for students, and laminated wall charts. The next stage of development includes interactive CD-ROMs and various computer applications.

One of the objectives of the Sound Approach to teaching English language is the development of a spelling recognition system for words expressed in a phonetic alphabet of forty-two symbols known as the *Sound Approach Phonetic Alphabet* (SA). The SA alphabet represents without ambiguity all sounds appearing in the pronunciation of English language words, and does so without using any special or unusual symbols or diacritical marks; SA only uses normal English letters that can be found on any keyboard but arranges them so that consistent combinations of letters always represent the same sound. Consequently, any spoken word can be uniquely expressed as a sequence of SA alphabet symbols, and pronounced properly when being read by a reader knowing the SA alphabet. Due to representational ambiguity and the insufficiency of English language characters to adequately and efficiently portray their sounds phonetically (i.e., there are between 15 and 20 English vowel sounds depending on regional dialect, but only five letters to represent them in traditional English orthography), the relationship between a word expressed in SA alphabet and its possible spellings is one to many. That is, each SA sequence of characters can be associated with a number of possible, homophonic sequences of English language characters. However, within a sentence usually only one spelling for a spoken word is possible. The major challenge in this context is the recognition of the proper spelling of a homophone/homonym given in SA language. Automated recognition of the spelling has the potential for development of SA-based phonetic text editors which would not require the user to know the spelling rules for the language but only being able to pronounce a word within a relatively generous margin of error and to express it in the simple phonetic SA-based form. Computerized text editors with this ability would tremendously simplify the English language training process, for example, by focusing the learner on the sound contents of the language and its representation in an unambiguous form using SA symbols, and in a wider sense, allow for more equal power in the use of English by any native or non-native speaker of English.

2 Approach

The approach adapted in this project would involve the application of the mathematical theory of rough sets in the development of a data-based word spelling recognizer. The theory of rough sets is a collection of mathematical tools mainly used in the processes of decision table derivation, analysis, decision table reduction and decision rules derivation from data (see, for instance references [4-9]). In the word spelling recognition problem, one of the difficulties is the fact that many spoken words given in SA form correspond to a number of English language words given in a standard alphabet. To resolve, or to reduce this ambiguity, the context information must be taken into account. That is, the recognition procedure should involve words possibly appearing before, and almost certainly after the word to be translated into standard English orthography. In the rough-set approach this will require the construction of a decision table for each spoken word. In the decision table, the possible information inputs would include context words surrounding the given word and other information such as the position of the word in the sentence, and so on. Identifying and minimizing the required number of information inputs in such decision tables would be one of the more labor-intensive parts of the project. In this part, the techniques of rough sets, supported by rough-set bas ed analytical software such as KDD-R [10-11], would be used in the analysis of the classificatory adequacy of the decision tables, and their minimization and extraction of classification (decision) rules to be used in the spelling recognition. It should be emphasized at this point, that the process of minimization and rule extraction would be automated to a large degree and adaptive in the sense that inclusion of new spoken word-context combinations would result in regeneration of the classification rules without human intervention. In this sense the system would have some automated learning ability allowing for continuous expansion as more and more experience is accumulated while being used.

3   Rough Sets

The theory of rough sets and their application methodology has been under continuous development for over 15 years now. The theory was originated by Zdzislaw Pawlak [4] in the 1970's as a result of long term fundamental research on logical properties of information systems, carried out by himself and a group of logicians from the Polish Academy of Sciences and the University of Warsaw, Poland. The methodology is concerned with the classificatory analysis of imprecise, uncertain or incomplete information or knowledge expressed in terms of data acquired from experience. The primary notions of the theory of rough sets are the approximation space and lower and upper approximations of a set. The approximation space is a classification of the domain of interest into disjointed categories. The classification formally represents our knowledge about the domain, i.e., knowledge is understood here as an ability to characterize all classes of the classification, for example, in terms of features of objects belonging to the domain. Objects belonging to the same category are not distinguishable which means that their membership status with respect to an arbitrary subset of the domain may not always be clearly definable. This fact leads to the definition of a set in terms of lower and upper approximations. The lower approximation characterizes domain objects about which it is known with certainty, or with a controlled degree of uncertainty [7-8] that they do belong to the subset of interest, whereas the upper approximation is a description of objects which possibly belong to the subset. Any subset defined through its lower and upper approximations is called a rough set. The main specific problems addressed by the theory of rough sets are:

- representation of uncertain, vague or imprecise information;

- empirical learning and knowledge acquisition from experience;

- decision table analysis;

- evaluation of the quality of the available information with respect to its consistency and presence or absence of repetitive data patterns;

- identification and evaluation of data dependencies;

- approximate pattern classification;

- reasoning with uncertainty;

- information-preserving data reduction.

A number of practical applications of this approach have been developed in recent years in areas such as medicine, drug research, process control and others. One of the primary applications of rough sets in artificial intelligence (AI) is for the purpose of knowledge analysis and discovery in data [6]. Several extensions of the original rough sets theory have been proposed in recent years to better handle probabilistic information occurring in empirical data, and in particular the variable precision rough sets (VPRS) model [7-8] which serves as a basis of the software system KDD-R to be used in this project. The VPRS model extends the original approach by using frequency information occurring in the data to derive classification rules.

In practical applications of rough sets methodology, the object of the analysis is a flat table whose rows represent some objects or observations expressed in terms of values of some features (columns) referred to as attributes. Usually, one column is selected as a decision or recognition target, called a decision attribute. The objective is to provide enough information in the table, in terms of attributes of a sufficient number and quality, and a sufficient number of observations, so that each value of the decision attribute could be precisely characterized in terms of some combinations of various features of observations. The methodology of rough sets provides a number of analytical techniques, such as dependency analysis, to asses the quality of the information accumulated in such table (referred to as a decision table). The decision table should be complete enough to enable the computer to correctly classify new observations or objects into one of the categories existing in the table (that is, matching the new observation vector by having identical values of conditional attributes). Also, it should be complete in terms of having enough attributes to make sure that no ambiguity would arise with respect to the predicted value of the target attribute (which is the spelling category in the case of this application). One of the advantages of the rough sets approach is its ability to optimize the representation of the classification information contained in the table by computing so-called reduct, that is, a minimal subset of conditional attributes preserving the prediction accuracy. Another useful aspect is the possibility of the extraction of the minimal length, or generalized decision rules from the decision table. Rules of this kind can subsequently be used for decision making, in particular for predicting the spelling category of an unknown sound.

CLASS	-5	-3	-1	spell
1	0	2	5	ade
2	a	1	2	ade
3	0	0	3	aid
4	0	0	5	aid
5	0	0	1	aid
6	0	0	b	aid
7	0	2	c	aid
8	0	1	2	aid
9	0	2	9	aid
10	0	0	2	ate
11	0	8	1	eight

Table 1. Classification training sentences by using grammatical categories

In the current preliminary testing of SA, a selection of homonyms were put into representative "training" sentences. For each group of "confusing" words one recognition table was constructed. For example, one decision table was developed to distinguish spelling of sounding similar words *ade*, *aid*, *ate* and *eight*. Some of the training sentences used in deriving the table were as follows:

"we need aid", *"she is a nurse's aid"*, *"we ate chicken for dinner"*, and so on. The relative word positions (relative to the target word) in the sentences were plying the role of attributes of each sentence. That is, attribute -1 represented the predecessor of the target word, attribute denoted by -2 was the next preceding word, and so on. Only up to five positions preceding the target word were used in the representation. The values of such defined attributes were grammatical categories of the words appearing on particular positions, eg. verb (value=1), noun (value=2), etc. These values were then used to synthesize decision tables by categorizing training sentences into a number of classes. The decision tables were subsequently the subject of dependency analysis and reduction to eliminate redundant inputs. For instance, an exemplary final reduced decision table obtained for words *ade*, *aid*, *ate* and *eight* is shown in Table 1.

In the preliminary experiments, it was found that using the decision tables the computer could accurately choose the correct spelling of non-dependent homonyms (i.e., those homonyms for which the simple grammatical protocol was unable to determine the correct spelling from the context) 83.3 percent of the time, as in the sentence, *The ayes/eyes have it*. With dependent homonyms, as in the sentence, *ate eight meals*, the computer could accurately choose the correct spelling more than 98 percent of the time.

4 Major Stages of the Initial Project

The initial project was divided into the following major stages which, depending on funding, could have significantly shortened time-frames:

1. Construction of decision tables for the selected number of English language homonyms or homophones. This part would involve research into possible

contexts surrounding the selected words in typical sentences and their representation in decision table format. This would also involve rough set analysis, optimization and testing (with respect to completeness and prediction accuracy) of the constructed tables using existing software systems Dataquest [12,13] or KDD-R. The related activity would be the extraction of classification rules from such tables. This is a very labor-intensive part of the project since the number of possible homonyms or homophones is in the range of approximately 3000. The time-frame for this part of the project is approximately two years.

2. Editor development using the tables constructed in Stage 1 as a main component of the spelling recognition system. The editor would have some learning capabilities in the sense of being able to automatically acquire new feedback word combinations in cases of unsuccessful recognitions. The editor will be constructed in a similar pattern to Japanese Romaji-Hiragana-Kanji word processing selection tables. The estimated time for this stage of the project is approximately one year to construct a working prototype system assuming two full-time programmers would be involved in the system development.

3. This stage would involve both system testing and refinement, going through multiple feedback loops until satisfactory system performance and user satisfaction is achieved. The system would be tested with English language students at Yamaguchi University and other international locations. The accumulated feedback would be used to retrain and enhance the system's spelling recognition capabilities and to refine the user's interface to make it as friendly as possible. It is also felt that using SA, it can be adapted to any regional pronunciation style (e.g., Australian, British Received, Indian, Irish, etc.) by offering the user their choice of keyboard's for their particular area. For example, in standard International Broadcast English the word *table* would be represented in SA by spelling it *teibul* , whereas in Australian English it could be represented in SA by spelling it *taibul* and the computer would still offer the standard orthographic representation of *table* in the spell-checking process in either keyboard format. At this stage, not only could it be used as an ordinary spell checker, but could be programmed for speech as well so that the user could have the word or passage read and spoken by the computer in either sound spelling or in regular spelling. As a normal spell checker, for example, it would be difficult to distinguish between the words *bother* and *brother*. However, with speech capacity, the user could potentially hear the difference and catch the mistake. This could also become an excellent teaching/learning device for practicing and learning correct pronunciation whether for native or for non-native English speakers.

5 Conclusions

In the initial study on the efficacy of the Sound Approach phonetic alphabet in meeting the requirements for the development of easily accessible and accurate computer word recognition capability conducted at the University of Regina in 1997, the rough set model was used to construct decision tables on a list of various English homonyms. It was found that the Sound Approach phonetic alphabet and the rough set model were quite compatible with each other in determining decision tables used in decision making for predicting the correct spelling of a word written either phonetically or in standard English orthography. It was found in preliminary experiments that even using a relatively unrefined grammatical protocol and decision tables, we were able to correctly identify the correct spelling of non-dependent homonyms 83.3 percent of the time. This accuracy rate rivals already extant forms of standard spelling recognition systems. When confronted with dependent homonyms, the computer could accurately choose the correct spelling more than 98 percent of the time.

It is felt that with further refining of the grammatical protocol and expansion of the sample sentences using the approximately 3000 English homonyms, a spelling recognition system could be constructed that would allow even non-native speakers of English to gain equal access and power in the language. Further, this would be but one of the necessary building blocks for the construction of a total voice recognition operating system, and a major step forward in computer speech technology. It is also considered that these advancements have considerable commercial possibilities that should be developed.

6 Acknowledgements

The first author would like to thank Professor Mitsuyasu Miyazaki for his kind and helpful advice concerning the format and for checking the Japanese abstract which follows this article. The research reported in this article was partially supported by a research grant awarded by Natural Sciences and Engineering Research Council of Canada to the second author.

References

1. Higgins, M.L., The Quest for a Universal Auxiliary Language: Addressing Pronunciation and Orthographic Barriers of English. Hawaii: University Microfilms, pp. 162, 1994.
2. Higgins, M.L., Higgins, M.S., and Shima, Y., Basic Training in Pronunciation and Phonics: A Sound Approach. The Language Teacher, vol. 19, number 4, April 1995, pp. 4-8.
3. Higgins, M.L. A Report On The Development Of The Yuzhno-Sakhalinsk International School: The First English Language Immersion Program in Russia. Journal of the Faculty of Liberal Arts (Humanities and Social Sciences). Yamaguchi University, Vol. 28. pp. 209-222.
4. Pawlak, Z., Rough Sets: Theoretical Aspects of Reasoning About Data. Kluwer Academic Publishers, Dordrecht, 1991.
5. Slowinski, R. (ed.) Intelligent Decision Support: Handbook of Applications and Advances of the Rough Sets Theory. Kluwer Academic Publishers, Dordrecht, 1992.

6. Ziarko, W The Discovery, Analysis and Representation of Data Dependencies in Databases. In Piatesky-Shapiro, G. and Frawley, W.J. (eds.) Knowledge Discovery in Databases, AAAI Press/MIT Press, 1991, pp. 177-195.
7. Ziarko, W. Variable Precision Rough Sets Model. Journal of Computer and Systems Sciences, vol. 46, no. 1, 1993, pp. 39-59.
8. Katzberg, J. and Ziarko, W. Variable Precision Extension of Rough Sets. Fundamenta Informaticae, Special Issue on Rough Sets, vol. 27, no. 2-3, 1996, pp. 223-234.
9. Ziarko, W. (ed.) Rough Sets, Fuzzy Sets and Knowledge Discovery. Springer Verlag, 1994.
10. Ziarko, W. and Shan, N. KDD-R: A Comprehensive System For Knowledge Discovery Using Rough Sets. Proceedings of the International Workshop on Rough Sets and Soft Computing, San Jose 1994, 164-173.
11. Ziarko, W. and Shan, N. On Discovery of Attribute Interactions and Domain Classifications. Intelligent Automation and Soft Computing, vol. 2, no. 2, 1996, pp. 211-218.
12. Reduct Systems Inc. Dataquest System, User Manual.
13. Levinson, L. Data Mining: Intelligent Technology Down to Business. PC-AI, Nov/Dec 1993, pp. 17-23.

An Adaptive Handwriting Recognition System[1]

Gaofeng Qian

GTE Enterprise Architecture, Planning and Integration
919 Hidden Ridge, M03C42
Irving, TX 75038
kgqian@yahoo.com

Abstract. This paper describes a recognition system for on-line cursive handwriting that requires very little initial training and that rapidly learns, and adapts to, the handwriting style of a user. Key features are a shape analysis algorithm that determines shapes in handwritten words, a linear segmentation algorithm that matches characters identified in handwritten words to characters of candidate words, and a learning algorithm that adapts to the user writing style. Using a lexicon with 10K words, the system achieved an average recognition rate of *81.3%* for top choice and *91.7%* for the top three choices.

1 Introduction

As more people use and depend on computers, it is important that computers become easier to use. Many systems for handwriting recognition have been developed in the past 35 years [1][4][5][6][7][8]. In contrast to those systems, the method proposed in this paper

* Dispenses with extensive training of the type required for Hidden Markov Models and Time Delay Neural Networks [6][7]. Initialization of the knowledge base consists of providing four samples for each character.
* Uses a shape analysis algorithm that not only supports the identification of characters but also allows efficient reduction of the lexicon to a small list of candidate words [1][4].
* Uses a linear-time segmentation technique that optimally matches identified characters of the handwritten word to characters of a candidate word, in the sense that the method completely avoids premature segmentation selections that may be made by some techniques [6][8].
* Learns not only from failure but also from correctly identified words, in contrast to other prior methods [5][7].
* The dictionary words need not be provided in script form. Thus, switching to a new vocabulary becomes very simple, requiring merely a new lexicon[6][7].

---

[1] This work is done as part of my Ph.D. study under Professor Klaus Truemper in the AI Lab of The University of Texas at Dallas, and funded by the Office of Naval Research.

552

2 Modules of The System

The system consists of three modules. The *preprocessing module* accepts as input the raw pixel sequence of a handwritten word recorded by a digitizing tablet and converts it to a sequence of feature vectors called the *basic code*. The *interpretation module* receives the basic code of a handwritten word as input, deduces word shapes, selects from a lexicon a list of candidate words, and from these candidates deduces by a matching process the interpretation. The *learning module* analyzes the correct word, which is either the output word of the interpretation module or the intended word supplied by the user, and locates opportunities for learning from misidentified letters and from identified letters with low match quality values. The insight so obtained results in addition, adjustment, or replacement of templates. The next three sections describe the preprocessing, interpretation and learning module respectively.

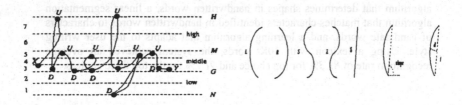

Fig. 1. An example of handwritten word 'help' with extracted features and regions; and possible shapes of strokes.

3 Preprocessing Module

An on-line handwriting recognition system accepts handwriting from a digitizer. Due to technical limitations of the tablet, the raw pixel sequence of a handwritten word includes imperfections and redundant information. We first delete duplicate pixels caused by a hesitation in writing and interpolate non-adjacent consecutive pixels caused by fast writing, to produce a continuous pixel sequence. We then identify pixels with particular characteristics such as *local maxima* and *local minima*. We also normalize the handwritten word and extract other features such as locations of extrema, shapes of strokes, slopes of strokes, curvatures of strokes, connections of strokes, and openings associated with maxima and minima. We organize these features into a sequence of *feature vectors* called *basic code* which is input of the *interpretation* module. The left part of Figure 1 shows an example of a handwritten word with extracted extrema and regions. The right part gives some sample shapes of strokes.

4 Interpretation Module

The interpretation module takes the basic code as input and interprets it as some word of a given lexicon. The module carries out that task as follows. It initially extracts the shape of the handwritten word, such as ascenders, descenders and their positions with respect to the baseline of the word. By using the shape information, it reduces a large reference lexicon to a list of candidates which have the same shape as the handwritten word. For each candidate, the module carries out the following steps. First, the module identifies letters of the candidate in the basic code using template matching and computes a match quality for each identified letter. We emphasize that the portions of the basic code corresponding to identified letters can, and often do, overlap. Second, for each contiguous segment of basic code connecting identified letters, a certain length is computed. Similarly, for the unidentified letters of the candidate, a certain length is determined as well. Third, a linear-time segmentation algorithm finds an optimal matching of identified characters of the handwritten word to the characters of the given candidate word, in the sense that the matching maximizes the sum of the match quality values for the identified letters minus the sum of the length differences for the unidentified letters. Once all candidate words have been processed, the optimal matching of each candidate word is scored and the candidate with the highest score is selected as the desired word.

5 Learning Module

The learning algorithm adapts the system to a specific writing style by learning user behavior and updating the template set. User-adaptive systems reported in the literature conduct their adaptive processes only when a word is not recognized correctly [5][7]. We employ a more elaborate adaptive learning strategy. The system learns the user's writing not only when the output word of the system is wrong, but also when it is correct. In the latter case, the system learns individual characters or sub-strings of the word that have not been recognized correctly.

With knowing the *correct* word of a handwritten word, which is either the output word of the interpretation module confirmed by the user, or the intended word supplied by the user, the learning module analyzes the identified segments and unidentified segments of the basic code to identify errors for learn. We do learning on the unidentified segments and the identified segments with low match quality. For each learning case, the learning module picks up one of the following three methods subsequently:

1. Adding the segment of basic code as a new template if the number of templates does not reach the maximum allowed in the systems.
2. Adjusting the parameters of a template so that the match quality is increased. Such a change may cause the template less often occurrences of other letters/strings. Hence, we evaluate the positive and negative impact of such adjustments to decide if we want to adjust a template or use the next method.
3. Replacing the least frequently used templates by the basic code segment.

6 Experimental Results

The handwriting data is collected using the Wacom ArtZ II tablet (140 samples per second and 100 lines per inch). The initial set of templates was collected from one writer who did not participate in the testing. Test data were collected from four writers. The user-independent system using preprocessing and interpretation modules had an average recognition rate of 65.5%, and the user-adaptive system using three modules reached 81.3%. Thus, the learning module improved the average system accuracy by 15.8%.

We have conducted experiments to analyze the error distribution. Table 1 shows the percentage of correct words appearing in different ranges using the user-adaptive system. The table shows that the system always determines the correct shape class. The screen process, which reduces the shape class to a small list of candidates, causes an average 4% error. The average performance at the top 1 choice is 81.3%. In the experiment of the top three choices, the average performance is improved to 91.7%. However, the average recognition rate of the top five choices is 92.5% which does not improve much on the top 3 choices.

Table 1. Recognition rates of the sytem on different criteria

Writer	Top 1	Top 3	Top 5	Candidate list	Shape Clase
A	84%	93%	93%	96%	100%
B	80%	91%	92%	97%	100%
C	75%	89%	91%	95%	100%
D	86%	94%	94%	97%	100%

7 Conclusions

This paper has presented a new approach for on-line handwriting recognition. The framework of our approach dispenses with elaborate training of the type required for statistical pattern recognition. Initialization of the system consists merely in providing four samples for each character, written in isolation by one writer. The dictionary words need not be provided in script form. Thus, even switching to a new vocabulary becomes very simple, requiring merely a new lexicon. While principles underlying the present approach are general enough, the techniques of segmentation and learning are particularly well suited for Roman scripts. Tests have shown that the method is robust because performance does not degrade significantly even when words written by one writer are interpreted using reference characters from another.

In creating a complete handwriting interpretation system, one must decide where effort can be most effectively applied to increase the performance. It is felt that in this system, the effort has been distributed with an emphasis on the work of the interpretation module. The preprocessing module could be improved upon, for example, by extracting a set of better features from the raw pixel sequence.

References

[1] Bramall, P. E. and Higgins, C. A., A Cursive Script Recognition System Based on Human Reading Models, *Machine Vision and Applications*, Vol 8, 1995, 224--231

[2] Higgins, C. A. and Ford, D. M. , On-Line Recognition of Connected Handwriting by Segmentation and Template Matching, *Proceedings of 11th IAPR International Conference on Pattern Recognition*, Vol 2 , 1992, 200—203

[3] Morasso, P., Limonceli, M. and Morchio, M, Incremental Learning experiments with SCRIPTOR: an Engine for On-line Recognition of Cursive Handwriting, *Machine Vision and Applications*, Vol 8, 1995, 206--214

[4] Powalka, R. K., Sherkat, N., Evett, L. J., and Whitrow, R. J., Multiple Word Segmentation with Interactive Look-up for Cursive Script Recognition, *Proceedings of the second International Conference on Document Analysis and Recognition*, 1993, 196--199

[5] Qian, G. The Kritzel System for On-line handwriting Interpretation, *Proceedings of the Fourteenth National Conference on Artificial Intelligence*, Portland, Oregon, 1996, 1403

[6] Schenkel, M., Guyon, I. and Henderson, D., On-line Cursive Script Recognition Using Time-Delay Neural Networks and Hidden Markov Models, *Machine Vision and Applications*, Vol 8, 1995, 215—223

[7] Schomaker, L., Using Stroke- or Character-based Self-Organizing Maps in the Recognition of On-line, Connected Cursive Script, *Pattern Recognition*, Vol 26, 1993, 443--450

[8] Tappert, C. C., Cursive Script Recognition by Elastic Matching, *IBM Journal of Research and Development*, Vol 26, 1982, 765--771

Author Index

Lecture Notes in Artificial Intelligence (LNAI)

Lecture Notes in Computer Science